Lecture Notes in Computer Science 8367

Christoph Beierle Carlo Meghini (Eds.)

Foundations of Information and Knowledge Systems

8th International Symposium, FoIKS 2014
Bordeaux, France, March 3-7, 2014
Proceedings

Volume Editors

Christoph Beierle
FernUniversität in Hagen
Fakultät für Mathematik und Informatik
Universitätstrasse 1
58084 Hagen, Germany
E-mail: christoph.beierle@fernuni-hagen.de

Carlo Meghini
Istituto di Scienza e Tecnologie della Informazione
Consiglio Nazionale delle Ricerche
Via G. Moruzzi 1
56124 Pisa, Italy
E-mail: carlo.meghini@isti.cnr.it

ISSN 0302-9743 e-ISSN 1611-3349
ISBN 978-3-319-04938-0 e-ISBN 978-3-319-04939-7
DOI 10.1007/978-3-319-04939-7
Springer Cham Heidelberg New York Dordrecht London

Library of Congress Control Number: 2014930887

CR Subject Classification (1998): G.2, F.4.1-2, I.2.3-4, D.3

LNCS Sublibrary: SL 3 – Information Systems and Application, incl. Internet/Web and HCI

Typesetting: Camera-ready by author, data conversion by Scientific Publishing Services, Chennai, India

Printed on acid-free paper

Springer is part of Springer Science+Business Media (www.springer.com)

Preface

This volume contains the articles that were presented at the 8th International Symposium on Foundations of Information and Knowledge Systems (FoIKS 2014), which was held in Bordeaux, France, March 3–7, 2014.

The FoIKS symposia provide a biennial forum for presenting and discussing theoretical and applied research on information and knowledge systems. The goal is to bring together researchers with an interest in this subject, share research experiences, promote collaboration, and identify new issues and directions for future research.

FoIKS 2014 solicited original contributions on foundational aspects of information and knowledge systems. This included submissions that apply ideas, theories, or methods from specific disciplines to information and knowledge systems. Examples of such disciplines are discrete mathematics, logic and algebra, model theory, information theory, complexity theory, algorithmics and computation, statistics, and optimization.

Previous FoIKS symposia were held in Kiel (Germany) in 2012, in Sofia (Bulgaria) in 2010, Pisa (Italy) in 2008, Budapest (Hungary) in 2006, Vienna (Austria) in 2004, Schloss Salzau near Kiel (Germany) in 2002, and Burg/Spreewald near Berlin (Germany) in 2000. FoIKS took up the tradition of the conference series Mathematical Fundamentals of Database Systems (MFDBS), which initiated East – West collaboration in the field of database theory. Former MFDBS conferences were held in Rostock (Germany) in 1991, Visegrad (Hungary) in 1989, and Dresden (Germany) in 1987.

The FoIKS symposia are a forum for intense discussions. Speakers are given sufficient time to present their ideas and results within the larger context of their research. Furthermore, participants are asked in advance to prepare a first response to a contribution of another author in order to initiate discussion.

Suggested topics for FoIKS 2014 included, but were not limited to:

- Database Design: formal models, dependencies and independencies
- Dynamics of Information: models of transactions, concurrency control, updates, consistency preservation, belief revision
- Information Fusion: heterogeneity, views, schema dominance, multiple source information merging, reasoning under inconsistency
- Integrity and Constraint Management: verification, validation, consistent query answering, information cleaning
- Intelligent Agents: multi-agent systems, autonomous agents, foundations of software agents, cooperative agents, formal models of interactions, logical models of emotions
- Knowledge Discovery and Information Retrieval: machine learning, data mining, formal concept analysis and association rules, text mining, information extraction

- Knowledge Representation, Reasoning and Planning: non-monotonic formalisms, probabilistic and non-probabilistic models of uncertainty, graphical models and independence, similarity-based reasoning, preference modeling and handling, argumentation systems
- Logics in Databases and AI: classical and non-classical logics, logic programming, description logic, spatial and temporal logics, probability logic, fuzzy logic
- Mathematical Foundations: discrete structures and algorithms, graphs, grammars, automata, abstract machines, finite model theory, information theory, coding theory, complexity theory, randomness
- Security in Information and Knowledge Systems: identity theft, privacy, trust, intrusion detection, access control, inference control, secure Web services, secure Semantic Web, risk management
- Semi-Structured Data and XML: data modeling, data processing, data compression, data exchange
- Social Computing: collective intelligence and self-organizing knowledge, collaborative filtering, computational social choice, Boolean games, coalition formation, reputation systems
- The Semantic Web and Knowledge Management: languages, ontologies, agents, adaptation, intelligent algorithms
- The WWW: models of Web databases, Web dynamics, Web services, Web transactions and negotiations

The call for papers resulted in the submission of 52 full articles. In a rigorous reviewing process, each submitted article was reviewed by at least three international experts. The 14 articles judged best by the Program Committee were accepted for long presentation. In addition, five articles were accepted for short presentation. This volume contains versions of these articles that were revised by their authors according to the comments provided in the reviews. After the conference, authors of a few selected articles were asked to prepare extended versions of their articles for publication in a special issue of the journal *Annals of Mathematics and Artificial Intelligence.*

We wish to thank all authors who submitted papers and all conference participants for fruitful discussions. We are grateful to Dov Gabbay, Cyril Gavoille, and Jeff Wijsen, who presented invited talks at the conference; this volume also contains articles for two of the three invited talks. We would like to thank the Program Committee members and additional reviewers for their timely expertise in carefully reviewing the submissions. We want to thank Markus Kirchberg for his work as publicity chair. The support of the conference provided by the European Association for Theoretical Computer Science (EATCS) and by CPU LABEX of the University of Bordeaux is greatfully acknowledged. Special thanks go to Sofian Maabout and his team for being our hosts and for the wonderful days in Bordeaux.

March 2014

Christoph Beierle
Carlo Meghini

Conference Organization

FoIKS 2014 was organized by the University of Bordeaux, France.

Program Committee Chairs

Christoph Beierle	University of Hagen, Germany
Carlo Meghini	ISTI-CNR Pisa, Italy

Program Committee

José Júlio Alferes	Universidade Nova de Lisboa, Portugal
Leila Amgoud	University of Toulouse, France
Peter Baumgartner	NICTA and The Australian National University
Salem Benferhat	Université d'Artois, Lens, France
Leopoldo Bertossi	Carleton University, Canada
Philippe Besnard	University of Toulouse, France
Joachim Biskup	University of Dortmund, Germany
Piero A. Bonatti	University of Naples Federico II, Italy
Gerd Brewka	University of Leipzig, Germany
François Bry	LMU München, Germany
Andrea Calì	Birkbeck, University of London, UK
Jan Chomicki	University at Buffalo, USA
Marina De Vos	University of Bath, UK
Michael I. Dekhtyar	Tver State University, Russia
James P. Delgrande	Simon Fraser University, Canada
Tommaso Di Noia	Technical University of Bari, Italy
Jürgen Dix	Clausthal University of Technology, Germany
Thomas Eiter	Vienna University of Technology, Austria
Ronald Fagin	IBM Almaden Research Center, San Jose, USA
Victor Felea	Al.I. Cuza University of Iasi, Romania
Flavio Ferrarotti	Victoria University of Wellington, New Zealand
Sergio Flesca	University of Calabria, Italy
Lluis Godo	Artificial Intelligence Research Institute (IIIA - CSIC), Spain
Gianluigi Greco	University of Calabria, Italy
Claudio Gutierrez	University of Chile, Chile
Marc Gyssens	Hasselt University, Belgium
Sven Hartmann	Clausthal University of Technology, Germany
Stephen J. Hegner	Umeå University, Sweden

Edward Hermann Haeusler	Pontifícia Universidade Católica, Brazil
Andreas Herzig	University of Toulouse, France
Pascal Hitzler	Wright State University, USA
Anthony Hunter	University College London, UK
Yasunori Ishihara	Osaka University, Japan
Gyula O. H. Katona	Alfréd Rényi Institute, Hungarian Academy of Sciences, Hungary
Gabriele Kern-Isberner	University of Dortmund, Germany
Sébastien Konieczny	University of Lens, France
Gerhard Lakemeyer	RWTH Aachen University, Germany
Jérôme Lang	University of Paris 9, France
Mark Levene	Birkbeck University of London, UK
Sebastian Link	University of Auckland, New Zealand
Weiru Liu	Queen's University Belfast, UK
Thomas Lukasiewicz	University of Oxford, UK
Carsten Lutz	University of Bremen, Germany
Sebastian Maneth	NICTA and University of New South Wales, Australia
Pierre Marquis	University of Artois, France
Wolfgang May	University of Göttingen, Germany
Thomas Meyer	CSIR Meraka and University of KwaZulu-Natal, South Africa
Leora Morgenstern	New York University, USA
Amedeo Napoli	LORIA Nancy, France
Bernhard Nebel	University of Freiburg, Germany
Wilfred S. H. Ng	Hong Kong University of Science and Technology, Hong Kong, SAR China
Henri Prade	University of Toulouse, France
Andrea Pugliese	University of Calabria, Italy
Sebastian Rudolph	University of Dresden, Germany
Attila Sali	Alfréd Rényi Institute, Hungarian Academy of Sciences, Hungary
Francesco Scarcello	University of Calabria, Italy
Klaus-Dieter Schewe	Software Competence Center Hagenberg, Austria
Dietmar Seipel	University of Würzburg, Germany
Nematollaah Shiri	Concordia University, Montreal, Canada
Gerardo I. Simari	University of Oxford, UK
Guillermo Ricardo Simari	Universidad Nacional del Sur, Argentina
Nicolas Spyratos	University of Paris-South, France
Umberto Straccia	ISTI-CNR Pisa, Italy
Letizia Tanca	Politecnico di Milano, Italy
Bernhard Thalheim	University of Kiel, Germany
Alex Thomo	University of Victoria, Canada
Miroslaw Truszczynski	University of Kentucky, USA

José María Turull-Torres	Massey University Wellington, New Zealand
Dirk Van Gucht	Indiana University, USA
Victor Vianu	University of California San Diego, USA
Peter Vojtáš	Charles University, Czech Republic
Qing Wang	Australian National University, Australia
Jef Wijsen	University of Mons-Hainaut, Belgium
Mary-Anne Williams	University of Technology, Sydney, Australia
Stefan Woltran	Vienna University of Technology, Austria

Additional Reviewers

Fernando Bobillo
Alain Casali
Giovanni Casini
Michelle Cheatham
Andrea Esuli
Fabrizio Falchi
Fabio Fassetti
Pietro Galliani
Alejandro Grosso
Szymon Klarman
Seifeddine Kramdi
Thomas Krennwallner
Francesco Lupia
Wenjun Ma
Roberto Mirizzi
Cristian Molinaro
Till Mossakowski
Michael Norrish
Antonino Rullo
Nicolas Schwind
Michaël Thomazo

Local Organization Chair

Sofian Maabout LaBRI, Bordeaux, France

Publicity Chair

Markus Kirchberg VISA Inc., Singapore

Sponsored By

CPU LABEX, Cluster of Excellence, University of Bordeaux, France
European Association for Theoretical Computer Science (EATCS)

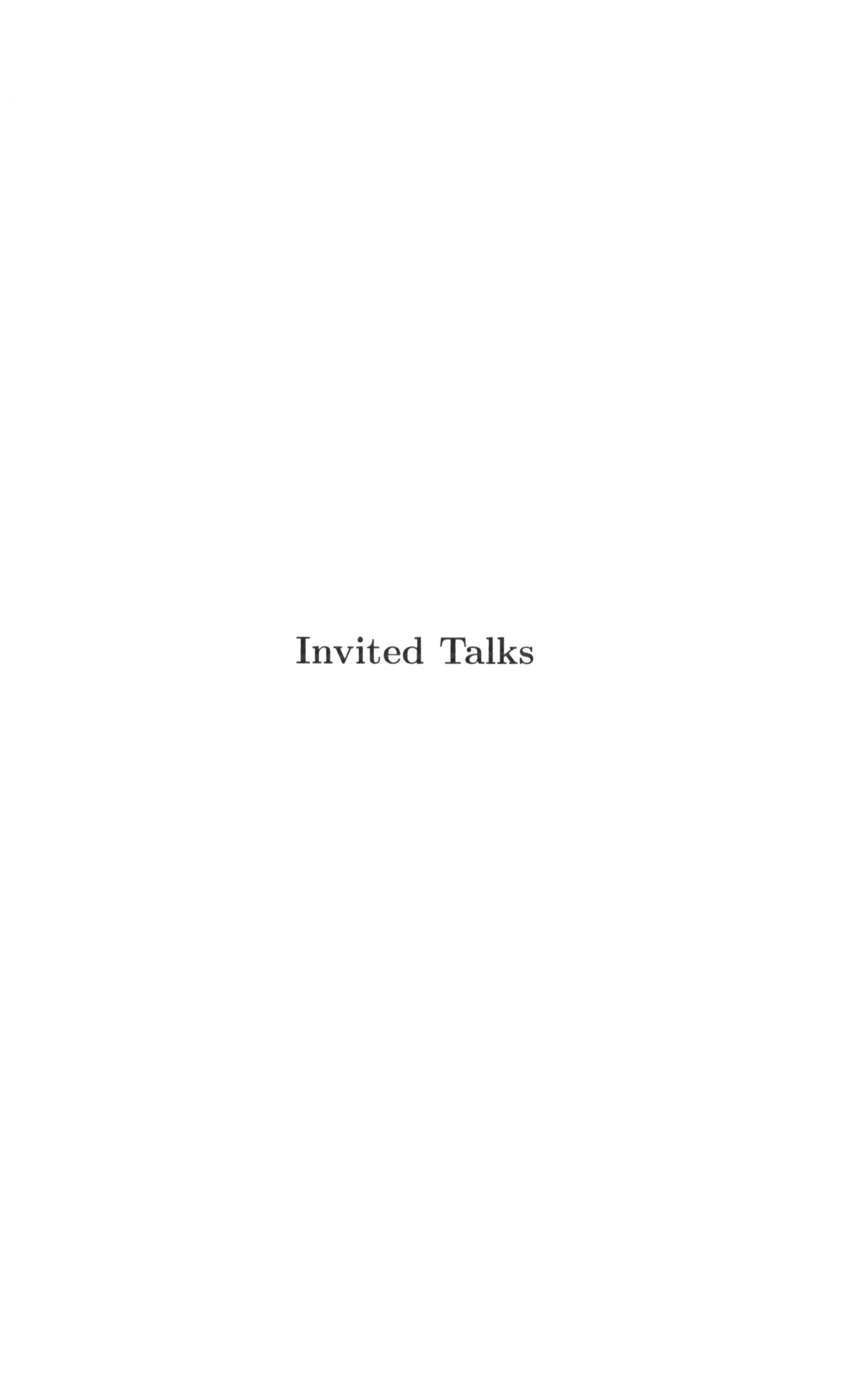

Invited Talks

The Equational Approach to Contrary-to-duty Obligations

Dov M. Gabbay

Bar-Ilan University, Ramat-Gan, Israel
King's College London, London, UK
University of Luxembourg, Luxembourg

We apply the equational approach to logic to define numerical equational semantics and consequence relations for contrary to duty obligations, thus avoiding some of the traditional known paradoxes in this area. We also discuss the connection with abstract argumentation theory. Makinson and Torre's input output logic and Governatori and Rotolo's logic of violation.

Data Structures for Emergency Planing

Cyril Gavoille

LaBRI - University of Bordeaux, Bordeaux, France

We present in this talk different techniques for quickly answer graph problems where some of the nodes may be turn off. Typical graph problems are such as connectivity or distances between pair of nodes but not only. Emergency planing for such problems is achieved by pre-processing the graphs and by virtually preventing all possible subsequent node removals. To obtain efficient data structures, the idea is to attach very little and localized information to nodes of the input graph so that queries can be solved using solely on these information. Contexts and solutions for several problems will be surveyed.

A Survey of the Data Complexity of Consistent Query Answering under Key Constraints

Jef Wijsen

Université de Mons, Mons, Belgium

This talk adopts a very elementary representation of uncertainty. A relational database is called uncertain if it can violate primary key constraints. A repair of an uncertain database is obtained by selecting a maximal number of tuples without selecting two distinct tuples of the same relation that agree on their primary key. For any Boolean query q, $\mathsf{CERTAINTY}(q)$ is the problem that takes an uncertain database **db** on input, and asks whether q is true in every repair of **db**. The complexity of these problems has been particularly studied for q ranging over the class of Boolean conjunctive queries. A research challenge is to solve the following complexity classification task: given q, determine whether $\mathsf{CERTAINTY}(q)$ belongs to complexity classes **FO**, **P**, or **coNP**-complete.

The counting variant of $\mathsf{CERTAINTY}(q)$, denoted $\sharp\mathsf{CERTAINTY}(q)$, asks to determine the exact number of repairs that satisfy q. This problem is related to query answering in probabilistic databases.

This talk motivates the problems $\mathsf{CERTAINTY}(q)$ and $\sharp\mathsf{CERTAINTY}(q)$, surveys the progress made in the study of their complexity, and lists open problems. We also show a new result comparing complexity boundaries of both problems with one another.

Table of Contents

Invited Talks

Regular Articles

The Equational Approach to Contrary-to-duty Obligations

Dov M. Gabbay

Bar-Ilan University, Ramat-Gan, Israel
King's College London, London, UK
University of Luxembourg, Luxembourg

Abstract. We apply the equational approach to logic to define numerical equational semantics and consequence relations for contrary to duty obligations, thus avoiding some of the traditional known paradoxes in this area. We also discuss the connection with abstract argumentation theory. Makinson and Torre's input output logic and Governatori and Rotolo's logic of violation.

1 Methodological Orientation

This paper gives equational semantics to contrary to duty obligations (CTDs) and thus avoids some of the known CTD paradoxes. The paper's innovation is on three fronts.

1. Extend the equational approach from classical logic and from argumentation [1,2] to deontic modal logic and contrary to duty obligations [5].
2. Solve some of the known CTD paradoxes by providing numerical equational semantics and consequence relation to CTD obligation sets.
3. Have a better understanding of argumentation semantics.

Our starting point in this section is classical propositional logic, a quite familiar logic to all readers. We give it equational semantics and define equational consequence relation. This will explain the methodology and concepts behind our approach and prepare us to address CTD obligations. We then, in Section 2, present some theory and problems of CTD obligations and intuitively explain how we use equations to represent CTD sets.

Section 3 deals with technical definitions and discussions of the equational approach to CTD obligations, Section 4 compares with input output logic, Section 5 compares with the logic of violation and and we conclude in Section 6 with general discussion and future research.

Let us begin.

1.1 Discussion and Examples

Definition 1. *Classical propositional logic has the language of a set of atomic propositions Q (which we assume to be finite for our purposes) and the connectives $\neg$ and $\wedge$. A classical model is an assignment $h : Q \mapsto \{0, 1\}$. h can be extended to all wffs by the following clauses:*

C. Beierle and C. Meghini (Eds.): FoIKS 2014, LNCS 8367, pp. 1–61, 2014.

- $h(A \wedge B) = 1$ *iff* $h(A) = h(B) = 1$
- $h(\neg A) = 1 - h(A)$

The set of tautologies are all wffs A such that for all assignments $h, h(A) = 1$.

The other connectives can be defined as usual

$$a \to b = \mathit{def.}\ \neg(a \wedge \neg b)$$
$$a \vee b = \neg a \to b = \neg(\neg a \wedge \neg b)$$

Definition 2.

1. *A numerical conjunction is a binary function* $\mu(x,y)$ *from* $[0,1]^2 \mapsto [0,1]$ *satisfying the following conditions*
   (a) μ *is associative and commutative*
   $\mu(x, \mu(y,z)) = \mu(\mu(x,y),z)$
   $\mu(x,y) = \mu(y,x)$
   (b) $\mu(x,1) = x$
   (c) $x < 1 \Rightarrow \mu(x,y) < 1$
   (d) $\mu(x,y) = 1 \Rightarrow x = y = 1$
   (e) $\mu(x,0) = 0$
   (f) $\mu(x,y) = 0 \Rightarrow x = 0$ *or* $y = 0$
2. *We give two examples of a numerical conjunction*

$$\mathbf{n}(x,y) = \min(x,y)$$
$$\mathbf{m}(x,y) = xy$$

For more such functions see the Wikipedia entry on t-norms [9]. However, not all t-norms satisfy condition (f) above.

Definition 3.

1. *Given a numerical conjunction* μ, *we can define the following numerical (fuzzy) version of classical logic.*
   (a) *An assignment is any function* $\mathbf{h}$ *from wff into* $[0,1]$.
   (b) $\mathbf{h}$ *can be extended to* $\mathbf{h}_\mu$ *defined for any formula by using* μ *by the following clauses:*
   - $\mathbf{h}_\mu(A \wedge B) = \mu(\mathbf{h}_\mu(A), \mathbf{h}_\mu(B))$
   - $\mathbf{h}_\mu(\neg A) = 1 - \mathbf{h}_\mu(A)$
2. *We call* μ*-tautologies all wffs A such that for all* $\mathbf{h}$, $\mathbf{h}_\mu(A) = 1$.

Remark 1. Note that on $\{0,1\}$, $\mathbf{h}_\mu$ is the same as h. In other words, if we assign to the atoms value in $\{0,1\}$, then $\mathbf{h}_\mu(A) \in \{0,1\}$ for any A. This is why we also refer to μ as "semantics".

The difference in such cases is in solving equations, and the values they give to the variables $0 < x < 1$.

Consider the equation arising from $(x \to x) \leftrightarrow \neg(x \to x)$. We want

$$\mathbf{h}_{\mathbf{m}}(x \to x) = \mathbf{h}_{\mathbf{m}}(\neg(x \to x))$$

We get

$$(1-\mathbf{m}(x))\mathbf{m}(x)=[1-\mathbf{m}(x)\cdot(1-\mathbf{m}(x))]$$

or equivalently

$$\mathbf{m}(x)^2-\mathbf{m}(x)+\tfrac{1}{2}=0.$$

Which is the same as

$$(\mathbf{m}(x)-\tfrac{1}{2})^2+\frac{1}{4}=0.$$

There is no real numbers solution to this equation.

However, if we use the **n** semantics we get

$$\mathbf{h_n}(x\to x)=\mathbf{h_n}(\neg(x\to x))$$

or

$$\min(\mathbf{n}(x),(1-\mathbf{n}(x))=1-\min(\mathbf{n}(x),1-\mathbf{n}(x))$$

$\mathbf{n}(x)=\frac{1}{2}$ is a solution.

Note that if we allow **n** to give values to the atoms in $\{0,\frac{1}{2},1\}$, then all formulas A will continue to get values in $\{0,\frac{1}{2},1\}$. I.e. $\{0,\frac{1}{2},1\}$ is closed under the function **n**, and the function $\nu(x)=1-x$.

Also all equations with **n** can be solved in $\{0,\frac{1}{2},1\}$.

This is not the case for **m**. Consider for the example the the equation corresponding to $x\equiv x\wedge\ldots\wedge x$, $(n+1$ times).

The equation is $x=x^{n+1}$. We have the solutions $x=0, x=1$ and all roots of unity of $x^n=1$.

Definition 4. *Let I be a set of real numbers $\{0,1\}\subseteq I\subseteq[0,1]$. Let μ be a semantics. We say that I supports μ iff the following holds:*

1. *For any $x,y\in I$, $\mu(x,y)$ and $\nu(x)=1-x$ are also in I.*
2. *By a μ expression we mean the following*
   (a) *x is a μ expression, for x atomic*
   (b) *If X and Y are μ expressions then so are $\nu(X)=(1-X)$ and $\mu(X,Y)$*
3. *We require that any equation of the form $E_1=E_2$, where E_1 and E_2 are μ expressions has a solution in I, if it is at all solvable in the real numbers.*

Remark 2. Note that it may look like we are doing fuzzy logic, with numerical conjunctions instead of t-norms. It looks like we are taking the set of values $\{0,1\}\subseteq I\subseteq[0,1]$ and allowing for assignments **h** from the atoms into I and assuming that I is closed under the application of μ and $\nu(x)$. For $\mu=\mathbf{n}$, we do indeed get a three valued fuzzy logic with the following truth table, Figure 1.

Note that we get the same system only because our requirement for solving equations is also supported by $\{0,\frac{1}{2},1\}$ for **n**.

The case for **m** is different. The values we need are all solutions of all possible equations. It is not the case that we choose a set I of truth values and close under **m**, and ν.

It is the case of identifying the set of zeros of certain polynomials (the polynomials arising from equations). This is an algebraic geometry exercise.

A	B	$\neg A$	$A \wedge B$	$A \vee B$	$A \to B$
0	0	1	0	0	1
0	$\frac{1}{2}$	1	0	$\frac{1}{2}$	1
0	1	1	0	1	1
$\frac{1}{2}$	0	$\frac{1}{2}$	0	$\frac{1}{2}$	$\frac{1}{2}$
$\frac{1}{2}$	$\frac{1}{2}$	$\frac{1}{2}$	$\frac{1}{2}$	$\frac{1}{2}$	$\frac{1}{2}$
$\frac{1}{2}$	1	$\frac{1}{2}$	$\frac{1}{2}$	1	1
1	0	0	0	1	0
1	$\frac{1}{2}$	0	$\frac{1}{2}$	1	$\frac{1}{2}$
1	1	0	1	1	1

Fig. 1.

Remark 3. The equational approach allows us to model what is considered traditionally inconsistent theories, if we are prepared to go beyond $\{0,1\}$ values. Consider the liar paradox $a \leftrightarrow \neg a$. The equation for this is (both for **m** for **n**) $a = 1 - a$ (we are writing 'a' for '$\mathbf{m}(a)$' or '$\mathbf{n}(a)$' f). This solves to $a = \frac{1}{2}$.

1.2 Theories and Equations

The next series of definitions will introduce the methodology involved in the equational point of view.

Definition 5

1. (a) *A classical equational theory has the form*

$$\Delta = \{A_i \leftrightarrow B_i \mid i = 1, 2, \ldots\}$$

   where A_i, B_i are wffs.

   (b) *A theory is called a B-theory*[1] *if it has the form*

$$x_i \leftrightarrow A_i$$

   where x_i are atomic, and for each atom y there exists at most one i such that $y = x_i$.

2. (a) *A function* **f**: *wff* $\to [0,1]$ *is an μ model of the theory if we have that* **f** *is a solution of the system of equations* $\mathbf{Eq}(\Delta)$.

$$\mathbf{h}_\mu(A_i) = \mathbf{h}_\mu(B_i), i = 1, 2, \ldots$$

   (b) *Δ is μ consistent if it has an μ model*

[1] B for Brouwer, because we are going to use Brouwer's fixed point theorem to show that theories always have models.

3. *We say that a theory Δ μ semantically (equationally) implies a theory Γ if every solution of* $\mathbf{Eq}(\Delta)$ *is also a solution of* $\mathbf{Eq}(\Gamma)$.
   We write

$$\Delta \vDash_{\mu} \Gamma.$$

   Let $\mathbb{K}$ be a family of functions from the set of wff to $[0,1]$. We say that $\Delta \vDash_{(\mu,\mathbb{K})} \Gamma$ if every μ solution $\mathbf{f}$ *of* $\mathbf{Eq}(\Delta)$ *such that $\mathbf{f} \in \mathbb{K}$ is also an μ solution of* $\mathbf{Eq}(\Gamma)$.
4. *We write*

$$A \vDash_{\mu} B$$

   iff the theory $\top \leftrightarrow A$ semantically (equationally) implies $\top \leftrightarrow B$.
   Similarly we write $A \vDash_{(\mu,\mathbb{K})} B$. In other words, if for all suitable solutions $\mathbf{f}$, $\mathbf{f}(A) = 1$ implies $\mathbf{f}(B) = 1$.

Example 1.

1. Consider $A \wedge (A \to B)$ does it $\mathbf{m}$ imply B? The answer is yes.
   Assume $\mathbf{m}(A \wedge (A \to B)) = 1$ then $\mathbf{m}(A)(1 - \mathbf{m}(A)(1 - \mathbf{m}(B))) = 1$. Hence $\mathbf{m}(A) = 1$ and $\mathbf{m}(A)(1 - \mathbf{m}(B)) = 0$. So $\mathbf{m}(B) = 1$.
   We now check whether we always have that $\mathbf{m}(A \wedge (A \to B) \to B) = 1$.
   We calculate $\mathbf{m}(A \wedge (A \to B) \to B) = [1 - \mathbf{m}(A \wedge (A \to B))(1 - \mathbf{m}(B))]$.

$$= [1 - \mathbf{m}(A)(1 - \mathbf{m}(A)(1 - \mathbf{m}(B))x(1 - \mathbf{m}(B))]$$

   Let $\mathbf{m}(A) = \mathbf{m}(B) = \frac{1}{2}$. we get

$$= [1 - \tfrac{1}{2}(1 - \tfrac{1}{2} \times \tfrac{1}{2}) \cdot \tfrac{1}{2} = 1 - \frac{3}{16} = \frac{13}{16}.$$

   Thus the deduction theorem does not hold. We have

$$A \wedge (A \to B) \vDash B$$

   but

$$\nvDash A \wedge (A \to B) \to B.$$

2. (a) Note that the theory $\neg a \leftrightarrow a$ is not $(\{0,1\}, \mathbf{m})$ consistent while it is $(\{0, \frac{1}{2}, 1\}, \mathbf{m})$ consistent.
   (b) The theory $(x \to x) \leftrightarrow \neg(x \to x)$ is not $([0,1], \mathbf{m})$ consistent but it is $(\{0, \frac{1}{2}, 1\}, \mathbf{n})$ consistent, but not $(\{0,1\}, \mathbf{n})$ consistent.

Remark 4. We saw that the equation theory $x \wedge \neg x \leftrightarrow \neg(x \wedge \neg x)$ has no solutions (no $\mathbf{m}$-models) in $[0,1]$. Is there a way to restrict $\mathbf{m}$ theories so that we are assured of solutions? The answer is yes. We look at B-theories of the form $x_i \leftrightarrow E_i$ where x_i is atomic and for each x there exists at most one clause in the theory of the form $x \leftrightarrow E$. These we called B theories. Note that if $x = \top$, we can have several clauses for it. The reason is that we can combine

$$\top \leftrightarrow E_1$$

$$\top \leftrightarrow E_2$$

into

$$\top \leftrightarrow E_1 \wedge E_2.$$

The reason is that the first two equations require

$$\mathbf{m}(E_i) = \mathbf{m}(\top) = 1$$

which is the same as

$$\mathbf{m}(E_1 \wedge E_2) = \mathbf{m}(E_1) \cdot \mathbf{m}(E_2) = 1.$$

If x is atomic different from $\top$, this will not work because

$$x \leftrightarrow E_i$$

requires $\mathbf{m}(x) = \mathbf{m}(E_i)$ while $x \leftrightarrow E_1 \wedge E_2$ requires $\mathbf{m}(x) = \mathbf{m}(E_1)\mathbf{m}(E_2)$.

The above observation is important because logical axioms have the form $\top \leftrightarrow A$ and so we can take the conjunction of the axioms and that will be a theory in our new sense.

In fact, as long as our μ satisfies

$$\mu(A \wedge B) = 1 \Rightarrow \mu(A) = \mu(B) = 1$$

we are OK.

Theorem 1. *Let Δ be a B-theory of the form*

$$x_i \leftrightarrow E_i.$$

Then for any continuous μ, Δ has a $([0,1], \mu)$ model.

Proof. Follows from Brouwer's fixed point theorem, because our equations have the form

$$\mathbf{f}(\boldsymbol{x}) = \mathbf{f}(\boldsymbol{E}(\boldsymbol{x}))$$

in $[0,1]^n$ where $\boldsymbol{x} = (x_1, \ldots, x_n)$ and $\boldsymbol{E} = (E_1, \ldots, E_n)$.

Remark 5. If we look at B-theories, then no matter what μ we choose, such theories have μ-models in $[0,1]$. We get that all theories are μ-consistent. A logic where everything is consistent is not that interesting.

It is interesting, therefore, to define classes of μ models according to some meaningful properties. For example the class of all $\{0,1\}$ models. There are other classes of interest. The terminology we use is intended to parallel semantical concepts used and from argumentation theory.

Definition 6. *Let Δ be a B-theory. Let* $\mathbf{f}$ *be a μ-model of Δ. Let A be a wff.*

1. *We say $\mathbf{f}(A)$ is crisp (or decided) if $\mathbf{f}(A)$ is either 0 or 1. Otherwise we say $\mathbf{f}(A)$ is fuzzy or undecided.*
2. *(a)* $\mathbf{f}$ *is said to be crisp if $\mathbf{f}(A)$ is crisp for all A.*

(b) *We say that* $\mathbf{f} \leq \mathbf{g}$, *if for all* A, *if* $\mathbf{f}(A) = 1$ *then* $\mathbf{g}(A) = 1$, *and if* $\mathbf{f}(A) = 0$ *then* $\mathbf{g}(A) = 0$.
We say $\mathbf{f} < \mathbf{g}$ *if* $\mathbf{f} \leq \mathbf{g}$ *and for some* $A, \mathbf{f}(A) \notin \{0, 1\}$ *but* $\mathbf{g}(A) \in \{0, 1\}$.

Note that the order relates to crisp values only.

3. *Define the* μ*-crisp (or* μ*-stable) semantics for* Δ *to be the set of all crisp* μ*-model of* Δ.
4. *Define the* μ*-grounded semantics for* Δ *to be the set of all* μ*-models* $\mathbf{f}$ *of* Δ *such that there is no* μ*-model* $\mathbf{g}$ *of* Δ *such that* $\mathbf{g} < \mathbf{f}$.
5. *Define the* μ*-preferred semantics of* Δ *to be the set of all* μ*-models* $\mathbf{f}$ *of* Δ *such that there is no* μ*-model* $\mathbf{g}$ *of* Δ *with* $\mathbf{f} < \mathbf{g}$.
6. *If* $\mathbb{K}$ *is a set of* μ *models, we therefore have the notion of* $\Delta \vDash_{\mathbb{K}} \Gamma$ *for two theories* Δ *and* Γ.

1.3 Generating B-theories

Definition 7. *Let* S *be a finite set of atoms and let* R_a *and* R_s *be two binary relations on* S. *We use* $\mathcal{A} = (S, R_a, R_s)$ *to generate a* B*-theory which we call the argumentation network theory generated on* S *from the attack relation* R_a *and the support relation* R_s.

For any $x \in S$, *let* $y_1, \ldots, y_m$ *be all the elements* y *of* S *such that* yR_ax *and let* $z_1, \ldots, z_n$ *be all the elements* z *of* S *such that* xR_sz *(of course* m, n *depend on* x*). Write the theory* $\Delta_{\mathcal{A}}$.

$$\{x \leftrightarrow \bigwedge z_j \wedge \bigwedge \neg y_i \mid x \in S\}$$

We understand the empty conjunction as $\top$.

These generate equations

$$x = \min(z_j, 1 - y_i)$$

using the $\mathbf{n}$ *function or*

$$x = (\Pi_j z_j)(\Pi_i(1 - y_i))$$

using the $\mathbf{m}$ *function.*

Remark 6.

1. If we look at a system with attacks only of the form $\mathcal{A} = (S, R_a)$ and consider the $\mathbf{n}(\min)$ equational approach for $[0, 1]$ then $\mathbf{n}$ models of the corresponding B-theory $\Delta_{\mathcal{A}}$ correspond exactly to the complete extensions of (S, R_a). This was extensively investigated in [1,2]. The semantics defined in Definition 6, the stable, grounded an preferred $\mathbf{n}$-semantics correspond to the same named semantics in argumentation, when restricted to B-theories arising from argumentation.
If we look at μ other than $\mathbf{n}$, example we look at $\mu = \mathbf{m}$, we get different semantics and extensions for argumentation networks. For example the network of Figure 2 has the $\mathbf{n}$ extensions $\{a = 1, b = 0\}$ and $\{a = b = \frac{1}{2}\}$
while it has the unique $\mathbf{m}$ extension $\{a = 1, b = 0\}$.

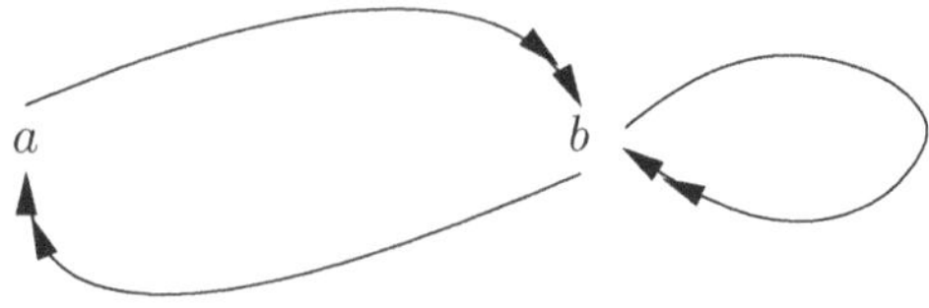

Fig. 2.

2. This correspondence suggests new concepts in the theory of abstract argumentation itself. Let $\Delta_\mathcal{A}, \Delta_\mathcal{B}$ be two B-theories arising from two abstract argumentation system $\mathcal{A} = (S, R_\mathcal{A})$ and $\mathcal{B} = (S, R_\mathcal{B})$ based on the same set S. Then the notion of $\Delta_\mathcal{A} \vDash_\mathbb{K} \Delta_\mathcal{B}$ as defined in Definition 5 suggest the following consequence relation for abstract argumentation theory.
   - $\mathcal{A} \vDash_\mathbb{K} \mathcal{B}$ iff any $\mathbb{K}$-extension ($\mathbb{K}$=complete, grounded, stable, preferred) of $\mathcal{A}$ is also a $\mathbb{K}$-extension of $\mathcal{B}$.

   So, for example, the network of Figure 3(a) semantically entails the network of Figure 3(b).

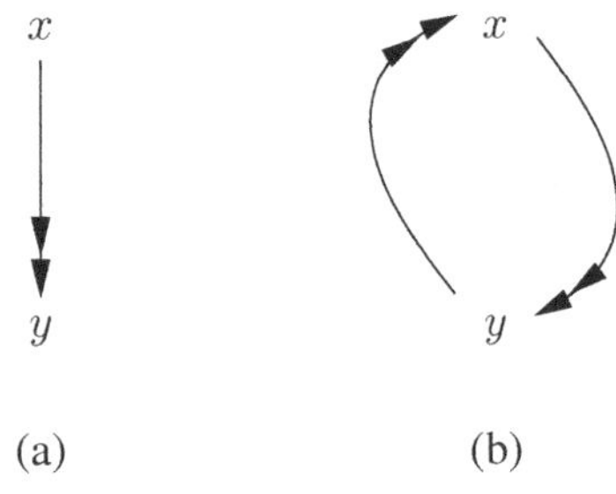

(a) (b)

Fig. 3.

Remark 7. We can use the connection of equational B-theories with argumentation networks to export belief revision and belief merging from classical logic into argumentation. There has been considerable research into merging of argumentation networks. Classical belief merging offers a simple solution. We only hint here, the full study is elsewhere [10].

Let $\mathcal{A}_i = (S, R_i), i = 1, \ldots, n$, be the argumentation networks to be merged based on the same S. Let Δ_i be the corresponding equational theories with the corresponding semantics, based on $\mathbf{n}$. Let $\mathbf{f}_i$ be respective models of Δ_i and let μ be a merging function, say $\mu = \mathbf{m}$.

Let $\mathbf{f} = \mu(\mathbf{f}_1, \ldots, \mathbf{f}_n)$. Then the set of all such $\mathbf{f}$s is the semantics for the merge result. Each such an $\mathbf{f}$ yields an extension.

Remark 8. The equational approach also allows us to generate more general abstract argumentation networks. The set S in (S, R_a) need not be a set of atoms. It can be a set of wffs.

Thus following Definition 7 and remark 6, we get the equations (for each A, B_j and where B_j are all the attackers of A:

$$\mathbf{f}(A) = \mu(\mathbf{f}(\neg B_1), \ldots,).$$

There may not be a solution.

2 Equational Modelling of Contrary to Duty Obligations

This section will use our μ-equational logic to model contrary to duty (CTD) sets of obligations. So far such modelling was done in deontic logic and there are difficulties involved. Major among them is the modelling of the Chisholm set [11].

We are going to use our equational semantics and consequence of Section 1 and view the set of contrary to duty obligations as a generator for an equational theory. This will give an acceptable paradox free semantics for contrary to duty sets.

We shall introduce our semantics in stages. We start with the special case of the generalised Chisholm set and motivate and offer a working semantical solution. Then we show that this solution does not work intuitively well for more general sets where there are loops. Then we indicate a slight mathematical improvement which does work. Then we also discuss a conceptual improvement.

The reader might ask why not introduce the mathematical solution which works right from the start? The answer is that we do not do this for reasons of conceptual motivation, so we do not appear to be pulling a rabbit out of a hat!

We need first to introduce the contrary to duty language and its modelling problems.

2.1 Contrary to Duty Obligations

Consider a semi-formal language with atomic variables $Q = \{p, q, r, \ldots\}$ the connective $\rightarrow$ and the unary operator $\bigcirc$. We can write statements like

1. $\bigcirc\neg$ fence
   You should not have a fence
2. fence $\rightarrow \bigcirc$ whitefence
   If you do have a fence it should be white.
3. Fact: fence

We consider a generalised Chisholm set of contrary to duty obligations (CTD) of the form

$$Oq_0$$

and for $i = 0, \ldots, n$ we have the CTD is

$$q_i \rightarrow Oq_{i+1}$$
$$\neg q_i \rightarrow O\neg q_{i+1}$$

and the facts $\pm q_j$ for some $j \in J \subseteq \{0, 1, \ldots, n+1\}$. Note that for the case of $n = 1$ and fact $\neg q_0$ we have the Chisholm paradox.

2.2 Standard Deontic Logic and Its Problems

A logic with modality $\Box$ is **KD** modality if we have the axioms

K0 All substitution instances of classical tautologies
K1 $\Box(p \wedge q) \equiv (\Box p \wedge \Box q)$
K2 $\vdash A \Rightarrow \vdash \Box A$
D $\neg\Box\bot$

It is complete for frames of the form (S, R, a) where $S \neq \varnothing$ is a set of possible worlds, $a \in S, R \subseteq S \times S$ and $\forall x \exists y(xRy)$.

Standard Deontic Logic **SDL** is a **KD** modality O. We read $u \vDash Op$ as saying p holds in all ideal worlds relative to u, i.e. $\forall t(uRt \Rightarrow t \vDash p)$. So the set of ideal worlds relative to u is the set $I(u) = \{t \mid uRt\}$.

The **D** condition says $I(x) \neq \varnothing$ for $x \in S$.

Following [8], let us quickly review some of the difficulties facing **SDL** in formalizing the Chisholm paradox.

The Chisholm Paradox

A. Consider the following statements:
   1. It ought to be that a certain man go to the assistance of his neighbour.
   2. It ought to be that if he does go he tell them he is coming.
   3. If he does not go then he ought not to tell them he is coming.
   4. He does not go.

   It is agreed that intuitively (1)–(4) of Chisholm set A are consistent and totally independent of each other. Therefore it is expected that their formal translation into logic **SDL** should retain these properties.

B. Let us semantically write the Chisholm set in semiformal English, where p and q as follows, p means HELP and q means TELL.
   1. Obligatory p.
   2. $p \to$ Obligatory q.
   3. $\neg p \to$ Obligatory $\neg q$.
   4. $\neg p$.

   Consider also the following:
   5. p.
   6. Obligatory q.
   7. Obligatory $\neg q$.

We intuitively accept that (1)–(4) of B are consistent and logically independent of each-other. Also we accept that (3) and (4) imply (7), and that (2) and (5) imply (6). Note that some authors would also intuitively expect to conclude (6) from (1) and (2).

Now suppose we offer a logical system **L** and a translation τ of (1), (2), (3), (4) of Chisholm into **L**.

For example **L** could be Standard Deontic Logic or **L** could be a modal logic with a dyadic modality $O(X/Y)$ (X is obligatory in the context of Y). We expect some coherence conditions to hold for the translation, as listed in Definition 8.

Definition 8. (Coherence conditions for representing contrary to duty obligations set in any logic)
We now list coherence conditions for the translation τ and for **L**.

We expect the following to hold.

- *(a)* *"Obligatory X" is translated the same way in (1), (2) and (3).*
  Say τ(Obligatory X)=$\varphi(X)$.
- *(b)* *(2) and (3) are translated the same way, i.e., we translate the form:*
  (23): $X \to$ Obligatory Y
  to be $\psi(X,Y)$ and the translation does not depend on the fact that we have (4) $\neg p$ as opposed to (5) p.
  Furthermore, we might, but not necessarily, expect $\psi(X/\top) = \varphi(X)$.
- *(c)* *if X is translated as $\tau(X)$ then (4) is translated as $\neg\tau(X)$, the form (23) is translated as $\psi(\tau(X), \tau(Y))$ and (1) is translated as $\varphi(\tau(X))$.*
- *(d)* *the translations of (1)–(4) remain independent in* **L** *and retain the connections that the translations of (2) and (5) imply the translation of (6), and the translations of (3) and (4) imply the translation of (7).*
- *(e)* *the translated system maintains its properties under reasonable substitution in* **L**.
  The notion of reasonable substitution is a tricky one. Let us say for the time being that if we offer a solution for one paradox, say $\Pi_1(p, q, r, \ldots)$ and by substitution for $p, q, r, \ldots$ we can get another well known paradox Π_2, then we would like to have a solution for Π_2. This is a reasonable expectation from mathematical reasoning. We give a general solution to a general problem which yields specific solutions to specific problems which can be obtained from the general problem.
- *(f)* *the translation is essentially linguistically uniform and can be done item by item in a uniform way depending on parameters derived from the entire database. To explain what we mean consider in classical logic the set*
  (1) p
  (2) $p \to q$.
  To translate it into disjunctive normal form we need to know the number of atoms to be used. Item (1) is already in normal form in the language of $\{p\}$ but in the language of $\{p, q\}$ its normal form is $(p \wedge q) \vee (p \wedge \neg q)$. If we had another item
  (3) r
  then the normal form of p in the language of $\{p, q, r\}$ would be
  $(p \wedge q \wedge r) \vee (p \wedge q \wedge \neg r) \vee (p \wedge \neg q \wedge r) \vee (p \wedge \neg q \wedge \neg r)$.
  The moral of the story is that although the translation of (1) is uniform algorithmically, we need to know what other items are in the database to set some parameters for the algorithm.

Jones and Pörn, for example, examine in [8] possible translations of the Chisholm (1)–(4) into **SDL**. They make the following points:

(1) If we translate according to, what they call, option a:
  (1a) Op
  (2a) $O(p \to q)$

(3a) $\neg p \to O\neg q$
(4a) $\neg p$
then we do not have consistency, although we do have independence

(2) If we translate the Chisholm item (2) according to what they call option b:
(2b) $p \to Oq$
then we have consistency but not independence, since (4a) implies logically (2b).

(3) If (3a) is replaced by
(3b) $O(\neg p \to \neg q)$
then we get back consistency but lose independence, since (1a) implies (3b).

(4) Further, if we want (2) and (5) to imply (6), and (3) and (4) to imply (7) then we cannot use (3b) and (2a).

The translation of the Chisholm set is a "paradox" because known translations into Standard Deontic Logic (the logic with O only) are either inconsistent or dependent.

All the above statements together are logically independent and are consistent. Each statement is independent of all the others. If we want to embed the (model them) in some logic, we must preserve these properties and correctly get all intuitive inferences from them.

Remark 9. We remark here that the Chisholm paradox has a temporal dimension to it. The $\pm$tell comes before the $\pm$go. In symbols, the $\pm q$ is temporally before the $\pm p$. This is not addressed in the above discussion.

Consider a slight variation:

1. It ought to be that a certain man go to the assistance of his neighbour.
2. It ought to be that if he does not go he should write a letter of explanation and apology.
3. If he does go, then he ought not write a letter of explanation and apology.
4. He does not go.

Here $p =$ he does go and $q =$ he does not write a letter. Here q comes after p.

It therefore makes sense to supplement the Chisholm paradox set with a temporal clause as follows:

1. p comes temporally before q.

In the original Chisholm paradox the supplement would be:

1. Tell comes temporally before go.

2.3 The Equational Approach to CTD

We are now ready to offer equational semantics for CTD. Let us summarise the tools we have so far.

1. We have μ semantics for the language of classical logic.
2. Theories are sets of equivalences of the form $E_1 \leftrightarrow E_2$.
3. We associate equations with such equivalences.
4. Models are solutions to the equations.

5. Using models, we define consequence between theories.
6. Axioms have the for $\top \leftrightarrow E$
7. B-theories have the form $x \leftrightarrow E$, where x is atomic and E is unique to x.
8. We always have solutions for equations corresponding to B-theories.

Our strategy is therefore to associate a B-theory $\Delta(\mathbb{C})$ with any contrary to duty set $\mathbb{C}$ and examine the associated μ-equations for a suitable μ. This will provide semantics and consequence for the CTD sets and we will discuss how good this representation is.

The perceptive reader might ask, if Obligatory q is a modality, how come we hope to successfully model it in μ classical logic? Don't we need modal logic of it? This is a good question and we shall address it later. Of course modal logic can be translated into classical logic, so maybe the difficulties and paradoxes are "lost in translation". See Remark 15.

Definition 9. *1. Consider a language with atoms, the semi-formal $\rightarrow$ and $\neg$ and a semi-formal connective O.*
A contrary to duty expression has the form $x \rightarrow Oy$ where x and y are literals, i.e. either atoms q or negations of atoms $\neg q$, and where we also allow for x not to appear. We might write $\top \rightarrow Oy$ in this case, if it is convenient.
2. Given a literal x and a set $\mathbb{C}$ of CTD expressions, then the immediate neighbourhood of x in $\mathbb{C}$. is the set $\mathbb{N}_x$ of all expressions from $\mathbb{C}$ of the form

$$z \rightarrow Ox$$

or the form

$$x \rightarrow Oy.$$

3. A set $\mathbb{F}$ of facts is just a set of literals.
4. A general CTD system is a pair $(\mathbb{C}, \mathbb{F})$
5. A Chisholm CTD set $\mathbb{CH}$ has the form

$$\begin{aligned} &x_i \rightarrow Ox_{i+1} \\ &\neg x_i \rightarrow O\neg x_{i+1} \\ &Ox_1 \end{aligned}$$

where $1 \leq i \leq m$ and x_i are literals (we understand that $\neg\neg x$ is x).

Example 2. Figure 4 shows a general CTD set

$$\mathbb{C} = \{a \rightarrow Ob, b \rightarrow O\neg a\}$$

Figure 5 shows a general Chisholm set. We added an auxiliary node x_0 as a starting point.

Figure 6 shows a general neighbourhood of a node x.

We employed in the figures the device of showing, whenever $x \rightarrow Oy$ is given, two arrows, $x \rightarrow y$ and $x \twoheadrightarrow \neg y$. The single arrow $x \rightarrow y$ means "from x go to y" and the double arrow $x \twoheadrightarrow \neg y$ means "from x do not go to $\neg y$".

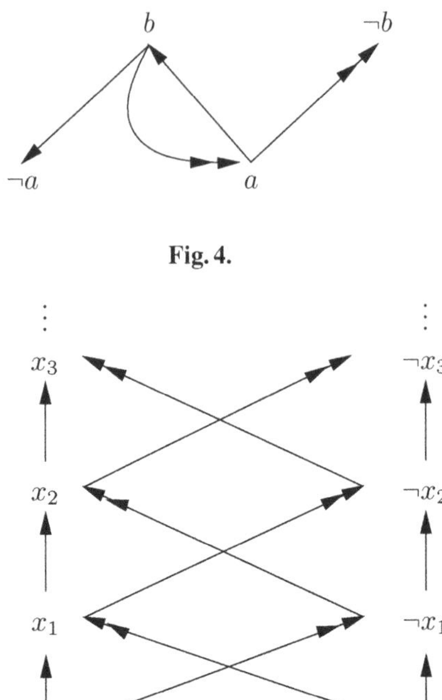

Fig. 4.

Fig. 5.

Remark 10. In Figures 4–6 we understand that an agent is at the starting point x_0 and he has to go along the arrows $\rightarrow$ to follow his obligations. He should not go along any double arrow, but if he does, new obligations (contrary to duty) appear.

This is a mathematical view of the CTD. The obligations have no temporal aspect to them but mathematically there is an obligation progression $(\pm x_0, \pm x_1, \pm x_2, \ldots)$.

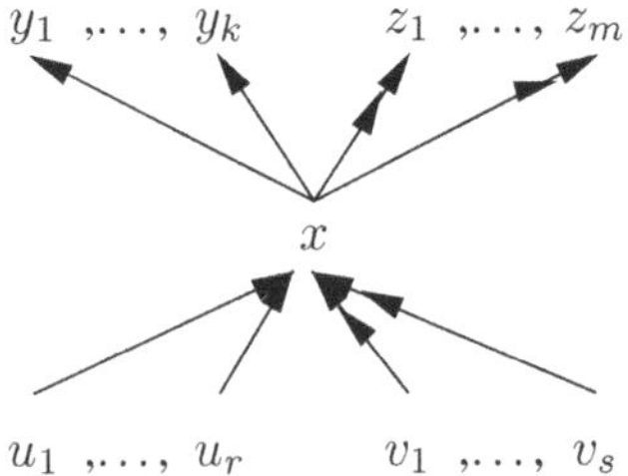

Fig. 6.

In the Chisholm example, the obligation progression is (± go, ±tell), while the practical temporal real life progression is (±tell, ±go). We are modelling the obligation progression.

To be absolutely clear about this we give another example where there is similar progression. Take any Hilbert axiom system for classical logic. The consequence relation $A \vdash B$ is timeless. It is a mathematical relation. But in practice to show $A \vdash B$ from the axioms, there is a progression of substitutions and uses of modus ponens. This is a mathematical progression of how we generate the consequence relations.

Remark 11. We want to associate equations with a given CTD set. This is essentially giving semantics to the set. To explain the methodology of what we are doing, let us take an example from the modal logic **S4**. This modal logic has wffs of the form $\Box q$. To give semantics for $\Box q$ we need to agree on a story for "$\Box$" which respects the logical theorems which "$\Box$" satisfies (completeness theorem). The following are possible successful stories about "$\Box$" for which there is completeness.

1. Interpret $\Box$ to mean provability in Peano arithmetic.
2. $\Box q$ means that q holds in all possible accessible situations (Kripke models).
3. $\Box$ means topological interior in a topological space.
4. $\Box$ means the English progressive tense:
   $\Box$ eat = "is eating"
5. $\Box$ means constructive provability.

For the case of CTD we need to adopt a story respecting the requirement we have on CTD.

Standard deontic logic **SDL** corresponds to the story that the meaning of OA in a world is that A holds in all accessible relative ideal worlds. It is a good story corresponding to the intuition that our obligations should take us to a better worlds. Unfortunately, there are difficulties with this story, as we have seen.

Our story is different. We imagine we are in states and our obligations tell us where we can and where we cannot go from our state. This is also intuitive. It is not descriptive as the ideal world story is, but it is operational , as real life is.

Thus in Figure 6 an agent at node x wants to say that he is a "good boy". So at x he says that he intends to go to one of $y_1, \ldots, y_k$ and that he did not come to x from $v_1, \ldots, v_k$, where the obligation was not to go to x.

Therefore the theory we suggest for node x is

$$x \leftrightarrow (\bigwedge_i y_i \wedge \bigwedge_j \neg v_j)$$

We thus motivated the following intuitive, but not final, definition.

Let $\mathbb{C}$ be a CTD set and for each x let $\mathbb{N}_x$ be its neighbourhood as in Figure 6. We define the theory $\Delta(\mathbb{C})$ to be

$$\{x \leftrightarrow (\bigwedge_i y_i \wedge \bigwedge_j \neg v_j) \mid \text{ for all } \mathbb{N}_x\}. \qquad (*1)$$

This definition is not final for technical reasons. We have literals "$\neg q$"and we do not want equivalences of the form $\neg q \leftrightarrow E$. So we introduce a new atom $\bar{q}$ to represent $\neg q$ with the theory $\bar{q} \leftrightarrow \neg q$.

So we take the next more convenient definition.

Definition 10.

1. *Let $\mathbb{C}$ be a CTD set using the atoms Q. Let $Q^* = Q \cup \{\bar{q} \mid q \in Q\}$, where $\bar{q}$ are new atoms.*
   Consider $\mathbb{C}^*$ gained from $\mathbb{C}$ by replacing any occurrence of $\neg q$ by $\bar{q}$, for $q \in Q$. Using this new convention Figure 5 becomes Figure 7.

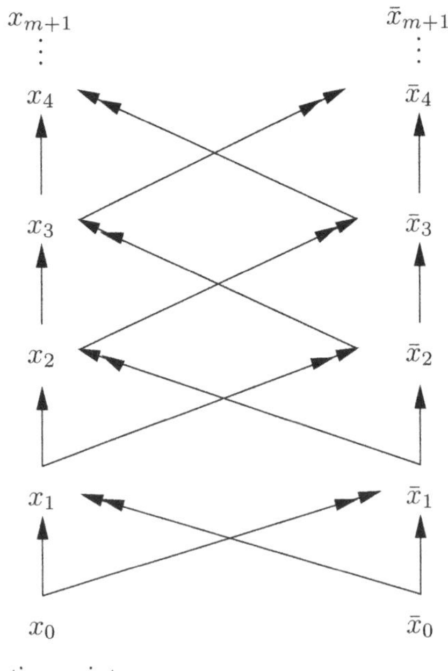

Fig. 7.

2. *The theory for the CTD set represented by Figure 7 is therefore*

$$
\begin{aligned}
&x_0 \leftrightarrow \top, \bar{x}_0 \leftrightarrow \bot \\
&x_0 \leftrightarrow x_1, \bar{x}_0 \leftrightarrow \bar{x}_1 \\
&x_i \leftrightarrow x_{i+1} \wedge \bar{x}_{i-1} \\
&\bar{x}_i \leftrightarrow \bar{x}_{i+1} \wedge x_{i-1} \\
&\bar{x}_i \leftrightarrow \neg x_i \\
&x_{m+1} \leftrightarrow \bar{x}_m \\
&\bar{x}_{m+1} \leftrightarrow x_m \\
&\textit{for } 1 \leq i \leq m
\end{aligned}
$$

   The above is not a B-theory. The variable $\bar{x}_i$ has two clauses associated with it. (x_0 is OK because the second equation is $\top$). So is $\bar{x}_0$.
   It is convenient for us to view clause $\bar{x}_i = \neg x_i$ as an integrity constraint. So we have a B-theory with some additional integrity constraints.
   Note also that we regard all x_i and $\bar{x}_i$ as different atomic letters. If some of them are the same letter, i.e. $x_i = x_j$ then we regard that as having further integrity constraints of the form $x_i \leftrightarrow x_j$.
3. *The equations corresponding to this theory are*

$$
\begin{aligned}
&x_0 = 1, \bar{x}_0 = 0 \\
&x_0 = x_1, \bar{x}_0 = \bar{x}_1 \\
&x_i = \min(x_{i+1}, 1 - \bar{x}_{i-1}) \\
&\bar{x}_i = \min(\bar{x}_{i+1}, 1 - x_{i-1}) \\
&\bar{x}_i = 1 - x_i \\
&x_{m+1} = 1 - \bar{x}_m \\
&\bar{x}_{m+1} = 1 - x_m \\
&\textit{for } 1 \leq i \leq m
\end{aligned}
$$

   Remember we regard the additional equation

$$\bar{x}_i = 1 - x_i$$

   as an integrity constraint.
   Note also that we regard all x_i and $\bar{x}_i$ as different atomic letters. If some of them are the same letter, i.e. $x_i = x_j$ then we regard that as having further integrity constraints of the form $x_i \leftrightarrow x_j$. The rest of the equations have a solution by Brouwer's theorem. We look at these solutions and take only those which satisfy the integrity constraints. There may be none which satisfy the constraints, in which case the system overall has no solution!
4. *The dependency of variables in the equations of Figure 7 is described by the relation $x \Rightarrow y$ reading (x depends on y), where*

$$x \Rightarrow y = \textit{def.}\ (x \to y) \vee (y \twoheadrightarrow x).$$

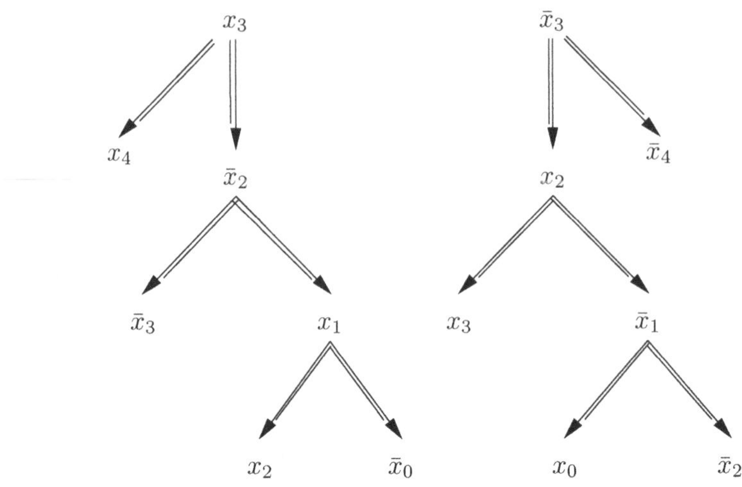

Fig. 8.

Figure 8 shows the variable dependency of the equations generated by Figure 7 up to level 3

Lemma 1.

1. *The equations associated with the Chisholm set of Figure 7 have the following unique solution, and this solution satisfies the integrity constraints:*

$$x_0 = 1, x_i = 1, \bar{x}_i = 0, \text{ for } 0 \leq i \leq m+1$$

2. *All the equations are independent.*

Proof.

1. By substitution we see the proposed solution is actually a solution. It is unique because $x_0 = 1$ and the variable dependency of the equations, as shown in Figure 8, is acyclic.
2. Follows from the fact that the variable dependency of the equations is acyclic. The variable x_i can depend only on the equations governing the variables below it in the dependency graph. Since it has the last equation in the tree, it cannot be derived from the equations below it.

Remark 12. We mentioned before that the theory (*1) and its equations above do not work for loops. Let us take the set $a \to \bigcirc\neg a$.

The graph for it, according to our current modelling would be Figure 9.

The equations for this figure would be

$$a = \min(1 - a, \bar{a})$$
$$a = 1 - \bar{a}$$

Fig. 9.

which reduces to

$$a = 1 - a$$
$$a = \tfrac{1}{2}$$

It does not have a consistent $\{0, 1\}$ solution.

We can fix the situation by generally including the integrity constraints $\bar{x} = 1 - x$ in the graph itself.

So Figure 9 becomes Figure 10, and the equations become

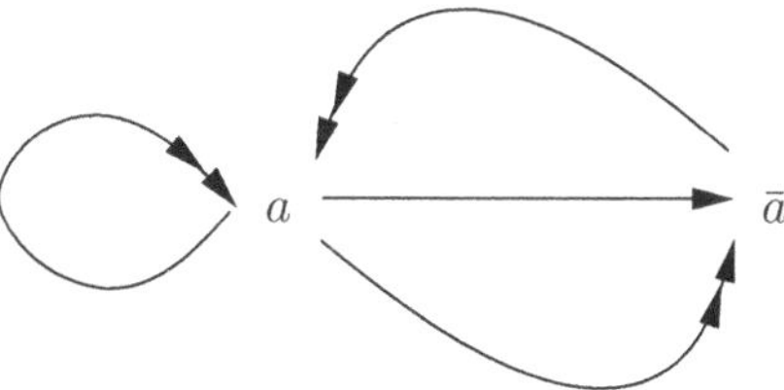

Fig. 10.

$$a = \min(\bar{a}, 1 - a, 1 - \bar{a})$$
$$\bar{a} = 1 - a$$

The two equations reduce to

$$a = \min(a, 1 - a)$$

which has the solution

$$a = 0, \bar{a} = 1$$

which fits our intuition.

Let us call this approach, (namely the approach where we do not view the equations $\bar{x} = 1 - x$ as integrity constraints but actually insert instead double arrow in the graph itself) the mathematical approach. What we have done here is to incorporate the integrity constraints $\bar{x} = 1 - x$ into the graph. Thus Figure 7 would become Figure 11, and the equations for the figure would become

$$\begin{aligned} x_i &= \min(x_{i+1}, 1 - \bar{x}_i, 1 - \bar{x}_{i-1}) \\ \bar{x}_i &= \min(\bar{x}_{i+1}, 1 - x_i, 1 - x_{i-1}) \\ x_0 &= 1, \bar{x}_0 = 0 \\ x_{m+1} &= \min(1 - \bar{x}_{m+1}, 1 - \bar{x}_m) \\ \bar{x}_{m+1} &= min(1 - x_{m+1}, 1 - x_m) \end{aligned}$$

for $1 \leq i \leq m$.

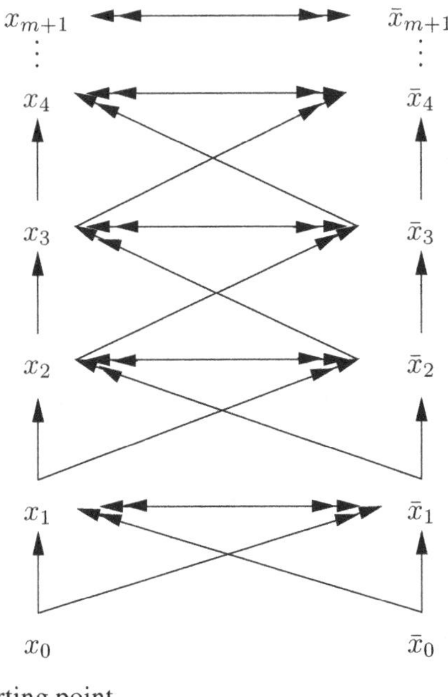

Fig. 11.

For the Chisholm set, we still get the same solution for these new equations, namely

$$\begin{aligned} x_0 = x_1 = \ldots x_{m+1} = 1 \\ \bar{x}_0 = \bar{x}_1 = \ldots = \bar{x}_{m+1} = 0 \end{aligned}$$

The discussion that follows in Definition 11 onwards applies equally to both graphs. We shall discuss this option in detail in Subsection 2.4.

The reader should note that we used here a mathematical trick. In Figure 11, there are two conceptually different double arrows. The double arrow $x_i \twoheadrightarrow x_{i+1}$ comes from an obligation $x_i \rightarrow \bigcirc x_{i+1}$, while the double arrows $x \twoheadrightarrow \bar{x}$ and $\bar{x} \twoheadrightarrow x$ come from logic (because $\bar{x} = \neg x$). We are just arbitratily mixing them in the graph!

Definition 11. *Consider Figure 7. Call this graph by $\mathbb{G}(m+1)$. We give some definitions which analyse this figure.*

First note that this figure can be defined analytically as a sequence of pairs

$$((x_0, \bar{x}_0), (x_1, \bar{x}_1), \ldots, (x_{m+1}, \bar{x}_{m+1})).$$

The relation $\to$ can be defined between nodes as the set of pairs $\{(x_i, x_{i+1})$ and $(\bar{x}_i, \bar{x}_{i+1})$ for $i = 0, 1, \ldots, m\}$. The relation $\twoheadrightarrow$ can be defined between nodes as the set of pairs $\{(x_i, \bar{x}_{i+1})$ and $(\bar{x}_i, x_{i+1})$ for $i = 0, 1, \ldots, m\}$. The starting point is a member of the first pair, in this case it is x_0, the left hand element of the first pair in the sequence, but we could have chosen $\bar{x}_0$ as the starting point.

1. *Let xRy be defined as $(x \to y) \vee (x \twoheadrightarrow y)$ and let R^* be the transitive and reflexive closure of R.*
2. *Let z be either x_i or $\bar{x}_i$. The truncation of $\mathbb{G}(m+1)$ at z is the subgraph of all points above z including z and $\bar{z}$ and all the arrow connections between them.*

   $$\mathbb{G}_z = \{y | zR^*y\} \cup \{\bar{z}\}$$

   We take z as the starting point of $\mathbf{G}(m+1)_z$. Note that $\mathbb{G}(m+1)_z$ is isomorphic to $\mathbb{G}(m+1-i)$. It is the same type of graph as $\mathbb{G}(m+1)$, only it starts at z. The corresponding equations for $\mathbb{G}_z$ will require $z = 1$.
3. *A path in the graph is a full sequence of points $(x_0, z_1, \ldots, z_{m+1})$ where z_i is $\bar{x}_i$ or x_i.*[2]
4. *A set of "facts" $\mathbb{F}$ in the graph is a set of nodes choosing at most exactly one of each pair $\{x_i, \bar{x}_i\}$.*
5. *A set of facts $\mathbb{F}$ restricts the possible paths by stipulating that the paths contain the nodes in the facts.*

Example 3. Consider Figure 7. The following is a path Π in the graph

$$\Pi = (x_0, x_1, x_2, x_3, \ldots, x_{m+1})$$

If we think in terms of an agent going along this path, then this agent committed two violations. Having gone to $\bar{x}_1$ instead of to x_1, he committed the first violation. From $\bar{x}_1$, the CTD says he should have gone to $\bar{x}_2$, but he went to x_2 instead. This is his second violation. After that he was OK.

Now look at the set of facts $= \{\bar{x}_1, x_2\}$. This allows for all paths starting with $(x_0, \bar{x}_1, x_2, \ldots)$. So our agent can still commit violations after x_2. We need more facts about his path.

---

[2] Note that the facts are sets of actual nodes. We can take the conjunction of the actual nodes as a formula faithfully representing the set of facts. Later on in this paper we will look at an arbitrary formula ϕ as generating the set of facts $\{y|y$ is either x_i or $\neg x_i$, unique for each i, such that $\phi| - y\}$.

According to this definition, $\phi = x_1 \vee x_2$, generates no facts. We will, however, find it convenient later in the paper, (in connection with solving the Miner's Paradox, Remark 20 below) to regard a disjunction as generating several possible sets of facts, one for each disjunct. See also Remark 19 below.

Suppose we add the fact $\bar{x}_4$.So our set is now $\mathbb{F} = \{\bar{x}_1, x_2, \bar{x}_4\}$.

We know now that the agent went from x_2 onto $\bar{x}_4$. The question is, did he pass through $\bar{x}_3$? If he goes to x_3, there is no violation and from there he goes to $\bar{x}_4$, and now there is violation.

If he goes to x_3, then the violation is immediate but when he goes from $\bar{x}_3$ to $\bar{x}_4$, there is no violation.

The above discussion is a story. We have to present it in terms of equations, if we want to give semantics to the facts.

Example 4. Let us examine what is the semantic meaning of facts. We have given semantic meaning to a Chisholm set $\mathbb{C}$ of contrary to duties; we constructed the graph, as in Figure 7 and from the graph we constructed the equations and we thus have equational semantics for $\mathbb{C}$.

We now ask what does a fact do semantically?

We know what it does in terms of our story about the agent. We described it in Example 3. What does a fact do to the graph? Let us take as an example the fact $\bar{x}_3$ added to the CTD set of Figure 7. What does it do? The answer is that it splits the figure into two figures, as shown in Figures 12 and 13.

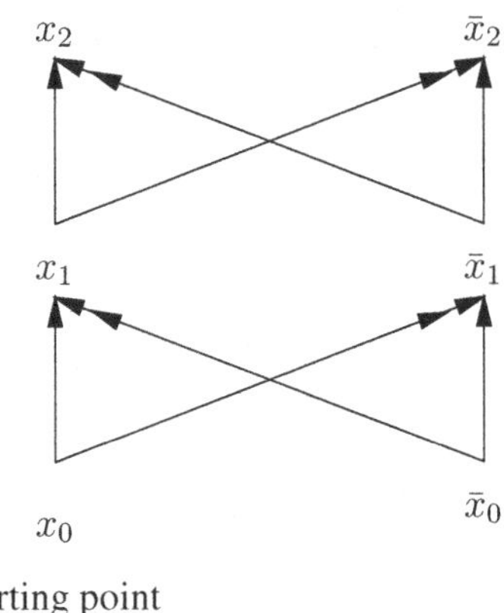

Fig. 12.

Note that Figure 13 is the truncation of Figure 7 at $\bar{x}_3$, and Figure 12 is the complement of this truncation.

Thus the semantical graphs and equations associated with $(\mathbb{C}, \{\bar{x}_3\})$ are the two figures, Figure 12 and Figure 13 and the equations they generate.

The "facts" operation is associative. Given another fact, say z it will be in one of the figures and so that figure will further split into two.

Definition 12. *Given a Chisholm system* $(\mathbb{C}, \mathbb{F})$ *as in Definition 9 we define its semantics in terms of graphs and equations. We associate with it with following system of*

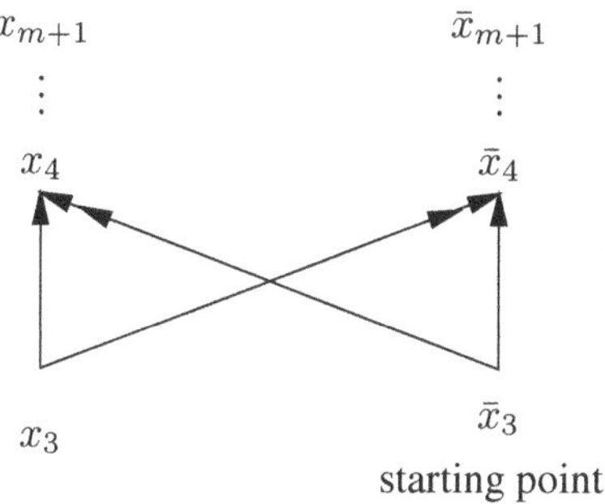

Fig. 13.

graphs (of the form of Figure 7) and these graphs will determine the equations, as in Definition 10.

The set $\mathbb{C}$ has a graph $\mathbb{G}(\mathbb{C})$. The set $\mathbb{F}$ can be ordered according to the relation R in the graph $\mathbb{G}(\mathbb{C})$ as defined in Definition 11. Let $(z_1, \ldots, z_k)$ be the ordering of $\mathbb{F}$.

We define by induction the following graphs:

1. (a) *Let $\mathbb{G}_k^+$ be $\mathbb{G}(\mathbb{C})_{z_k}$, (the truncation of $\mathbb{G}(\mathbb{C})$ at z_k). item Let $\mathbb{G}_k^-$ be $\mathbb{G}(\mathbb{C}) - \mathbb{G}_k^+$ (the remainder graph after deleting from it the top part $\mathbb{G}_k^+$).*
   (b) *The point z_{k-1} is in the graph $\mathbb{G}_k^-$.*
2. *Assume that for $z_i, 1 < i \leq k$ we have defined $\mathbb{G}_i^+$ and $\mathbb{G}_i^-$ and that $\mathbb{G}_i^+$ is the truncation of $\mathbb{G}_{i+1}^-$ at point z_i, and that $\mathbb{G}_i^- = \mathbb{G}_{i+1}^- - \mathbb{G}_i^+$. We also assume that z_{i-1} is in $\mathbb{G}_i^-$.*
   Let $\mathbb{G}_{i-1}^+ = (\mathbb{G}_i^-)_{z_{-1}}$, (i.e. the truncation of $\mathbb{G}_i^-$ at point z_{i-1}).
   Let $\mathbb{G}_{i-1}^- = \mathbb{G}_i^- - \mathbb{G}_{i-1}^+$.
3. *The sequence of graphs $\mathbb{G}, \mathbb{G}_1^-, \mathbb{G}_1^+, \mathbb{G}_2^+, \ldots, \mathbb{G}_k^+$ is the semantical object for $(\mathbb{C}, \mathbb{F})$. They generate equations which are the equational semantics for $(\mathbb{C}, \mathbb{F})$.*

Example 5. Consider a system $(\mathbb{C}, \mathbb{F})$ where $\mathbb{F}$ is a maximal path, i.e. $\mathbb{F}$ is the sequence $(z_1, \ldots, x_{m+1})$. The graph system for it will be as in Figure 14.

starting point	z_{m+1}	$\bar{z}_{m+1}$	graph $\mathbb{G}_{m+1}^+$
	$\vdots$	$\vdots$	$\vdots$
starting point	z_1	$\bar{z}_1$	graph $\mathbb{G}_1^+$
starting point	x_0	$\bar{x}_0$	graph $\mathbb{G}_1^-$

Fig. 14.

Remark 13. The nature of the set of facts $\mathbb{F}$ is best understood when the set $\mathbb{C}$ of Chisholm CTDs is represented as a sequence. Compare with Definition 12.

$\mathbb{C}$ has the graph $\mathbb{G}(\mathbb{C})$. The graph can be represented as a sequence

$$\mathbf{E} = ((x_0, \bar{x}_0), (x_1, \bar{x}_1), \ldots, (x_{m+1}, \bar{x}_{m+1}))$$

together with the starting point (x_0).

When we get a set of facts $\mathbb{F}$ and arrange it as a sequence $(z_1, \ldots, z_k)$ in accordance with the obligation progression, we can add x_0 to the sequence and look at $\mathbb{F}$ as

$$\mathbb{F} = (x_0, z_1, \ldots, z_k).$$

We also consider $(\mathbb{E}, \mathbb{F})$ as a pair, one the sequence $\mathbb{E}$ and the other as a multiple sequence of starting points. The graph $\mathbb{G}_i$ is no more and no less than the subsequence $\mathbb{E}_i$, beginning from the pair $(z_i, \bar{z}_i)$ up to the pair $(z_{i+1}, \bar{z}_{i+1})$ but *not* including $(z_{i+1}, \bar{z}_{i+1})$.

This way it is easy to see how $\mathbb{G}$ is the sum of all the $\mathbb{G}_i$, strung together in the current progression order. Furthermore, we can define the concept of "the fact z_j is in violation of the CTD of z_i", for $i < j$. To find out if there was such a violation, we solve the equations for

$$\mathbb{E}_i = ((z_i, \bar{z}_i), \ldots, (x_{m+1}, \bar{x}_{m+1}))$$

and if the equation solves with $z_j = 0$ then putting $z_j = 1$ is a violation.

Remark 14. Let us check whether our equational modelling of the Chisholm CTD set satisfies the conditions set out in Definition 8.

Consider Figure 15 (a) and (b):

(a) Obligatory x must be translated the same way throughout.
This holds because we use a variable x in a neighbourhood generated equation.

(b) The form $X \to OY$ must be translated uniformly no matter whether $X = q$ or $X = \neg q$.
This is is true of our model.

(c) This holds because "X" is translated as itself.

(d) The translation of the clauses must be all independent.
Indeed this holds by Lemma 1.
It is also true that (see Figure 15(a))

2. $p \to Oq$

and

5. p

imply

6. Oq

This holds because (5) p is a fact. So this means that Figure 15(b) truncated at the point p.
The truncated figure is indeed what we would construct for Oq.
A symmetrical argument shows that (4) and (3) imply (7).

1. Op	5. p
2. $p \to Oq$	6. Oq
3. $\neg p \to Oq$	7. $O\neg q$
4. $\neg p$	

(a)

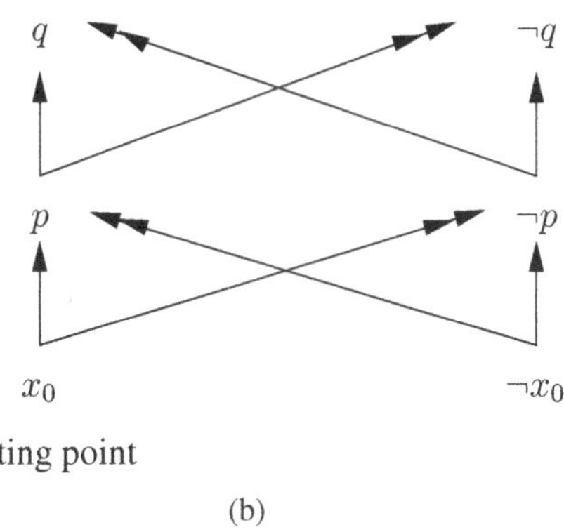

(b)

Fig. 15.

(e) The system is required to be robust with respect to substitution.
This condition arose from criticism put forward in [7] against the solution to the Chisholm paradox offered in [8]. [8] relies on the fact that p, q are independent atoms.
The solution does not work when $q \vdash p$, e.g. substituting for "q" the wff "$r \wedge p$" (like $p =$ fence and $q =$ white fence).
In our case we use equations and if we substitute "$r \wedge p$" for "q" we get the equations

$$r \wedge p = 1 - p$$
$$p = r \wedge p$$

Although this type of equation is not guaranteed a solution, there is a solution in this case;$p = r = 1$.
If we add the fact $\neg p$, i.e. $1 - p = 1, p = 0$, (there is no fence) the equation solves to $\neg q = \neg p \vee \neg r$, which is also $= 1$ because of $\neg p$. So we have no problem with such substitution. In fact we have no problem with any substitution because the min function which we use always allows for solutions.

(f) The translation must be uniform and it to be done item by item.
Yes. Indeed, this is what we do!

Remark 15. We can now explain how classical logic can handle CTD, even though the CTD $x \to Oy$ involves a modality. The basic graph representation such as Figure 7 can be viewed as a set of possible worlds where the variables x and y act as nominals (i.e. atoms naming worlds by being true exactly at the world they name). x is a world, y is a world and $x \to y$ means y is ideal for x. $x \twoheadrightarrow \bar{y}$ means that $\bar{y}$ is sub ideal for x. Let $\Box_1$ be the modality for $\to$ and $\Box_2$ the modality for $\twoheadrightarrow$. Then we have a system with two disjoint modalities and we can define

$$OA \equiv \Box_1 A \wedge \Box_2 \neg A.$$

Now this looks familiar and comparable to [8], and especially to [12]. The perceptive reader might ask, if we are so close to modal logic, and in the modal logic formulation there are the paradoxes, why is it that we do not suffer from the paradoxes in the equational formulation?

The difference is because of how we interpret the facts! The equational approach spreads and inserts the facts into different worlds according to the obligation progression. Modal logic cannot do that because it evaluates formulas in single worlds. With equations, each variable is a nominal for a different world but is also is natural to substitute values to several variables at the same time!

Evaluating in several possible worlds at the same time in modal logic would solve the paradox but alas, this is not the way it is done.

Another difference is that in modal logic we can iterate modalities and write for example

$$O(x \to Oy).$$

We do not need that in Chisholm sets. This simplifies the semantics.

2.4 Looping CTDs

So far we modelled the Chisholm set only. Now we want to expand the applicability of the equational approach and deal with looping CTDs, as in the set in Figure 4. Let us proceed with a series of examples.

Example 6. Consider the CTD set of Figure 4. If we write the equations for this example we get

1. $a = \min(b, 1 - b)$
2. $b = \neg a$
3. $\neg b = 1 - a$

and the constants

4. $\neg b = 1 - b$
5. $\neg a = 1 - a$.

The only solution here is $a = b = \frac{1}{2}$. In argumentation and in classical logic terms this means the theory of Figure 4 is $\{0, 1\}$ inconsistent.

This is mathematically OK, but is this the correct intuition? Consider the set $\{b, \neg a\}$. The only reason this is not a solution is because we have $a \twoheadrightarrow \neg b$ and if $a = 0$, we get $\neg b = 1$ and so we cannot have $b = 1$.

However, we wrote $a \twoheadrightarrow \neg b$ because of the CTD $a \to Ob$, which required us to go from a to b (i.e $a \to b$) and in this case we put in the graph $a \twoheadrightarrow \neg b$ to stress "do not go to $\neg b$".

However, if $a = 0$, why say anything? We do not care in this case whether the agent goes from a to b!

Let us look again at Figure 6. We wrote the following equation for the node x

$$x = \min(u_i, 1 - v_j).$$

The rationale behind it was that we follow the rules, so we are going to u_i as our obligations say, and we came to x correctly, not from v_j, because $v_j \to O\neg x$ is required. Now if $v_j = 1$ (in the final solution) then the equation is correct. But if $v_j = 0$, then we do not care if we come to x from v_j, because $v_j \to O\neg x$ is not activated. So somehow we need to put into the equation that we care about v_j only when $v_j = 1$.

Remark 16. Let us develop the new approach mentioned in Example 6 and call it the soft approach. We shall compare it with the mathematical approach of Remark 12.

First we need a δ function as follows:

$$\delta(w) = \bot \text{ if } w = \bot$$

and

$$\delta(w) = \top \text{ if } w \neq \bot.$$

$\delta(w) = w$, if we are working in two valued $\{0, 1\}$ logic. Otherwise it is a projective function

$$\delta(0) = 0 \text{ and } \delta(w) = 1 \text{ for } w > 0.$$

We can now modify the equivalences (*1) (based on figure 6) as follows:

Let $_1, \ldots, v_s$ be as in Figure 6. Let $J, K \subseteq \{1, \ldots, s\}$ be such that $J \cap K = \varnothing$ and $J \cup K = \{1, \ldots, s\}$. Consider the expression

$$\varphi_{J,K} = \bigwedge_{j \in J} \delta(v_j) \wedge \bigwedge_{k \in K} \neg\delta(v_k).$$

This expression is different from 0 (or $\bot$), exactly when K is the set of all indices k for which $v_j = \bot$.

Replace (*1) by the following group of axioms for each pair J, K and for each x

$$x \wedge \varphi_{J,K} \leftrightarrow \varphi_{J,K} \wedge \bigwedge_r u_r \wedge \bigwedge_{j \in J} \neg v_j. \qquad (*2)$$

Basically what (*2) says is that the value of x should be equal to

$$\min\{u_r, 1 - v_j \text{ for those } j \text{ whose value is } \neq 0\}.$$

Note that this is an implicit definition for the solution of the equations. It is clear when said in words but looks more complicated when written mathematically. Solutions may not exist.

Example 7. Let us now look again at Figure 4.

The soft equations discussed in Remark 16 are

$$\begin{aligned}\delta(a) \wedge \bar{b} &= \delta(a)(1-a)\\ \delta(b) \wedge a &= \delta(b)\min(b, 1-b)\\ b &= \bar{a}\\ \bar{b} &= 1-b\\ \bar{a} &= 1-a.\end{aligned}$$

For these equations $\bar{a} = 1, \bar{b} = a = 0, b = 1$ is a solution.

Note that $\bar{a} = \bar{b} = 1$ and $a = b = 0$ is *not* a solution!

Let us now examine and discuss the mathematical approach alternative, the one mentioned in Remark 12. The first step we take is to convert Figure 4 into the right form for this alternative approach by adding double arrows between all x and $\bar{x}$. We get Figure 16.

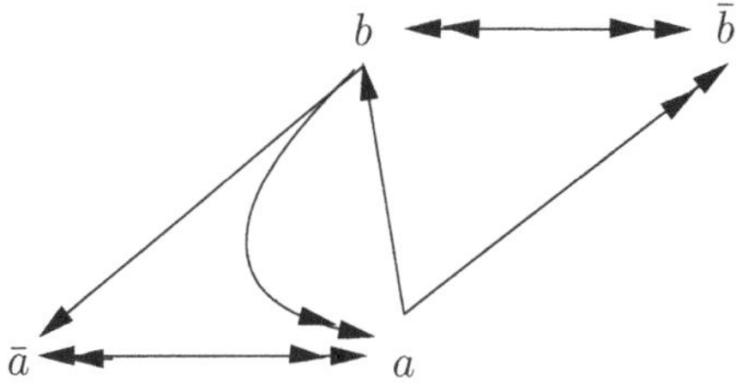

Fig. 16.

The equations are the following:

$$\begin{aligned}a &= \min(b, 1-\bar{a}, 1-b)\\ \bar{a} &= 1-a\\ b &= \min(\bar{a}, 1-\bar{b})\\ \bar{b} &= \min(1-a, 1-b).\end{aligned}$$

Let us check whether $a = \bar{b} = 0$ and $b = \bar{a} = 1$ is a solution. We get respectively by substitution

$$\begin{aligned}0 &= \min(1, 0, 0)\\ 1 &= 1-0\\ 1 &= \min(1, 1-0)\\ 0 &= min(1-0, 1-1).\end{aligned}$$

Indeed, we have a solution. Let us try the solution $\bar{b} = \bar{a} = 1$ and $a = b = 0$. Substitute in the equations and get

$$0 = \min(0, 0, 1)$$
$$1 = 1 - 0$$
$$0 = \min(1, 1 - 1)$$
$$1 = \min(1 - 0, 1 - 0).$$

Again we have a solution.

This solution also makes sense. Note that this is not a solution of the previous soft approach!

We need to look at more examples to decide what approach to take, and which final formal definition to give.

Example 8. Consider the following two CTD sets, put forward by two separate security advisors D and F.

D1: you should have a dog
Od

D2: If you do not have a dog, you should have a fence
$\neg d \to Of$

D3: If you have a dog you should not have a fence
$d \to O\neg f$

F1: You should have a fence
Of

F2: If you do not have a fence you should have a dog
$\neg f \to Od$

F3: If you do have a fence you should not have a dog.
$f \to O\neg d$

If we put both sets together we have a problem. They do not agree, i.e. {D1, D2, D3, F1, F2, F3}. However, we can put together both D1, D2 and F1, F2. They do agree, and we can have both a dog and a fence.

The mathematical equational modelling of D1 and D2 also models D3, i.e. D1, D2 $\vDash$ D3 and similarly F1, F2 $\vDash$ F3. So according to this modelling {D1, D2, F1, F2} cannot be consistently together. Let us check this point. Consider Figure 17

The equations for Figure 17 are:

$$x_0 = 1$$
$$x_0 = d$$
$$\bar{x}_0 = 1 - x_0$$
$$d = 1 - \bar{d}$$
$$\bar{d} = \min(1 - d, 1 - x_0)$$
$$\bar{d} = f$$
$$f = 1 - \bar{f}$$
$$\bar{f} = \min(1 - f, 1 - \bar{d})$$

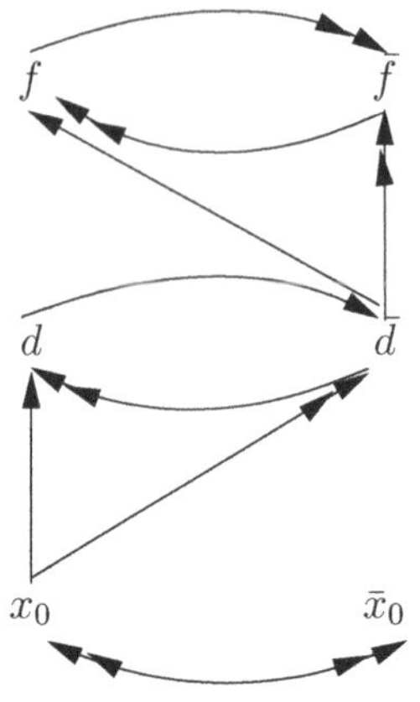

Fig. 17.

The only solution is

$$x_0 = d = \bar{f} = 1$$
$$\bar{x}_0 = \bar{d} = f = 0.$$

The important point is that $\bar{f} = 1$, i.e. no fence.

Thus D1,D2 $\vdash \bar{f}$.

By complete symmetry beget that F1,F2 $\vdash \bar{d}$. Thus we cannot have according to the mathematical approach that having both a dog and a fence is consistent with $\{$D1,D2, F1,F2$\}$.

Let us look now at the soft approach. Consider Figure 18

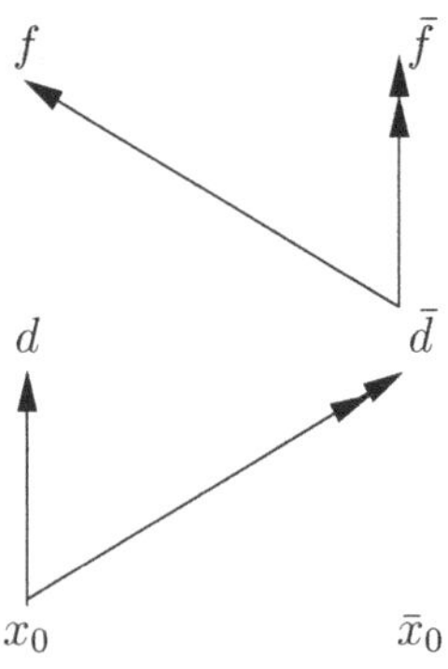

Fig. 18.

The soft equations for Figure 18 are:

$$\begin{aligned}
&x_0 = 1\\
&x_0 = d\\
&\min(x_0, \bar{d}) = \min(x_0, 1 - x_0)\\
&\min(\bar{d}, \bar{f}) = \min(\bar{d}, 1 - \bar{d})
\end{aligned}$$

There are two solutions

$$x_0 = 1, d = 1, \bar{d} = 0, \bar{f} = 1, f = 0$$

and

$$x_0 = 1, d = 1, \bar{d} = 0, \bar{f} = 0, f = 1.$$

The conceptual point is that since $\bar{d} = 0$, we say nothing about $\bar{f}$.

Now similar symmetrical solution is available for {F1,F2} Since D1,D2 allow for $f = 1$ and F1,F2 allow for $d = 1$, they are consistent together. In view of this example we should adopt the soft approach.

Remark 17. Continuing with the previous Example 8, let us see what happens if we put together in the same CTD set the clauses {D1,D2,E1,E2} and draw the graph for them all together, in contrast to what we did before, where we were looking at two separate theories and seeking a joint solution. If we do put them together, we get the graph in Figure 19.

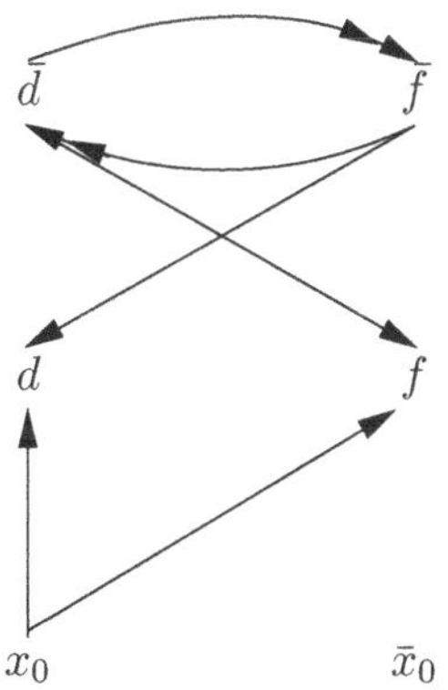

Fig. 19.

If we use the mathematical equations, there will be no solution. If we use the soft approach equations, we get a unique solution

$$d = f = 1, \bar{d} = \bar{f} = 0$$

The reason for the difference, I will stress again, is in the way we write the equations for $\bar{d}$ and $\bar{f}$. In the mathematical approach we write

$$\begin{aligned}\bar{d} &= \min(f, 1 - \bar{f})\\ \bar{f} &= \min(d, 1 - \bar{d})\\ \bar{d} &= 1 - d\\ \bar{f} &= 1 - f\end{aligned}$$

In the soft approach we write

$$\begin{aligned}\min(\bar{d}, \bar{f}) &= \min(f, \bar{d}, 1 - \bar{f})\\ \min(\bar{f}, \bar{d}) &= \min(d, \bar{f}, 1 - d)\end{aligned}$$

This example also shows how to address a general CTD set, where several single arrows can come out of a node (in our case x_0). The equations for x_0 in our example are:

$$\begin{aligned}x_0 &= 1\\ x_0 &= min(f, d)\end{aligned}$$

which forces $d = f = 1$. We will check how to generalise these ideas in the next section.

2.5 Methodological Discussion

Following the discussions in the previous sections, we are now ready to give general definitions for the equational approach to general CTD sets. However, before we do that we would like to have a methodological discussion. We aleady have semantics for CTD. It is the soft equations option discussed in the previous subsection. So all we need to do now is to define the notion of a general CTD set (probably just a set of clauses of the form $\pm x \to O \pm y$) and apply the soft equational semantics to it. This will give us a consequence relation and a consistency notion for CTD sets and the next step is to find proof theory for this consequence and prove a completeness theorem.

We need to ask, however, to what extent is the soft semantics going to be intuitive and compatible with our perception of how to deal with conflicting CTD sets? So let us have some discussion about what is intuitive first, before we start with the technical definitions in the next section. Several examples will help.

Example 9. Consider the following CTD set:

1. You should not have a dog
   $O\neg d$
2. If you have a dog you must keep it
   $d \to Od$
3. d: you have a dog

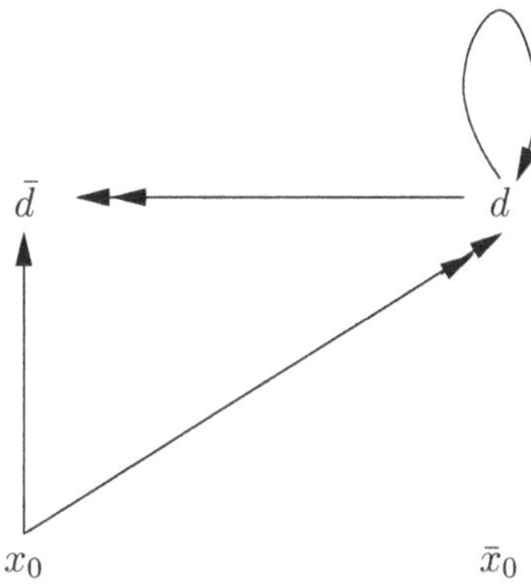

Fig. 20.

Here we have a problem. Is (1), (2), (3) a consistent set? In **SDL** we can derive from (2) and (3) $O\neg d$ and get a contradiction $Od \wedge O\neg d$.

However, in our semantics we produce a graph and write equations and if we have solutions, then the set is consistent. Let us do this.

The original graph for clauses (1)–(2) is Figure 20. This graph generates equations.

The fact d splits the graph and we get the two graphs in Figures 21 and 22.

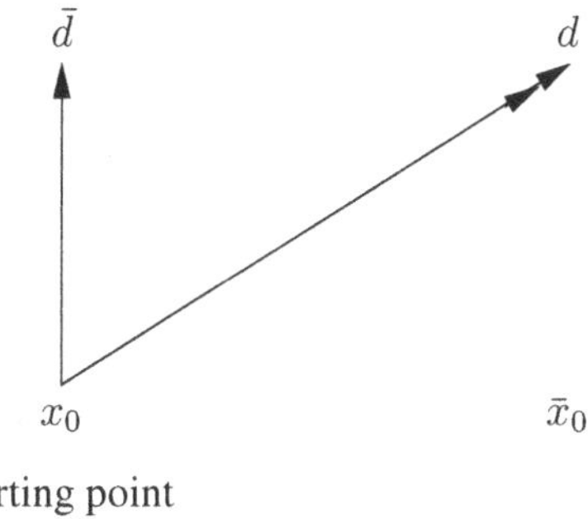

Fig. 21.

The solution of the soft equations for the original graph (without the fact d) is $x_0 = \bar{d} = 1, d = 0$.

The solution for the two split graphs, after the fact d gives $d = 1$ for Figure 21 and $\bar{d} = 0$ for Figure 22.

There is no mathematical contradiction here. We can identify a violation from the graphs. However we may say there is something unintuitive, as the CTD proposal for a remedy for the violation $O\neg d$, namely $d \to Od$ violates the original obligation $O\neg d$, and actually perpetuates this violation. This we see on the syntactical level. No problem in the semantics.

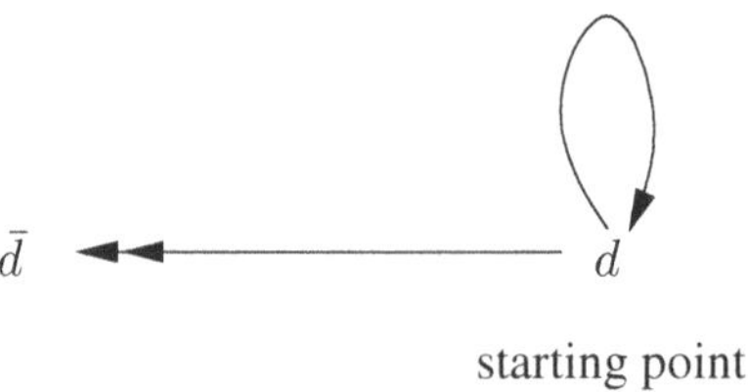

Fig. 22.

We can explain and say that since the fact d violates $O\neg d$ then a new situation has arisen and $O\neg d$ is not "inherited" across a CTD. In fact, in the case of a dog it even makes sense. We should not have a dog but if we violate the obligation and get it, then we must be responsible for it and keep it.

The next example is more awkward to explain.

Example 10. This example is slightly more problematic. Consider the following.

1. You should not have a dog
   $O\neg d$
2. you should not have a fence
   $O\neg f$
3. If you do have a dog you should have a fence
   $d \to Of$

The graph for (1)–(3) is Figure 23.

The solution is $\bar{d} = \bar{f} = 1, d = f = 1$.

Let us add the new fact

4. d: You have a dog

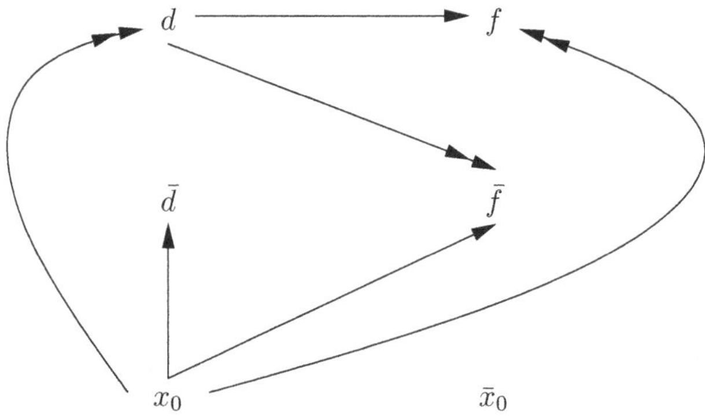

Fig. 23.

The graph of Figure 23 splits into two graphs, Figure 24 and 25.

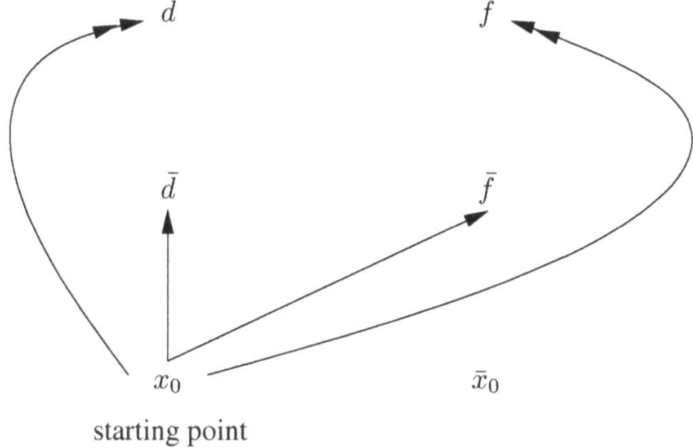

Fig. 24.

The equations for Figure 24 solve to $\bar{d} = \bar{f} = 1, d = f = 0$. The equations for Figure 25 solve to $d = f = 1, \bar{d} = \bar{f} = 0$.

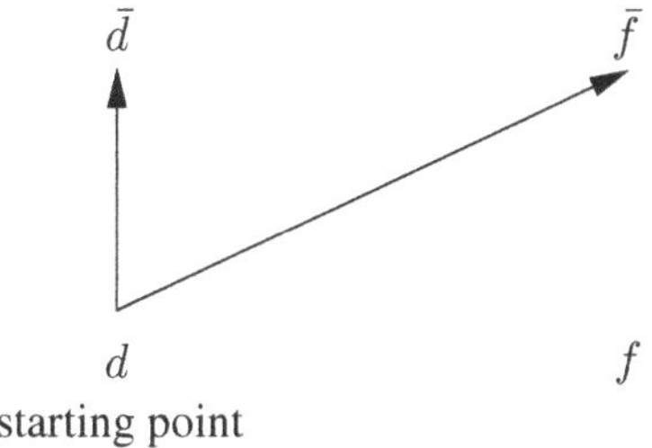

Fig. 25.

There is no mathematical contradiction here because we have three separate graphs and their solutions. We can, and do, talk about violations, not contradictions.

Note that in **SDL** we can derive Of and $O\neg f$ from (1), (2), (3) and we do have a problem, a contradiction, because we are working in a single same system.

Still, even for the equational approach, there is an intuitive difficulty here. The original $O\neg f$ is contradicted by $d \to Of$. The "contradiction" is that we offer a remedy for the violation d namley Of by violating $O\neg f$.

You might ask, why offer the remedy Of? Why not say keep the dog chained? Oc? The Oc remedy does not violate $O\neg f$.

The explanation that by having a dog (violating $O\neg d$) we created a new situation is rather weak, because having a fence is totally independent from having a dog, so we would expect that the remedy for having a dog will not affect $O\neg f$!

The important point is that the equational approach can identify such "inconsistencies" and can add constraints to avoid them if we so wish.

Remark 18. Let us adopt the view that once a violation is done by a fact then any type of new rules can be given. This settles the problems raised in Example 10. However, we have other problems. We still have to figure out a technical problem, namlely how to deal with several facts together. In the case of the Chisholm set there were no loops and so there was the natural obligation progression. We turned the set of facts into a sequence and separated the original graph (for the set of CTD wihout the facts) into a sequence of graphs, and this was our way of modelling the facts. When we have loops there is a problem of definition, how do we decompose the original graph when we have more than one fact? The next example will illustrate.

Example 11. This example has a loop and two facts. It will help us understand our modelling options in dealing with facts. Consider the following clauses. This is actually the Reykjavik paradox, see for example [13]:

1. There should be no dog
   $O\neg d$
2. There should be no fence
   $O\neg f$
3. If there is a dog then there should be a fence
   $d \to Of$
4. If there is a fence then there should be a dog
   $f \to Od.$

The figure for these clauses is Figure 26.

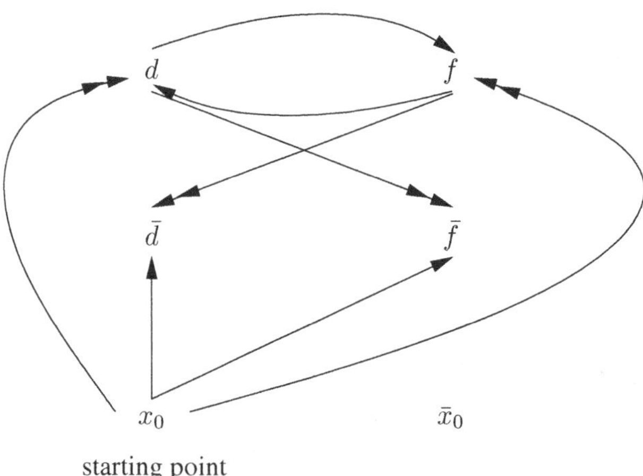

Fig. 26.

The soft equations solve this figure into $x_0 = \bar{d} = \bar{f} = 1$. $f = d = 0$. We now add the input that there is a dog and a fence.

5. d: dog, f: fence

The question is how to split Figure 26 in view of this input.

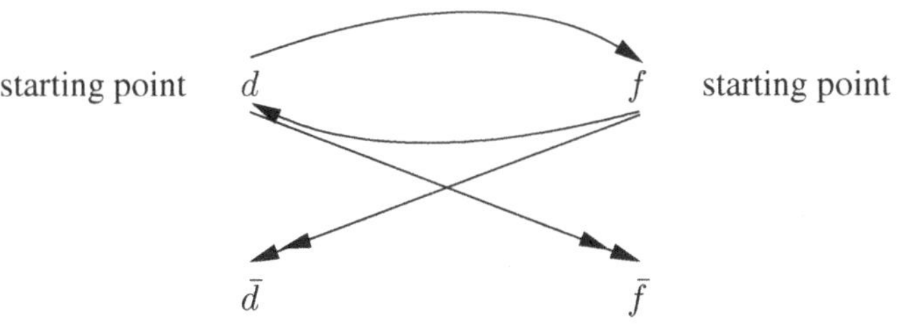

Fig. 27.

If we substitute $d = 1$ and $f = 1$ together and split, we get Figure 27, with two starting points.

Comparison with the original figure shows two violations of $O\neg d$ and $O\neg f$.

Let us now first add the fact d and then add the fact f.

When we add the fact d, Figure 26 split (actually is modified) into Figure 28. This figure happens to look just like Figure 27 with only d as a starting point. (Remember that any starting point x gets the equation $x = 1$.)

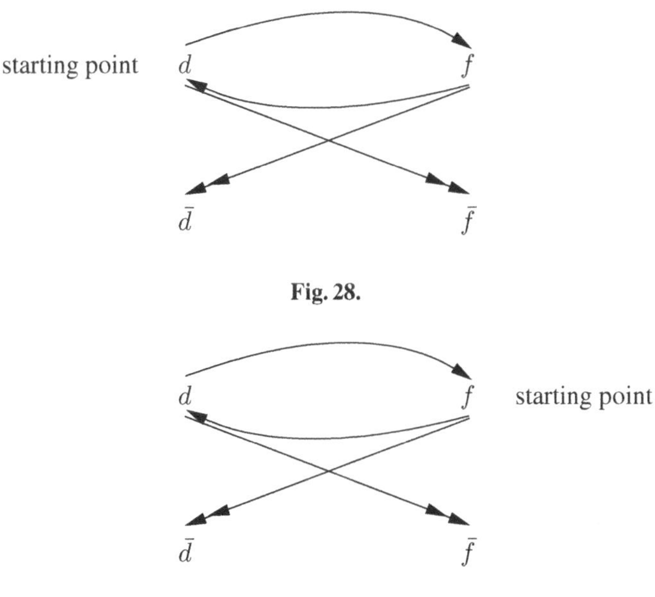

Fig. 28.

Fig. 29.

Adding now the additional fact f changes Figure 28 into Figure 29. In fact we would have got Figure 29 first, had we introduced the fact f first, and then added the fact d, we would have got Figure 29.

The difference between the sequencing is in how we perceive the violations.

The following is a summary.

Option 1. Introduce facts $\{d, f\}$ simultaneously. Get Figure 27, with two starting points. There are two violations, one of $O\neg f$ and one of $O\neg d$. This is recognised by comparing the solutions for the equations of Figure 26 with those of Figure 27.

Option 2. Introduce the fact d first. Figure 26 changes into Figure 28. Solving the equations for these two figures shows a violationof $O\neg d$ and a vioation of $O\neg f$, because f also gets $f = 1$ in the equations of Figure 28.

Option 2df. We now add to option 2d the further fact f. We get that Figure 28 becomes Figure 29. The solutions of the two figures are the soame, $f = d = 1$. So adding f gives no additional violation.

We thus see that adding $\{d, f\}$ together or first d and then f or (by symmetry) first f and then d all essentially agree and there is no problem. So where is the problem with simultaneous facts? See the next Example 12.

Example 12 (Example 11 continued). We continue the previous Example 11:

Let us try to add the facts $\{d, \neg f\}$ to the CTD set of Figure 26. Here we have a problem because we get Figure 30. In this figure both d and $\bar{f}$ are starting points. These two must solve to $d = \bar{f} = 1$. This is impossible in the way we set up the system. This means that it is inconsistent from the point of view of our semantics to add the facts $\{d, \neg f\}$ simultaneously in the semantics, or technically to have two starting points!

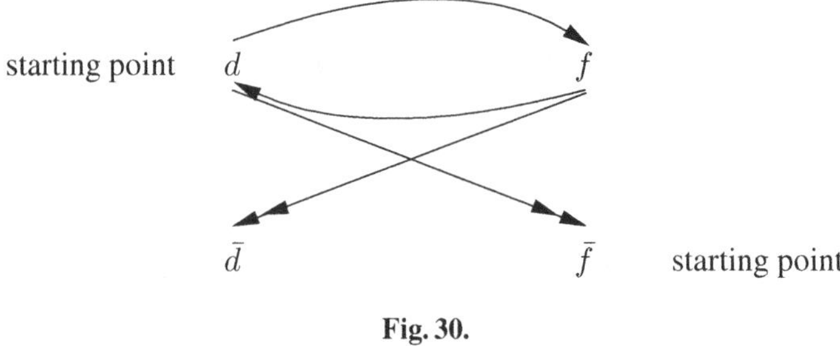

Fig. 30.

But we know that it is consisent and possible in reality to have a dog and no fence. So where did we go wrong in our semantic modelling? Mathematically the problem arises with making two nodes starting points. This means that we are making two variables equal to 1 at the same time. The equations cannot adjust and have a solution.[3]

The obvious remedy is to add the facts one at a time. Option 3d first adds d and then takes option 3d $\neg f$ and add $\neg f$ and in parallel, option 4$\neg f$ first adds the fact $\neg f$ and then take option 4$\neg fd$ and add the fact d. Let us see what we get doing these options and whether we can make sense of it.

[3] Remember when we substitute a fact we split the graph into two and so the equations change. We are not just substituting values into equations (in which case the order simultaneous or not does not matter), we are also changing the equations.

Recall what you do in Physics: If we have, for example, the equation $y = \sin x$ and we substitute for x a very small positive value, then we change the equation to $y = x$.

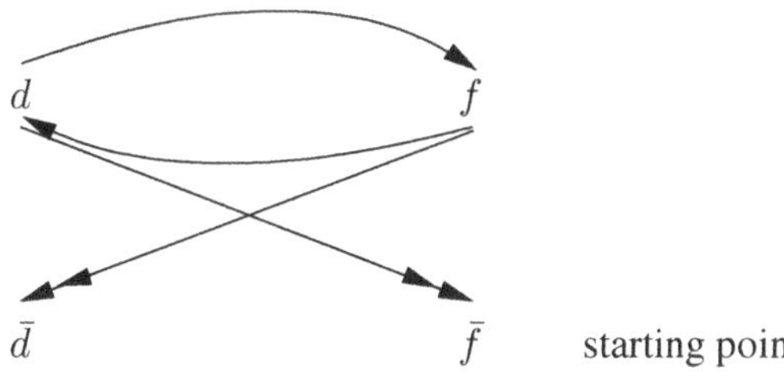

Fig. 31.

Option 3$d\neg f$. Adding the fact d would give us Figure 28 from Figure 26. We now add fact $\neg f$. This gives us Figure 31 from Figure 28.

Figure 31 violates Figure 28.

Option 3$\neg fd$. If we add the fact $\neg d$ first, we get Figure 31 from the original Figure 26.

If we now add the fact d, we get Figure 28.

The solution to the equations of this figure is $d = 1, f = 1, \bar{f} = 0$, but we already have the fact $\neg d$, so the $f = 1$ part cannot be accepted.

Summing up:

- facts $\{d, \neg f\}$ cannot be modelled simultaneously.
- First d then $\neg f$, we get that $\neg f$ violates $d \to Of$.
- first $\neg f$ then d, we get that $d \to Of$ cannot be implemented.

So the differences in sequencing the facts manifests itself as differences in taking a point of view of the sequencing of the violations.

The two views, when we have as additional data both d and $\neg f$, are therefore the following:

we view $d \to Of$ as taking precedent and $\neg f$ is violating it

or

we view $O\neg f$ as as taking precedence over $d \to Of$ and hence $d \to Of$ cannot be implemented.

3 Equational Semantics for General CTD Sets

We now give general definitions for general equational semantics for general CTD sets.

Definition 13.

1. *Let Q be a set of distinct atoms. Let $\bar{Q}$ be $\{\bar{a} | a \in Q\}$. Let $Q^* = Q \cup \bar{Q} \cup \{\top, \bot\}$. For $x \in \bar{Q}$, let $\bar{x}$ be x (i.e. $\bar{\bar{x}} = x$). Let $\bar{\top} = \bot$ and $\bar{\bot} = \top$.*

2. *A general CTD clause has the form* $x \to Oy$, *where* $x, y \in Q^*$ *and* $x \neq \bot, y \neq \bot, \top$.
3. *Given a set* $\mathbb{C}$ *of general CTD clauses let* $Q^*(\mathbb{C})$ *be the set* $\{x, \bar{x} | x$ *appears in a clause of* $\mathbb{C}\}$.
4. *Define two relations on* $Q^*(\mathbb{C})$, $\to$ *and* $\twoheadrightarrow$ *as follows:*
   - $x \to y$ *if the clause* $x \to Oy$ *is in* $\mathbb{C}$
   - $x \twoheadrightarrow y$ *if the clause* $x \to O\bar{y}$ *is in* $\mathbb{C}$
5. *Call the system* $\mathbb{G}(\mathbb{C}) = (Q^*(\mathbb{C}), \to, \twoheadrightarrow)$ *the graph of* $\mathbb{C}$.
6. *Let* $x \in Q^*(\mathbb{C})$. *Let*

$$E(x \to) = \{y | x \to y\}$$
$$E(\twoheadrightarrow x) = \{y | y \twoheadrightarrow x\}$$

Definition 14.

1. *Let* $\mathbb{C}$ *be a CTD set and let* $\mathbb{G}(\mathbb{C})$ *be its graph. Let* x *be a node in the graph. Let* $\mathbf{f}$ *be a function from* $Q^*(\mathbb{C})$ *into [0,1].*
   Define
   $$E^+(\twoheadrightarrow x, \mathbf{f}) = \{y | y \twoheadrightarrow x \text{ and } \mathbf{f}(x) > 0\}.$$
2. *Let* $\mathbf{f}, x$ *be as in (1). We say* $\mathbf{f}$ *is a model of* $\mathbb{C}$ *if the following holds*
   (a) $\mathbf{f}(\top) = 1, \mathbf{f}(\bot) = 0$
   (b) $\mathbf{f}(\bar{x}) = 1 - \mathbf{f}(x)$
   (c) $\mathbf{f}(x) = \min(\{\mathbf{f}(y) | x \to y\} \cup \{1 - \mathbf{f}(z) | z \in E^+(\twoheadrightarrow x, \mathbf{f})\})$
3. *We say* $\mathbf{f}$ *is a* $\{0,1\}$ *model of* $\mathbb{C}$ *if* $\mathbf{f}$ *is a model of* $\mathbb{C}$ *and* $\mathbf{f}$ *gives values in* $\{0,1\}$.

Example 13. Consider the set

1. $\top \to a$
2. $a \to \bar{a}$

This set has no models. However (2) alone has a model $\mathbf{f}(a) = 0, \mathbf{f}(\bar{a}) = 1$. The equations for (2) are: $a = 1 - \bar{a}, a = \min(1 - a, \bar{a})$.

The graph for (1) and (2) is Figure 32

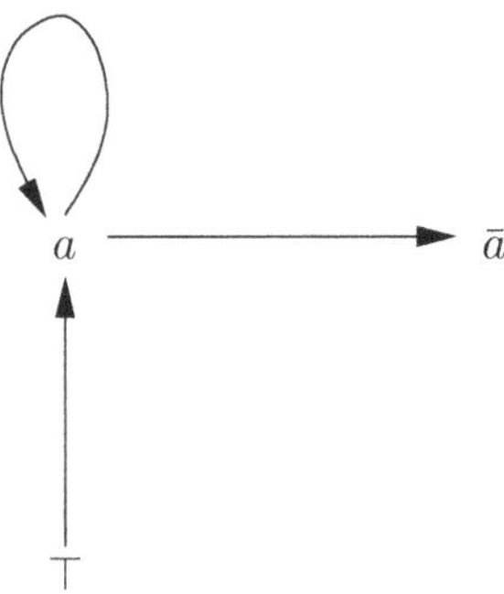

Fig. 32.

The graph for (2) alone is Figure 32 without the node $\top$.

Definition 15. *Let $\mathbb{C}$ be a CTD set. Let $\mathbb{G}(\mathbb{C}) = (Q^*, \rightarrow, \twoheadrightarrow)$ be its graph. Let $x \in Q^*$. We define the truncation graph $\mathbb{G}(\mathbb{C})_x$, as follows.*

1. *Let R^* be the reflexive and transitive closure of R where*

$$xRy = \mathrm{def}(\mathrm{x} \rightarrow \mathrm{y}) \vee (\mathrm{x} \twoheadrightarrow \mathrm{y}).$$

2. *Let Q^*_x be the set*

$$\{z | xR^*z \vee \bar{x}R^*z\} \cup \{\top, \bot\}$$

   Let $\rightarrow_x = (\rightarrow \upharpoonright Q_x) \cup \{\top \rightarrow x\}$.
   Then

$$\mathbb{G}(\mathbb{C})_x = (Q^*_x, \rightarrow_x, \twoheadrightarrow \upharpoonright Q^*_x).$$

3. *In words: the truncation of the graph at x is obtained by taking the part of the graph of all points reachable from x or $\bar{x}$ together with $\top$ and $\bot$ and adding $\top \rightarrow x$ to the graph.*

Example 14. Consider a m level Chisholm set as in Figure 11. The truncation of this figure at point $\bar{x}_3$ is essentially identicl with Figure 13. It is Figure 33. The difference is that we write "$\top \rightarrow \bar{x}_3$" instead of "$\bar{x}_3$ starting point". These two have the same effect on the equations namely that $\bar{x}_3 = 1$.

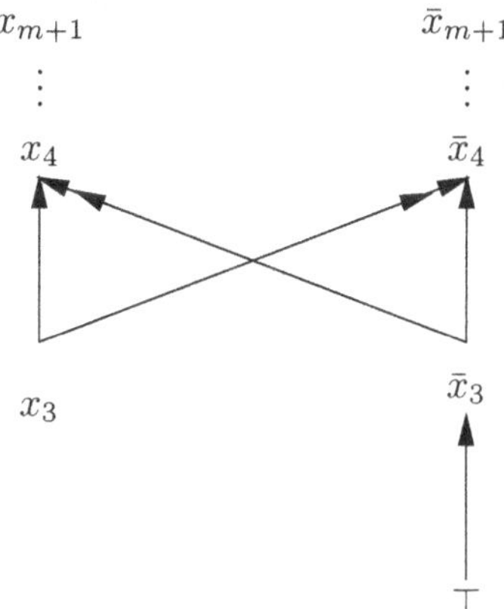

Fig. 33.

Definition 16. *Let $\mathbb{C}$ be a CTD set. Let $\mathbb{F}$ be a set of facts. We offer equational semantics for $(\mathbb{C}, \mathbb{F})$.*

1. *Let Ω be any ordering of $\mathbb{F}$.*

$$\Omega = (f_1, f_2, \ldots, f_k).$$

2. *Let $\mathbb{G}(\mathbb{C})$ be the graph of $\mathbb{C}$ and consider the following sequence of graphs and their respective equations.*

$$(\mathbb{G}(\mathbb{C}), \mathbb{G}(\mathbb{C})_{f_1}, \mathbb{G}(\mathbb{C})_{f_1 f_2}, \ldots, \mathbb{G}(\mathbb{C})_{f_1,\ldots,f_k})$$

We call the above sequence

$$\mathbb{G}(\mathbb{C})_{\Omega}.$$

We consider Ω as a "point of view" of how to view the violation sequnce arising from the facts.

3. *The full semantics for $(\mathbb{C}, \mathbb{F})$ is the family of all sequences $\{\mathbb{G}(\mathbb{C})_{\Omega}\}$ for all Ω orderings of $\mathbb{F}$.*

Remark 19. The CTD sets considered so far had the form $\pm x \to O \pm y$, where x and y are atomic. This remark expands our language allowing for x, y to be arbitrary propositional formulas. Our technical machinery of graphs and equations works just the same for this case. We can write in the graph $A, \bar{A}$ and then write the appropriate equations, and we use $\mathbf{h}_\mu$. For example the equation $\bar{A} = 1 - A$ becomes $\mathbf{h}_\mu(\bar{A}) = 1 - \mathbf{h}_\mu(A)$. Starting points A must satisfy $\mathbf{h}_\mu(A) = 1$, all the same as before. The only difference is that since the equations become implicit on the atoms, we may not have a solution.

In practice the way we approach such a CTD set is as follows: Let $\mathbb{C}_1$ be a set of CTD obligations of the form $\{A_i \to OB_i\}$. We pretend that A_i, B_i are all atomic. We do this by adding a new atomic constant $y(A)$, associated with every wff A. The set $\mathbb{C}_1 = \{A_i \to OB_i\}$ becomes the companion set $\mathbb{C}_2\{y(A_i) \to Oy(B_i)\}$. We now apply the graphs and equational approach to $\mathbb{C}_2$ and get a set of equations to be solved.

We add to this set of equations the further constraint equations

$$y(A_i) = \mathbf{h}_\mu(A_i)$$
$$y(B_i) = \mathbf{h}_\mu(B_i).$$

We now solve for the atomic propositions of the language.

We need to clarify one point in this set-up. What do we mean by facts $\mathbb{F}$? We need to take $\mathbb{F}$ as a propositional theory, it being the conjunction of some of the A_i. If we are given a set $\mathbb{C}_1$ of contrary to duty clauses of the form $A \to OB$ and facts $\mathbb{F}_1$, we check whether $\mathbb{F}_1 \vDash A$ in classical logic, (or in any other logic we use as a base. Note that if A_i are all atomic then it does not matter which logic we use as a base the consequence between conjunctions of atoms is always the same). If yes, then to the companion set $\mathbb{C}_2$ we add the fact $y(A)$. We thus get the companion set of facts $\mathbb{F}_2$ and we can carry on. This approach is perfectly compatible with the previous system where A, B were already atomic. The theory $\mathbb{F}$ is the conjunction of all the $\pm$ atoms in $\mathbb{F}$.

There is a slight problem here. When the formulas involved were atomic, a set of facts was a set of atoms $\mathbb{F}$, obtained by choose one of each pair $\{+x, -x\}$. So $\mathbb{F}$ was consistent. When the formulas involved are not atomic, even if we choose one of each pair $\{A, \neg A\}$, we may end up with a set $\mathbb{F}$ being inconsistent. We can require that we choose only consistent sets of facts and leave this requirement as an additional constraint.

This remark is going to be important when we compare our approach to that of Makinson and Torre's input output logic approach.

Example 15. To illustrate what we said in Remark 19, let us consider Figure 17.

The equations are listed in Example 8. Let us assume that in the figure we replace d by $y(D)$, where $D = d \vee c$. We get the equations involving $y(D)$ instead of D and get the solution as in Example 8, to be

$$x_0 = y(D) = \bar{f} = 1$$
$$\bar{x}_0 = y(\bar{D}) = f = 0.$$

Now we have the additional equation

$$\begin{aligned} y(D) &= \mathbf{h}_\mu(D) \\ &= \mathbf{h}_\mu(c \vee d) \\ &= \max(c, d) \end{aligned}$$

So we get $\max(c, d) = 1$ and we do have the solution with $d = 1, c = 1, ord = 0, c = 1, ord = 1, c = 0$.

Remark 20 (Miner Paradox [21,22]). We begin with a quote from Malte Willer in [22]

> Every adequate semantics for conditionals and deontic ought must offer a solution to the miners paradox about conditional obligations..... Here is the miners paradox. Ten miners are trapped either in shaft A or in shaft B, but we do not know which one. Water threatens to flood the shafts. We only have enough sand bags to block one shaft but not both. If one shaft is blocked, all of the water will go into the other shaft, killing every miner inside. If we block neither shaft, both will be partially flooded, killing one miner. [See Figure 34

Action	if miners in A	if miners in B
Block A	All saved	All drowned
Block B	All drowned	All saved
Block neither shaft	One drowned	One drowned

Fig. 34.

> Lacking any information about the miners exact whereabouts, it seems to say that
>
> 1. We ought to block neither shaft.
>
> However, we also accept that
>
> 2. If the miners are in shaft A, we ought to block shaft A,
> 3. If the miners are in shaft B, we ought to block shaft B.

But we also know that

4. Either the miners are in shaft A or they are in shaft B.

And (2)-(4) seem to entail

5. Either we ought to block shaft A or we ought to block shaft B, which contradicts (1).

Thus we have a paradox.

We formulate the Miners paradox as follows:

1. $\top \to O\neg\text{Block}A$
   $\top \to O\neg\text{Block}B$
2. Miners in $A \to O\text{Block}A$
3. Miners in $B \to O\text{Block}B$
4. Facts: Miners in $A \vee$ Miners in B.

The graph for (1)–(3) is Figure 35

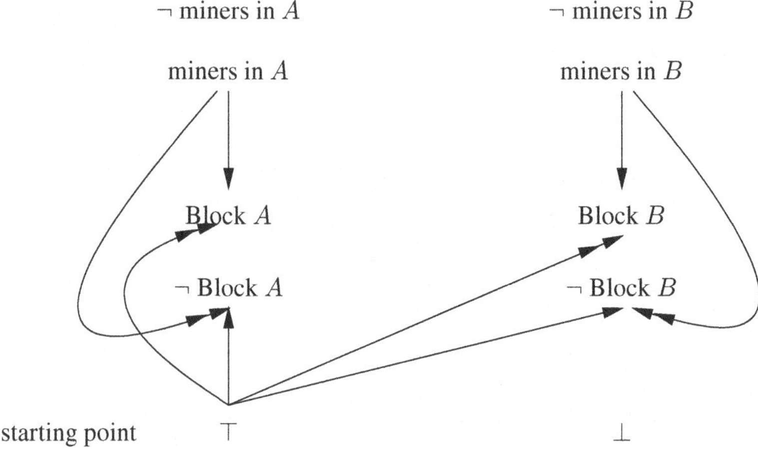

Fig. 35.

The Miners paradox arises because we want to detach using (2), (3) and (4) and get (5).

5. $O\text{Block}A \vee O\text{Block}B$

which contradicts (1).

However, according to our discussion, facts simply choose new starting points in the figure. The fact (4) is read as two possible sets of facts. Either the set of the fact that miner in A or the other possibility, the set containing miner in B. We thus get two possible graphs, Figure 36 and Figure 37.

We can see that there is no paradox here.

We conclude with a remark that we can solve the paradox directly using H. Reichenbach [24] reference points, without going through the general theory of this paper. See [23].

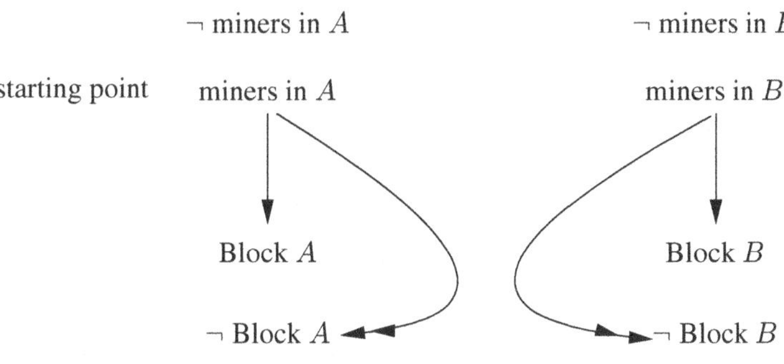

Fig. 36.

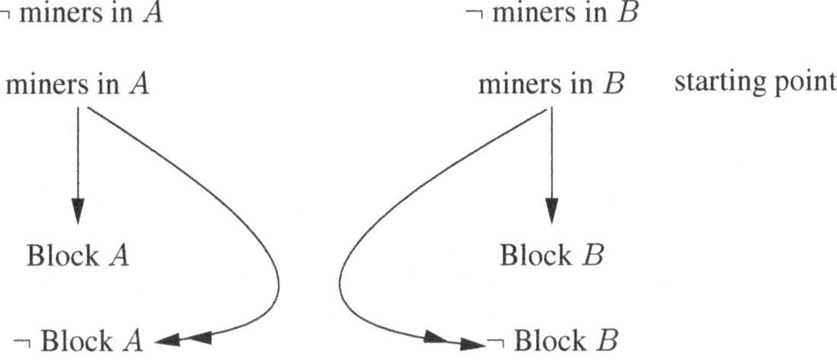

Fig. 37.

4 Proof Theory for CTDs

Our analysis in the previous sections suggest proof theory for sets of contrary to duty obligations. We use Gabbay's framework of labelled deductive systems [25].

We first explain intuitively our approach before giving formal definitions. Our starting point is Definition 13. The contrary to duty obligations according to this definition have the form $x \to Oy$, where x, y are atoms q or their negation $\neg q$ and x may be $\top$ and y is neither $\top$ nor $\bot$.

For our purpose we use the notation $x \Rightarrow y$. We also use labels annotating the obligations, and we write

$$t : x \Rightarrow y.$$

The label we use is the formula itself

$$t = (x \Rightarrow y).$$

Thus our CTD data for the purpose of proof theory has the form

$$(x \Rightarrow y) : x \Rightarrow y.$$

Given two CTD data items of the form

$$t : x \Rightarrow y; s : y \Rightarrow z$$

we can derive a new item

$$t * s : x \Rightarrow z$$

where $*$ is a concatenation of sequences. (Note that the end letter of t is the same as the beginning letter of s, so we can chain them.)

So we have the rule

$$\frac{(x \Rightarrow y) : x \Rightarrow y; (y \Rightarrow z) : y \Rightarrow z}{(x \Rightarrow y, y \Rightarrow z) : x \Rightarrow z}$$

It may be that we also have $(x \Rightarrow z) : x \Rightarrow z$ (i.e., the CTD set contains $x \to Oy, y \to Oz$ and $x \to Oz$), in which case $x \Rightarrow z$ will have two different labels, namely

$$t_1 = (x \Rightarrow y, y \Rightarrow z)$$
$$t_2 = (x \Rightarrow z).$$

We thus need to say that the proof theory allows for lables which are sets of chained labels (we shall give exact definitions later). So the label for $x \Rightarrow z$ would be $\{t_1, t_2\}$. There may be more labels $t_3, t_4, \ldots$ for $x \Rightarrow z$ depending on the original CTD set.

Suppose that in the above considerations $x = \top$. This means that our CTD set described above has the form $\{Oy$ (being $\top \to Oy), y \to Oz$ and $Oz\}$. By using the chaining rule we just described (and not mentioning any labels) we also get

$$\top \Rightarrow z, \top \Rightarrow y.$$

We can thus intuitively detach with $\top$ and get that our CTD set proves $\{y, z\}$. Notations $\vdash \{y, z\}$.

Alternatively, even if x were arbitrary, not necessarily $\top$, we can detach with x and write $x \vdash \{y, z\}$.

Of course when we use labels we will write

$$t : x \vdash \{s_1 : y, s_2 : z\}$$

the labels s_1, s_2 will contain in them the information of how y, z were derived from x.

To be precise, if for example,

$$\mathbb{C} = \{(x \Rightarrow y) : x \Rightarrow y, (y \Rightarrow z) : y \Rightarrow z, (x \Rightarrow z) : x \Rightarrow z\}.$$

We get

$$(x) : x \vdash_{\mathbb{C}} \{(x, x \Rightarrow y, y \Rightarrow z) : z, (x, x \Rightarrow z) : z, (x, x \Rightarrow y) : y\}.$$

Definition 17. *1. Let Q be a set of atoms. Let $\neg$ be a negation and let $\Rightarrow$ be a CTD implication symbol.*
A clause has the form $x \Rightarrow y$, where x is either $\top$ or atom q or $\neg q$ and y is either atom a or $\neg a$.

2. *A basic label is either $(\top)$ or (q) or $(\neg q)$. (q atomic) or a clause $(x \Rightarrow y)$.*

3. *A chain label is a sequence of the following form*

$$(x_0 \Rightarrow x_1, x_1 \Rightarrow x_2, \ldots, x_n \Rightarrow x_{n+1})$$

where $x_i \Rightarrow x_{i+1}$ are clauses. x_0 is called the initial element of the sequence and x_{n+1} is the end element.
4. *A set label is a set of chain labels.*
5. *A labelled CTD dataset $\mathbb{C}$ is a set of elements of the form $(x \Rightarrow y) : (x \Rightarrow y)$ where $x \Rightarrow y$ is a clause and $(x \Rightarrow y)$ is a basic label.*
6. *A fact has the form $(x) : x$ where x is either atom q or $\neg q$ or $\top$.*
   A fact set $\mathbb{F}$ is a set of facts.

Definition 18. *Let $\mathbb{C}$ be a CTD dataset. We define the notions of*

$$\mathbb{C} \vdash_n t : x \Rightarrow y$$

where $n \geq 0, t$ a basic or chain label. This we do by induction on n.

We note that we may have $\mathbb{C} \vdash_n t : x \Rightarrow y$ hold for several different ns and different ts all depending on $\mathbb{C}$.

Case $n = 0$
$\mathbb{C} \vdash_0 t : x \Rightarrow y$ *if* $t = (x \Rightarrow y)$ *and* $(x \Rightarrow y) : x \Rightarrow y \in \mathbb{C}$.

Case $n = 1$
$\mathbb{C} \vdash_1 t : x \Rightarrow y$ *if for some* $x \Rightarrow w$ *we have* $(x \Rightarrow w) : x \Rightarrow w$ *in* $\mathbb{C}$ *and* $(w \Rightarrow y) : (w \Rightarrow y)$ *in* $\mathbb{C}$ *and* $t = (x \Rightarrow w, w \Rightarrow y)$.

Note that the initial element of t is x and the end element is y.

Case $n = m + 1$
Assume that $\mathbb{C} \vdash_m t : x \Rightarrow y$ has been defined and that in such cases the end element of t is y and the initial element of t is x.

Let $\mathbb{C} \vdash_{m+1} t : x \Rightarrow y$ hold if for some $t' : x \Rightarrow w$ we have $\mathbb{C} \vdash_m t' : x \Rightarrow w$ (and therefore the end element of t' is w and the initial element of t is x) and $(w \Rightarrow y) : w \Rightarrow y \in \mathbb{C}$ and $t = t' * (w \Rightarrow y)$, where $*$ is concatenation of sequences.

Definition 19. *Let $\mathbb{C}$ be a dataset and let $(x) : x$ be a fact. We write $\mathbb{C} \vdash^x_{n+1} t : y$ if for some $t : x \Rightarrow y$ we have $\mathbb{C} \vdash_n t : x \Rightarrow y$.*

We may also use the clearer notation

$$\mathbb{C} \vdash_{n+1} (x, t) : y.$$

Example 16. Let $\mathbb{C}$ be the set

$$(x \Rightarrow y) : x \Rightarrow y$$
$$(y \Rightarrow z) : y \Rightarrow z$$
$$(z \Rightarrow y) : z \Rightarrow y$$

Then

$$\begin{aligned}
&\mathbb{C} \vdash_0 (x \Rightarrow y) : x \Rightarrow y \\
&\mathbb{C} \vdash_2 (x \Rightarrow y, y \Rightarrow z, z \Rightarrow y) : x \Rightarrow y \\
&\mathbb{C} \vdash_1^x (x \Rightarrow y) : y \\
&\mathbb{C} \vdash_3^x (x \Rightarrow y, y \Rightarrow z, z \Rightarrow y) : y
\end{aligned}$$

or using the clearer notation

$$\begin{aligned}
&\mathbb{C} \vdash_1 (x, (x \Rightarrow y)) : y \\
&\mathbb{C} \vdash_3 (x, (x \Rightarrow y, y \Rightarrow z, z \Rightarrow y)) : y.
\end{aligned}$$

Note also that y can be proved with different labels in different ways.

Definition 20. *Let $\mathbb{C}$ be a dataset and let $\mathbb{F}$ be a set of facts. We define the notion of $\mathbb{C}, \mathbb{F} \vdash_n (z, t) : x$ where x, z are atomic or negation of atomic and z also possibly $z = \top$, as follows:*

Case $n = 0$
$\mathbb{C}, \mathbb{F} \vdash_0 (z, t) : x$ *if* $(x) : x \in \mathbb{F}$ *and* $(z, t) = (x)$.

Case $n = m + 1$
$\mathbb{C}, \mathbb{F} \vdash_{m+1} (z, t) : x$ *if* $\mathbb{C} \vdash_n t : z \Rightarrow x$ *and* $z = \top$ *or* $(z) : z \in \mathbb{F}$.

Example 17. We continue Example 16. We have

$$\begin{aligned}
&\mathbb{C}, \{(z) : z\} \vdash_0 (z) : z \\
&\mathbb{C} \vdash_2 (z, (z \Rightarrow y)) : y \\
&\mathbb{C} \vdash_4 (x, (x \Rightarrow y, y \Rightarrow z, z \Rightarrow y)) : y
\end{aligned}$$

Example 18. To illustrate the meaning of the notion of $\mathbb{C}, \mathbb{F} \vdash t : x$ let us look at the CTD set of Figure 26 (this is the Reykajavic set) with $d =$ dog and $f =$ fence:

1. $O\neg d$
   written as $(\top \Rightarrow \neg d) : \top \Rightarrow \neg d$.
2. $O\neg f$
   written as $(\top \Rightarrow \neg f) : \top \Rightarrow \neg f$.
3. $d \rightarrow Of$
   written as $(d \Rightarrow f) : d \Rightarrow f$.
4. $f \rightarrow Od$ written as $(f \Rightarrow d) : f \Rightarrow d$.

The above defines $\mathbb{C}$. Let the facts $\mathbb{F}$ be $(d) : d$ and $(\neg f) : \neg f$. We can equally write the facts as

$$\begin{aligned}
&(\top \Rightarrow d) : \top \Rightarrow d) \\
&(\top \Rightarrow \neg f) : \top \Rightarrow \neg f.
\end{aligned}$$

(a) CTD point of view
    Let us first look at the contrary to duty set and the facts intuitively from the deontic point of view. The set says that we are not allowed to have neither a dog d nor a

fence f. So good behaviour must "prove" from $\mathbb{C}$ the two conclusions $\{\neg d, \neg f\}$. This is indeed done by

$$\mathbb{C} \vdash_1 (\top, (\top \Rightarrow \neg d)) : \neg d$$
$$\mathbb{C} \vdash_1 (\top, (\top \Rightarrow \neg f) : \neg f$$

The facts are that we have a dog (in violation of $\mathbb{C}$) and not a fence

$$\mathbb{F} = \{(d) : d, (\neg f) : \neg f\}.$$

So we can prove

$$\mathbb{C}, \mathbb{F} \vdash_0 (d) : d$$
$$\mathbb{C}, \mathbb{F}) \vdash_0 (\neg f) : \neg f$$

but we also have

$$\mathbb{C}, \mathbb{F} \vdash_1 (d, (d \Rightarrow f)) : f.$$

We can see that we have violations, and the labels tell us what violates what.
Let us take the facts as a sequence. First we have a dog and then not a fence.
Let $\mathbb{F}_d = \{(d) : d\}$ and $\mathbb{F}_{\neg f} = \{(\neg f) : \neg f\}$.
then

$$\mathbb{C}, \mathbb{F}_d \vdash_0 (d) : d$$

$$\mathbb{C}, \mathbb{F}_d \vdash_1 (d, (d \Rightarrow f)) : f$$

which violates

$$\mathbb{C} \vdash_0 (\top, (\top \Rightarrow \neg f)) : \neg f$$

but $\mathbb{F} = \mathbb{F}_d \cup \mathbb{F}_{\neg f}$, and so $\mathbb{F}$ viewed in this sequece (first d then $\neg f$) gives us a choice of points of view. Is the addition $\neg f$ a violation of the CTD dog $\to O$ fence or is it in accordance with the original $O\neg f$?
The problem here is that the remedy for the violation of $O\neg d$ by the fact d is $d \to Of$, which is a violation of another CTD namely $O\neg f$. One can say the remedy wins or one can say this rememdy is wrong, stick to $O\neg f$.
The important point about the proof system $\mathbb{C}, \mathbb{F} \vdash_n t : A$ is that we can get exactly all the information we need regarding facts and violations.

(b) Modal point of view
To emphasise the mechanical uniterpreted nature of the proof system let us give it a modal logic interpretaiton. We regard the labels as possible worlds and regard $*$ as indicating accessibility. We read $\mathbb{C}, \mathbb{F} \vdash_n t : A$ as $t \vDash A$, in the model $\mathbf{m}$ defined by $\mathbb{C}, \mathbb{F}$ i.e. $\mathbf{m} = \mathbf{m}(\mathbb{C}, \mathbb{F})$). The model of (a) above is shown in Figure 38
What holds at node t in Figure 38 is the end element of the sequence t.
The facts give us no contradiction, because they are true at different worlds. At $(\top, (\top \Rightarrow d))$ we have dog and so at $(\top, (\top \Rightarrow d, d \Rightarrow f))$ we have a fence while at $(\top, (\top \Rightarrow \neg f))$, we have no fence.
Inconsistency can only arise if we have $(x) : z$ and $(x) : \neg z$ or $t : x \Rightarrow y$ and $t : x \Rightarrow \neg y$ but we cannot express that in our language.

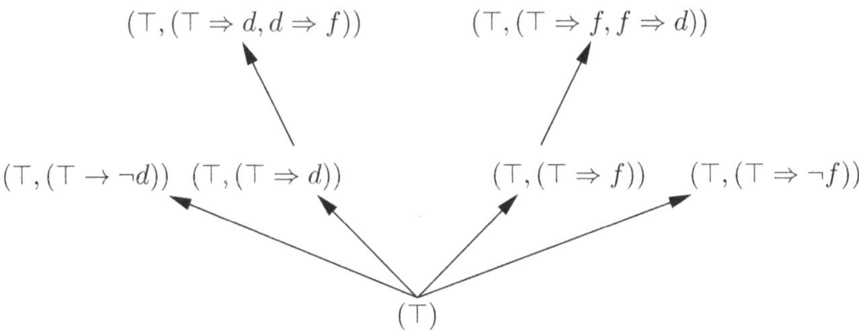

Fig. 38.

(c) The deductive view
This is a labelled deductive system view. We prove all we can from the system and to the extent that we get both A and $\neg A$ with different labels, we collect all labels and implement a *flattening policy*, to decide whether to adopt A or adopt $\neg A$.
Let us use, by way of example, the following flattening policy:
FP1 Longer labels win over shorter labels. (This means in CTD intrepretation that once an obligation is violated, the CTD has precedence.)
FP2 In case of same length labels, membership in $\mathbb{F}$ wins. This means we must accept the facts!
So according to this policy we have

$$(\top, (\top \Rightarrow d, d \Rightarrow f)) : f \text{ wins over } (\top, (\top \Rightarrow \neg f)) : \neg f$$

and

$$(\top, (\top \Rightarrow d)) : d \text{ wins over } (\top, (\top \Rightarrow \neg d)) : \neg d.$$

So we get the result $\{d, f\}$.

We can adopt the input-output policy of Makinson-Torre. We regard $\mathbb{C}$ as a set of pure mathematical input output pairs.

We examine each rule in $\mathbb{C}$ against the input $\mathbb{F}$. If it yields a contradictory output, we drop the rule.

The final result is obtained by closing the input under the remaining rules. So let us check:

$$\text{Input} : \{d, \neg f\}.$$

Rules in $\mathbb{C}$
$\top \Rightarrow \neg f$, OK
$\top \Rightarrow \neg d$, drop rule
$d \Rightarrow f$, drop rule
$f \Rightarrow d$, not applicable.

Result of closure: $\{d, \neg f\}$.

The input-output approach is neither proof theory not CTD. To see this add another rule

$$\neg d \Rightarrow b$$

where b is something completely different, consistent with $\pm d, \pm f$. This rule is not activated by the input $\{d, \neg f\}$. In the labelled approach we get b in our final set.

Problems of this kind have already been addressed by our approach of *compromise revision*, in 1999, see [25].

Example 19. Let us revisit the miners paradox of Remark 20 and use our proof theory.

We have the following data:

1. $\top \Rightarrow \neg$ Block A
2. $\top \Rightarrow \neg$ Block B
3. Miners in $A \Rightarrow$ Block A
4. Miners in $B \Rightarrow$ Block B
5. Fact: Miners in $A\vee$ miners in B

using ordinary logic.

We get from (2), (3) and (4)

5. Block $A\vee$ Block B

(5) contradicts (1).

Let us examine how we do this in our labelled system.

We have

1*. $(\top \Rightarrow$ Block $A) : \top \Rightarrow$ Bock A
$(\top \Rightarrow$ Block $B) : \top \Rightarrow$ Block B
2*. (miners in $A \Rightarrow$ Block A): miners in $A \Rightarrow$ Block A.
3*. (miners in $B \Rightarrow$ Block B): miners in $B \Rightarrow$ Block B
4*. (miners in $A\vee$ miners in B)': miners in $A\vee$ miners in B.

To do labelled proof theory we need to say how to chain the labels of disjunctions.

We do the obvious, we chain each disjunct. So if t is a label with end element $x \vee y$ and we have two rules $x \Rightarrow z$ and $y \Rightarrow w$ then we can chain

$$(t, (x \Rightarrow z, y \Rightarrow w))$$

So we have the following results using such chaining:
(1*), (2*), (3*), (4*) $\vdash_1 (\top, (\top \Rightarrow \neg$ Block $A)) : \neg$ Block A
(1*), (2*), (3*), (4*) $\vdash_1 (\top, (\top \Rightarrow \neg$ Block $B)) : \neg$ Block B
(1*), (2*), (3*), (4*) $\vdash_3$ ((miners$A \vee$ miners B), ((miners $A \Rightarrow$ Block A), (miners $B \Rightarrow$ Block B))) : Block $A \vee$ Block B

Clearly we have proofs of $\neg$ Block A, $\neg$ Block B and Block $A\vee$ Block B but with different labels! The labels represent levels of knowledge. We can use a flattening process on the labels, or we can leave it as is.

There is no paradox, because the conclusions are on different levels of knowledge.

This can be seen also if we write a classical logic like proof.

To prove Block $A \vee$ Block B from (2), (3), (4), we need to use subproofs. Whenever we use a subproof we regard the subproof as a higher level of knowledge. To see this consider the attempt to prove

$$\frac{(l1.1) \quad A \Rightarrow B}{(l1.2) \quad \neg B \Rightarrow \neg A}$$

We get $\neg B \Rightarrow \neg A$ from the subproof in Figure 39.

Let us now go back to the miners problem. The proof rules we have to use are

MP $$\frac{A, A \Rightarrow B}{B}$$

Outer Box

(11.2.1) Assume $\neg B$, show $\neg A$

To show $\neg A$ use subproof in Inner Box

Inner Box

(1 1.2.1.1) Assume A, show $\bot$

(11.2.1.2) We want to reiterate (11.1) $A \Rightarrow B$ and bring it here to do modus ponens and get B

(1 1.2.1.3) We want to reiterate (1 1.2.1) $\neg B$ and bring it here to get a contradiction.

To do these actions we need proof theoretic permissions and procedures, because moving assumptions across levels of knowledge, from outer box to inner box

Such procedures are part of the definition of the logic.

Fig. 39.

DE $$\frac{A \vee B \quad A \text{ proves } C \quad B \text{ proves } D}{C \vee D}$$

RI We can reiterate positive (but not negative) wffs into subproofs.

The following is a proof using these rules:

Level 0

0.1a $\neg$ Block A. This is a negative assumption
0.1b $\neg$ Block B, negative assumption
0.2 miners in $A \Rightarrow$ Block A, assumption
0.3 miners in $B \Rightarrow$ Block B, assumption
0.4 miners in $A \vee$ miners in B
0.5 Block $A \vee$ Block B, would have followed from the proof in Figure 40, if there were no restriction rule RI. As it is the proof is blocked.

Box 1

1.1	Miners in $A \vee$ miners in B, reiteration of 0.4 into Box 1
1.2	Miners in $A \Rightarrow$ Block A, reiteration of 0.2 into Box 1.
1.3	Miners in $B \Rightarrow$ Block B, reiteration of 0.3 into Box 1
1.4	Block $A \vee$ Block B, from 1.1, 1.2, and 1.3 using DE
1.5	To get a contradiction we need to bring 0.1a $\neg$ Block A 0.1.b $\neg$ Block B as reiterations into Box 1. However, we cannot do so because these are negative information assumptions and cannot be reiterated

Fig. 40.

5 Comparing with Makinson and Torre's Input Output Logic

This section compares our work with Input Output logic , I/O, of Makinson and Torre. Our starting point is [19]. Let us introduce I/O using Makinson and Torre own words from [19].

> BEGIN QUOTE 1
> Input/output logic takes its origin in the study of conditional norms. These may express desired features of a situation, obligations under some legal, moral or practical code, goals, contingency plans, advice, etc. Typically they may be expressed in terms like: In such-and-such a situation, so-and-so should be the case, or ...should be brought about, or ...should be worked towards, or ...should be followed — these locutions corresponding roughly to the kinds of norm mentioned. To be more accurate, input/output logic has its source in a tension between the philosophy of norms and formal work of deontic logicians...
>
> Like every other approach to deontic logic, input/output logic must face the problem of accounting adequately for the behaviour of what are called 'contrary-to-duty' norms. The problem may be stated thus: given a set of norms to be applied, how should we determine which obligations are operative in a situation that already violates some among them. It appears that input/output logic provides a convenient platform for dealing with this problem by imposing consistency constraints on the generation of output.
>
> We do not treat conditional norms as bearing truth-values. They are not embedded in compound formulae using truth-functional connectives. To avoid all confusion, they are not even treated as formulae, but simply as ordered pairs (a,x) of purely boolean (or eventually first-order) formulae.
> Technically, a normative code is seen as a set G of conditional norms, i.e. a set of such ordered pairs (a, x). For each such pair, the body a is thought of as an input, representing some condition or situation, and the head x is thought of as an output, representing what the norm tells us to be desirable, obligatory or whatever in that situation. The task of logic is seen as a modest one. It is not to create or determine a distinguished set of norms, but rather to prepare information before it goes in as input to such a set G, to unpack output as it emerges and, if needed, coordinate the two in certain ways. A set G of conditional norms is thus seen as a transformation device, and the task of logic is to act as its 'secretarial assistant'.

Makinson and Torre adapt an example from Prakken and Sergot [4] to illustrate their use of input/output logic. We shall use the same example to compare their system with ours.

Example 20. We have the following two norms:

1. The cottage should not have a fence or a dog;
   $O\neg(f \vee d)$
   or equivalently

   (a) $O\neg f$
   (b) $O\neg d$
2. If it has a dog it must have both a fence and a warning sign.
   $d \to O(f \wedge w)$
   or equivalently
   (c) $d \to Of$
   (d) $d \to Ow$
   In the notation of input/output logic the above data is written as
   (e) $(\top, \neg(f \vee d))$
   (f) $(d, f \wedge w)$.
   Suppose further that we are in the situation that the cottage has a dog, in other words we have the fact:
3. Fact: d
   thus violating the first norm.

The question we ask is: what are our current obligations? or in other words, how are we going to model this set? We know from our analysis in the previous section that a key to the problem is modelling the facts and that deontic logic gets into trouble because it does not have the means to pay attention to what we called the obligation progression.

Figures 41–43 describe our model, which is quite straight forward.

Let us see how Makinson and Torre handle this example.

The input output model will apply the data as input to the input output rules (f) and (e). This is the basic idea of Makinson and Torre for handling CTD obligations with facts.

Makinson and Torre realise that, and I quote again

> BEGIN QUOTE 2
> Unrestricted input/output logic gives
> f: the cottage has a fence
> and
> w: the cottage has a warning sign.
> Less convincingly, because unhelpful in the supposed situation, it also gives
> $\neg d$: the cottage does not have a dog.
> Even less convincingly, it gives
> $\neg f$: the cottage does not have a fence,
> which is the opposite of what we want. These results hold even for simple-minded output, ...

Makinson and Torre propose as a remedy to use constraints, namely to apply to the facts only those I/O rules which outputs are consistent with the facts. They say, and I quote again:

> BEGIN QUOTE 3
> Our strategy is to adapt a technique that is well known in the logic of belief change cut back the set of norms to just below the threshold of making the current situation contrary-to-duty. In effect, we carry out a contraction on the set G of given norms. Specifically, we look at the maximal subsets G' of G such

> that out(G', A) is consistent with input A. To illustrate this consider the cottage example, where $G = \{(t, \neg(f \vee d), (d, f \wedge w)\}$, with the contrary-to-duty input d. Using just simple minded output, G' has just one element $(d, f \wedge w)$ and so the output is just $f \wedge w$.

We note that this output corresponds to our Figure 43.

Makinson and Torre continue to say, a key paragraph showing the difference between our methods and theirs:

> BEGIN QUOTE 4
> Although the ... strategy is designed to deal with contrary-to-duty norms, its application turns out to be closely related to belief revision and nonmonotonic reasoning when the underlying input/output operation authorizes throughput More surprisingly, there are close connections with the default logic of Reiter, falling a little short of identity...

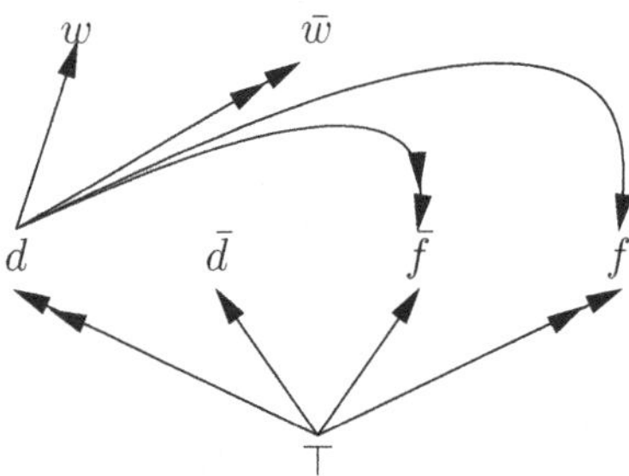

Fig. 41.

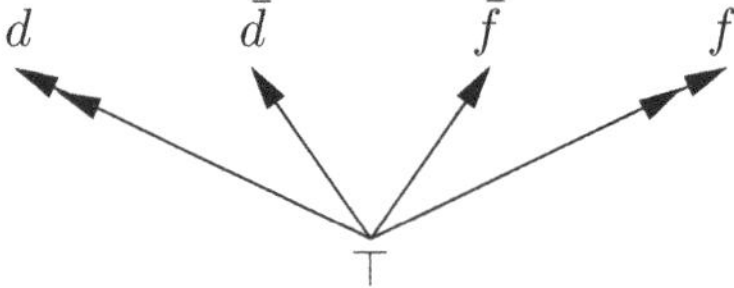

Fig. 42.

Let us, for the sake of comparison, consider the CTD sets of Figure 26 (this is actually the Reykajavik set of CTDs) and the facts as considered in Example 12. We have the following CTD (or equivalently the input output rules):

1. $(\top, \neg(d \vee f))$
2. (d, f)
3. (f, d)

The input is $A = d \wedge \neg f$.

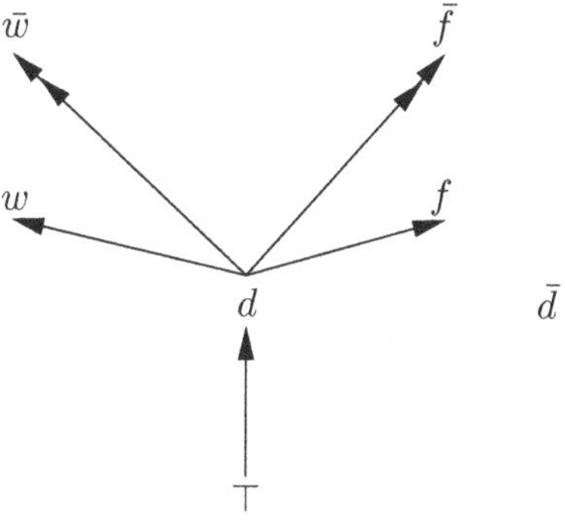

Fig. 43.

In this example the only rules (x, y) for which $A \vDash x$ are rules (1) and (2), but neither of their output is consistent with the input. So nothing can be done here. This corresponds to the lack of solution of our equations where we want to make both d and $\neg f$ the starting points. Our analysis in Example 12 however, gives a different result, because we first input d then $\neg f$ and in parallel, put $\neg f$ and then d.

So apart from the difference that input output logic is based on classical semantics for classical logic and we use equational semantics, there is also the difference that input output logic puts all the input in one go and detaches with all CTD rules whose output does not contradict it, while we use all possible sequencing of the input, inputting them one at a time. (To understand what 'one at a time' means, recall Remark 19 and Example 15.) There is here a significant difference in point of view. We take into account the obligation progression and given a set of facts as inputs, we match them against the obligation progression. In comparison, Input Output logic lumps all CTD as a set of input output engines and tries to plug the inputs into the engines in different ways and see what you get. The CTD clauses lose their Deontic identity and become just input output engines. See our analysis and comparison in part (c) of Example 18, where this point is clearly illustrated.

Let us do a further comparison. Consider the looping CTD set of Figure 4 which is analysed in Example 7. This has two input output rules

1. (a, b)
2. $(b, \neg a)$

Consider the two possible inputs

$$A = \neg a \wedge b$$

and

$$B = \neg a \wedge \neg b$$

A was a solution according to the soft approach option. B was a solution according to the mathematical approach option.

Using the input output approach, we can use $(b, \neg a)$ for A and we cannot use anything for B.

So there is compatibility here with the soft approach.

Let us summarise the comparison of our approach with the input output approach.

Com 1. We use equational sematics, I/O uses classical semantics.

Com 2. We rely on the obligation progression, breaking the input into sequence and modelling it using graphs. I/O does not do that, but uses the input all at once and taking maximal sets of CTDs (x, y) such that the input proves the xs and is consistent with the ys. The question whether it is possible to define violation progression from this is not clear. The I/O approach is a consequence relation/consistency approach. Our graph sequences and input facts sequences can also model action oriented/temporal (real time or imainary obligation progression 'time'). So for example we can model something like

$$f \to O\neg f$$

If you have a fence you should take it down.

Com 3. We remain faithful to the contrary to duty spirit, keeping our graphs and equations retain the CTD structure. I/O brought into their system significant AGM revision theory and turn I/O into a technical tool for revision theory and other nonmonotonic systems. See their quoted text 4.

Com 4. The connections are clear enough for us to say we can give equational semantics directly to input output logic, as it is, and never mind its connections with contrary to duty. Makinson and Torre defined input output logic, we have our equational approach, so we apply our approach to their logic directly. This is the subject of a separate paper.

6 Comparing with Governatori and Rotolo's Logic of Violations

We now compare with Governatori and Rotolo's paper [13]. This is an important paper which deserves more attention.

Governatori and Rotolo present a Gentzen system for reasoning with contrary-to-duty obligations. The intuition behind the system is that a contrary-to-duty is a special kind of normative exception. The logical machinery to formalise this idea is taken from substructural logics and it is based on the definition of a new non-classical connective capturing the notion of reparational obligation.

Given in our notation the following sequence of CTDs

$$A_1, \ldots, A_n \Rightarrow OB_1$$
$$\neg B_1 \Rightarrow OB_2$$
$$\neg B_1 \Rightarrow OB_3$$

They introduce a substructural connective and consider it as a sub-structural consequence relation without the structural rules of contraction, duplication, and exchange, and write the above sequence as

$$A_1, \ldots, A_n \Rightarrow B_1, \ldots, B_m.$$

The meaning is: the sequence $A_1, \ldots, A_n$ comports that B_1 is the case; but if B_1 is not satisfied, then B_2 should be the case; if both B_1 and B_2 are not the case, then B_3 should be satisfied, and so on. In a normative context, this means that the content of the obligation determined by the conditions $A_1, \ldots, A_n$. The As are the facts and they are not ordered. So in this respect Governatori and Rotolo approach are like the I/O approach.

Govenatori and Rotolo give proof theoretical rules for manipulating such sequents.

This approach is compatible with our approach in the sense that it relies on the obligation progression. It is also compatible with our proof theory of Section 4.

For the purpose of comparison, we need not go into details of their specific rules. it is enough to compare one or two cases.

Consider the CTD set represented by Figure 5. Since this set and figure is acyclic, Governatori and Rotolo can represent it by a theory containing several of their sequents . Each sequent will represent a maximal path in the figure. I don't think however that they can represent all possible paths. So the graph representation is a more powerful representation and we could and plan to present proof theory on graphs in a subsequent paper.

From my point of view , Governatori and Rotolo made a breakthrough in 2005 in the sense that they proposed to respect what I call the obligation (or violation) progression and their paper deserves more attention.

They use Gentzen type sequences which are written linearly, and are therefore restricted. We use plannar graphs (think of them as planar two dimensional Gentzen sequents) which are more powerful.

I am not sure how Governatori and Rotolo will deal with loops in general. They do find a way to deal with some loops for example I am sure they can handle the CTD of Figure 9,or of Figure 4, but I am not sure how they would deal with a general CTD set.

By the way, we used ordered sequences with hierarchical consequents in [18].

Governatori and Rotolo do not offer semantics for their system.

We offer equational semantics.

This means that we can offer equational semantics to their Gentzen system and indeed offer equational semantics to substructural logics in general.

This is a matter for another future paper.

Let us quote how they deal with the Chisholm paradox

> **Chisholm's Paradox.** The basic scenario depicted in Chisholm's paradox corresponds to the following implicit normative system:
>
> $$\{\vdash_O h, h \vdash_O i, \neg h \vdash_O \neg i\}$$
>
> plus the situation $s = \{\neg h\}$. First of all, note that the system does not determine in itself any normative contradiction. This can be checked by making explicit the normative system. In this perspective, a normative system consisting of the above norms can only allow for the following inference:
>
> $$\frac{\vdash_O h, \quad \neg h \vdash_O \neg i}{\vdash_O (h, \neg i)}$$

Thus, the explicit system is nothing but

$$\{h \vdash_O i, \vdash_O (h, \neg i)]\}.$$

It is easy to see that s is ideal (my words: i.e. no violations) wrt the first norm . On the other hand, while s is not ideal wrt $\vdash_O (h, \neg i)$, we do not know if it is sub-ideal (i.e. there are some violations but they are compensated by obeying the respective CTD) wrt such a norm. Then, we have to consider the two states of affairs $s_1 = \{\neg h, i\}$ and $s_2 = \{\neg h, \neg i\}$. It is immediate to see that s_1 is non-ideal (i.e. all violations throughout, no compensation) in the system, whereas s_2 is sub-ideal.

If so, given s, we can conclude that the normative system says that $\neg i$ ought to be the case.

7 Conclusion

We presented the equational approach for classical logic and presented graphs for General CTD sets which gave rise to equations . These equations provided semantics for general CTD sets.

The two aspects are independent of one another, though they are well matched.

We can take the graph representation and manipulate it using syntactical rules and this would proof theoretically model CTD's. Then we can give it semantics, either equational semantics or possible world semantics if we want.

We explained how we relate to Makinson and Torre's input output approach and Governatory and Rotolo's logic of violations approach.

The potential "output" from this comparison are the following possible future papers:

1. Equational semantics for input output logic
2. Equational semantics for substructural logics
3. Development of planar Gentzen systems (that would be a special case of labelled deductive systems)
4. Planar proof theory for input output logic (again, a special case of labelled deductive systems).
5. Proof theory and equational semantics for embedded CTD clauses of the form $x \to O(y \to Oz))$.

References

1. Gabbay, D.: An Equational Approach to Argumentation Networks. Argument and Computation 3(2-3), 87–142 (2012)
2. Gabbay, D.: Meta-Logical Investigations in Argumentation Networks. Research Monograph College publications, 770 p. (2013)
3. Gabbay, D.: Temporal deontic logic for the generalised Chisholm set of contrary to duty obligations. In: Ågotnes, T., Broersen, J., Elgesem, D. (eds.) DEON 2012. LNCS, vol. 7393, pp. 91–107. Springer, Heidelberg (2012)

4. Prakken, H., Sergot, M.J.: Contrary-to-duty obligations. Studia Logica 57(1), 91–115 (1996)
5. Carmo, J., Jones, A.J.I.: Deontic Logic and Contrary-to-duties. In: Gabbay, D.M., Guenthner, F. (eds.) Handbook of Philosophical Logic, vol. 8, pp. 265–343. Springer, Heidelberg (2002)
6. Prakken, H., Sergot, M.: Dyadic deontic logic and contrary to duty obligations. In: Nute, D. (ed.) Defeasible Deontic Logic, Synthese Library, pp. 223–262. Kluwer (1997)
7. de Boer, M., Gabbay, D., Parent, X., Slavkova, M.: Two dimensional deontic logic. Synthese 187(2), 623–660 (2012), doi:10.1007/s11229-010-9866-4
8. Jones, A.J.I., Pörn, I.: Ideality, sub-ideality and deontic logic. Synthese 65 (1985)
9. `http://en.wikipedia.org/wiki/T-norm`
10. Gabbay, D., Rodrigues, O.: Voting and Fuzzy Argumentation Networks. Submitted to Journal of Approximate Reasoning; Short version to appear in Proceedings of CLIMA 2012. Springer. New revised version now entitled: Equilibrium States on Numerical Argumentation Networks
11. Chisholm, R.M.: Contrary-to-duty imperatives and deontic logic. Analysis 24 (1963)
12. Gabbay, D., Gammaitoni, L., Sun, X.: The paradoxes of permission. An action based solution. To appear in Journal of Applied Logic
13. Governatori, G., Rotolo, A.: Logic of violations: a Gentzen system for reasoning with contrary-to-duty obligations. Australasian Journal of Logic 4, 193–215 (2005)
14. Governatori, G., Rotolo, A.: A Gentzen system for reasoning with contrary-to-duty obligations, a preliminary study. In: Jones, A.J.I., Horty, J. (eds.) Deon 2002, London, pp. 97–116 (May 2002)
15. Makinson, D., van der Torre, L.: Input/output logics. Journal of Philosophical Logic 29(4), 383–408 (2000)
16. Makinson, D., van der Torre, L.: Constraints for input/output logics. Journal of Philosophical Logic 30(2), 155–185 (2001)
17. Parent, X., Gabbay, D., van der Torre, L.: Intuitionistic Basis for Input/Output Logic. To appear springer volume in honour of David Makinson
18. Gabbay, D.M., Schlechta, K.: A Theory of Hierarchical Consequence and Conditionals. Journal of Logic, Language and Information 12(1), 3–32 (2010)
19. Makinson, D., van der Torre, L.: What is Input/Output Logic? In: ESSLLI 2001 (2001)
20. Gabbay, D., Strasser, C.: A dynamic proof theory for reasoning with conditional obligations on the basis of reactive graphs. Paper 477
21. Kolodny, N., MacFarlane, J.: Ifs and oughts. Journal of Philosophy 107(3), 115–143 (2010)
22. Willer, M.: A remark on iffy oughts. Journal of Philosophy 109(7), 449–461 (2012)
23. Baniasad, Z., Gabbay, D., Robaldo, L., van der Torre, L.: A solution to the miner paradox: A Reichenbach reference points approach (2013)
24. Reichenbach, H.: Elements of Symbolic Logic. Free Press, New York (1947)
25. Gabbay, D.M.: Labelled Deductive Systems. Oxford Logic Guides 33 (1996)
26. Gabbay, D.M.: Compromise update and revision: A position paper. In: Fronhoffer, B., Pareschi, R. (eds.) Dynamic Worlds. Applied Logic Series, vol. 12, pp. 111–148. Springer (1999)

A Survey of the Data Complexity of Consistent Query Answering under Key Constraints

Jef Wijsen

Université de Mons, Mons, Belgium
jef.wijsen@umons.ac.be

Abstract. This paper adopts a very elementary representation of uncertainty. A relational database is called uncertain if it can violate primary key constraints. A repair of an uncertain database is obtained by selecting a maximal number of tuples without selecting two distinct tuples of the same relation that agree on their primary key. For any Boolean query q, $\mathsf{CERTAINTY}(q)$ is the problem that takes an uncertain database **db** on input, and asks whether q is true in every repair of **db**. The complexity of these problems has been particularly studied for q ranging over the class of Boolean conjunctive queries. A research challenge is to solve the following complexity classification task: given q, determine whether $\mathsf{CERTAINTY}(q)$ belongs to complexity classes **FO**, **P**, or **coNP**-complete.

The counting variant of $\mathsf{CERTAINTY}(q)$, denoted $\sharp\mathsf{CERTAINTY}(q)$, asks to determine the exact number of repairs that satisfy q. This problem is related to query answering in probabilistic databases.

This paper motivates the problems $\mathsf{CERTAINTY}(q)$ and $\sharp\mathsf{CERTAINTY}(q)$, surveys the progress made in the study of their complexity, and lists open problems. We also show a new result comparing complexity boundaries of both problems with one another.

1 Motivation

Uncertainty shows up in a variety of forms and representations. In this paper, we consider a very elementary representation of uncertainty. We model uncertainty in the relational database model by primary key violations. A *block* is a maximal set of tuples of the same relation that agree on the primary key of that relation. Tuples of a same block are mutually exclusive alternatives for each other. In each block, only one (and exactly one) tuple can be true, but we do not know which one. We will refer to databases as "uncertain databases" to stress that such databases can violate primary key constraints.

Primary keys are underlined in the conference planning database of Fig. 1. Blocks are separated by dashed lines. There is uncertainty about the city of ICDT 2016 (Rome or Paris), about the rank of KDD (A or B), and about the frequency of ICDT (biennial or annual).

There can be several reasons why a database is uncertain. On the positive side, it allows one to represent several possible future scenarios. In Fig. 1, the relation **C** represents that there are still two candidate cities for hosting ICDT

C. Beierle and C. Meghini (Eds.): FoIKS 2014, LNCS 8367, pp. 62–78, 2014.

C	**conf**	**year**	**city**	**country**
	ICDT	2016	Rome	Italy
	ICDT	2016	Paris	France
	KDD	2017	Rome	Italy

R	**conf**	**rank**	**frequency**
	ICDT	A	biennial
	ICDT	A	annual
	KDD	A	annual
	KDD	B	annual
	DBPL	B	biennial
	BDA	B	annual

Fig. 1. Uncertain database

2016. On the reverse side, inconsistency may be an undesirable but inescapable consequence of data integration. The relation **R** may result from integrating data from different web sites that contradict one another.

A *repair* (or possible world) of an uncertain database is obtained by selecting exactly one tuple from each block. In general, the number of repairs of an uncertain database **db** is exponential in the size of **db**. For instance, if an uncertain database contains n blocks with two tuples each, then it contains $2n$ tuples and has 2^n repairs.

There are three natural semantics for answering Boolean queries q on an uncertain database. Under the *possibility semantics*, the question is whether the query evaluates to true on some repair. Under the *certainty semantics*, the question is whether the query evaluates to true on every repair. More generally, under the *counting semantics*, the question is to determine the number of repairs on which the query evaluates to true. In this paper, we consider the certainty and counting semantics. The certainty semantics adheres to the paradigm of *consistent query answering* [1,3], which introduces the notion of database repair with respect to general integrity constraints. In this work, repairing is exclusively with respect to primary key constraints, one per relation.

Example 1. The uncertain database of Fig. 1 has eight repairs. The Boolean first-order query $\exists x \exists y \exists z \exists w \left(\mathbf{C}(\underline{x, y}, \text{'Rome'}, z) \wedge \mathbf{R}(\underline{x}, \text{'A'}, w)\right)$ (Will Rome host some A conference?) is true in six repairs.

For any Boolean query q, the decision problem $\mathsf{CERTAINTY}(q)$ is the following.

PROBLEM:	$\mathsf{CERTAINTY}(q)$
INPUT:	uncertain database **db**
QUESTION:	Does every repair of **db** satisfy q?

Two comments are in place. First, the Boolean query q is not part of the input. Every Boolean query q gives thus rise to a new problem. Since the input to $\mathsf{CERTAINTY}(q)$ is an uncertain database, the problem complexity is *data complexity*. Second, we will assume that every relation name in q or **db** has a fixed known arity and primary key. The primary key constraints are thus implicitly present in all problems.

The complexity of $\mathsf{CERTAINTY}(q)$ has gained considerable research attention in recent years. A challenging question is to distinguish queries q for which the problem $\mathsf{CERTAINTY}(q)$ is tractable from queries for which the problem is intractable. Further, if $\mathsf{CERTAINTY}(q)$ is tractable, one may ask whether it is first-order definable. We will refer to these questions as the *complexity classification task for* $\mathsf{CERTAINTY}(q)$.

For any Boolean query q, the counting problem $\sharp\mathsf{CERTAINTY}(q)$ is defined as follows.

PROBLEM:	$\sharp\mathsf{CERTAINTY}(q)$
INPUT:	uncertain database **db**
QUESTION:	How many repairs of **db** satisfy q?

The complexity classification task for $\sharp\mathsf{CERTAINTY}(q)$ is then to determine the complexity of $\sharp\mathsf{CERTAINTY}(q)$ for varying q.

In this paper, we review known results in the aforementioned complexity classification tasks. We also contribute a new result relating the complexity classifications for $\mathsf{CERTAINTY}(q)$ and $\sharp\mathsf{CERTAINTY}(q)$. We discuss variations and extensions of the basic problems, and review existing systems that implement algorithms for consistent query answering under primary keys.

This paper is organized as follows. Section 2 introduces the basic concepts and terminology. Section 3 discusses *consistent first-order rewriting*, which consists in solving $\mathsf{CERTAINTY}(q)$ in first-order logic. Section 4 reviews known dichotomy theorems for $\mathsf{CERTAINTY}(q)$ and $\sharp\mathsf{CERTAINTY}(q)$. Section 5 contains our new result. From Section 6 on, we present a number of variations and extensions of the basic framework. Section 6 introduces the notion of *nucleus* of an uncertain database **db** relative to a class $\mathcal{C}$ of Boolean queries. Intuitively, a nucleus is a new (consistent) database that "summarizes" all repairs of **db** such that it returns certain answers to all queries in $\mathcal{C}$. Section 7 relates $\sharp\mathsf{CERTAINTY}(q)$ to query evaluation in probabilistic databases. Section 9 discusses practical implementations. Finally, Section 10 lists some questions for future research.

2 Preliminaries

In this section, we first introduce basic notions and terminology. We then recall a number of complexity classes that will occur in the complexity classification tasks mentioned in Section 1.

2.1 Data and Query Model

We assume disjoint sets of *variables* and *constants*. If $\boldsymbol{x}$ is a sequence containing variables and constants, then $\mathsf{vars}(\boldsymbol{x})$ denotes the set of variables that occur in $\boldsymbol{x}$. A *valuation* over a set U of variables is a total mapping θ from U to the set of constants. Such a valuation θ is extended to be the identity on constants and on variables not in U.

Atoms and Key-Equal Facts. Each *relation name* R of arity n, $n \geq 1$, has a unique *primary key* which is a set $\{1, 2, \ldots, k\}$ where $1 \leq k \leq n$. We say that R has *signature* $[n, k]$ if R has arity n and primary key $\{1, 2, \ldots, k\}$. Elements of the primary key are called *primary-key positions*, while $k+1$, $k+2$, ..., n are *non-primary-key positions*. For all positive integers n, k such that $1 \leq k \leq n$, we assume denumerably many relation names with signature $[n, k]$.

If R is a relation name with signature $[n, k]$, then $R(s_1, \ldots, s_n)$ is called an R-*atom* (or simply atom), where each s_i is either a constant or a variable ($1 \leq i \leq n$). Such an atom is commonly written as $R(\underline{\boldsymbol{x}}, \boldsymbol{y})$ where the primary key value $\boldsymbol{x} = s_1, \ldots, s_k$ is underlined and $\boldsymbol{y} = s_{k+1}, \ldots, s_n$. A *fact* is an atom in which no variable occurs. Two facts $R_1(\underline{\boldsymbol{a}_1}, \boldsymbol{b}_1), R_2(\underline{\boldsymbol{a}_2}, \boldsymbol{b}_2)$ are *key-equal* if $R_1 = R_2$ and $\boldsymbol{a}_1 = \boldsymbol{a}_2$.

We will use letters F, G, H for atoms. For an atom $F = R(\underline{\boldsymbol{x}}, \boldsymbol{y})$, we denote by $\mathsf{key}(F)$ the set of variables that occur in $\boldsymbol{x}$, and by $\mathsf{vars}(F)$ the set of variables that occur in F, that is, $\mathsf{key}(F) = \mathsf{vars}(\boldsymbol{x})$ and $\mathsf{vars}(F) = \mathsf{vars}(\boldsymbol{x}) \cup \mathsf{vars}(\boldsymbol{y})$.

Uncertain Database, Blocks, and Repairs. A *database schema* is a finite set of relation names. All constructs that follow are defined relative to a fixed database schema.

An *uncertain database* is a finite set **db** of facts using only the relation names of the schema. We write **adom**(**db**) for the active domain of **db** (i.e., the set of constants that occur in **db**). A *block* of **db** is a maximal set of key-equal facts of **db**. An uncertain database **db** is *consistent* if it does not contain two distinct facts that are key-equal (i.e., if every block of **db** is a singleton). A *repair* of **db** is a maximal (with respect to set containment) consistent subset of **db**.

Boolean Conjunctive Query. A *Boolean query* is a mapping q that associates a Boolean (true or false) to each uncertain database, such that q is closed under isomorphism [22]. We write $\mathbf{db} \models q$ to denote that q associates true to **db**, in which case **db** is said to *satisfy* q. A *Boolean first-order query* is a Boolean query that can be defined in first-order logic. A *Boolean conjunctive query* is a finite set $q = \{R_1(\underline{\boldsymbol{x}_1}, \boldsymbol{y}_1), \ldots, R_n(\underline{\boldsymbol{x}_n}, \boldsymbol{y}_n)\}$ of atoms. By $\mathsf{vars}(q)$, we denote the set of variables that occur in q. The set q represents the first-order sentence

$$\exists u_1 \cdots \exists u_k \left(R_1(\underline{\boldsymbol{x}_1}, \boldsymbol{y}_1) \wedge \cdots \wedge R_n(\underline{\boldsymbol{x}_n}, \boldsymbol{y}_n)\right),$$

where $\{u_1, \ldots, u_k\} = \mathsf{vars}(q)$. This query q is satisfied by uncertain database **db** if there exists a valuation θ over $\mathsf{vars}(q)$ such that for each $i \in \{1, \ldots, n\}$, $R_i(\underline{\boldsymbol{a}}, \boldsymbol{b}) \in \mathbf{db}$ with $\boldsymbol{a} = \theta(\boldsymbol{x}_i)$ and $\boldsymbol{b} = \theta(\boldsymbol{y}_i)$.

If q is a Boolean conjunctive query, $\boldsymbol{x} = \langle x_1, \ldots, x_\ell \rangle$ is a sequence of distinct variables that occur in q, and $\boldsymbol{a} = \langle a_1, \ldots, a_\ell \rangle$ is a sequence of constants, then $q_{[\boldsymbol{x} \mapsto \boldsymbol{a}]}$ denotes the query obtained from q by replacing all occurrences of x_i with a_i, for all $1 \leq i \leq \ell$.

Computational Problems. The decision problem $\mathsf{CERTAINTY}(q)$ and the counting problem $\sharp\mathsf{CERTAINTY}(q)$ have been defined in Section 1.

2.2 Restrictions on Conjunctive Queries

The class of Boolean conjunctive queries can be further restricted by adding syntactic constraints.

Acyclicity. A Boolean conjunctive query q is *acyclic* if it has a *join tree* [2]. A *join tree* for q is an undirected tree whose vertices are the atoms of q such that for every variable x in $\mathsf{vars}(q)$, the set of vertices in which x occurs induces a connected subtree.

No Self-joins. We say that a Boolean conjunctive query q has a *self-join* if some relation name occurs more than once in q. If q has no self-join, then it is called *self-join-free.*

Restrictions on Signatures. Let R be a relation name with signature $[n, k]$. The relation name R is *simple-key* if $k = 1$. The relation name R is *all-key* if $n = k$.

We introduce names for some classes of special interest:

- BCQ denotes the class of Boolean conjunctive queries;
- SJFBCQ denotes the class of self-join-free Boolean conjunctive queries; and
- ACYSJFBCQ denotes the class of acyclic self-join-free Boolean conjunctive queries.

2.3 Complexity Classes

The following complexity classes will occur in the complexity classification tasks for $\mathsf{CERTAINTY}(q)$ and $\sharp\mathsf{CERTAINTY}(q)$.

- **FO**, the class of first-order definable problems. In particular, for a given Boolean query q, $\mathsf{CERTAINTY}(q)$ is in **FO** if there exists a Boolean first-order query φ such that for every uncertain database **db**, every repair of **db** satisfies q if and only if **db** satisfies φ. Such a φ, if it exists, is called a *consistent first-order rewriting of* q.
- **P**, the class of decision problems that can be solved in deterministic polynomial time.
- **NP**, the class of decision problems whose "yes" instances have succinct certificates that can be verified in deterministic polynomial time.
- **coNP**, the class of decision problems whose "no" instances have succinct disqualifications that can be verified in deterministic polynomial time. In particular, $\mathsf{CERTAINTY}(q)$ is in **coNP** for every Boolean first-order query q, because if q is not true in every repair of **db**, then a succinct disqualification is a repair of **db** that falsifies q. Indeed, repair checking (i.e., given **rep** and **db**, check whether **rep** is a repair of **db**) is in polynomial time, and so is the data complexity of first-order queries.

- **FP**, the class of function problems that can be solved in deterministic polynomial time. In particular, for a given Boolean query q, $\sharp$CERTAINTY(q) is in **FP** if there exists a polynomial-time algorithm that takes any uncertain database **db** on input, and returns the number of repairs of **db** that satisfy q.
- $\sharp$**P**, the class of counting problems associated with decision problems in **NP**. Given an instance of a decision problem in **NP**, the associated counting problem instance asks to determine the number of succinct certificates of its being a "yes" instance. The following problem is obviously in **NP** for every Boolean first-order query q: given an uncertain database **db** on input, determine whether some repair of **db** satisfies q. Its associated counting problem, called $\sharp$CERTAINTY(q), is thus in $\sharp$**P**.

Concerning the latter item, notice that the decision variant of $\sharp$CERTAINTY(q) is *not* CERTAINTY(q), which is a decision problem in **coNP**. The problem $\sharp$CERTAINTY(q) might better have been named $\sharp$POSSIBILITY(q) or so, because its decision variant asks, given an uncertain database on input, whether some repair satisfies q.

3 Consistent First-Order Rewriting

The detailed investigation of CERTAINTY(q) was pioneered by Fuxman and Miller [14,15]. This initial research focused on determining classes of Boolean conjunctive queries q for which CERTAINTY(q) is in **FO**, and hence solvable by a single Boolean first-order query, which is called a consistent first-order rewriting of q. The practical significance is that such a consistent first-order rewriting can be directly implemented in SQL. A concrete example is given next.

Example 2. Let q_0 be the query $\exists z\mathbf{R}(\text{'ICDT'}, \text{'A'}, z)$ (Is ICDT a conference of rank A?). Clearly, q_0 is true in every repair of some uncertain database **db** if and only if **db** contains an **R**-fact stating that ICDT has rank A, and contains no **R**-fact mentioning a different rank for ICDT. These conditions are expressed by the following Boolean first-order query (call it φ_0):

$$\exists z\mathbf{R}(\text{'ICDT'}, \text{'A'}, z) \wedge \forall y \forall z \left(\mathbf{R}(\text{'ICDT'}, y, z) \rightarrow y = \text{'A'}\right).$$

If φ_0 evaluates to true on **db**, then every repair of **db** satisfies q_0; if φ_0 evaluates to false on **db**, then some repair of **db** falsifies q_0. Thus, φ_0 a consistent first-order rewriting of q_0, and solves CERTAINTY(q_0) without any need for enumerating repairs.

Fuxman and Miller defined a class $\mathcal{C}_{forest} \subsetneq$ SJFBCQ such that for every $q \in \mathcal{C}_{forest}$, CERTAINTY($q$) is in **FO**. Their results were improved by Wijsen [34], as follows.

Theorem 1 ([34]). *Given $q \in$ ACYSJFBCQ, it is decidable (in quadratic time in the size of q) whether* CERTAINTY(q) *is in* **FO**. *If* CERTAINTY(q) *is in* **FO**, *then a consistent first-order rewriting of q can be effectively constructed.*

To be precise, the class $\mathcal{C}_{forest}$ contains some conjunctive queries that are cyclic in some very restricted way (see [9, Theorem 2.4]). On the other hand, there exist queries $q \in$ AcySjfBcq such that CERTAINTY(q) is in **FO** but $q \notin \mathcal{C}_{forest}$. For Boolean conjunctive queries with self-joins, some sufficient conditions for the existence of consistent first-order rewritings appear in [32].

4 Complexity Dichotomy Theorems

A further research challenge is to distinguish Boolean conjunctive queries q for which the problem CERTAINTY(q) is tractable from queries for which the problem is intractable.

In general, we say that a class $\mathcal{P}$ of decision problems *exhibits a* **P**-**coNP**-*dichotomy* if all problems in $\mathcal{P}$ are either in **P** or **coNP**-hard. We say that $\mathcal{P}$ *exhibits an effective* **P**-**coNP**-*dichotomy* if in addition it is decidable whether a given problem in $\mathcal{P}$ is in **P** or **coNP**-hard. Likewise, we say that a class $\mathcal{P}$ of function problems *exhibits an* **FP**-♯**P**-*dichotomy* if all problems in $\mathcal{P}$ are either in **FP** or ♯**P**-hard (under polynomial-time Turing reductions). We say that $\mathcal{P}$ *exhibits an effective* **FP**-♯**P**-*dichotomy* if in addition it is decidable whether a given problem in $\mathcal{P}$ is in **FP** or ♯**P**-hard. We use the term *complexity dichotomy theorem* to refer to a theorem that establishes a **P**-**coNP**-dichotomy in a class of decision problems, or an **FP**-♯**P**-dichotomy in a class of counting problems. By Ladner's theorem [21], if $\mathbf{P} \neq \mathbf{NP}$, or $\mathbf{FP} \neq \sharp\mathbf{P}$, then no such complexity dichotomy theorems exist for **coNP** or ♯**P**.

Let $\mathcal{C}$ be a class of Boolean queries. We write CERTAINTY[$\mathcal{C}$] to denote the class of decision problems that contains CERTAINTY(q) for every $q \in \mathcal{C}$. Likewise, we write ♯CERTAINTY[$\mathcal{C}$] for the class of counting problems that contains ♯CERTAINTY(q) for every $q \in \mathcal{C}$.

The following result by Kolaitis and Pema was chronologically the first complexity dichotomy theorem for consistent query answering under primary keys.

Theorem 2 ([17]). *Let $\mathcal{C}$ be the class of self-join-free Boolean conjunctive queries that contain at most two atoms. Then,* CERTAINTY[$\mathcal{C}$] *exhibits an effective* **P**-**coNP**-*dichotomy.*

Clearly, every Boolean conjunctive query with at most two atoms, is acyclic. The following generalization of Theorem 2 was conjectured in both [26] and [35]. What is remarkable is that both works have independently conjectured exactly the same boundary between tractable and intractable problems in the class CERTAINTY[AcySjfBcq]. The exposition of this boundary is involved and will not be given here.

Conjecture 1. The problem class CERTAINTY[AcySjfBcq] exhibits an effective **P**-**coNP**-dichotomy.

Recently, Koutris and Suciu showed the following complexity dichotomy theorem.

Theorem 3 ([19,20]). *Let $\mathcal{C}$ be the class of self-join-free Boolean conjunctive queries in which each relation name is either simple-key or all-key. Then,* $\mathsf{CERTAINTY}[\mathcal{C}]$ *exhibits an effective* **P**-**coNP**-*dichotomy.*

In summary, for Boolean conjunctive queries q, the complexity classification task for $\mathsf{CERTAINTY}(q)$ is far from accomplished. More is known about the complexity classification of $\sharp\mathsf{CERTAINTY}(q)$, as becomes clear from the following two theorems.

Theorem 4 ([24]). *The problem class* $\sharp\mathsf{CERTAINTY}[\text{SJFBCQ}]$ *exhibits an effective* **FP**-$\sharp$**P**-*dichotomy.*

The following is one of the few complexity dichotomies known for conjunctive queries with self-joins.

Theorem 5 ([25]). *Let $\mathcal{C}$ be the class of Boolean conjunctive queries in which all relation names are simple-key. Then,* $\sharp\mathsf{CERTAINTY}[\mathcal{C}]$ *exhibits an effective* **FP**-$\sharp$**P**-*dichotomy.*

5 Comparing Complexity Boundaries

Theorems 1-5 contribute to the complexity classification tasks for $\mathsf{CERTAINTY}(q)$ and $\sharp\mathsf{CERTAINTY}(q)$. These theorems guarantee the existence of effective procedures to classify problems in different complexity classes. We will not expose these effective procedures in detail, but provide some examples for queries in SJFBCQ.

1. For $q_1 = \{R(\underline{x}, y), S(\underline{y}, z)\}$, we have that $\mathsf{CERTAINTY}(q_1)$ is in **FO**, and $\sharp\mathsf{CERTAINTY}(q_1)$ is in **FP**. A consistent first-order rewriting of q_1 is $\exists x \exists y \left(R(\underline{x}, y) \wedge \forall y \left(R(\underline{x}, y) \rightarrow \exists z S(\underline{y}, z)\right)\right)$.
2. For $q_2 = \{R(\underline{x}, y), S(\underline{y}, a)\}$, where a is a constant, $\mathsf{CERTAINTY}(q_2)$ is in **FO**, but $\sharp\mathsf{CERTAINTY}(q_2)$ is already $\sharp$**P**-hard. A consistent first-order rewriting of q_2 is $\exists x \exists y \left(R(\underline{x}, y) \wedge \forall y \left(R(\underline{x}, y) \rightarrow \left(S(\underline{y}, a) \wedge \forall z \left(S(\underline{y}, z) \rightarrow z = a\right)\right)\right)\right)$.
3. For $q_3 = \{R(\underline{x}, y), S(\underline{y}, x)\}$, we have that $\mathsf{CERTAINTY}(q_3)$ is in **P** $\setminus$ **FO** [33], and $\sharp\mathsf{CERTAINTY}(q_3)$ is $\sharp$**P**-hard.
4. For $q_4 = \{R(\underline{x}, y), S(\underline{z}, y)\}$, we have that $\mathsf{CERTAINTY}(q_4)$ is **coNP**-complete, and $\sharp\mathsf{CERTAINTY}(q_4)$ is $\sharp$**P**-hard.

We have no example of a query $q \in$ SJFBCQ such that $\mathsf{CERTAINTY}(q)$ is in **P** $\setminus$ **FO** and $\sharp\mathsf{CERTAINTY}(q)$ is in **FP**. The following theorem implies that no such query exists unless **FP** $=$ $\sharp$**P**. The proof is in the Appendix.

Theorem 6. *For every $q \in$ SJFBCQ, if* $\mathsf{CERTAINTY}(q)$ *is not in* **FO**, *then* $\sharp\mathsf{CERTAINTY}(q)$ *is* $\sharp$**P**-*hard.*

Clearly, the number of repairs of an uncertain database **db** can be computed in polynomial time in the size of **db**, by multiplying the cardinalities of all blocks. Therefore, for every Boolean query q, if $\sharp\mathsf{CERTAINTY}(q)$ is in **FP**, then

$\mathbf{R}'$	<u>conf</u>	rank	frequency
	ICDT	A	ℓ_1
	KDD	ℓ_2	annual
	DBPL	B	biennial
	BDA	B	annual
	ℓ_3	B	ℓ_1
	ℓ_4	ℓ_2	ℓ_1

Fig. 2. BCQ-nucleus for the relation **R** of Fig. 1. The symbols ℓ_1, ℓ_2, ℓ_3, ℓ_4 are new distinct constants that cannot be used in queries.

CERTAINTY(q) must be in **P**. That is, tractability of CERTAINTY(q) could in principle be established from tractability of ♯CERTAINTY(q). We notice, however, that Theorem 4 does not give us new tractable cases of CERTAINTY(q) in this way for $q \in$ ACYSJFBCQ (i.e., Theorem 4 does not help to prove Conjecture 1). By Theorem 1, we can already distinguish in ACYSJFBCQ the queries q that have a consistent first-order rewriting from those that have not. If a query $q \in$ ACYSJFBCQ has no consistent first-order rewriting, then ♯CERTAINTY(q) is ♯**P**-hard by Theorem 6.

6 Nucleus

Solving CERTAINTY(q) consists in developing an algorithm that takes any uncertain database **db** on input, and checks whether every repair of **db** satisfies the Boolean query q. As indicated in Section 3, for some q, such an algorithm can be expressed in first-order logic. A natural question is whether algorithms for solving CERTAINTY(q) could benefit from some database preprocessing. An approach proposed in [29] consists in "rewriting" an uncertain database **db** into a new database **db**$'$ such that for all queries q in some query class, the answer to CERTAINTY(q) on input **db** is obtained by executing q on **db**$'$. This is formalized next.

Nucleus. Let $\mathcal{C}$ be a class of Boolean queries. A $\mathcal{C}$-*nucleus* of an uncertain database **db** is a database **db**$'$ such that for every query $q \in \mathcal{C}$, every repair of **db** satisfies q if and only if **db**$'$ satisfies q.

It follows from [29] that every uncertain database **db** has a BCQ-nucleus. To be precise, this result supposes the existence of some special constants, called *labeled nulls*, that can occur in uncertain databases, but not in queries.

Example 3. A BCQ-nucleus for the relation **R** of Fig. 1 is shown in Fig. 2, where ℓ_1, ℓ_2, ℓ_3, ℓ_4 are distinct labeled nulls. The atom $\mathbf{R}'(\underline{\ell_3}, \mathrm{B}, \ell_1)$, for example, expresses that in every repair, some conference of rank B has the same frequency as ICDT. The query $\exists z \mathbf{R}(\text{'ICDT'}, \text{'A'}, z)$ evaluates to true on every repair of **R**,

R	<u>**conf**</u>	**rank**	**frequency**	P
	ICDT	A	biennial	0.3
	ICDT	A	annual	0.6
	KDD	A	annual	0.5
	KDD	B	annual	0.5
	DBPL	B	biennial	0.7
	BDA	B	annual	1.0

Fig. 3. Representation of a BID probabilistic database

and evaluates to true on the BCQ-nucleus. The query $\exists y \mathbf{R}(\text{'ICDT'}, y, \text{'annual'})$ evaluates to false on some repair of **R**, and evaluates to false on the BCQ-nucleus.

The notion of $\mathcal{C}$-nucleus is closely related to the notion of *universal repair* in [28]. Clearly, since CERTAINTY(q) is **coNP**-hard for some $q \in$ BCQ, any algorithm that takes an uncertain database **db** on input and computes a BCQ-nucleus of **db**, must be exponential-time (unless **P** = **coNP**).

7 Probabilistic Databases

For any fixed Boolean query q, the problem ♯CERTAINTY(q) is a special case of probabilistic query answering. Let N be the total number of repairs of a given uncertain database **db**. If a fact A (or, by extension, a Boolean query) evaluates to true in m repairs, then its probability, denoted $P(A)$, is m/N. For example, in Fig. 1, the probability of the fact **R**(<u>ICDT</u>, A, biennial) is 4/8, because it belongs to 4 repairs out of 8. It can now be easily verified that for all distinct facts A, B of **db**, the following hold:

- If the facts A and B belong to a same block, then $P(A \wedge B) = 0$. In probabilistic terms, distinct facts of the same block represent *disjoint* (i.e., exclusive) events.
- If the facts A and B belong to distinct blocks, then $P(A \wedge B) = P(A) \cdot P(B)$. In probabilistic terms, facts of distinct blocks are *independent*.

Probabilistic databases satisfying the above two properties have been coined *block-independent-disjoint* (BID) by Dalvi, Ré, and Suciu [6]. BID probabilistic databases can be represented by listing the probability of each fact, as illustrated in Fig. 3. The main differences between uncertain databases and BID probabilistic databases are twofold:

- In an uncertain database, all facts of a same block have the same probability. In BID probabilistic databases, facts of a same block need not have the same probability. For example, in the BID probabilistic database of Fig. 3, the two facts about ICDT have distinct probabilities (0.3 and 0.6).

- In an uncertain database, the probabilities of facts in a same block sum up to 1. In BID probabilistic databases, this sum can be strictly less than 1. The BID probabilistic database of Fig. 3 admits a possible world with non-zero probability in which ICDT does not occur.

A detailed comparison of both data models can be found in [35]. The difference between both data models is further diminished in [23], where some positive integer multiplicity is associated to every tuple of an uncertain database.

The tractability/intractability frontier of evaluating SJFBCQ queries on BID probabilistic databases has been revealed by Dalvi et al. [7]. Theorem 4 settles this frontier for uncertain databases. For conjunctive queries with self-joins, no analogue of Theorem 5 is currently known for BID probabilistic databases.

The situation is different for tuple-independent probabilistic databases. In such a database, there is no notion of block and all tuples represent independent events. The tractability/intractability frontier of evaluating unions of conjunctive queries (possibly with self-joins) on tuple-independent probabilistic databases has been revealed by Dalvi and Suciu [8].

8 Integrity Constraints on Uncertain Databases

Integrity constraints allow to restrict the set of legal databases. Although uncertainty is modeled by primary key violations in our approach, this does not mean that constraints should be given up altogether. Some constraints, including some primary keys, could still be enforced.

Example 4. The uncertain database of Fig. 1 satisfies the functional dependency $\mathbf{R} : \mathbf{city} \rightarrow \mathbf{country}$, the inclusion dependency $\mathbf{C}[\mathbf{conf}] \subseteq \mathbf{R}[\mathbf{conf}]$, and the join dependency $\mathbf{C} : \bowtie [\{\mathbf{conf}, \mathbf{rank}\}, \{\mathbf{conf}, \mathbf{frequency}\}]$. This join dependency expresses that, given a conference, the uncertainties in rank and frequency are independent [31].

The problem $\mathsf{CERTAINTY}(q)$ has been generalized to account for constraints that are satisfied by the uncertain databases that are input to the problem [16], as follows. Let q be a Boolean query, and let Σ be a set of first-order constraints referring exclusively to the relation names in **db**. Then $\mathsf{CERTAINTY}(q, \Sigma)$ is the following decision problem.

PROBLEM:	$\mathsf{CERTAINTY}(q, \Sigma)$
INPUT:	uncertain database **db** that satisfies Σ
QUESTION:	Does every repair of **db** satisfy q?

Clearly, if $\Sigma = \emptyset$, then $\mathsf{CERTAINTY}(q, \Sigma)$ is the same as $\mathsf{CERTAINTY}(q)$. At another extreme, if Σ contains all primary key constraints, then the input to $\mathsf{CERTAINTY}(q, \Sigma)$ is restricted to consistent databases without primary key violations.

Concerning the following theorem, a join dependency $R : \bowtie [K_1, \ldots, K_\ell]$ is called a *key join dependency* (KJD) if for all $1 \leq i < j \leq \ell$, the intersection

$K_i \cap K_j$ is exactly the primary key of R. The join dependency in Example 4 is a key join dependency.

Theorem 7 ([16]). *Given a query $q \in$ ACYSJFBCQ and a set Σ of functional dependencies and KJDs containing at most one KJD for every relation name in q, it is decidable whether* CERTAINTY(q, Σ) *is in* **FO**. *If* CERTAINTY(q, Σ) *is in* **FO**, *then a first-order definition of it can be effectively constructed.*

9 Non-boolean Queries and Implemented Systems

In this section, we discuss some systems that have implemented consistent query answering under primary key constraints. In practice, non-Boolean queries are more prevalent than Boolean queries, on which we have focused so far. Nevertheless, most results can be easily extended to non-Boolean queries, as follows.

Henceforth, we will write $q(x_1, \ldots, x_\ell)$ to indicate that q is a (domain independent) first-order query with free variables $x_1, \ldots, x_\ell$. The function problem CERTAINTY$(q(\boldsymbol{x}))$ takes on input an uncertain database **db**, and asks to return all certain answers to q, i.e., all sequences $\boldsymbol{a}$ of constants (of the same length as $\boldsymbol{x}$) such that $q(\boldsymbol{a})$ is true in every repair of **db**. A first-order formula $\varphi(\boldsymbol{x})$ is called a *consistent first-order rewriting of* $q(\boldsymbol{x})$ if for every uncertain database **db**, for all sequences $\boldsymbol{a}$ of constants, $q(\boldsymbol{a})$ is true in every repair of **db** if and only if $\varphi(\boldsymbol{a})$ is true in **db**.

Notice that the number of sequences $\boldsymbol{a}$ that consist exclusively of constants in **db**, is polynomially bounded in the size of **db**. Therefore, the non-Boolean case is at most polynomially more difficult than the Boolean one. Further, $q(\boldsymbol{x})$ has a consistent first-order rewriting if and only if the Boolean query $q_{[\boldsymbol{x} \mapsto \boldsymbol{c}]}$ has a consistent first-order rewriting, where $\boldsymbol{c}$ is a sequence of new constants.

ConQuer [13] and EQUIP [18] are two systems for solving CERTAINTY$(q(\boldsymbol{x}))$ where $q(\boldsymbol{x})$ is a conjunctive query. ConQuer applies to a class of conjunctive queries $q(\boldsymbol{x})$ for which CERTAINTY$(q(\boldsymbol{x}))$ is known to be in **FO**.[1] ConQuer rewrites such a query $q(\boldsymbol{x})$ into a new SQL query Q that gives the certain answers on any uncertain database. The query Q can then be executed in any commercial DBMS. Notice that Q does not depend on the data.

EQUIP applies to all conjunctive queries $q(\boldsymbol{x})$. When an uncertain database **db** is given as the input of the problem CERTAINTY$(q(\boldsymbol{x}))$, EQUIP transforms the database and the query into a Binary Integer Program (BIP) that computes the certain answers. The BIP can then be executed by any existing BIP solver. Since the BIP depends on the database **db**, a new BIP has to be generated whenever the database changes.

Extensive experiments [18,26] show that if CERTAINTY$(q(\boldsymbol{x}))$ is in **FO**, then encoding the problem in SQL (like in ConQuer) is always preferable to binary integer programming. This is not surprising, because binary integer programming is **NP**-hard, while the data complexity of "first-order" SQL is **FO**. A main

[1] ConQuer also deals with aggregation, but we will not consider queries with aggregation here.

conclusion of [26] is that *consistent first-order rewriting should be used whenever possible*. In this respect, Theorems 1 and 7 are of practical importance, because they tell us when exactly consistent first-order rewriting is possible, i.e., when the problem can be solved in SQL. The viability of consistent first-order rewriting was also demonstrated in [10].

10 Open Problems

Notwithstanding active research, the complexity classification of $\mathsf{CERTAINTY}(q)$ for Boolean conjunctive queries q is far from completed. The case of self-joins remains largely unexplored. Furthermore, existing research has exclusively focused on complexity classes **FO**, **P**, and **coNP**-complete. A more fine-grained classification could be pursued. For example, can we characterize Boolean conjunctive queries q for which $\mathsf{CERTAINTY}(q)$ is **P**-complete?

Fontaine [12] has established a number of results relating the complexities of consistent query answering and the constraint satisfaction problem (CSP). One result is the following. Let DisBcq be the class of Boolean first-order queries that can be expressed as disjunctions of Boolean conjunctive queries (possibly with constants and self-joins). Then, a **P-coNP** dichotomy in $\mathsf{CERTAINTY}[\text{DisBcq}]$ implies Bulatov's dichotomy theorem for conservative CSP [4]. Since the proof of the latter theorem is highly involved, it is a major challenge to establish a **P-coNP** dichotomy in $\mathsf{CERTAINTY}[\text{DisBcq}]$. A further natural question is whether complexities in $\sharp\mathsf{CERTAINTY}[\text{DisBcq}]$ can be related to the effective dichotomy theorem for counting CSP proved in [5,11]. Conversely, can complexity results for CSP be used in the complexity classification of $\mathsf{CERTAINTY}(q)$ and $\sharp\mathsf{CERTAINTY}(q)$?

The concept of nucleus has not yet been studied in depth. Can we determine a large class of queries $\mathcal{C} \subseteq \text{Bcq}$ such that a $\mathcal{C}$-nucleus of any uncertain database **db** can be computed in polynomial time in the size of **db**? Some preliminary results appear in [30].

Currently, no dichotomy is known in the complexity of evaluating conjunctive queries with self-joins on BID probabilistic databases. Can the dichotomy of Theorem 5 be extended to BID probabilistic databases?

It remains an open task to gain a better understanding of the role of Σ in the complexity of $\mathsf{CERTAINTY}(q, \Sigma)$. Currently, we have only studied the case where Σ is a set of functional dependencies and join dependencies, the latter of a restricted form. The set Σ of satisfied constraints could be used, for example, to limit the amount of uncertainty by restricting the number of tuples per block.

References

1. Arenas, M., Bertossi, L.E., Chomicki, J.: Consistent query answers in inconsistent databases. In: PODS, pp. 68–79. ACM Press (1999)
2. Beeri, C., Fagin, R., Maier, D., Yannakakis, M.: On the desirability of acyclic database schemes. J. ACM 30(3), 479–513 (1983)

3. Bertossi, L.E.: Database Repairing and Consistent Query Answering. In: Synthesis Lectures on Data Management. Morgan & Claypool Publishers (2011)
4. Bulatov, A.A.: Complexity of conservative constraint satisfaction problems. ACM Trans. Comput. Log. 12(4), 24 (2011)
5. Bulatov, A.A.: The complexity of the counting constraint satisfaction problem. J. ACM 60(5), 34 (2013)
6. Dalvi, N.N., Ré, C., Suciu, D.: Probabilistic databases: diamonds in the dirt. Commun. ACM 52(7), 86–94 (2009)
7. Dalvi, N.N., Re, C., Suciu, D.: Queries and materialized views on probabilistic databases. J. Comput. Syst. Sci. 77(3), 473–490 (2011)
8. Dalvi, N.N., Suciu, D.: The dichotomy of probabilistic inference for unions of conjunctive queries. J. ACM 59(6), 30 (2012)
9. Decan, A.: Certain Query Answering in First-Order Languages. PhD thesis, Université de Mons (2013)
10. Decan, A., Pijcke, F., Wijsen, J.: Certain conjunctive query answering in SQL. In: Hüllermeier, E., Link, S., Fober, T., Seeger, B. (eds.) SUM 2012. LNCS, vol. 7520, pp. 154–167. Springer, Heidelberg (2012)
11. Dyer, M.E., Richerby, D.: An effective dichotomy for the counting constraint satisfaction problem. SIAM J. Comput. 42(3), 1245–1274 (2013)
12. Fontaine, G.: Why is it hard to obtain a dichotomy for consistent query answering? In: LICS, pp. 550–559. IEEE Computer Society (2013)
13. Fuxman, A., Fazli, E., Miller, R.J.: ConQuer: Efficient management of inconsistent databases. In: Özcan, F. (ed.) SIGMOD Conference, pp. 155–166. ACM (2005)
14. Fuxman, A.D., Miller, R.J.: First-order query rewriting for inconsistent databases. In: Eiter, T., Libkin, L. (eds.) ICDT 2005. LNCS, vol. 3363, pp. 337–351. Springer, Heidelberg (2005)
15. Fuxman, A., Miller, R.J.: First-order query rewriting for inconsistent databases. J. Comput. Syst. Sci. 73(4), 610–635 (2007)
16. Greco, S., Pijcke, F., Wijsen, J.: Certain query answering in partially consistent databases. PVLDB 7(5) (2014)
17. Kolaitis, P.G., Pema, E.: A dichotomy in the complexity of consistent query answering for queries with two atoms. Inf. Process. Lett. 112(3), 77–85 (2012)
18. Kolaitis, P.G., Pema, E., Tan, W.-C.: Efficient querying of inconsistent databases with binary integer programming. PVLDB 6(6), 397–408 (2013)
19. Koutris, P., Suciu, D.: A dichotomy on the complexity of consistent query answering for atoms with simple keys. CoRR, abs/1212.6636 (2012)
20. Koutris, P., Suciu, D.: A dichotomy on the complexity of consistent query answering for atoms with simple keys. In: Schweikardt [27]
21. Ladner, R.E.: On the structure of polynomial time reducibility. J. ACM 22(1), 155–171 (1975)
22. Libkin, L.: Elements of Finite Model Theory. Springer (2004)
23. Maslowski, D., Wijsen, J.: Uncertainty that counts. In: Christiansen, H., De Tré, G., Yazici, A., Zadrozny, S., Andreasen, T., Larsen, H.L. (eds.) FQAS 2011. LNCS, vol. 7022, pp. 25–36. Springer, Heidelberg (2011)
24. Maslowski, D., Wijsen, J.: A dichotomy in the complexity of counting database repairs. J. Comput. Syst. Sci. 79(6), 958–983 (2013)
25. Maslowski, D., Wijsen, J.: Counting database repairs that satisfy conjunctive queries with self-joins. In: Schweikardt ed. [27]
26. Pema, E.: Consistent Query Answering of Conjunctive Queries under Primary Key Constraints. PhD thesis, University of California Santa Cruz (2013)

27. Schweikardt, N. (ed.): 17th International Conference on Database Theory, ICDT 2014, Athens, Greece, March 24-28. ACM (2014)
28. ten Cate, B., Fontaine, G., Kolaitis, P.G.: On the data complexity of consistent query answering. In: Deutsch, A. (ed.) ICDT, pp. 22–33. ACM (2012)
29. Wijsen, J.: Database repairing using updates. ACM Trans. Database Syst. 30(3), 722–768 (2005)
30. Wijsen, J.: On condensing database repairs obtained by tuple deletions. In: DEXA Workshops, pp. 849–853. IEEE Computer Society (2005)
31. Wijsen, J.: Project-join-repair: An approach to consistent query answering under functional dependencies. In: Larsen, H.L., Pasi, G., Ortiz-Arroyo, D., Andreasen, T., Christiansen, H. (eds.) FQAS 2006. LNCS (LNAI), vol. 4027, pp. 1–12. Springer, Heidelberg (2006)
32. Wijsen, J.: On the consistent rewriting of conjunctive queries under primary key constraints. Inf. Syst. 34(7), 578–601 (2009)
33. Wijsen, J.: A remark on the complexity of consistent conjunctive query answering under primary key violations. Inf. Process. Lett. 110(21), 950–955 (2010)
34. Wijsen, J.: Certain conjunctive query answering in first-order logic. ACM Trans. Database Syst. 37(2), 9 (2012)
35. Wijsen, J.: Charting the tractability frontier of certain conjunctive query answering. In: Hull, R., Fan, W. (eds.) PODS, pp. 189–200. ACM (2013)

A Appendix: Proof of Theorem 6

We first expose the tractability/intractability boundary of Theorem 4.

Complex Part of a Boolean Conjunctive Query. Let q be a Boolean conjunctive query. A variable $x \in \mathsf{vars}(q)$ is called a *liaison variable* if x has at least two occurrences in q.[2] The *complex part* of a Boolean conjunctive query q, denoted $[\![q]\!]$, contains every atom $F \in q$ such that some non-primary-key position in F contains a liaison variable or a constant.

Example 5. The variable y is the only liaison variable in $q = \{R(\underline{x}, y), R(\underline{y}, z), S(\underline{y}, u, a)\}$, in which a is a constant. The complex part of q is $[\![q]\!] = \{R(\underline{x}, y), S(\underline{y}, u, a)\}$. The complex part of $\{R(\underline{y}, w), R(\underline{x}, u), T(\underline{x, y})\}$, where T is all-key, is empty.

If some atom $F = R(\underline{\boldsymbol{x}}, y_1, \ldots, y_\ell)$ of a Boolean conjunctive query q does *not* belong to q's complex part, then $y_1, \ldots, y_\ell$ are distinct variables that have only one occurrence in q. Intuitively, such variables can be disregarded when evaluating the query q, because they do not impose any join condition.

Function IsSafe takes a query $q \in$ SJFBCQ on input, and always terminates with either **true** or **false**. The function is recursive. The base rules (SE0a and SE0b) apply if q consists of a single fact, or if the complex part of q is empty.

[2] Liaison variables are sometimes called "join variables" in the literature. Notice nevertheless that in the singleton query $\{R(\underline{x}, x)\}$, which is not a genuine join, the variable x is a liaison variable.

```
Function. IsSafe(q) Determine whether q is safe
Input: A query q in SJFBCQ.
Result: Boolean in {true, false}.
begin
    SE0a: if |q| = 1 and vars(q) = ∅ then
        return true;
    SE0b: if ⟦q⟧ = ∅ then
        return true;
    SE1: if q = q1 ∪ q2 with q1 ≠ ∅ ≠ q2, vars(q1) ∩ vars(q2) = ∅ then
        return IsSafe(q1) ∧ IsSafe(q2);
    /* a is an arbitrary constant                                      */
    SE2: if ⟦q⟧ ≠ ∅ and ⋂_{F∈⟦q⟧} key(F) ≠ ∅ then
        select x ∈ ⋂_{F∈⟦q⟧} key(F);
        return IsSafe(q_[x↦a]);
    SE3: if there exists F ∈ q such that key(F) = ∅ ≠ vars(F) then
        select F ∈ q such that key(F) = ∅ ≠ vars(F);
        select x ∈ vars(F);
        return IsSafe(q_[x↦a]);
    if none of the above then
        return false;
```

The recursive rule SE1 applies if q can be partitioned into two subqueries which have no variables in common. The recursive rule SE2 applies if for some variable x, all atoms in the complex part of q contain x at some of their primary-key positions. The recursive rule SE3 applies if all primary-key positions of some atom are occupied by constants and some non-primary-key position contains a variable.

A query $q \in$ SJFBCQ is called *safe* if Function IsSafe returns **true** on input q; otherwise q is *unsafe*. The following result refines Theorem 4.

Theorem 8 (**[24]**). *For every* $q \in$ SJFBCQ,

1. *if* q *is safe, then* $\sharp\mathsf{CERTAINTY}(q)$ *is in* **FP**; *and*
2. *if* q *is unsafe, then* $\sharp\mathsf{CERTAINTY}(q)$ *is* $\sharp$**P**-*hard.*

We use the following helping lemma.

Lemma 1. *For every* $q \in$ SJFBCQ, *if* q *is safe, then* $\mathsf{CERTAINTY}(q)$ *is in* **FO**.

Proof. Let $q \in$ SJFBCQ such that q is safe. The proof runs by induction on the execution of Function IsSafe. Some rule among SE0a, SE0b, SE1, SE2, or SE3 must apply to q.

Case SE0a Applies. If q consists of a single fact, then $\mathsf{CERTAINTY}(q)$ is obviously in **FO**.

Case SE0b Applies. If $[\![q]\!] = \emptyset$, then for any given uncertain database **db**, we have that q evaluates to true on every repair of **db** if and only if q evaluates to true on **db**. It follows that $\mathsf{CERTAINTY}(q)$ is in **FO**.

Case SE1 Applies. Let $q = q_1 \cup q_2$ with $q_1 \neq \emptyset \neq q_2$ and $\mathsf{vars}(q_1) \cap \mathsf{vars}(q_2) = \emptyset$. Since q is safe, q_1 and q_2 are safe by definition of safety. By the induction hypothesis, there exists a consistent first-order rewriting φ_1 of q_1, and a consistent first-order rewriting φ_2 of q_2. Obviously, $\varphi_1 \wedge \varphi_2$ is a consistent first-order rewriting of q.

Case SE2 Applies. Assume variable x such that for every $F \in [\![q]\!] \neq \emptyset$, $x \in \mathsf{key}(F)$. We first show that for every uncertain database **db**, the following are equivalent:

1. every repair of **db** satisfies q; and
2. for some constant a, every repair of **db** satisfies $q_{[x \mapsto a]}$.

$\boxed{2 \Longrightarrow 1}$ Trivial. $\boxed{1 \Longrightarrow 2}$ Proof by contraposition. Assume that for every constant a, there exists a repair $\mathbf{rep}_a$ of **db** such that $\mathbf{rep}_a \not\models q_{[x \mapsto a]}$. Assume without loss of generality that $[\![q]\!] = \{R_1(\underline{x, \boldsymbol{x}_1}, \boldsymbol{y}_1), \ldots, R_n(\underline{x, \boldsymbol{x}_n}, \boldsymbol{y}_n)\}$. For every $a \in \mathbf{adom}(\mathbf{db})$, let $\mathbf{rep}'_a$ be the subset of $\mathbf{rep}_a$ that contains each R_i-fact whose leftmost position is occupied by a, for all $1 \leq i \leq n$. Let $\mathbf{db}'$ be the subset of **db** that contains each fact F whose relation name is not among $R_1, \ldots, R_n$. Let $\mathbf{rep}'$ be a repair of $\mathbf{db}'$. It can now be easily seen that $\mathbf{rep}' \cup \left(\bigcup_{a \in \mathbf{adom}(\mathbf{db})} \mathbf{rep}'_a\right)$ is a repair of **db** that falsifies q.

By definition of safety, $q_{[x \mapsto a]}$ is safe. By the induction hypothesis, the problem $\mathsf{CERTAINTY}(q_{[x \mapsto a]})$ is in **FO**. Let φ be a consistent first-order rewriting of $q_{[x \mapsto c]}$, where we assume without loss of generality that c is a constant that does not occur in q. Let $\varphi(x)$ be the first-order formula obtained from φ by replacing each occurrence of c with x. By the equivalence shown in the previous paragraph, $\exists x \varphi(x)$ is a consistent first-order rewriting of q.

Case SE3 Applies. Assume $F \in q$ such that $\mathsf{key}(F) = \emptyset$ and $\mathsf{vars}(F) \neq \emptyset$. Let $\boldsymbol{x}$ be a sequence of distinct variables such that $\mathsf{vars}(\boldsymbol{x}) = \mathsf{vars}(F)$. Let $\boldsymbol{a} = \langle a, a, \ldots, a \rangle$ be a sequence of length $|\boldsymbol{x}|$. By definition of safety, $q_{[\boldsymbol{x} \mapsto \boldsymbol{a}]}$ is safe. By the induction hypothesis, $\mathsf{CERTAINTY}(q_{[\boldsymbol{x} \mapsto \boldsymbol{a}]})$ is in **FO**. From Lemma 8.6 in [34], it follows that $\mathsf{CERTAINTY}(q)$ is in **FO**. This concludes the proof of Lemma 1. □

The proof of Theorem 6 can now be given.

Proof (of Theorem 6). Assume that $\mathsf{CERTAINTY}(q)$ is not in **FO**. By Lemma 1, q is unsafe. By Theorem 8, $\sharp\mathsf{CERTAINTY}(q)$ is $\sharp$**P**-hard. □

Arguments Using Ontological and Causal Knowledge

Philippe Besnard[1,4], Marie-Odile Cordier[2,3,4], and Yves Moinard[3,4]

[1] CNRS (UMR IRIT), Université de Toulouse, France
[2] Université de Rennes 1
[3] Inria
[4] IRISA, Campus de Beaulieu, 35042 Rennes, France
besnard@irit.fr,
cordier@irisa.fr,
yves.moinard@inria.fr

Abstract. We explore an approach to reasoning about causes via argumentation. We consider a causal model for a physical system, and we look for arguments about facts. Some arguments are meant to provide explanations of facts whereas some challenge these explanations and so on. At the root of argumentation here, are causal links ($\{A_1, \cdots, A_n\}$ *causes* B) and also ontological links (c_1 *is_a* c_2). We introduce here a logical approach which provides a candidate explanation ($\{A_1, \cdots, A_n\}$ *explains* $\{B_1, \cdots, B_m\}$) by resorting to an underlying causal link substantiated with appropriate ontological links. Argumentation is then at work from these various explanation links. A case study is developed: a severe storm Xynthia that devastated a county in France in 2010, with an unaccountably high number of casualties.

1 Introduction and Motivation

Looking for explanations is a frequent operation, in various domains, from judiciary to mechanical fields. We consider the case where we have a (not necessarily exhaustive) description of some mechanism, or situation, and we are looking for explanations of some facts. The description contains logical formulas, together with some *causal* and *ontological* formulas (or links). Indeed, it is well-known that, although there are similarities between *causation* and *implication*, causation cannot be rendered by a simple logical implication. Moreover, confusing causation and *co-occurrence* could lead to undesirable relationships. This is why we resort here to a *causal formalism* such that some *causal links* and *ontological links* are added to classical logical formulas. Then, our causal formalism will produce various *explanation links* [2].

Each causal link gives rise to explanation links, and each explanation link must appeal to at least one causal link. The ontology gives rise to further explanation links, although only in the case that these come from explanation links previously obtained: no explanation link can come only from ontological information. In fact, the ontology determines a new connective (it can be viewed as a strong implication) which can induce these further explanation links, whereas classical implication cannot. Indeed, given an explanation link, logical implication is not enough to derive from it other explanation links (apart from trivially equivalent ones).

C. Beierle and C. Meghini (Eds.): FoIKS 2014, LNCS 8367, pp. 79–96, 2014.

Despite these restrictions, if the situation described is complex enough, there will be a large number of explanation links, i.e., possible explanations, and *argumentation* is an appealing approach to distinguish between all these explanations.

We introduce in section 2 an explicative model, built from a causal model and an ontological model. Section 4 shows how the explicative model produces explanations. Section 5 deals with argumentation about explanations and we conclude in section 6. Section 3 presents a case study: a severe storm called Xynthia, that resulted in 26 deaths in a single group of houses in La Faute sur Mer, a village in Vendée during a night in 2010.

2 Explicative Model = Causal Model + Ontological Model

The model that is used to obtain explanations and support argumentation, called the explicative model, is built from a causal model relating literals in causal links and from an ontological model where classes (which denote types of object as usual in the literature about ontology) are related by specialization/generalization links. Data consist of causal links and "is_a" relationships (specifying a hierarchy of classes, see the ontological model in section 2.3) and background knowledge (formulae of a sorted logic whose sorts are the classes of the aforementioned hierarchy).

From these data, tentative explanations are inferred according to principles using the so-called ontological deduction links obtained in the explicative model.

2.1 Closed Literals

By a *closed literal*, we mean a propositional literal or a formula

$$\exists\, x : class\ \neg^{\{0,1\}} P(x) \qquad \text{or} \qquad \forall\, x : class\ \neg^{\{0,1\}} P(x)$$

where x is a variable and P is a unary predicate symbol, preceded or not by negation. Throughout, a closed literal of the form $\exists\, x : classP(x)$ is abbreviated as $\exists\ P(class)$ and $\forall\, x : classP(x)$ is abbreviated as $\forall\ P(class)$ and similarly for $\neg P$ instead of P. From now on, when we write simply *literal*, we mean a closed literal.

Lastly, extension to n-ary predicates is unproblematic except for heavy notation. Henceforth, it is not considered in this paper for the sake of clarity.

2.2 The Causal Model

By a causal model [11], we mean a representation of a body of causal relationships to be used to generate arguments that display explanations for a given set of facts. Intuitively, a causal link expresses that a bunch of facts causes some effect.

Notation 1. A *causal link* is of the form

$$\{\alpha_1, \alpha_2, \cdots, \alpha_n\}\ causes\ \beta$$

where $\alpha_1, \alpha_2, \cdots, \alpha_n, \beta$ are literals.

These causal links will be used in order to obtain *explanation links* in section 4.

Example 1. *My being a gourmet with a sweet tooth causes me to appreciate some cake can be represented by*

$$\{\texttt{sweet\_tooth\_gourmet}\} \quad causes \quad \exists\, X : \texttt{cake Ilike(X)}$$

Similarly, my being greedy causes me to appreciate all cakes can be represented by

$$\{\texttt{IamGreedy}\} \quad causes \quad \forall\, X : \texttt{cake Ilike(X)}$$

In the figures (e.g., part of the causal model for Xynthia in Fig. 2), each plain black arrow represents a causal link.

2.3 The Ontological Model

Our approach assumes an elementary ontology in which specialization/generalization links between classes are denoted $c_n \xrightarrow{\text{ISA}} c_m$.

Notation 2. An $\xrightarrow{\text{ISA}}$ link has the form $c_1 \xrightarrow{\text{ISA}} c_2$ where c_1 and c_2 are sorts in our logical language that denote classes such that c_1 is a subclass of c_2 in the ontology.

Thus, $\xrightarrow{\text{ISA}}$ denotes the usual specialization link between classes. E.g., we have `Hurri` $\xrightarrow{\text{ISA}}$ `SWind` and `House1FPA` $\xrightarrow{\text{ISA}}$ `HouseFPA` and `HouseFPA` $\xrightarrow{\text{ISA}}$ `BFPA`[1]: the class `Hurri` of hurricane is a specialization of the class `SWind` of strong wing, the class `House1FPA` of typical Vendée low houses with a single level in flood-prone area is a specialization of the class `HouseFPA` of houses in this area, which itself is a specialization of the class `BFPA` of buildings in this area.

In the figures (e.g., part of the ontological model for Xynthia in Fig. 1), each white-headed arrow labelled with *is-a* denotes an $\xrightarrow{\text{ISA}}$ link.

$$\text{The relation} \xrightarrow{\text{ISA}} \text{is required to be transitive and reflexive.} \quad (1)$$

Reflexivity is due to technical reasons simplifying various definitions and properties.

2.4 The Explicative Model

The resulting model (causal model + ontological deduction link) is the *explicative model*, from which explanation links can be inferred.

Notation 3. An *ontological deduction link* has the form $\Phi_1 \xrightarrow{\text{DEDO}} \Phi_2$ where Φ_1 and Φ_2 are two sets of literals.

[1] `FPA` stands for some precise flood-prone area, `BFPA` for the buildings in this area, `HouseFPA` for the houses in this area and `House1FPA` for the one floor low houses in this area.

If $\Phi_1 = \{\varphi_1\}$ and $\Phi_2 = \{\varphi_2\}$ are singletons, we may actually omit curly brackets in the link $\{\varphi_1\} \overset{\text{DEDO}}{\longrightarrow} \{\varphi_2\}$ to abbreviate it as $\varphi_1 \overset{\text{DEDO}}{\longrightarrow} \varphi_2$.

Such a link between literals $\varphi_1 \overset{\text{DEDO}}{\longrightarrow} \varphi_2$ actually requires that φ_1 and φ_2 are two literals built on the same predicate, say P. If $\varphi_1 = \exists\, P(c_1)$ and $\varphi_2 = \exists\, P(c_2)$, then $\exists\, P(c_1) \overset{\text{DEDO}}{\longrightarrow} \exists\, P(c_2)$ simply means that $\exists\, P(c_2)$ can be deduced from $\exists\, P(c_1)$ due to specialization/generalization links (namely here, the $c_1 \overset{\text{ISA}}{\longrightarrow} c_2$ link in the ontological model that relate the classes c_1 and c_2 mentioned in φ_1 and φ_2).

Technically, the $\overset{\text{DEDO}}{\longrightarrow}$ links between literals are generated through a single principle:

$$
\begin{array}{lr}
\text{If in the ontology is the link} & class_1 \overset{\text{ISA}}{\longrightarrow} class_2, \\
\text{then, in the explicative model is the link} & \exists\, P(class_1) \overset{\text{DEDO}}{\longrightarrow} \exists\, P(class_2) \\
\text{as well as the link} & \forall\, P(class_2) \overset{\text{DEDO}}{\longrightarrow} \forall\, P(class_1) \\
\text{Also, the following links are added} & \forall\, P(class_i) \overset{\text{DEDO}}{\longrightarrow} \exists\, P(class_i).
\end{array}
\qquad (2)
$$

The same principle holds for P replaced by $\neg P$. That is, from $class_1 \overset{\text{ISA}}{\longrightarrow} class_2$, both $\exists\, \neg P(class_1) \overset{\text{DEDO}}{\longrightarrow} \exists\, \neg P(class_2)$ and $\forall\, \neg P(class_2) \overset{\text{DEDO}}{\longrightarrow} \forall\, \neg P(class_1)$ ensue, and the links $\forall\, \neg P(class_i) \overset{\text{DEDO}}{\longrightarrow} \exists\, \neg P(class_i)$ are added whenever necessary.

Let us provide an example from Xynthia, with a predicate `Occ` so that $\exists$ `Occ(Hurri)` intuitively means: some hurricane occurs.

$$
\begin{array}{lr}
\text{By means of the } \overset{\text{ISA}}{\longrightarrow} \text{ link} & \texttt{Hurri} \overset{\text{ISA}}{\longrightarrow} \texttt{SWind}, \\
\text{we obtain the } \overset{\text{DEDO}}{\longrightarrow} \text{ link} & \exists\,\texttt{Occ(Hurri)} \overset{\text{DEDO}}{\longrightarrow} \exists\,\texttt{Occ(SWind)}.
\end{array}
$$

The $\overset{\text{DEDO}}{\longrightarrow}$ links between literals introduced in (2) are extended to a relation among sets of literals (as announced in Notation 3), which is done as follows:

$$
\begin{array}{c}
\text{Let } \Phi \text{ and } \Psi \text{ be two sets of literals,} \\
\text{we add to the explicative model } \Phi \overset{\text{DEDO}}{\longrightarrow} \Psi \\
\text{if for each } \psi \in \Psi, \text{ there exists } \varphi \in \Phi \text{ such that } \varphi \overset{\text{DEDO}}{\longrightarrow} \psi \\
\text{and for each } \varphi \in \Phi, \text{ there exists } \psi \in \Psi \text{ such that } \varphi \overset{\text{DEDO}}{\longrightarrow} \psi.
\end{array}
\qquad (3)
$$

From (1), we obtain that $\psi \overset{\text{DEDO}}{\longrightarrow} \psi$ is in the explicative model for each literal ψ. Accordingly,

$$\Psi \overset{\text{DEDO}}{\longrightarrow} \Psi \text{ is in the explicative model for each set of literals } \Psi. \qquad (4)$$

Back to the hurrican illustration (`Hurri` $\overset{\text{ISA}}{\longrightarrow}$ `SWind`), the explicative model contains:

$\{\exists\,\texttt{Occ(Hurri)}, \texttt{ItRains}\} \overset{\text{DEDO}}{\longrightarrow} \{\exists\,\texttt{Occ(SWind)}, \texttt{ItRains}\}$

but it does not contain

$\{\exists\,\texttt{Occ(Hurri)}, \texttt{ItRains}\} \overset{\text{DEDO}}{\longrightarrow} \{\exists\,\texttt{Occ(SWind)}$

because, in the latter case, `ItRains` contributes nothing in the consequent.

Definition 4. Items (2)-(3) give all and only the ontological deduction links $\Phi \overset{\text{DEDO}}{\longrightarrow} \Psi$ (introduced in Notation 3) comprised in the explicative model.

Summing up, the explicative model consists of the causal links (in the causal model) together with the ontological deduction links (obtained from the ontological model). The explicative model contains all the ingredients needed to derive explanations as is described in section 4.

2.5 Background Knowledge

In addition to the explicative model, background knowledge is used for consistency issues when defining explanation links, as will be seen in Section 4, Notation 5. Background knowledge consists of logical formulas of sorted logic. Part of this knowledge is freely provided by the user. Moreover, we take causal and ontological deduction links to entail classical implication:

$$\begin{array}{rll} \{\alpha_1, \cdots, \alpha_n\} \ causes\ \beta & \text{entails} & (\bigwedge_{i=1}^{n} \alpha_i) \rightarrow \beta. \\ \alpha \overset{\text{DEDO}}{\longrightarrow} \gamma & \text{entails} & \alpha \rightarrow \gamma. \end{array} \tag{5}$$

Consequently, the rightmost logical formulas $(\bigwedge_{i=1}^{n} \alpha_i) \rightarrow \beta$ and $\alpha \rightarrow \gamma$ from (5), are necessarily included in the background knowledge.

3 The Xynthia Example

In this section, we consider as an example a severe storm, called Xynthia, which made 26 deaths in a single group of houses in La Faute sur Mer, a village in Vendée during a night in February 2010. It was a severe storm, with strong winds, low pressure, but it had been forecast. Since the casualties were excessive with respect to the strength of the meteorological phenomenon, various investigations have been ordered. This showed that various factors combined their effects. The weather had its role, however, other factors had been involved: recent houses and a fire station had been constructed in an area known as being susceptible of getting submerged. Also, the state authorities did not realize that asking people to stay at home was inappropriate in case of flooding given the traditionally low Vendée houses. From various enquiries, including one from the French parliament[2] and one from the Cour des Comptes (a national juridiction responsible for monitoring public accounts in France)[3] and many articles on the subject, we have plenty of information about the phenomenon and its dramatic consequences. We have extracted a small part from all this information as an illustration of our approach.

3.1 Classes and Predicates for the Xynthia Example

The classes we consider in the causal model are the following ones: `Hurri`, `SWind`, `BFPA`, `House1FPA`, `HouseFPA`, and `BFPA` have already been introduced in §2.3, together with a few $\overset{\text{ISA}}{\longrightarrow}$ links. Among the buildings in the flood-prone area `FPA`, there

[2] `http://www.assemblee-nationale.fr/13/rap-info/i2697.asp`

[3] `www.ccomptes.fr/Publications/Publications/Les-enseignements-des-inondations-de-2010-sur-le-littoral-atlantique-Xynthia-et-dans-le-Var`

is also a fire station FireSt. Besides Hurri, we consider two other kinds of natural disasters NatDis: tsunami Tsun and flooding Flooding. As far as meteorological phenomena are concerned, we restrict ourselves to very low pressure VLP, together with the aforementioned Hurri and SWind, and add high spring tide HST to our list of classes.

Two kinds of alerts Alert may be given by the authorities, Alert-Evacuate AlertE and Alert-StayAtHome AlertS. Also, PeopleS expresses that people stay at home. There exists an anemometer (able to measure wind strength) with a red light, described by OK_Anemo meaning that it is in a normal state and Red_Anemo meaning that its light is on, which is caused by strong wind, while during a hurricane the anemometer is in abnormal state.

The following predicates are introduced: Flooded and Vict_I, applied to a group of building, respectively meaning that flooding occurs over this group, and that there were victims in this group ($I \in \{1, 2, 3\}$ is a degree of gravity, e.g. Vict_1, Vict_2 and Vict_3 respectively mean, in % of the population of the group: at least a small number, at least a significant number and at least a large number of victims).

Remember that Occ means that some fact has occurred (a strong wind, a disaster, ...), similarly a unary predicate Exp means that some fact is expected to occur.

3.2 The Causal and Ontological Models for the Xynthia Example

The classes and the ontological model are given in Fig. 1.

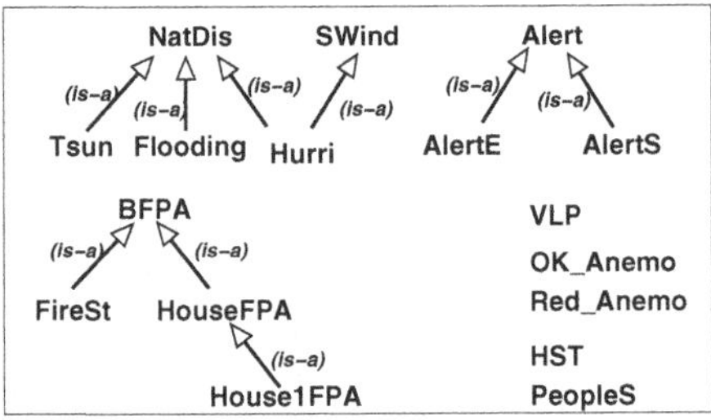

Fig. 1. Ontological model for Xynthia ($\stackrel{\text{ISA}}{\longrightarrow}$ links and constants)

Part of the causal model is given in Table 1 and in Fig. 2. It represents causal relations between (sets of) literals. It expresses that an alert occurs when a natural disaster is expected, or when a natural disaster occurs. Also, people stay at home if alerted to stay at home, and then having one level home flooded results in many victims, and even more victims if the fire station itself is flooded,...

3.3 The Explicative Model for the Xynthia Example

The explicative model can be built, which contains various $\stackrel{\text{DEDO}}{\longrightarrow}$ links between literals. For instance, from Hurri $\stackrel{\text{ISA}}{\longrightarrow}$ SWind, the links

Table 1. Part of the causal model for Xynthia

$\exists$ `Exp(VLP)`	*causes*	$\exists$ `Exp(SWind)`,
$\exists$ `Occ(Hurri)`	*causes*	$\neg$ `OK_Anemo`,
{$\exists$ `Occ(SWind)`, `OK_Anemo`}	*causes*	`Red_Anemo`,
{$\neg\exists$ `Occ(SWind)`, `OK_Anemo`}	*causes*	$\neg$ `Red_Anemo`,
$\exists$ `Occ(NatDis)`	*causes*	$\exists$ `Occ(Alert)`,
$\exists$ `Exp(NatDis)`	*causes*	$\exists$ `Occ(Alert)`,
{$\exists$ `Occ(VLP)`, $\exists$ `Occ(SWind)`, $\exists$ `Occ(HST)`}	*causes*	$\exists$ `Occ(Flooding)`,
$\exists$ `Occ(Flooding)`	*causes*	$\forall$ `Flooded(BFPA)`,
$\forall$ `Flooded(BFPA)`	*causes*	$\forall$ `Vict_1(BFPA)`,
$\exists$ `Occ(AlertS)`	*causes*	$\exists$ `Occ(PeopleS)`,
{$\exists$ `Occ(PeopleS)`, $\forall$ `Flooded(House1FPA)`}	*causes*	$\forall$ `Vict_2(House1FPA)`,
{$\forall$ `Vict_2(House1FPA)`, $\forall$ `Flooded(FireSt)`}	*causes*	$\forall$ `Vict_3(House1FPA)`.

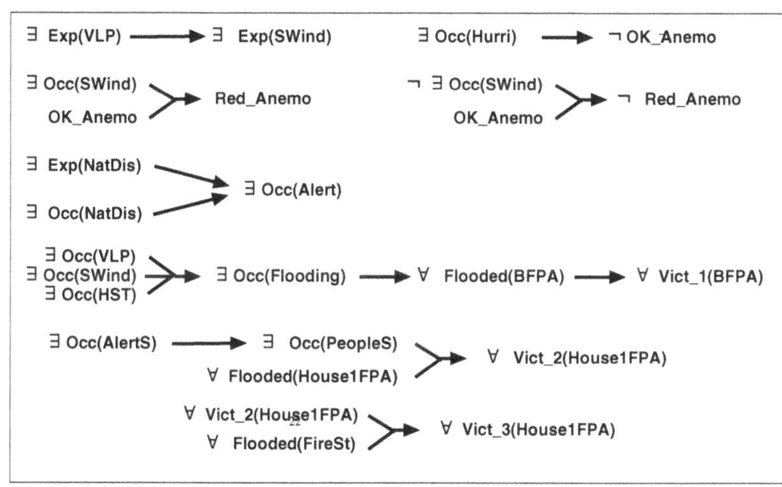

Fig. 2. Part of the causal model for Xynthia

$\exists$ `Occ(Hurri)` $\overset{\text{DEDO}}{\longrightarrow}$ $\exists$ `Occ(SWind)`, and
$\exists$ `Exp(Hurri)` $\overset{\text{DEDO}}{\longrightarrow}$ $\exists$ `Exp(SWind)` are obtained.

And similarly, from {`HouseFPA` $\overset{\text{ISA}}{\longrightarrow}$ `BFPA`, `House1FPA` $\overset{\text{ISA}}{\longrightarrow}$ `HouseFPA`}, we obtain `House1FPA` $\overset{\text{ISA}}{\longrightarrow}$ `BFPA` by (1), from which we consequently get the link $\forall$ `Flooded(BFPA)` $\overset{\text{DEDO}}{\longrightarrow}$ $\forall$ `Flooded(House1FPA)`.

Fig. 3 represents some causal and $\overset{\text{DEDO}}{\longrightarrow}$ links which are part of the explicative model. In the figures, white-headed arrows represent $\overset{\text{DEDO}}{\longrightarrow}$ links. Remember that each black-headed arrow represents a causal link, from a literal or, in the case of a forked entry, from a set of literals.

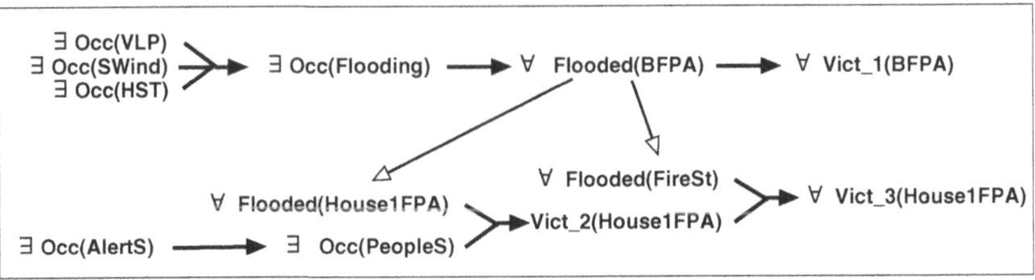

Fig. 3. Part of the explicative model: data used to explain why there were numerous victims in low houses in the flood-prone area House1FPA

4 Explanations

4.1 Introducing Explanation Links

The explicative model (it consists of causal and ontological links) allows us to infer *explanation links*. We want to exhibit candidate reasons that can explain a fact by means of at least one causal link. We disregard "explanations" that would involve only links which are either classical implications ($\rightarrow$) or $\overset{\text{DEDO}}{\longrightarrow}$ links: some causal information is necessary for an "explanation" to hold. Here is how causal and ontological links are used in order to obtain (tentative) explanations in our formalism.

Notation 5. Let Φ, Δ and Ψ be sets of literals. An *explanation link*

$$\Phi \textit{ explains } \Delta \textit{ unless } \neg\,\Psi$$

is intended to mean that Φ explains Δ provided that, given Φ, the set Ψ is possible: if adding $\Phi \cup \Psi$ to available data (i.e., background knowledge and formulas from (5)) leads to an inconsistency, then the explanation link cannot be used to explain Δ by Φ.

Ψ is called the *provision set* of the explanation link.

When the set Ψ is empty, we may omit the " *unless* $\neg\,\emptyset$" (i.e., "unless $\bot$") part.

Throughout the text, we write as usual $\bigwedge \Phi$ for $\bigwedge_{\varphi \in \Phi} \varphi$ and $\neg\Psi$ for $\neg \bigwedge \Psi$.

We set the following equivalences between explanation links, so that the leftmost link can under any circumstance be substituted for the rightmost link and vice-versa:

$$\begin{array}{ll} \Phi \textit{ explains } \Delta & \text{is equivalent to } \Phi \textit{ explains } \Delta \textit{ unless } \neg\,\Phi. \\ \Phi \textit{ explains } \Delta \textit{ unless } \neg\,\Psi & \text{is equivalent to } \Phi \textit{ explains } \Delta \textit{ unless } \neg\,(\Phi \cup \Psi). \end{array} \quad (6)$$

Let us now describe how explanation links are inferred from the explicative model. First is the case that Δ is a singleton set.

4.2 Explaining a Singleton from a Set of Literals

The basic case consists in taking it that a direct causal link Φ *causes* β between a set of literals Φ and a literal β provides an explanation such that the cause explains the (singleton set of) effect: see (7a).

If $\beta = \exists\, P(c)$ or $\beta = \neg \exists\, P(c)$, a more interesting case arises. Take $\beta = \exists\, P(c)$ for instance. Since the causal link expresses that the effect of Φ is $\exists\, P(c)$, it means that for any subclass c' of c, $\exists\, P(c')$ could be caused by Φ (unless a logical inconsistency would indicate that $\exists\, P(c')$ cannot be the case in the presence of Φ and background knowledge and all the formulas from (5)). Accordingly, Φ can be viewed as explaining $\exists\, P(c')$. This is the reason for (7b).

$$\begin{array}{ll} \{\Phi \textit{ causes } \beta\} \text{ yields: } \Phi \textit{ explains } \{\beta\}. & (a) \\ \left\{\begin{array}{l}\Phi \textit{ causes } \exists\,\beta \\ \exists\,\delta \overset{\text{DEDO}}{\longrightarrow} \exists\,\beta\end{array}\right\} \text{ yields: } \Phi \textit{ explains } \{\exists\,\delta\}, \textit{ unless } \neg\,\{\exists\,\delta\}. & (b) \end{array} \qquad (7)$$

(5) yields $\bigwedge \Phi \to \beta$ (case (7a)) as well as $\bigwedge \Phi \to \exists\,\beta$ and $\exists\,\delta \to \exists\,\beta$ (case (7b)), hence adding β (case (7a)) or $\exists\,\beta$ (case (7b)) to the provision set makes no difference, thereby justifying the equivalences in (6).

If $\Phi = \{\varphi\}$ is a singleton set, we may abbreviate Φ *explains* $\{\beta\}$ as φ *explains* β.

Here are a couple of examples from the Xynthia case. First, that flooding occurred can be explained by the conjunction of very low pressure, strong wind, as well as high spring tide. In symbols,

$$\left\{\begin{array}{c}\exists\,\texttt{Occ(VLP)},\\ \exists\,\texttt{Occ(SWind)},\\ \exists\,\texttt{Occ(HST)}\end{array}\right\} \textit{ causes } \exists\,\texttt{Occ(Flooding)}$$

yields

$$\left\{\begin{array}{c}\exists\,\texttt{Occ(VLP)},\\ \exists\,\texttt{Occ(SWind)},\\ \exists\,\texttt{Occ(HST)}\end{array}\right\} \textit{ explains } \exists\,\texttt{Occ(Flooding)}$$

Second, expecting a hurricane can be explained from expecting very low pressure:

$$\left\{\begin{array}{l}\exists\,\texttt{Exp(VLP)} \textit{ causes } \exists\,\texttt{Exp(SWind)}\\ \exists\,\texttt{Exp(Hurri)} \overset{\text{DEDO}}{\longrightarrow} \exists\,\texttt{Exp(SWind)}\end{array}\right\}$$

yields

$$\exists\,\texttt{Exp(VLP)} \textit{ explains } \exists\,\texttt{Exp(Hurri)}$$

Third, that all buildings in the flood-prone area are flooded can be explained by flooding:

$$\exists\,\texttt{Occ(Flooding)} \textit{ causes } \forall\,\texttt{Flooded(BFPA)}$$

yields

$$\exists\,\texttt{Occ(Flooding)} \textit{ explains } \forall\,\texttt{Flooded(BFPA)}$$

In the figures, dotted arrows represent explanation links (to be read *explains*), these arrows being sometimes labelled with the corresponding provision set.

We now introduce explanation links between sets of literals, extending the notion of explanation links from sets of literals to literals presented so far. Since it is an extension, we keep the same name *explanation link*.

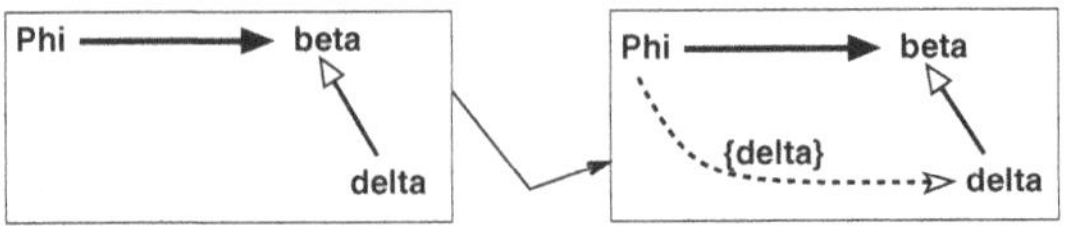

Fig. 4. The schema of the explanation link from (7)

4.3 Explaining a Set of Literals from a Set of Literals

The patterns (7) inducing an explanation for a single observation (a singleton set) are now extended so that they can be used to obtain an explanation for a set of observations:

Let $\Phi_1, \Phi_2, \Delta, \Psi_1$ and Ψ_2 be sets of literals and β be a literal.

$$\begin{array}{ll} \text{If we have} & \Phi_1 \textit{ explains } \Delta \textit{ unless } \neg\Psi_1, \quad \text{and} \\ & \Phi_2 \textit{ explains } \{\delta\} \textit{ unless } \neg\Psi_2, \\ \text{then we get} & \Phi_1 \cup \Phi_2 \textit{ explains } \Delta \cup \{\delta\} \textit{ unless } \neg(\Psi_1 \cup \Psi_2). \end{array} \tag{8}$$

Notice that the condition in (8) is that $\Psi_1 \cup \Psi_2$ must be possible (it is not enough that Ψ_1 be possible ***and*** that Ψ_2 be possible —the same applies to (10) below).

Still further explanation links can be generated from these, *by following* $\overset{\text{DEDO}}{\longrightarrow}$ *links*:

$$\text{If we have} \left\{ \begin{array}{c} \Phi \textit{ explains } \Delta \textit{ unless } \neg\Psi \\ \Phi_0 \overset{\text{DEDO}}{\longrightarrow} \Phi \\ \Delta \overset{\text{DEDO}}{\longrightarrow} \Delta_1 \end{array} \right\} \tag{9}$$

$$\text{then we get} \quad \Phi_0 \textit{ explains } \Delta_1 \textit{ unless } \neg\Psi.$$

Let us return to our example. Applying (7a), that all the buildings in the flood-prone area are flooded can be explained by flooding (this is shown at the end of section 4.2). This explanation link (Φ is $\{\exists\,\texttt{Occ(Flooding)}\}$ and Δ is $\{\forall\,\texttt{Flooded(BFPA)}\}$) can be exploited through (9), letting $\Phi_0 = \Phi$ and $\Delta_1 = \{\forall\,\texttt{Flooded(HouseFPA)}\}$. I.e., that all houses in the flood-prone area are flooded can also be explained by flooding:

$$\left\{ \begin{array}{c} \exists\,\texttt{Occ(Flooding)}\ \textit{causes}\ \forall\,\texttt{Flooded(BFPA)} \\ \forall\,\texttt{Flooded(BFPA)} \overset{\text{DEDO}}{\longrightarrow} \forall\,\texttt{Flooded(HouseFPA)} \end{array} \right\}$$

yields

$$\exists\,\texttt{Occ(Flooding)}\ \textit{explains}\ \forall\,\texttt{Flooded(HouseFPA)}$$

The last, but not least, way by which explanation links induce further explanation links is *transitivity* (of a weak kind because provision sets are unioned).

$$\text{If} \quad \left\{ \begin{array}{c} \Phi \textit{ explains } \Delta \textit{ unless } \neg\Psi_1 \\ \Gamma \cup \Delta \textit{ explains } \Theta \textit{ unless } \neg\Psi_2 \end{array} \right\} \tag{10}$$

$$\text{then} \quad \Phi \cup \Gamma \textit{ explains } \Theta \textit{ unless } \neg(\Psi_1 \cup \Psi_2).$$

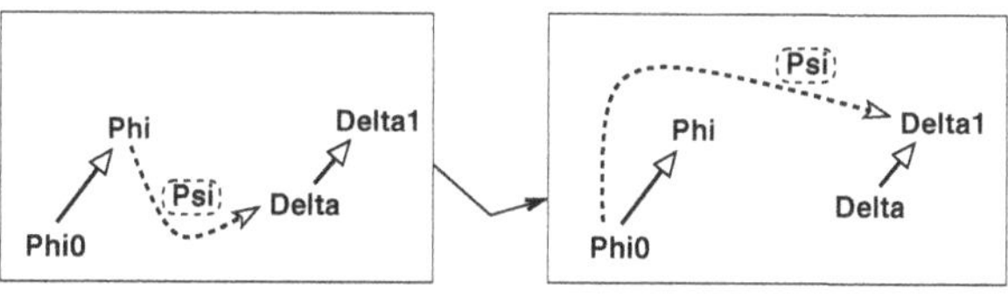

Fig. 5. Explanation links follow $\stackrel{\text{DEDO}}{\longrightarrow}$ links [cf (9)]

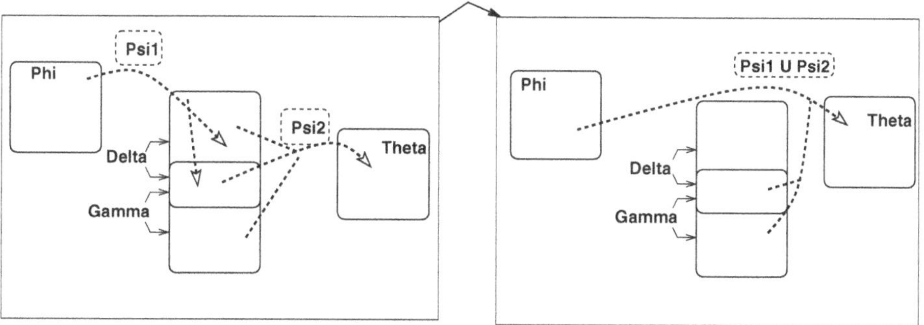

Fig. 6. Transitivity of explanations among sets of literals (10)

Now, we have defined the notion introduced in Notation 5:

Definition 6. The explanation links Φ *explains* Δ *unless* $\neg\,\Psi$ introduced in Notation 5 arising from the explicative model are those and only those resulting from applications of (7), (8), (9) and (10).

The reader should keep in mind that Φ must always be included in the set to be checked for consistency, as is mentioned in Notation 5 (cf (6)).

Definition 6 is such that we can neither explain Φ by Φ itself nor explain Φ by Φ_0 if all we know is $\Phi_0 \stackrel{\text{DEDO}}{\longrightarrow} \Phi$. Intuitively, providing such "explanations" would be cheating, given the nature of an explanation: some causal information is required.

4.4 More Examples Detailed

Let us start with an example from Xynthia illustrating the use of the patterns (7b) and (9) depicted in Fig. 4 and 5.

In the causal model for Xynthia, we focus on the causal link

$$\exists\,\texttt{Exp(VLP)}\ \textit{causes}\ \exists\,\texttt{Exp(SWind)}$$

In the ontological model for Xynthia, we consider the following ontological links

$$\left\{\begin{array}{l}\texttt{Hurri} \stackrel{\text{ISA}}{\longrightarrow} \texttt{SWind} \\ \texttt{Hurri} \stackrel{\text{ISA}}{\longrightarrow} \texttt{NatDis}\end{array}\right\}$$

which give rise, in the explicative model, to the $\overset{\text{DEDO}}{\longrightarrow}$ links below

$$\left\{\begin{array}{lll} \exists\,\texttt{Exp(Hurri)} & \overset{\text{DEDO}}{\longrightarrow} & \exists\,\texttt{Exp(SWind)} \\ \exists\,\texttt{Exp(Hurri)} & \overset{\text{DEDO}}{\longrightarrow} & \exists\,\texttt{Exp(NatDis)} \end{array}\right\}$$

We are looking for Exp(NatDis) to be explained by Exp(VLP) hence we consider

$$\left\{\begin{array}{lll} \exists\,\texttt{Exp(VLP)} & \mathit{causes} & \exists\,\texttt{Exp(SWind)} \\ \exists\,\texttt{Exp(Hurri)} & \overset{\text{DEDO}}{\longrightarrow} & \exists\,\texttt{Exp(SWind)} \end{array}\right\}$$

and we apply (7b) in order to obtain, as a first step,

$$\exists\,\texttt{Exp(VLP)}\ \mathit{explains}\ \exists\,\texttt{Exp(Hurri)}\ \mathit{unless}\ \neg\exists\,\mathit{Exp(Hurri)}$$

over which we apply (9) using the ontological deduction link obtained above, that is,

$$\exists\,\texttt{Exp(Hurri)} \overset{\text{DEDO}}{\longrightarrow} \exists\,\texttt{Exp(NatDis)}$$

in order to arrive at

$$\exists\,\texttt{Exp(VLP)}\ \mathit{explains}\ \exists\,\texttt{Exp(NatDis)}\ \mathit{unless}\ \neg\exists\,\mathit{Exp(Hurri)}$$

Please observe that applying (9) actually requires $\exists\,\texttt{Exp(VLP)} \overset{\text{DEDO}}{\longrightarrow} \exists\,\texttt{Exp(VLP)}$ which is obtained by using (4).

That a natural disaster occurs can be explained from the fact that very low pressure is expected. However, if $\neg\exists$ Exp(Hurri) holds (it is impossible that some hurricane be expected), then this explanation no longer stands (because the effect of the causal link underlying it is strong wind and the explanation chain here identifies hurricane as the kind of strong wind expected).

Let us now turn to an example showing how a chain of explanations can be constructed by means of transitivity (10) applied over explanations already detailed above. The fact that

$$\left\{\begin{array}{c} \exists\,\texttt{Occ(VLP)}, \\ \exists\,\texttt{Occ(SWind)}, \\ \exists\,\texttt{Occ(HST)} \end{array}\right\}\ \mathit{causes}\ \exists\,\texttt{Occ(Flooding)}$$

is in the explicative model allowed us to conclude

$$\left\{\begin{array}{c} \exists\,\texttt{Occ(VLP)}, \\ \exists\,\texttt{Occ(SWind)}, \\ \exists\,\texttt{Occ(HST)} \end{array}\right\}\ \mathit{explains}\ \exists\,\texttt{Occ(Flooding)} \qquad \text{(i)}$$

and the fact that

$$\left\{\begin{array}{lll} \exists\,\texttt{Occ(Flooding)} & \mathit{causes} & \forall\,\texttt{Flooded(BFPA)} \\ \forall\,\texttt{Flooded(BFPA)} & \overset{\text{DEDO}}{\longrightarrow} & \forall\,\texttt{Flooded(HouseFPA)} \end{array}\right\}$$

is in the explicative model allowed us to conclude

$$\exists\,\texttt{Occ(Flooding)}\ \mathit{explains}\ \forall\,\texttt{Flooded(HouseFPA)}. \qquad \text{(ii)}$$

Hence chaining the explanations (i) and (ii) through (10) by letting $\Gamma = \emptyset$ yields

$$\left\{\begin{array}{c}\exists\,\texttt{Occ(VLP)}\\ \exists\,\texttt{Occ(SWind)},\\ \exists\,\texttt{Occ(HST)}\end{array}\right\} \textit{ explains } \forall\,\texttt{Flooded(HouseFPA)} \qquad \text{(iii)}$$

Let us now suppose that we have multiple observations

$$\{\forall\,\texttt{Flooded(BFPA)},\texttt{Red\_Anemo}\}.$$

From $\{\exists\,\texttt{Occ(SWind)},\texttt{OK\_Anemo}\}$ *causes* $\texttt{Red\_Anemo}$,
we get $\{\exists\,\texttt{Occ(SWind)},\texttt{OK\_Anemo}\}$ *explains* $\texttt{Red\_Anemo}$.

Then, from (iii), using (8) we get

$$\left\{\begin{array}{c}\exists\,\texttt{Occ(VLP)}\\ \exists\,\texttt{Occ(SWind)},\\ \exists\,\texttt{Occ(HST)},\\ \texttt{OK\_Anemo}\end{array}\right\} \textit{ explains } \{\forall\,\texttt{Flooded(HouseFPA)},\texttt{Red\_Anemo}\} \qquad \text{(iv)}$$

Also, from $\texttt{Hurri} \xrightarrow{\text{ISA}} \texttt{SWind}$ we get $\exists\,\texttt{Occ(Hurri)} \xrightarrow{\text{DEDO}} \exists\,\texttt{Occ(SWind)}$
Thus from (iii), using (9), we get

$$\left\{\begin{array}{c}\exists\,\texttt{Occ(VLP)}\\ \exists\,\texttt{Occ(Hurri)},\\ \exists\,\texttt{Occ(HST)}\end{array}\right\} \textit{ explains } \forall\,\texttt{Flooded(HouseFPA)} \qquad \text{(v)}$$

However, since $\exists\,\texttt{Occ(Hurry)}$ *causes* $\neg\,\texttt{Red\_Anemo}$, we do not get

$$\left\{\begin{array}{c}\exists\,\texttt{Occ(VLP)}\\ \exists\,\texttt{Occ(Hurri)},\\ \exists\,\texttt{Occ(HST)},\\ \texttt{OK\_Anemo}\end{array}\right\} \textit{ explains } \{\forall\,\texttt{Flooded(HouseFPA)},\texttt{Red\_Anemo}\}.$$

Fig. 7 displays another example from Xynthia of various possible explanations (represented by dotted lines) labelled as $1, 1a, \ldots$ The sets of literals, from which the explanation links start, are framed and numbered (1) to (4). These sets are not disjoint, some literals are then duplicated for readability and the copies are annotated with *(bis)*. Transitivity of explanations is again at work, e.g.,

- $\texttt{set}\,1$ *explains* $\forall\,\texttt{Vict\_1(BFPA)}$ (label $1{+}1a{+}1b$)
  It is obtained by transitivity over explanation links 1, $1a$ and $1b$.
- $\texttt{set}\,4$ *explains* $\forall\,\texttt{Vict\_2(House1FPA)}$ (label 1+1a+2)
  It follows from explanations 1, $1a$ and 2. The latter results from explanation 1+1a together with the $\forall\,\texttt{Flooded(BFPA)} \xrightarrow{\text{DEDO}} \forall\,\texttt{Flooded(House1FPA)}$ link.
- $\texttt{set}\,4$ *explains* $\forall\,\texttt{Vict\_3(House1FPA)}$ (label 1+1a+2+3)
  Explanation 3 results from the $\forall\,\texttt{Flooded(BFPA)} \xrightarrow{\text{DEDO}} \forall\,\texttt{Flooded(FireSt)}$ link together with explanations 1+1a+2.

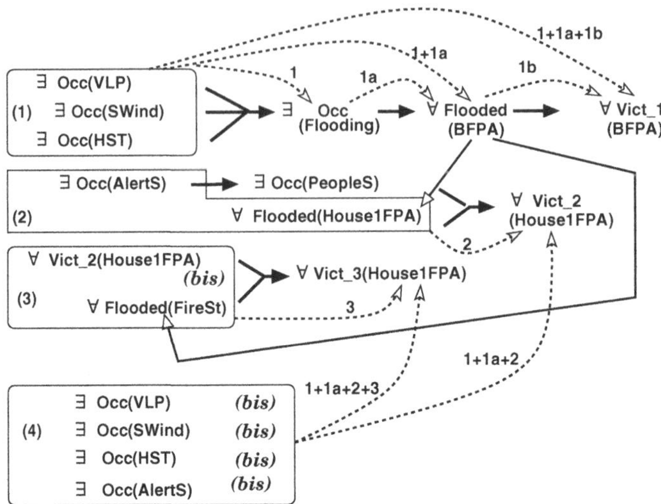

Fig. 7. A few explanations for victims

5 Argumentation

The explicative causal model allows us to infer explanations for a set of statements and these explanations might be used in an argumentative context [3,4]. Let us first provide some motivation from our case study, Xynthia.

An explanation for the flooded buildings is the conjunction of the bad weather conditions (very low pressure and strong wind) and high spring tide (see Fig. 2). Let us take this explanation as an argument. It can be attacked by noticing: a strong wind is supposed to trigger the red alarm of the anemometer and no alarm was shown. However, this counter-argument may itself be attacked by remarking that, in the case of a hurricane, that is a kind of strong wind, the anemometer is no longer operating, which explains that a red alarm cannot be observed.

Let us see how to consider formally argumentation when relying on an explicative model and explanations as described in sections 2 and 4. Of course, we begin with introducing arguments.

5.1 Arguments

An argument is a tuple $(\Phi, \Delta, \Psi, \Theta)$ such that Θ yields that

$$\Phi \text{ explains } \Delta, \text{ unless } \neg\Psi$$

is an explanation link according to Definition 6. The components of the argument are:

- Φ, *the explanation*, a set of literals.
- Δ, *the statements being explained*, a set of literals.
- Ψ, *the provision* of the explanation (see Section 4), a set of formulas.

- Θ, *the evidence*, comprised of formulas (e.g., $\bigwedge\Phi \to \gamma$), causal links (e.g., Φ *causes* β), and ontological deduction links (e.g., $\Delta \overset{\text{DEDO}}{\longrightarrow} \{\beta\}$).

Back to (iii) in the example from Xynthia in section 4.4, that the FPA houses are flooded is explained by the set of literals

$$\Phi = \left\{ \begin{array}{c} \exists\, \texttt{Occ(VLP)} \\ \exists\, \texttt{Occ(SWind)} \\ \exists\, \texttt{Occ(HST)} \end{array} \right\}$$

on the grounds of the following set consisting of two causal links and one ontological deduction link

$$\Theta = \left\{ \begin{array}{l} \exists\, \texttt{Occ(Flooding)} \;\; \textit{causes} \;\; \forall\, \texttt{Flooded(BFPA)}, \\ \left\{ \begin{array}{c} \exists\, \texttt{Occ(VLP)} \\ \exists\, \texttt{Occ(SWind)} \\ \exists\, \texttt{Occ(HST)} \end{array} \right\} \;\; \textit{causes} \;\; \exists\, \texttt{Occ(Flooding)}, \\ \forall\, \texttt{Flooded(BFPA)} \;\; \overset{\text{DEDO}}{\longrightarrow} \;\; \forall\, \texttt{Flooded(HouseFPA)} \end{array} \right\}$$

That is, (iii) gives rise to the argument $(\Phi, \{\delta\}, \Psi, \Theta)$ where

- *The explanation* is Φ.
- *There is a single statement being explained*, i.e., $\delta = \forall\, \texttt{Flooded(HouseFPA)}$.
- *The provision* of the explanation is empty.
- *The evidence* is Θ.

As for the argumentation part, our approach is concerned with sense-making. I.e., there is complex information that needs to be made sense of, and our approach is meant to provide a way to organize that information so that the key points are identified. This is a primary task in argumentation, as argumentation (even in computational guise) is much more than evaluating arguments, and in any case, does not begin with evaluating arguments [4]. Accordingly, our approach does *not* aim at evaluating a collection of arguments and counterarguments (as in the sense of determining extensions or identifying warranted arguments).

5.2 Counter-Arguments

A counter-argument for an argument $(\Phi, \Delta, \Psi, \Theta)$ is an argument $(\Phi', \Delta', \Psi', \Theta')$ which questions

1. either Φ (e.g., an argument exhibiting an explanation for $\neg\Phi$)
2. or Δ (e.g., an argument exhibiting an explanation for $\neg\Delta$)
3. or Ψ (e.g., an argument exhibiting an explanation for $\neg\Psi$)
4. or any item in Θ (e.g., an argument exhibiting an explanation for the negation of some θ occurring in Θ)
5. or does so by refutation: i.e an argument exhibiting an explanation for a statement known to be false and using any of Φ, Θ, Ψ and Δ. In this case, at least one of Φ', Θ', Ψ' intersects one of Φ, Δ, Θ, or Ψ.

Type (5) counter-arguments do not directly oppose an item in the argument being challenged. They rather question such an item by using it to provide an argument whose conclusion is wrong. The presence of such an item is ensured by checking that the challenged argument and the counter-argument indeed share something in common, i.e., that the intersection is not empty. Otherwise, in the case that the intersection is empty, then the two arguments have nothing in common, hence none can be viewed as a counter-argument to the other.

These counter-arguments have the form of an argument. They explain something that contradicts something in the challenged argument.

Dispute.
Let us consider the illustration at the start of this section: The argument (that the houses in the flood-prone area are flooded is partly explained by a strong wind) is under attack on the grounds that the anemometer did not turn red – indicating that no strong wind occurred. The latter is a counter-argument of type 5 in the above list. Indeed, the statement explained by the counter-argument is Red_Anemo that has been observed to be false. The explanation uses ∃ Occ(SWind), i.e., an item used by the explanation and then belonging to Φ in the attacked argument.

Taking Red_Anemo to be a falsehood, the counter-argument $(\Phi', \Delta', \emptyset, \Theta')$ results from Θ' yielding that

$$\left\{ \begin{array}{c} \exists\,\texttt{Occ(SWind)} \\ \texttt{OK\_Anemo} \end{array} \right\} \quad \text{explains} \quad \{\texttt{Red\_Anemo}\}$$

where

- *The explanation* is

$$\Phi' = \{\exists\,\texttt{Occ(SWind)}, \texttt{OK\_Anemo}\}$$

- *The statement being explained* is

$$\Delta' = \{\texttt{Red\_Anemo}\}$$

- *The evidence* is

$$\Theta' \quad = \quad \left\{ \left\{ \begin{array}{c} \exists\,\texttt{Occ(SWind)} \\ \texttt{OK\_Anemo} \end{array} \right\} \textit{causes}\ \texttt{Red\_Anemo} \right\}$$

Notice that Φ' does intersect Φ.

This is a counter-argument because, taking the anemometer being red as falsity, it is an argument which uses the occurring of a strong wind to conclude the anemometer being red. As explained above in the general case, such a type (5) counter-argument uses an item (a strong wind occurring) from the argument being challenged, in order to conclude a falsity (the anemometer being red).

Dispute (continued)
This counter-argument has in turn a counter-argument (of type 1.). It explains the misbehavior of the anemometer by the occurrence of an hurricane (that is a strong wind), and then explains the negation of an item OK_Anemo of the explanation Φ' of the counter-argument. The anemoter not getting red, instead of being explained by the absence of a

strong wind, is explained by the fact that the wind was so strong (an hurricane) that the anemometer misbehaved.

So, the counter-counter-argument is: $(\Phi'', \Delta'', \emptyset, \Theta'')$ resulting from Θ'' yielding that:

$$\{\exists\,\texttt{Occ(Hurri)}\} \text{ explains } \{\neg\texttt{OK\_Anemo}\}$$

where

- *The explanation* is
$$\Phi'' = \{\exists\,\texttt{Occ(Hurri)}\}$$
- *The statement being explained* is
$$\Delta'' = \{\neg\texttt{OK\_Anemo}\}$$
- *The evidence* is
$$\Theta'' = \{\exists\,\texttt{Occ(Hurri)}\ \mathit{causes}\ \neg\texttt{OK\_Anemo}\}$$

The dispute can extend to a counter-counter-counter-argument and so on as the process iterates.

6 Conclusion

The contribution of our work is firstly to propose a new logic-based formalism where explanations result from both causal and ontological links. It is important to stress that our approach reasons from causal relationships which are given, in contrast to a number of models for causality that aim at finding causal relationships (e.g., [8,9]). This causal-based approach for explanations, already defended in [1], is relatively different from other work on explanations that rely on expert knowledge and are considered as useful functionalities for expert systems and recommender systems (for a synthetic view on explanations in these domains, see [5,10,14]. We then show how these explanation links may be interestingly used as building blocks in an argumentative context [3]. It has similarities with the work by [12], who argue that, in the context of knowledge-based decision support systems, integrating explanations and argumentation capabilities is a valuable perspective.

Although explanation and argumentation have long been identified as distinct processes [13], it is also recognized that the distinction is a matter of context, hence they both play a role [7] when it comes to eliciting an answer to a "why" question. This is exactly what is attempted in this paper, as we are providing "possible" explanations, that thus can be turned into arguments. The argument format has some advantages inasmuch as its uniformity allows us to express objection in an iterated way: "possible" explanations are challenged by counter-arguments that happen to represent rival, or incompatible, "possible" explanations. Some interesting issues remain to be studied. Among others, comparing competing explanations according to minimality, preferences, and generally a host of criteria.

We have designed a system in answer set programming that implements the explicative proposal introduced above. Indeed, answer set programming [6] is well fitted for this kind of problem. One obvious reason is that rules such as (5), (8) or (9) can be translated literally and efficiently. Also ASP is known to be good for working with graphs such as the one depicted in the figures of this text. We plan to include our system in an argumentative framework and think it will be a good basis for a really practical system, able to manage with as a rich and tricky example as the Xynthia example.

Acknowledgements. It is our pleasure to thank the reviewers for their detailed and constructive remarks.

References

1. Besnard, P., Cordier, M.-O., Moinard, Y.: Deriving explanations from causal information. In: Ghallab, M., Spytopoulos, C.D., Fakotakis, N., Avouris, N.M. (eds.) ECAI 2008, pp. 723–724. IOS Press (2008)
2. Besnard, P., Cordier, M.-O., Moinard, Y.: Ontology-based inference for causal explanation. Integrated Computer-Aided Engineering 15, 351–367 (2008)
3. Besnard, P., Hunter, A.: Elements of Argumentation. MIT Press, Cambridge (2008)
4. Dung, P.M.: On the acceptability of arguments and its fundamental role in nonmonotonic reasoning, logic programming and n-person games. Artificial Intelligence 77, 321–357 (1995)
5. Friedrich, G., Zanker, M.: A taxonomy for generating explanations in recommender systems. AI Magazine 32(3), 90–98 (2011)
6. Gebser, M., Kaminski, R., Kaufmann, B., Schaub, T.: Answer Set Solving in Practice. Synthesis Lectures on Artificial Intelligence and Machine Learning. Morgan and Claypool Publishers (2012)
7. Giboney, J.S., Brown, S., Nunamaker Jr., J.F.: User acceptance of knowledge-based system recommendations: Explanations, arguments, and fit. In: 45th Annual Hawaii International Conference on System Sciences (HICSS'45), pp. 3719–3727. IEEE Computer Society (2012)
8. Halpern, J.Y., Pearl, J.: Causes and Explanations: A Structural-Model Approach. Part I: Causes. In: Breese, J.S., Koller, D. (eds.) 17th Conference in Uncertainty in Artificial Intelligence (UAI 2001), pp. 194–202. Morgan Kaufmann (2001)
9. Halpern, J.Y., Pearl, J.: Causes and Explanations: A Structural-Model Approach. Part II: Explanations. In: Nebel, B. (ed.) 17th International Joint Conference on Artificial Intelligence (IJCAI 2001), pp. 27–34. Morgan Kaufmann (2001)
10. Lacave, C., Díez, F.J.: A review of explanation methods for heuristic expert systems. The Knowledge Engineering Review 19, 133–146 (2004)
11. Mellor, D.H.: The Facts of Causation. Routledge, London (1995)
12. Moulin, B., Irandoust, H., Bélanger, M., Desbordes, G.: Explanation and argumentation capabilities: Towards the creation of more persuasive agents. Artif. Intell. Rev. 17(3), 169–222 (2002)
13. Walton, D.: Explanations and arguments based on practical reasoning. In: Roth-Berghofer, T., Tintarev, N., Leake, D.B. (eds.) Workshop on Explanation-Aware Computing at IJCAI 2009, Pasadena, CA, U.S.A, pp. 72–83 (July 2009)
14. Ye, L.R., Johnson, P.E.: The impact of explanation facilities in user acceptance of expert system advice. MIS Quarterly 19(2), 157–172 (1995)

Reasoning on Secrecy Constraints under Uncertainty to Classify Possible Actions*

Joachim Biskup, Gabriele Kern-Isberner,
Patrick Krümpelmann, and Cornelia Tadros

Fakultät für Informatik, Technische Universität Dortmund, Germany
{joachim.biskup,gabriele.kern-isberner,patrick.kruempelmann,
cornelia.tadros}@cs.tu-dortmund.de

Abstract. Within a multiagent system, we focus on an intelligent agent $\mathcal{D}$ maintaining a view on the world and interacting with another agent $\mathcal{A}$. Defending its own interests, $\mathcal{D}$ wants to protect sensitive information according to secrecy constraints in a secrecy policy. Hereby, a secrecy constraint intuitively expresses the desire that $\mathcal{A}$, seen as attacking these interests, should not believe some target sentence by reasoning on the world. For deciding on its actions, $\mathcal{D}$ has to interpret secrecy constraints under uncertainty about the epistemic state of $\mathcal{A}$. To this end, we equip $\mathcal{D}$ with a secrecy reasoner which classifies the agent's possible actions according to their compliance with its secrecy policy. For this classification task, we introduce principles to guide the reasoning based on postulates about $\mathcal{A}$. In particular, these principles give guidance on how to deal with the uncertainty about $\mathcal{A}$ and, if in the face of other desires, $\mathcal{D}$ considers an action which is potentially violating secrecy constraints, how to mitigate the effect of potential violations. Moreover, we design a secrecy reasoner by presenting a constructive approach for the classification task and verify that the designed reasoner adheres to the principles.

Keywords: action, agent view, a priori knowledge, attacker postulates, belief, credulity, epistemic state, multiagent system, secrecy constraint, secrecy reasoner, uncertainty.

1 Introduction

We consider a system of interacting and intelligent agents of the following kind: the agents are situated in a dynamic, non-deterministic and inaccessible environment [18]. Each agent maintains views and, based on that, beliefs on its environment while interpreting observations and deciding on its own actions. An agent's environment consists of some non-explicitly represented world and the other agents. Our long-term goal is to enable an agent to reason on secrecy constraints, declared by a security/knowledge engineer in a secrecy policy, in order

* This work has been supported by the Deutsche Forschungsgemeinschaft (German Research Council) under grant SFB 876/A5 within the framework of the Collaborative Research Center "Providing Information by Resource-Constrained Data Analysis".

C. Beierle and C. Meghini (Eds.): FoIKS 2014, LNCS 8367, pp. 97–116, 2014.

to classify its actions, capturing its secrecy constraints, to the end of decision making. This reasoning should follow the underlying intention to protect sensitive information against other agents in a best possible way – even under the uncertainty inherent in the agent's assumption about another agent and even under consideration of potential violations of secrecy constraints.

In this article, we first present a model of an intelligent agent for treating that goal of secrecy reasoning. Thereby, we focus on a scenario of two intelligent agents – each of them is capable to act as a *defending agent* $\mathcal{D}$ which can choose to execute an *inform-action* to its ends, i.e., tell another agent that some sentence is true; but doing this should comply with its secrecy constraints in its secrecy policy in a best possible way. In turn, the other agent is postulated as an *attacking agent* $\mathcal{A}$ with respect to $\mathcal{D}$'s secrecy constraints. This small scenario allows us to elaborate on a fundamental task to be solved for an agent as defender $\mathcal{D}$: fixing a point in time, how should $\mathcal{D}$ reason on its secrecy constraints to classify possible inform-actions according to their degree of compliance with $\mathcal{D}$'s secrecy constraints? We restrict to comparable degrees expressed by natural numbers where smaller numbers indicate better compliance.

Figure 1 shows the main components of an intelligent agent, both as an attacking and as a defending agent, in the focused scenario. We describe the agent's components by abstract concepts, leaving out details that are irrelevant for our purposes. This way, as we exemplify, our results can be applied to several instances of these concepts. On the one hand, an agent seen as an attacker $\mathcal{A}$ reasons about the world. On the other hand, an agent seen as a defender $\mathcal{D}$ considers the other agent's reasoning about the world, on the basis of *postulates* about the other agent's capabilities. In the following, we consider that one of the agents has the fixed role of a defender and the other agent a fixed role of an attacker.

A secrecy constraint of agent $\mathcal{D}$ towards the other agent $\mathcal{A}$ roughly expresses $\mathcal{D}$'s desire that "agent $\mathcal{A}$ *should not believe* the information expressed by some logical target sentence ϕ by reasoning with an operator Bel". To interpret such a constraint, the defending agent first needs the postulates about $\mathcal{A}$'s capabilities.

Under the uncertainty about $\mathcal{A}$ inherent in these postulates, the agent $\mathcal{D}$ also needs guidance by some appropriate *norm*. In our approach, as further discussed in Section 5 on related work, such a norm is induced by the specific declarations of $\mathcal{D}$'s engineer – the secrecy constraints and a belief operator family with credulity order – together with three principles for the classification task.

The *belief operator family with credulity order* enables $\mathcal{D}$ to compare belief operators according to the credulity of inferred belief. This way, if $\mathcal{D}$ considers that $\mathcal{A}$ believes a target sentence ϕ using operator Bel and thus *considers a respective secrecy constraint potentially violated*, $\mathcal{D}$ is enabled to mitigate the potential violation by considering less credulous reasoning than reasoning with Bel.

The principles present requirements on the classification in account of the uncertainty about $\mathcal{A}$ and the credulity of $\mathcal{A}$'s reasoning about a target sentence and for minimality of classification. Minimality should prevent that $\mathcal{D}$'s options

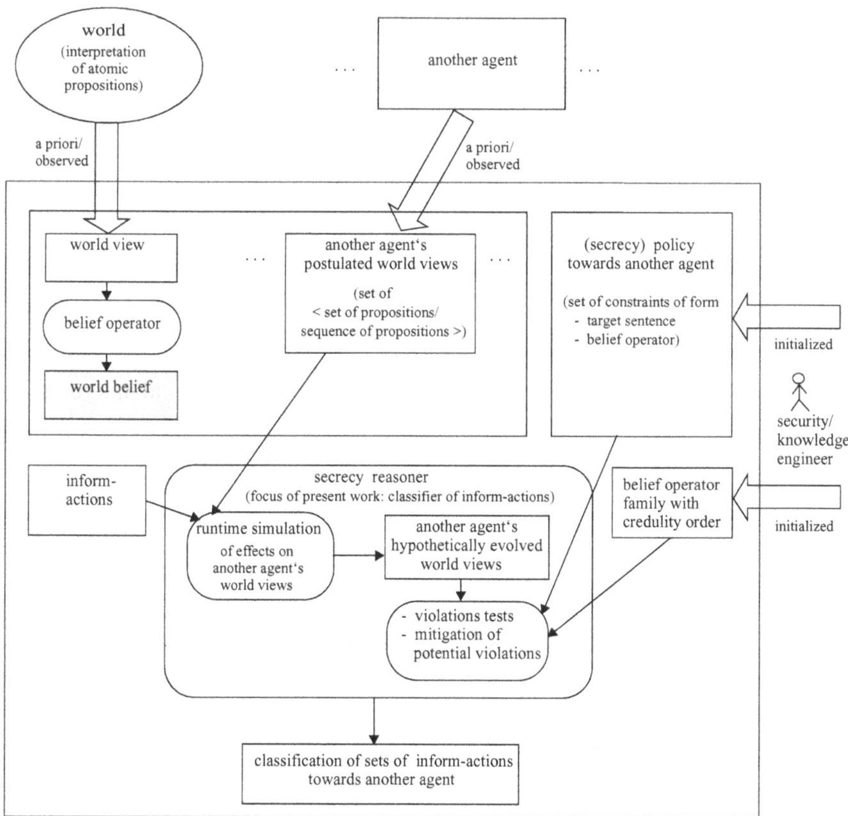

Fig. 1. Main components of an intelligent agent both as an attacking agent reasoning about the world and as a defending agent reasoning on secrecy constraints under uncertainty, focusing on the classification of actions

to execute inform-actions are unnecessarily restricted to respect other desires of $\mathcal{D}$ such as information sharing.

Our contribution is the specification and design of a reasoner on secrecy constraints. First we specify the reasoner declaratively by the formalization of the principles. Then, we develop an algorithm for the secrecy reasoner. Finally, we verify that the design and algorithm adheres to the declarative specification. In the following sections, we proceed as follows. In Section 2, we detail an intelligent agent's components essential for secrecy reasoning. In Section 3 we elaborate and formalize the principles for a secrecy reasoner as declarative requirements on its classification results. In Section 4, we design a secrecy reasoner and verify its adherence to the declarative requirements. In Section 5, we relate our contributions to other works. Finally, in the concluding Section 6, we discuss issues purposely left open in our abstractions and suggest lines for further research. Proofs of statements are either omitted or shifted to the appendix.

2 Epistemic Agent Model for Secrecy Reasoning

In this section, we first introduce the components of an epistemic agent being essential for $\mathcal{A}$'s reasoning about the world as shown in Figure 1. Then, we introduce the essential components for $\mathcal{D}'s$ reasoning on secrecy constraints including its postulates about $\mathcal{A}$'s means of reasoning to form belief and about the effects of $\mathcal{D}$'s actions on that belief.

Actual Situation of an Attacking Agent. Common to both agents, there is a (non-explicitly represented) *world* saying "what is really the case". The actual case is semantically specified by an interpretation of atomic propositions in the alphabet $\mathcal{A}t$ which forms the basic vocabulary of both agents. Syntactically, the agents communicate about the world in the shared propositional language $\mathcal{L}$ over $\mathcal{A}t$ with the standard propositional connectives, or an appropriate selection thereof. An attacking agent $\mathcal{A}$ has information about some aspects of the world, resulting in its *world view*, which may have resulted from a priori knowledge about the world in general or from observations during interactions. The world view is syntactically represented as a sequence $\langle B; \phi_1, \ldots, \phi_n \rangle$ comprised of two parts: the *background knowledge* B as a set of sentences from a language $\mathcal{L}_\mathcal{B}$, which is an extension of $\mathcal{L}$ to express rules, and its *observations* $\phi_i \in \mathcal{L}$.

Further, agent $\mathcal{A}$ can form *belief* about a sentence $\phi \in \mathcal{L}$ by means of a *belief operator* $Bel : 2^{\mathcal{L}_\mathcal{B}} \times \mathcal{L}^* \rightarrow 2^{\mathcal{L}}$. The agent believes ϕ using Bel if $\phi \in Bel(\langle B; \phi_1, \ldots, \phi_n \rangle)$. The operator is *chosen* by the agent to its ends, depending on the target ϕ of its reasoning, out of a family Ξ of such operators. The task of the belief operator is to derive propositions in $\mathcal{L}$ the agent accepts as true by reasoning on its observations using its background knowledge. Examples of the language $\mathcal{L}_\mathcal{B}$ for background knowledge are simply $\mathcal{L}$, extended logic programs [9] or conditionals [12].

Our intent is to consider general families Ξ of belief operators for modeling different kinds of reasoning such as skeptical and credulous reasoning [16]. Yet, we want to make sensible restrictions on what $\mathcal{A}$ may accept as true by using an operator from Ξ: $\mathcal{A}$ cannot accept as true a formula and its negation at the same time and it has to accept all propositional consequences in $\mathcal{L}$ of any formula it accepts as true. Moreover, belief operators in Ξ with different kind of reasoning usually can be compared with respect to the credulity of reasoning. We formalize this aspect by a *credulity order* on Ξ with the intuition that a belief operator is more *credulous* than another one if with the former the agent accepts more propositions as its belief.

Definition 1 (Belief Operator Family with Credulity Ordering). *A belief operator family with credulity order is a pair* $(\Xi, \preceq_{cred})$ *consisting of a set* Ξ *of belief operators of the form* $Bel : 2^{\mathcal{L}_\mathcal{B}} \times \mathcal{L}^* \rightarrow 2^{\mathcal{L}}$ *and a credulity order* $\preceq_{cred}$ *on* Ξ *fulfilling the following properties.*

First, for each belief operator $Bel \in \Xi$ *it holds:*

- *Consistency: There does not exist* $\phi \in \mathcal{L}$ *and* $W \in 2^{\mathcal{L}_\mathcal{B}} \times \mathcal{L}^*$ *such that* $\phi \in Bel(W)$ *and* $\neg\phi \in Bel(W)$.

– *Propositional Consequence: For all $\phi, \psi \in \mathcal{L}$ such that ψ is a propositional consequence of ϕ, i.e., $\phi \vdash_{pl} \psi$, for all $W \in 2^{\mathcal{L}_\mathcal{B}} \times \mathcal{L}^*$ it holds $\phi \in Bel(W)$ implies $\psi \in Bel(W)$.*

Second, the credulity order $\preceq_{cred}$ satisfies the following credulity *property: If $Bel \preceq_{cred} Bel'$, then for all $W \in 2^{\mathcal{L}_\mathcal{B}} \times \mathcal{L}^*$ it holds that $Bel(W) \subseteq Bel'(W)$. We read $Bel \preceq_{cred} Bel'$ as Bel' is at least as credulous as Bel.*

Example 1. We use an instantiation of the abstract concepts, taking $\mathcal{L}_\mathcal{B} = \mathcal{L}$ as the propositional language with all standard connectives and assuming a finite alphabet $\mathcal{At}$. Then, we define a set of belief operators $\Xi^{RW} = \{Bel_p \mid p \in (0.5, 1]\}$ indexed by threshold parameter p. Each operator calculates the agent's certainty in the truth of a formula $\phi \in \mathcal{L}$ as the ratio of its models among the models of the agent's background knowledge and its observations in its world view $\langle B; \phi_1, \ldots, \phi_n \rangle$ like in the *random worlds* approach [2]. Then, the agent's reasoning can be seen as accepting every formula as true that holds at least in the "majority" of the considered models, more precisely in at least $p \cdot 100$ percent of those models (given by the model operator Mod):

$$Bel_p(\langle B; \phi_1, \ldots, \phi_n \rangle) = \{\phi \in \mathcal{L} \mid \mathsf{r}(\phi, \langle B; \phi_1, \ldots, \phi_n \rangle) \geq p\}$$
$$\text{with } \mathsf{r}(\phi, \langle B; \phi_1, \ldots, \phi_n \rangle) = \frac{|\mathsf{Mod}(\phi) \cap \mathsf{Mod}(B \cup \{\phi_1, \ldots, \phi_n\})|}{|\mathsf{Mod}(B \cup \{\phi_1, \ldots, \phi_n\})|} \quad (1)$$

The credulity order is given as $Bel_p \preceq^{RW}_{cred} Bel_{p'}$ iff $p' \leq p$. We can easily verify that the pair $(\Xi^{RW}, \preceq^{RW}_{cred})$ is a belief operator family with credulity order of Def. 1. For simplicity, we neglect the cases where $B \cup \{\phi_1, \ldots, \phi_n\}$ is inconsistent and thus consider only consistent sets in the following examples.

Upon receiving an inform-action $inform(\phi) \in \mathcal{Act} := \{inform(\phi) \mid \phi \in \mathcal{L}\}$ from $\mathcal{D}$, interpreted by $\mathcal{A}$ as conveying the claim of truth of a sentence $\phi \in \mathcal{L}$ in the world, agent $\mathcal{A}$ appends ϕ to its observations in its world view. This is formalized by the operator $+ : 2^{\mathcal{L}_\mathcal{B}} \times \mathcal{L}^* \times \mathcal{Act} \to 2^{\mathcal{L}_\mathcal{B}} \times \mathcal{L}^*$ with $\langle B; \phi_1, \ldots, \phi_n \rangle + inform(\phi) := \langle B; \phi_1, \ldots, \phi_n, \phi \rangle$. In this context, we treat an inform-action as an abstraction that only cares about the information conveyed to $\mathcal{A}$, whether explicitly by communicating a statement ϕ or implicitly as consequences of some notification. All other aspects of an action, e.g., being a reaction on some previous request or urging the recipient to activities, are either abstracted into $\mathcal{D}$'s assumptions about $\mathcal{A}$ or simply neglected.

A Defending Agent's Postulates About an Attacking Agent. Being unable to determine an attacking agent $\mathcal{A}$'s actual situation, a defending agent $\mathcal{D}$ has to make assumptions about that situation to evaluate the effect of a considered inform-action on $\mathcal{A}$'s belief with respect to $\mathcal{D}$'s secrecy constraints. In particular, $\mathcal{D}$ is challenged by its uncertainty about $\mathcal{A}$'s world view and about $\mathcal{A}$'s choice of the operator Bel for a particular target sentence ϕ. We speak of $\mathcal{D}$'s *postulates* rather than of assumptions to stress that we do not only express what

$\mathcal{D}$ reasonably assumes about $\mathcal{A}$'s actual situation, but also what $\mathcal{D}$ presumes as a matter of precaution for the sake of secrecy in the face of $\mathcal{D}$'s uncertainty.

In this sense, agent $\mathcal{D}$ keeps $\mathcal{A}$'s *postulated world views* which is a non-empty set $\mathcal{W} \subseteq 2^{\mathcal{L}_\mathcal{B}} \times \mathcal{L}^*$ of world views. This way, $\mathcal{D}$ postulates that each $W \in \mathcal{W}$ could be held by $\mathcal{A}$, but that each $W \in (2^{\mathcal{L}_\mathcal{B}} \times \mathcal{L}^*) \setminus \mathcal{W}$ is *not* held by $\mathcal{A}$.

In this set, agent $\mathcal{D}$ might have incorporated a priori assumptions or observations of $\mathcal{A}$ as having been observed by $\mathcal{D}$ in turn. For example, $\mathcal{D}$ might only have partially accessed the contents of an inform-action from another agent to $\mathcal{A}$ since that agent might have encrypted parts of the contents for the sake of its secrecy constraints.

Example 2. Consider that agent $\mathcal{A}$ reasons about a particular person P who can have several properties $\mathcal{A}t = \{p_1, \ldots, p_4, s_1, s_2\}$. Considering $\mathcal{A}$'s relationship to P, agent $\mathcal{D}$ reasonably assumes that $\mathcal{A}$ knows whether P has properties p_1 and p_2, whereas $\mathcal{D}$ does not know this about P. Moreover, $\mathcal{D}$ postulates that $\mathcal{A}$ has the following background knowledge:

$$B = \{p_1 \wedge p_2 \wedge p_4 \Rightarrow s_1,\ p_2 \wedge p_3 \Rightarrow s_1,\ p_1 \wedge p_3 \Rightarrow s_2\}. \tag{2}$$

Thus, $\mathcal{D}$ postulates that the following world views could be held by $\mathcal{A}$: $W_1 = (B; p_1, p_2)$, $W_2 = (B; p_1, \neg p_2)$, $W_3 = (B; \neg p_1, p_2)$ and $W_4 = (B; \neg p_1, \neg p_2)$.

Further, the defending agent postulates which belief operator $\mathcal{A}$ would choose to reason about a target sentence ϕ in the context of a specific secrecy constraint. *Intuitively*, such a secrecy constraint should represent agent $\mathcal{D}$'s desire to avoid that agent $\mathcal{A}$ believes ϕ when using the belief operator Bel on its world view W, i.e., $\phi \in Bel(W)$ is undesired.

Definition 2 (Secrecy Policy). *A* secrecy constraint *is a tuple* $(\phi, Bel) \in \mathcal{L}_S = \mathcal{L} \times \Xi$ *consisting of a target sentence* $\phi \in \mathcal{L}$ *and a belief operator* $Bel \in \Xi$*. Further, a secrecy policy is a set* $\mathcal{S} \subseteq \mathcal{L}_S$ *of secrecy constraints.*

Note that $\mathcal{D}$ may have several secrecy constraints with the same target sentence to account for several possible choices of $\mathcal{A}$.

Example 3. Agent $\mathcal{D}$ considers the properties s_1 and s_2 as target sentences of secrecy constraints. Concerned with P's privacy, $\mathcal{D}$ thus has the following secrecy constraints towards $\mathcal{A}$: $\mathcal{S} = \{(s_1, Bel_{0.7}),\ (s_2, Bel_{0.6})\}$. This way, $\mathcal{D}$ postulates a more credulous reasoning of $\mathcal{A}$ about property s_2 than $\mathcal{A}$'s reasoning about property s_1 ($Bel_{0.6} \succ^{RW}_{cred} Bel_{0.7}$).

Taking all this into consideration, we define the epistemic state of a defending agent $\mathcal{D}$ as follows.

Definition 3 (Epistemic State). *The epistemic state* $\mathcal{K}_\mathcal{D}$ *of agent* $\mathcal{D}$ *(for the given point in time), focused on attacker* $\mathcal{A}$*, is determined by the following state operators. The world view is given by* $\mathsf{V}^W(\mathcal{K}_\mathcal{D}) \in 2^{\mathcal{L}_\mathcal{B}} \times \mathcal{L}^*$*. The set of postulated world views of* $\mathcal{A}$ *is given by* $\mathsf{V}^{PW}(\mathcal{K}_\mathcal{D}) \subseteq 2^{\mathcal{L}_\mathcal{B}} \times \mathcal{L}^*$ *with* $\mathsf{V}^{PW}(\mathcal{K}_\mathcal{D}) \neq \emptyset$*. The secrecy policy is given by* $\mathcal{S}(\mathcal{K}_\mathcal{D}) \subseteq \mathcal{L}_S$*.*

Generally, each agent enjoys the same structure, i.e., the state operators can be applied for any agent. Moreover, generally, the epistemic state contains postulated world views for each other agent in the environment. We denote the set of all epistemic states by $\mathcal{Es}$.

3 Secrecy Reasoner: Declarative Principles

Agent $\mathcal{D}$'s secrecy policy is interpreted by $\mathcal{D}$'s secrecy reasoner yielding a classification of possible actions. This way, a secrecy reasoner defines a *formal semantics* of $\mathcal{D}$'s secrecy policy. A low classification of an action indicates that the execution of that *action complies with $\mathcal{D}$'s secrecy policy* possibly well. Typically, $\mathcal{D}$ considers a finite subset $\mathcal{Act}' \subset_{fin} \mathcal{Act}$ of all actions as its options. Thus, formally, a *classification* is a function $\mathsf{cl} : \mathcal{Act}' \to \mathbb{N}_0$ such that $\mathsf{cl}^{-1}(0) \neq \emptyset$ where a classification rank of 0 means best compliance. The infinite range of classifications should enable the agent to mitigate potential violations as suggested by [14]. Roughly, the reason is that, if the agent should be enabled to do so, actions cannot be simply classified into two kinds, one complying with its secrecy policy and the others not complying. We denote the set of all classifications of arbitrary finite subsets $\mathcal{Act}'$ of actions by $\mathcal{Cl}$.

In this section, we define how a secrecy reasoner should classify actions in a declarative way by principles taking $\mathcal{D}$'s uncertainty about $\mathcal{A}$'s situation into account. Above all, the classification of an action $inform(\phi)$ depends on a runtime simulation of its effect on $\mathcal{D}$'s postulated world views $\mathcal{W}$. These *hypothetically evolved world views* are determined by the operator $\oplus$ as follows:

$$\oplus(\mathcal{W}, inform(\phi)) = \{W + \phi \mid W \in \mathcal{W}\}. \tag{3}$$

The principles for a secrecy reasoner are explained in the following and summarized by definitions at the end of this section.

Principle I.1 (*Avoid potential violations*) intuitively expresses that agent $\mathcal{D}$ desires to avoid that in some of its postulated world views more secrecy constraints become violated. We make this precise using the following definition.

Definition 4 (Violation Sets). *Let $\mathcal{S} \subseteq \mathcal{L}_S$ be a secrecy policy, $W \in 2^{\mathcal{L}_B} \times \mathcal{L}^*$ a world view, $\mathcal{W}$ postulated world views and $a \in \mathcal{Act}$ an inform-action. Then, the secrecy constraints in $\mathcal{S}$ violated under world view W are defined as the set*

$$\mathsf{vio}(\mathcal{S}, W) = \{(\phi, Bel) \in \mathcal{S} \mid \phi \in Bel(W)\}.$$

Further, combinations of secrecy constraints from $\mathcal{S}$ potentially violated under postulated world views $\mathcal{W}$ after action a are defined by

$$\mathsf{vioAfter}(\mathcal{S}, \mathcal{W}, a) = \{\mathsf{vio}(\mathcal{S}, W) \mid W \in \oplus(\mathcal{W}, a)\}.$$

In words, each of these combinations is a set of constraints that are jointly violated under one and the same $W \in \mathcal{W}$.

In the context of an epistemic state $\mathcal{K}$, we often use the notation $\mathsf{vio}(\mathcal{K}, W)$ for $\mathsf{vio}(\mathcal{S}(\mathcal{K}), W)$ and $\mathsf{vioAfter}(\mathcal{K}, a)$ for $\mathsf{vioAfter}(\mathcal{S}(\mathcal{K}), \mathsf{V}^{PW}(\mathcal{K}), a)$. Moreover, we say that $\mathcal{D}$ *considers a constraint potentially violated after action* a if the constraint occurs in some set in $\mathsf{vioAfter}(\mathcal{K}, a)$. Furthermore, we base several definitions on the set-inclusion maximal sets of a set X of sets which we denote by $\max_{\subseteq} X := \{S \in X \mid \text{ there is no } S' \in X \text{ such that } S \subset S'\}$.

Example 4. Agent $\mathcal{D}$ now wants to decide whether to reveal that P has property p_3; or to reveal less by saying $p_1 \vee p_2 \vee p_3$; or to intentionally lie to $\mathcal{A}$ about p_3. Thus, $\mathcal{D}$ considers one of the options $\mathcal{A}ct' = \{inform(p_3), inform(p_1 \vee p_2 \vee p_3), inform(\neg p_3)\}$ to inform $\mathcal{A}$, referred to as a_1, a_2 and a_3, respectively. Concerned with secrecy, $\mathcal{D}$ further considers the combinations of its secrecy constraints potentially violated after each action resulting in: $\mathsf{vioAfter}(\mathcal{K}_D, a_1) = \mathsf{vioAfter}(\mathcal{K}_D, a_2) = \{\{(s_2, Bel_{0.6}), (s_1, Bel_{0.7})\}, \{(s_2, Bel_{0.6})\}, \{(s_1, Bel_{0.7})\}, \emptyset\}$ and $\mathsf{vioAfter}(\mathcal{K}_D, a_3) = \{\emptyset\}$. Hereby, $\mathcal{D}$'s postulated world views and its secrecy policy are given by Ex. 2 and Ex. 3, respectively. Agent $\mathcal{D}$ uses them to evaluate the effect of the optional actions on the respective world views as follows (violations are marked):

	$\mathsf{r}(s_1, W_i)$	$\mathsf{r}(s_2, W_i)$
$W_1 + a_2$	0.75	0.625
$W_2 + a_2$	0.5	$0.\overline{6}$
$W_3 + a_2$	$0.\overline{6}$	0.5
$W_4 + a_2$	0.5	0.5

	$\mathsf{r}(s_1, W_i)$	$\mathsf{r}(s_2, W_i)$
$W_1 + a_1$	1	1
$W_2 + a_1$	1	0.5
$W_3 + a_1$	0.5	1
$W_4 + a_1$	0.5	0.5

The effects of action a_3 are not shown in the table since after that action no secrecy constraint is potentially violated.

In terms of Def. 4, Principle I.1 (*Avoid potential violations*) instructs the secrecy reasoner to classify an action b higher (worse) than another action a given an epistemic state $\mathcal{K}$ if after the former action more secrecy constraints are potentially violated in combination. This is formalized by the following relation: $\mathsf{vioAfter}(\mathcal{K}, b) \sqsupset \mathsf{vioAfter}(\mathcal{K}, a)$ iff

1. for all $S_a \in \mathsf{vioAfter}(\mathcal{K}, a)$ there exists $S_b \in \mathsf{vioAfter}(\mathcal{K}, b)$ such that $S_a \subseteq S_b$ and
2. there exists some $S_a \in \max_{\subseteq} \mathsf{vioAfter}(\mathcal{K}, a)$ such that there exists some $S_b \in \mathsf{vioAfter}(\mathcal{K}, b)$ with $S_a \subset S_b$.

Example 5. Taking up Ex. 4, we find that $\mathsf{vioAfter}(\mathcal{K}_D, a_1) \sqsupset \mathsf{vioAfter}(\mathcal{K}_D, a_3)$ and $\mathsf{vioAfter}(\mathcal{K}_D, a_2) \sqsupset \mathsf{vioAfter}(\mathcal{K}_D, a_3)$ hold, but $\mathsf{vioAfter}(\mathcal{K}_D, a_1)$ and $\mathsf{vioAfter}(\mathcal{K}_D, a_2)$ are not ordered by $\sqsupset$.

Using $\sqsupset$, a secrecy reasoner is not instructed to classify two actions differently if it lacks the information in $\mathcal{K}$ to compare combinations of constraints potentially violated after these actions. This is the case if in $\mathcal{K}$ for neither of two actions a and b the first condition for the relation $\sqsupset$ is satisfied. In this case, there are constraints potentially violated after a in combination, but which are not so after b; however, others are, but are not so after a. In such a case, a secrecy reasoner cannot decide the execution of which action is worse, considering all the constraints. In the presence of additional information, such as priorities among secrecy constraints, further decisions might be possible.

Example 6. As examples of the above case, the following pairs of violation sets are not ordered by $\sqsupset$: (1) $\{\{(s_1, Bel_{0.7})\}, \emptyset\}$ and $\{\{(s_2, Bel_{0.6})\}, \emptyset\}$; (2) $\{\{(s_1, Bel_{0.7})\}, \emptyset\}$ and $\{\{(s_2, Bel_{0.6}), (s_3, Bel_{0.5})\}, \emptyset\}$.

Principle I.2 (*Mitigate potential violations*) addresses interests of $\mathcal{D}$ conflicting with secrecy. An example of such an interest is that $\mathcal{D}$ is cooperative and generally motivated to share information. Pursuing such interests, $\mathcal{D}$ may choose an action which it considers potentially violating more constraints in combination than other actions. In this case, Principle I.2 says that, if $\mathcal{D}$ so chooses, it should at least mitigate the effect of its action on potentially violated secrecy constraints. Mitigation means that, if a target sentence could not be protected against inferences with Bel as desired, it should at least be protected against inferences with operators more skeptical than Bel. To this end, a secrecy reasoner considers actions with the same effect concerning combinations of potentially violated constraints in an epistemic state $\mathcal{K}$, i.e., actions $a, b \in \mathcal{A}ct'$ such that

$$\mathsf{vioAfter}(\mathcal{K}, a) \sim \mathsf{vioAfter}(\mathcal{K}, b) \text{ defined as } \max\nolimits_{\subseteq} \mathsf{vioAfter}(\mathcal{K}, a) = \max\nolimits_{\subseteq} \mathsf{vioAfter}(\mathcal{K}, b). \quad (4)$$

Such pairs of actions might mitigate the effect of one another as formalized in the following relation.

Definition 5 (Mitigation of Potential Violations). *Let $\mathcal{K}$ be an epistemic state and $a, b \in \mathcal{A}ct'$ such that $\mathsf{vioAfter}(\mathcal{K}, a) \sim \mathsf{vioAfter}(\mathcal{K}, b)$. Action a mitigates potential violation of action b, written as a mvio b, if the following two conditions hold:*

1. *There exists $S \in \max_{\subseteq} \mathsf{vioAfter}(\mathcal{K}, b)$ such that there exist $(\phi, Bel_1) \in S$ and $Bel'_1 \in \Xi$ with $Bel'_1 \prec_{cred} Bel_1$ such that the constraint (ϕ, Bel'_1) is potentially violated after action b in $\mathcal{K}$, but not after action a.*
2. *There are no $(\psi, Bel_2) \in S$ and $Bel'_2 \in \Xi$ with $Bel'_2 \prec_{cred} Bel_2$ such that the constraint (ψ, Bel'_2) is potentially violated after action a in $\mathcal{K}$, but not after action b.*

Example 7. We can see from the table in Ex. 4 that a_2 mvio a_1, but not a_1 mvio a_2. This is the case since $\mathsf{vioAfter}(\mathcal{K}_D, a_1) \sim \mathsf{vioAfter}(\mathcal{K}_D, a_2)$ holds on the one hand. On the other hand, for the constraint $(s_2, Bel_{0.6})$ in $\{(s_2, Bel_{0.6}), (s_1, Bel_{0.7})\} \in \mathsf{vioAfter}(\mathcal{K}_D, a_1)$ and the belief operator $Bel_{0.7}$ we obtain that

$$\mathsf{vioAfter}(\{(s_2, Bel_{0.7})\}, \mathsf{V}^{PW}(\mathcal{K}_D), a_1) = \{\{(s_2, Bel_{0.7})\}\} \text{ and}$$
$$\mathsf{vioAfter}(\{(s_2, Bel_{0.7})\}, \mathsf{V}^{PW}(\mathcal{K}_D), a_2) = \{\emptyset\}.$$

Further, for the constraint $(s_1, Bel_{0.7})$ in $\{(s_2, Bel_{0.6}), (s_1, Bel_{0.7})\}$ and the belief operator $Bel_{0.8}$ we obtain that

$\mathsf{vioAfter}(\{(s_1, Bel_{0.8})\}, \mathsf{V}^{PW}(\mathcal{K}_D), a_1) = \{\{(s_1, Bel_{0.8})\}\}$ and
$\mathsf{vioAfter}(\{(s_1, Bel_{0.8})\}, \mathsf{V}^{PW}(\mathcal{K}_D), a_2) = \{\emptyset\}$.

Thus, there are no $(\psi, Bel_2) \in S$ and $Bel_2' \in \Xi$ with $Bel_2' \prec_{cred} Bel_2$ such that (ψ, Bel_2') is potentially violated after action a_2 in $\mathcal{K}$, but not after action a_1.

If action a mitigates a potential violation after action b it should be classified lower at best. However, the relation mvio is not acyclic and if a mvio b is part of a cycle then there is no justification to classify a lower than b. In this case, we call a and b *conflicting*. Principle I.2 demands that if a mitigates potential violation of b and a and b are not conflicting, then a should be classified lower than b. If a mvio b and a and b are conflicting, then the principle accounts for this local mitigation by requiring that a should be classified at least as low as b.

Principle II (*Minimize classification*) stipulates that a classification should be as little restrictive as possible with regard to $\mathcal{D}$'s possible other desires such as cooperative information sharing. The lower the classification rank of an action the less $\mathcal{D}$ is admonished to refrain from executing that action. In particular, a secrecy reasoner does not pose any restriction on all actions with a rank of 0.

Definition 6 (Secrecy Reasoner For Action Classification). *A secrecy reasoner is a function* $\mathsf{sr} : 2^{\mathcal{A}ct}_{fin} \times \mathcal{E}s \rightarrow \mathcal{C}l$ *being parameterized with a family* $(\Xi, \preceq_{cred})$ *of belief operators with credulity order. It takes as input a finite subset* $\mathcal{A}ct'$ *of* $\mathcal{A}ct$ *and an epistemic state* $\mathcal{K}$ *and outputs a classification* cl *of* $\mathcal{A}ct'$*. Moreover, the function* sr *has to fulfill the following principles:*

Principle I.1: Avoid potential violations *Let* $a, b \in \mathcal{A}ct'$ *be actions and* $\mathcal{K}$ *an epistemic state such that* $\mathsf{vioAfter}(\mathcal{K}, b) \sqsupset \mathsf{vioAfter}(\mathcal{K}, a)$*. Further, let* $\mathsf{cl} = \mathsf{sr}(\mathcal{A}ct', \mathcal{K})$ *be the reasoner's classification of* $\mathcal{A}ct'$ *in* $\mathcal{K}$*. Then, it follows* $\mathsf{cl}(b) > \mathsf{cl}(a)$.

Principle I.2: Mitigate potential violations *Let* $\mathcal{K}$ *be an epistemic state and* $a, b \in \mathcal{A}ct'$ *such that* a mvio b*, then:*

1. *Conflict Free Mitigation: If there do not exist actions* $a_1, \ldots, a_n \in \mathcal{A}ct'$ *such that* $a_1 = b$*,* $a_n = a$ *and* a_i mvio a_{i+1} *for all* $i \in \{1, \ldots, n-1\}$*, then it follows* $\mathsf{cl}(b) > \mathsf{cl}(a)$ *with* $\mathsf{cl} = \mathsf{sr}(\mathcal{A}ct', \mathcal{K})$.
2. *Local Mitigation: Otherwise, it follows* $\mathsf{cl}(b) \geq \mathsf{cl}(a)$.

Principle II: Minimize classification *Given* $(\Xi, \preceq_{cred})$ *as parameter, let* sr' *be another function* $\mathsf{sr}' : 2^{\mathcal{A}ct}_{fin} \times \mathcal{E}s \rightarrow \mathcal{C}l$ *fulfilling Principles I.1 and I.2 Then, for all* $\mathcal{A}ct' \subset_{fin} \mathcal{A}ct$*, for all* $\mathcal{K} \in \mathcal{E}s$ *and for all* $a \in \mathcal{A}ct'$ *it holds* $\mathsf{cl}'(a) \geq \mathsf{cl}(a)$ *with* $\mathsf{cl}' = \mathsf{sr}'(\mathcal{A}ct', \mathcal{K})$ *and* $\mathsf{cl} = \mathsf{sr}(\mathcal{A}ct', \mathcal{K})$.

Beside the core Principles I.1, I.2 and II we define two supplemental principles that express interesting and desirable properties for a secrecy reasoner. However, the supplemental principles can be shown to be consequences of the core principles. Both principles consider the cautiousness of a secrecy reasoner in classifying an action with best compliance (classification rank 0) and this way for unrestricted use of $\mathcal{D}$.

Principle III (*Be cautious towards credulous reasoners*) follows the intuition that the more credulous $\mathcal{A}$ is postulated to reason about a target sentence, the easier that sentence might be inferred because $\mathcal{A}$ may accept more propositions as true and believe them. Thus, the more credulous the belief operator is postulated in a secrecy constraint, the more cautious agent $\mathcal{D}$ has to be while acting.

Principle IV (*Be more cautious the more uncertain*) bases on the general idea that being uncertain leads to cautious behavior. Here, the more world views could be held by $\mathcal{A}$ according to the postulated world views, the more uncertain $\mathcal{D}$ is about $\mathcal{A}$'s actual situation.

Principle III: Be cautious towards credulous reasoners
Let $\mathcal{K}$ and $\mathcal{K}'$ be epistemic states with equal components possibly except for the secrecy policies which are of the following form:

$$\mathcal{S}(\mathcal{K}) = \{(\phi_1, Bel_1), \dots (\phi_n, Bel_n)\}, \mathcal{S}(\mathcal{K}') = \{(\phi_1, Bel_1'), \dots (\phi_n, Bel_n')\}$$

with $Bel_i \succeq_{cred} Bel_i'$ for all $i \in \{1, \dots, n\}$.
If there exist actions $a, b \in \mathcal{A}ct'$ such that $\mathsf{vioAfter}(\mathcal{K}, a) = \{\emptyset\}$ and $\mathsf{vioAfter}(\mathcal{K}', b) = \{\emptyset\}$, then for the classifications $\mathsf{cl} = \mathsf{sr}(\mathcal{A}ct', \mathcal{K})$ and $\mathsf{cl}' = \mathsf{sr}(\mathcal{A}ct', \mathcal{K}')$ it holds that $\{a \in \mathcal{A}ct' \mid \mathsf{cl}(a) = 0\} \subseteq \{a \in \mathcal{A}ct' \mid \mathsf{cl}'(a) = 0\}$.

Principle IV: Be more cautious the more uncertain Let $\mathcal{K}$ and $\mathcal{K}'$ be epistemic states with equal components except for $\mathsf{V}^{PW}(\mathcal{K}) \supseteq \mathsf{V}^{PW}(\mathcal{K}')$.
If there exist actions $a, b \in \mathcal{A}ct'$ such that $\mathsf{vioAfter}(\mathcal{K}, a) = \{\emptyset\}$ and $\mathsf{vioAfter}(\mathcal{K}', b) = \{\emptyset\}$, then for the classifications $\mathsf{cl} = \mathsf{sr}(\mathcal{A}ct', \mathcal{K})$ and $\mathsf{cl}' = \mathsf{sr}(\mathcal{A}ct', \mathcal{K}')$ it holds that $\{a \in \mathcal{A}ct' \mid \mathsf{cl}(a) = 0\} \subseteq \{a \in \mathcal{A}ct' \mid \mathsf{cl}'(a) = 0\}$.

As we show in the appendix, the supplemental principles are consequences of the core principles.

Proposition 1. *Any function* $\mathsf{sr} : 2^{\mathcal{A}ct}_{fin} \times \mathcal{E}s \to \mathcal{C}l$ *that satisfies Principles I.1, I.2 and II also satisfies Principle III and Principle IV.*

4 Secrecy Reasoner: Constructive Design

We present an algorithm, given below in Procedure 1, which implements a secrecy reasoner as defined in Def. 6. In particular, it implements a function $\mathsf{sr} : 2^{\mathcal{A}ct}_{fin} \times \mathcal{E}s \to \mathcal{C}l$ being parameterized with a family $(\Xi, \preceq_{cred})$ of belief operators with credulity order that takes as input a finite subset $\mathcal{A}ct'$ of $\mathcal{A}ct$ and an epistemic state $\mathcal{K}$ and outputs a classification cl of $\mathcal{A}ct'$.

The main idea of the algorithm is to keep track of all not yet classified actions *unclass* while it iteratively assigns the currently considered (classification) rank to the actions of the input set of actions $\mathcal{A}ct'$. To this end, starting with a classification rank of 0, all actions for which there is no reason not to classify them with the current rank are assigned the current rank as their classification. Intuitively, a reason not to classify an action a with the current rank is given if more constraints are potentially violated in combination after a than after another unclassified action b (Principle I.1) or another unclassified action a' mitigates a potential violation of a without conflict (Principle I.2).

[1] We define $\max(\emptyset) := -1$ which is needed in the first iteration only.

Procedure 1. Secrecy Reasoner

Input: $\mathcal{Act}', \mathcal{K}, (\Xi, \preceq_{cred})$
Output: Array cl of classification ranks for actions $a \in \mathcal{Act}'$

```
unclass := Act'
for each a ∈ Act' do
   cl[a] := 0
end for
repeat
   best := {a ∈ unclass | there is no b ∈ unclass
                          such that vioAfter(K, a) ⊐ vioAfter(K, b)}
   eqbest := best/∼
   for each A ∈ eqbest do
      rank[A] := max¹{ cl[a] | a ∈ Act' \ unclass and
                          there is b ∈ A such that vioAfter(K, b) ⊐ vioAfter(K, a)} + 1
   end for
   for each A ∈ eqbest do
      conflictSets := conflictSets(A, K)
      repeat
         classSets := ∅
         for each CS ∈ conflictSets do
            if there is no CS' ∈ conflictSets with CS' ≠ CS such that a' ∈ CS'
            and a ∈ CS exist with a' mvio a then
               cl(a) := rank[A] for all a ∈ CS
               classSets := classSets ∪ {CS}
            end if
         end for
         conflictSets := conflictSets \ classSets
         rank[A] := rank[A] + 1
      until conflictSets = ∅
   end for
   unclass := unclass \ best
until unclass = ∅
```

In addition to the definitions already introduced, the algorithm makes use of auxiliary sets to test the preconditions of Principle I.2 defined in the following. For a given epistemic state $\mathcal{K}$ the algorithm determines the set of equivalence classes with respect to the equivalence relation $\sim$ of (4) for an inspected set of actions A, formally: $A/\sim$. Only pairs taken from such an equivalence class might satisfy the precondition of Principle I.2 by definition. Further, the algorithm has to be able to test *conflict free mitigation* of a pair of actions a, b with a mvio b. For this, all (maximal) conflict sets $\mathsf{conflictSets}(A, \mathcal{K})$ are computed for a given equivalence class A. Formally:

$$\mathsf{conflictSets}(A, \mathcal{K}) = \max_{\subseteq} \{A' \subseteq A \mid \text{for all } a, b \in A',\ a \neq b \text{ exist } a_1, \ldots, a_n \in A' \text{ such that } a_1 = a,\ a_n = b \text{ and for all } i \in \{1, \ldots, n-1\}\text{: } a_i \ \mathsf{mvio}\ a_{i+1}.\}$$

Note that, for any pair of actions $a, b \in \mathsf{conflictSets}(A, \mathcal{K})$ for some A and $\mathcal{K}$ with a mvio b the condition *conflict free mitigation* of Principle I.2 is violated.

Then, roughly speaking, the algorithm consists of two nested repeat-until loops. The outer one determines in each execution the set of not yet classified actions for which Principle I.1 does not give a reason not to classify them with the currently considered classification rank. The inner one determines in each execution the set of actions out of the selected actions from the outer loop for which Principle I.2 does not give a reason not to classify them with the currently considered classification rank and classifies them.

The outer loop first determines the subset of currently unclassified actions *best* the effects of which are not worse concerning combinations of potentially violated constraints than that of other unclassified actions (Principle I.1). Then, it constructs the auxiliary sets to check of the precondition for Principle I.2. In particular, it creates a partitioning *eqbest* of the set *best* consisting of the equivalence classes wrt. $\sim$. Then an array $rank[]$ is created which holds for each equivalence class its currently considered classification rank. Each $rank[A]$ is initialized in line 9 by the minimal classification rank for which Principle I.1 does not give a reason to classify any action of A higher than any action which is already classified. In the for-loop from line 11 to line 24 each current equivalence class A is first partitioned into its conflict sets.

The for-loop from line 15 to line 20 is intended to determine all actions of the current equivalence class for which no other action in the same equivalence class exists such that this pair satisfies the precondition of Principle I.2. This is done by comparing the conflict sets and either classifying all elements of the conflict set or none. This way all elements of a conflict set are classified with the same classification rank. The classified conflict sets of the currently considered equivalence class are stored in *classSets* and removed from the current set of conflict sets after the termination of the for-loop over all conflict sets in line 21.

If all actions in *best* are classified the condition in line 23 is true and they are removed from the set of unclassified actions in line 25. If all input actions are classified the condition in line 26 is true and the algorithm terminates.

Example 8. We consider the execution of Procedure 1 for our running example. Initially we have $unclass := \{a_1, a_2, a_3\}$ in line 1. As shown in Ex. 4 it holds that $\mathsf{vioAfter}(\mathcal{K}_D, a_1) = \mathsf{vioAfter}(\mathcal{K}_D, a_2)$ and $\mathsf{vioAfter}(\mathcal{K}_D, a_1) \sqsupset \mathsf{vioAfter}(\mathcal{K}_D, a_3)$ such that $best := \{a_3\}$ and therefore also $eqbest = \{\{a_3\}\}$. Since $\mathcal{A}ct' \setminus unclass = \emptyset$ it holds for all A that $rank[A] = 0$. In line 12 we get $conflictSets = \{\{a_3\}\}$. The conflict set $\{a_3\}$ trivially satisfies the condition in line 16 such that $cl(a_3) = 0$. Then $\{a_3\}$ is removed from $conflictSets$ such that $conflictsSets = \emptyset$. In line 22 $rank[\{a_3\}] := 1$, which does not have any effect in this special case, and the inner repeat-until loop terminates.

For the second execution of the outer repeat-until loop $unclass = \{a_1, a_2\}$. As shown in Ex. 4 $best := \{a_1, a_2\}$ and $eqbest = \{\{a_1, a_2\}\}$. In line 9 we get $rank[\{a_1, a_2\}] := 1$.

We already showed in Ex. 7 that it holds that a_2 mvio a_1 and that it does not hold that a_1 mvio a_2. Therefore $conflictSets = \{\{a_1\}, \{a_2\}\}$ in line 12. The condition in line 16 is satisfied for $\{a_2\}$ but not for $\{a_1\}$ such that $cl(a_2) := 1$ in line 17 and $conflictSets := \{\{a_1\}\}$ in line 21. Then $rank[\{a_1, a_2\}] := 2$.

The next execution of inner repeat-until loop begins. The only remaining conflict set $\{a_1\}$ trivially satisfies the condition of line 16 such that $cl(a_1) = 2$ and $conflictSets := \emptyset$ in line 21. The inner repeat-until loop terminates. In line 25 $unclass := \emptyset$ so that the outer repeat-until loop and thus the algorithm terminates. The output classification is $\mathsf{cl}(a_1) = 2, \mathsf{cl}(a_2) = 1, \mathsf{cl}(a_3) = 0$.

Proposition 2. *If all elementary operations of Procedure 1 are computable, the algorithm always terminates and returns a complete classification.*

Theorem 1. *Procedure 1 satisfies the principles of Def. 6.*

5 Related Work

In previous works, e.g. in [4,5], we focus on procedures for the defending agent $\mathcal{D}$ to control its reactions to the attacking agent $\mathcal{A}$'s update/revise-actions and query-actions to the end of effective preservation of secrecy (there: confidentiality). As an involved task, those procedures prevent implicit conveyance of information by $\mathcal{D}$'s reactions resulting from $\mathcal{A}$'s reasoning about the cause of these reactions. The effectiveness of control essentially bases on the postulate for security engineering that $\mathcal{D}$ is certain about $\mathcal{A}$'s epistemic state. The secrecy reasoning in this work is closely related to normative reasoning as "a norm defines principles of right action binding upon the members of a group and serving to guide, control, or regulate proper and acceptable behavior" [7]. In our case, a norm is defined by the semantics of the defending agent's secrecy policy as being defined by the declarative requirements on secrecy reasoning in Def. 6.

As one aspect, normative reasoning deals with how an agent's obligations may be derived from a norm, in particular, *contrary-to-duty (CTD) obligations* that "are in force just in case some other norm is violated" [7]. The idea is that a CTD obligation mitigates the effects of the violation of another obligation.
Similarly to our focused scenario, the works of [3,14] consider a scenario of an agent about to choose an action at a fixed point in time while being subject to obligations and CTD obligations. More specifically, Bartha in [3] advocates that a CTD obligation should be represented in the form $\mathsf{O}([\mathcal{X} : \phi] \Rightarrow [\mathcal{X} : \psi])$ and discusses how to derive further obligations from it. Here, the formula $[\mathcal{X} : \phi]$ means that agent $\mathcal{X}$'s choice of an action has the effect that ϕ (a propositional formula) becomes true in the world. Thus, the CTD obligation means if $\mathcal{X}$ chooses an action with a "bad" effect ϕ then it should mitigate that effect by ensuring that its action also has the effect ψ.[2]

In our work, with a similar idea, in Principle I.2 of Def. 6 the relation mvio compares actions with "equivalent" potential violations as defined by the relation $\sim$ to mitigate these violations. As suggested by [14], to express different

[2] A well-known example in the literature is that ϕ means "murder" and ψ means "gentle murder" to be found, e.g., in [14].

degrees of compliance due to potential violations and their mitigation we use an infinite range of classification ranks for actions. Other aspects of normative reasoning such as its relation to defeasible reasoning [15,17] and actions of policy change [7,1] for the defending agent are worthwhile lines for future research.

Closely related to our goal of secrecy reasoning, the work [1] specifically treats secrecy (there: privacy) in the context of normative reasoning. A sender can do inform-actions with a recipient towards whom the sender has obligations for privacy. These obligations are expressed in a policy language with modalities O for obligation and K for knowledge. The sender's aim of privacy is to prevent the recipient from knowing the truth of target sentences expressed in the modal language. The major differences to our work are as follows. The semantics of the policy is defined by Kripke structures with accessibility relations on possible worlds each for the interpretation of a modality. In particular, the semantic does not handle the sender's uncertainty about the recipient. Moreover, target sentences should only be protected towards the recipient's knowledge.

6 Conclusion and Issues Left Open

Having in mind our long-term goal of enriching an intelligent agent with sophisticated reasoning on secrecy constraints to support its decision making, in this work we focused on the fundamental task to define and implement semantics of secrecy constraints as a classification of possible inform-actions for a fixed point in time. In this context, a classification rank assigned to an inform-action expresses a degree of compliance with the constraints, in particular considering the uncertainty about an attacker's epistemic state and the possible desire to perform particular actions although some constraints might be violated. We presented a list of declaratively expressed principles for the semantics, designed an enforcement algorithm, and formally verified the satisfaction of the principles.

For the present work, we aimed at being as general as seen by us to be still meaningful, in order to cover a wide range of more concrete situations in the future. Moreover, we deliberately left open several issues for deeper and more refined studies. In the following we briefly discuss some of these issues. (1) An intelligent agent performs an ongoing loop of observation, reasoning with decision making and action; thus semantics of secrecy constraints should deal with sequences of actions rather than a single action for a fixed point in time. (2) An intelligent agent may dispose of a rich collection of possible actions; thus semantics of secrecy constraints should cover all possible actions rather than only inform-actions, essentially by identifying the information explicitly or implicitly conveyed by each of the actions. (3) Considering sequences of diverse actions, an action might be a reaction in form of an answer or a notification on a previously perceived request from another agent; thus semantics of secrecy constraints should seriously consider meta-inferences by the other agent based on the knowledge of the range of the functionally expected reactions. (4) While in this work a defending agent maintains its own world view and world belief, it does not consider the actual state of the world; thus we could explore the impact of the world on the secrecy reasoning. (5) A multi-agent system might be formed by several and diverse agents; thus all considerations should be extended to multi-side communications, then facing many problems known from other contexts to be highly challenging, like transitive information flows or hidden side channels. (6) The security/knowledge engineer of an agent must initialize the postulates about other agents appropriately.

Although each issue is intricate, related works suggest directions how to deal with issues (3) and (6) as examples for future research. Addressing (3), meta-inference by $\mathcal{A}$ is captured by other notions of secrecy, e.g., in [10,4]. These notions base on postulates for security engineering referring to $\mathcal{A}$'s reasoning about $\mathcal{D}$'s internal functionality, contrasting with the postulates in this article referring to $\mathcal{A}$'s reasoning about the world. In future research, our approach should be complemented by postulates of the former kind, related notions of secrecy and appropriate mechanisms of enforcement. It remains a challenging question whether the strategies of enforcement followed in those works can be applied to the proposed secrecy reasoning. Such a strategy in the work of controlled interaction execution, e.g., in [4,5,6], is to represent every information conveyed to $\mathcal{A}$ in $\mathcal{D}$'s view of $\mathcal{A}$ (here: the postulated world views) and its closure under an associated entailment relation, e.g., propositional entailment (here: a belief operator in a respective secrecy constraint). More precisely, the strategy is to prevent that $\mathcal{A}$ acquires any other information than that represented in $\mathcal{D}$'s view by $\mathcal{A}$'s capabilities postulated for security engineering. Then, a formal proof is provided that secrecy is enforced by this strategy. A major challenge is to represent meta-information such as $\mathcal{D}$'s reasoning class [5] or information conveyed by $\mathcal{D}$'s refusal reactions [6]. Addressing (6), within her initialization task, the knowledge engineer has to estimate the other agent's knowledge about the world and to choose a belief operator for each secrecy constraint. The estimation might benefit from insights in adversarial reasoning [13] while the choice of an operator depends on several factors such as the treatment of meta-inference and the definition of the credulity order. Lastly, concrete situations arise from the requirements of particular applications such as the integration of intelligent agents into business processes [8] or e-commerce [11].

In conclusion, this work presents an extendible model for the secrecy reasoning of an intelligent agent and exemplarily studies semantics of secrecy constraints for a fundamental task. As indicated above, our contributions can be seen as a first step of a challenging project to be performed in the future.

References

1. Aucher, G., Boella, G., van der Torre, L.: A dynamic logic for privacy compliance. Artificial Intelligence and Law 19(2), 187–231 (2011)
2. Bacchus, F., Grove, A.J., Halpern, J.Y., Koller, D.: From statistical knowledge bases to degrees of belief. In: CoRR, cs.AI/0307056 (2003)
3. Bartha, P.: Conditional obligation, deontic paradoxes, and the logic of agency. Annals of Mathematics and Artificial Intelligence 9(1-2), 1–23 (1993)

4. Biskup, J.: Inference-usability confinement by maintaining inference-proof views of an information system. International Journal of Computational Science and Engineering 7(1), 17–37 (2012)
5. Biskup, J., Tadros, C.: Preserving confidentiality while reacting on iterated queries and belief revisions. In: Annals of Mathematics and Artificial Intelligence (2013), doi:10.1007/s10472-013-9374-6
6. Biskup, J., Weibert, T.: Keeping secrets in incomplete databases. International Journal of Information Security 7(3), 199–217 (2008)
7. Broersen, J., van der Torre, L.: Ten problems of deontic logic and normative reasoning in computer science. In: Bezhanishvili, N., Goranko, V. (eds.) ESSLLI 2010 and ESSLLI 2011. LNCS, vol. 7388, pp. 55–88. Springer, Heidelberg (2012)
8. Burmeister, B., Arnold, M., Copaciu, F., Rimassa, G.: BDI-agents for agile goal-oriented business processes. In: Berger, M., Burg, B., Nishiyama, S. (eds.) AAMAS (Industry Track), pp. 37–44. International Foundation for Autonomous Agents and Multiagent Systems, Richland, Richland (2008)
9. Gelfond, M., Leone, N.: Logic programming and knowledge representation: the A-Prolog perspective. Artificial Intelligence 138 (2002)
10. Halpern, J.Y., O'Neill, K.R.: Secrecy in multiagent systems. ACM Transactions on Information and System Security 12(1) (2008)
11. He, M., Jennings, N.R., Leung, H.-F.: On agent-mediated electronic commerce. IEEE Transactions on Knowledge and Data Engineering 15(4), 985–1003 (2003)
12. Kern-Isberner, G. (ed.): Conditionals in Nonmonotonic Reasoning and Belief Revision - Considering Conditionals as Agents. LNCS, vol. 2087. Springer, Heidelberg (2001)
13. Kott, A., McEneaney, W.M.: Adversarial Reasoning: Computational Approaches to Reading the Opponent's Mind. Chapman & Hall/CRC, Boca Raton (2007)
14. Kuijer, L.B.: Sanction semantics and contrary-to-duty obligations. In: Ågotnes, T., Broersen, J., Elgesem, D. (eds.) DEON 2012. LNCS, vol. 7393, pp. 76–90. Springer, Heidelberg (2012)
15. Makinson, D.: Five faces of minimality. Studia Logica 52(3), 339–379 (1993)
16. Makinson, D.: General patterns in nonmonotonic reasoning. In: Handbook of Logic in Artificial Intelligence And Logic Programming, Vol. III. Clarendon Press, Oxford (1994)
17. van der Torre, L., Tan, Y.-H.: The many faces of defeasibility in defeasible deontic logic. In: Nute, D. (ed.) Defeasible Deontic Logic. Synthese Library, vol. 263, pp. 79–121. Springer, Netherlands (1997)
18. Wooldridge, M.: Intelligent agents. In: Weiss, G. (ed.) Multiagent Systems: A Modern Approach to Distributed Artificial Intelligence. MIT Press, Cambridge (1999)

A Appendix: Selected Proofs

Proof (of Proposition 1). Assume a secrecy reasoner sr which satisfies Principle I.1, Principle I.2 and Principle II. Let $\mathcal{Act}'$ be a finite set of actions and $\mathcal{K}$ an epistemic state. Let $\mathsf{cl} = \mathsf{sr}(\mathcal{Act}', \mathcal{K})$ be the classification determined by sr.

Claim 1. We first show that if there exists $b \in \mathcal{Act}'$ with $\mathsf{vioAfter}(\mathcal{K}, b) = \{\emptyset\}$, then for all actions $a \in \mathcal{Act}'$ it holds $\mathsf{cl}(a) = 0$ iff $\mathsf{vioAfter}(\mathcal{K}, a) = \{\emptyset\}$.

If $\mathsf{vioAfter}(\mathcal{K}, a) = \{\emptyset\}$, then $\mathsf{cl}(a) = 0$:
Let $a \in \mathcal{Act}'$ be an action with $\mathsf{vioAfter}(\mathcal{K}, a) = \{\emptyset\}$.

1. By the definitions of $\mathsf{vioAfter}$ and $\sqsupset$, there cannot exist an action $a' \in \mathcal{Act}'$ such that $\mathsf{vioAfter}(\mathcal{K}, a') \sqsubset \mathsf{vioAfter}(\mathcal{K}, a)$ holds.
2. Hence, for no $a' \in \mathcal{Act}'$ the precondition of Principle I.1 is satisfied. Therefore, there does not exist an action $a' \in \mathcal{Act}'$ such that $\mathsf{cl}(a') < \mathsf{cl}(a)$ is demanded by this principle.
3. There are no constraint $(\phi, Bel) \in \mathcal{S}(\mathcal{K})$ and no belief operator $Bel' \in \Xi$ with $Bel' \prec_{cred} Bel$ such that the constraint (ϕ, Bel') is potentially violated after action a.
   The reason is that, if there was a potential violation of (ϕ, Bel') after a, then by definition there exists $W \in \mathsf{V}^{PW}(\mathcal{K}) \oplus a$ such that $\phi \in Bel'(W)$. Since $Bel' \prec_{cred} Bel$ holds, the credulity property of the order $\preceq_{cred}$ implies that $Bel(W) \supseteq Bel'(W) \ni \phi$. Hence, it holds $(\phi, Bel) \in \mathsf{vio}(\mathcal{K}, W)$ and $\mathsf{vio}(\mathcal{K}, W) \in \mathsf{vioAfter}(\mathcal{K}, a)$ by Def. 4. This cannot happen since $\mathsf{vioAfter}(\mathcal{K}, a) = \{\emptyset\}$ holds.
4. Therefore, for each action $a' \in \mathcal{Act}'$ by Def. 5 the relation a' mvio a does not hold. Hence, for no $a' \in \mathcal{Act}'$ $\mathsf{cl}(a') < \mathsf{cl}(a)$ is demanded by Principle I.2.
5. By Points 2 and 4, there exists a classification cl' with $\mathsf{cl}'(a) = 0$ which satisfies Principles I.1 and I.2.
6. From the assumed satisfaction of Principle II by the secrecy reasoner sr, it indeed outputs a classification cl such that $\mathsf{cl}(a) = 0$.

If $\mathsf{cl}(a) = 0$, then $\mathsf{vioAfter}(\mathcal{K}, a) = \{\emptyset\}$:
We show the implication by contraposition. Let $a \in \mathcal{Act}'$ be an action such that $\mathsf{vio}(a, \mathcal{K}) \neq \{\emptyset\}$ holds.

1. By presupposition of Claim 1 there exists $b \in \mathcal{Act}'$ with $\mathsf{vioAfter}(\mathcal{K}, b) = \{\emptyset\}$.
2. Hence, it holds $\mathsf{vioAfter}(\mathcal{K}, a) \sqsubset \mathsf{vioAfter}(\mathcal{K}, b)$ by definition..
3. By Principle I.1, it follows that $\mathsf{cl}(a) > \mathsf{cl}(b) \geq 0$.

Satisfaction of Principle IV
Let $\mathcal{K}$ and $\mathcal{K}'$ be epistemic states with equal components possibly except for $\mathsf{V}^{PW}(\mathcal{K}) \supseteq \mathsf{V}^{PW}(\mathcal{K}')$. Assume there exist actions $a, b \in \mathcal{Act}'$ such that $\mathsf{vioAfter}(\mathcal{K}, a) = \{\emptyset\}$ and $\mathsf{vioAfter}(\mathcal{K}', b) = \{\emptyset\}$.

We show that for each $c \in \mathcal{Act}'$ if $\mathsf{cl}(c) = 0$ holds then $\mathsf{cl}'(c) = 0$ holds with $\mathsf{cl} = \mathsf{sr}(\mathcal{Act}', \mathcal{K})$ and $\mathsf{cl}' = \mathsf{sr}(\mathcal{Act}', \mathcal{K}')$. Let c be an action $c \in \mathcal{Act}'$ such that $\mathsf{cl}(c) = 0$.

1. Since $a \in \mathcal{Act}'$ satisfies $\mathsf{vioAfter}(\mathcal{K}, a) = \{\emptyset\}$ it holds that $\mathsf{vioAfter}(\mathcal{K}, c) = \{\emptyset\}$ by Claim 1.
2. From the definition of $\mathsf{vioAfter}$ it follows directly that for $\mathsf{V}^{PW}(\mathcal{K}) \supseteq \mathsf{V}^{PW}(\mathcal{K}')$ it holds that

$$\mathsf{vioAfter}(\mathcal{S}(\mathcal{K}), \mathsf{V}^{PW}(\mathcal{K}), c) \supseteq \mathsf{vioAfter}(\mathcal{S}(\mathcal{K}), \mathsf{V}^{PW}(\mathcal{K}'), c) = \mathsf{vioAfter}(\mathcal{S}(\mathcal{K}'), \mathsf{V}^{PW}(\mathcal{K}'), c).$$

3. It follows that $\mathsf{vioAfter}(\mathcal{S}, \mathsf{V}^{PW}(\mathcal{K}'), c) = \{\emptyset\}$.
4. Since action $b \in \mathcal{Act}'$ satisfies $\mathsf{vioAfter}(\mathcal{K}', b) = \{\emptyset\}$ it follows that $\mathsf{cl}'(c) = 0$ by Claim 1.

Satisfaction of Principle III
Let $\mathcal{K}$ and $\mathcal{K}'$ be epistemic states with equal components possibly except for the secrecy policies which are of the following form:

$$\begin{aligned} \mathcal{S}(\mathcal{K}) &= \{(\phi_1, Bel_1), \dots (\phi_n, Bel_n)\}, \\ \mathcal{S}(\mathcal{K}') &= \{(\phi_1, Bel_1{}'), \dots (\phi_n, Bel_n{}')\} \\ \text{with} \quad & Bel_i \succeq_{cred} Bel_i{}' \text{ for all } i \in \{1, \dots, n\}. \end{aligned} \tag{5}$$

Assume there exist actions $a, b \in \mathcal{A}ct'$ such that $\mathsf{vioAfter}(\mathcal{K}, a) = \{\emptyset\}$ and $\mathsf{vioAfter}(\mathcal{K}', b) = \{\emptyset\}$.
We show that for all actions $c \in \mathcal{A}ct'$, if $\mathsf{cl}(c) = 0$ holds, then $\mathsf{cl}'(c) = 0$ holds with $\mathsf{cl} = \mathsf{sr}(\mathcal{A}ct', \mathcal{K})$ and $\mathsf{cl}' = \mathsf{sr}(\mathcal{A}ct', \mathcal{K}')$.
Let c be an action $c \in \mathcal{A}ct'$ such that $\mathsf{cl}(c) = 0$.

1. Since by assumption, action $a \in \mathcal{A}ct'$ satisfies $\mathsf{vioAfter}(\mathcal{K}, a) = \{\emptyset\}$ it holds that $\mathsf{vioAfter}(\mathcal{K}, c) = \{\emptyset\}$ by Claim 1.
2. From the definition of vioAfter it follows directly that for all $W \in \oplus(\mathsf{V}^{PW}(\mathcal{K}), c)$ and for all $(\phi, Bel) \in \mathcal{S}(\mathcal{K})$ it holds that $\phi \notin Bel(W)$.
3. By presupposition, the secrecy policies are of the form in (5) so that for all $W \in 2^{\mathcal{L}_\mathcal{B}} \times \mathcal{L}^*$ it holds $Bel_i'(W) \subseteq Bel_i(W)$ by the credulity property of the order $\preceq_{cred}$. Hence, it follows that for all $W \in \oplus(\mathsf{V}^{PW}(\mathcal{K}), c)$ and for all $(\phi, Bel') \in \mathcal{S}(\mathcal{K}')$ it holds that $\phi \notin Bel'(W)$ by Point 2.
4. It follows that

$$\{\emptyset\} = \mathsf{vioAfter}(\mathcal{S}(\mathcal{K}'), \mathsf{V}^{PW}(\mathcal{K}), c) = \mathsf{vioAfter}(\mathcal{S}(\mathcal{K}'), \mathsf{V}^{PW}(\mathcal{K}'), c).$$

5. Since by assumption, action $b \in \mathcal{A}ct'$ satisfies $\mathsf{vioAfter}(\mathcal{K}', b) = \{\emptyset\}$ it follows that $\mathsf{cl}'(c) = 0$ by Claim 1. □

Proof (of Theorem 1, satisfaction of Principle II: Minimize classification). Given $(\Xi, \preceq_{cred})$ as parameter, let cl be the classification function as defined by Procedure 1. Further, let sr′ be another function $\mathsf{sr}' : 2^{\mathcal{A}ct}_{fin} \times \mathcal{E}s \to \mathcal{C}l$ fulfilling Principles I.1 and I.2. We prove that, for all $\mathcal{A}ct' \subseteq \mathcal{A}ct$, for all $\mathcal{K} \in \mathcal{E}s$ and for all $a \in \mathcal{A}ct'$ it holds $\mathsf{cl}'(a) \geq \mathsf{cl}(a)$ with $\mathsf{cl}' = \mathsf{sr}'(\mathcal{A}ct', \mathcal{K})$.

We proceed by induction on the classification rank r in the range of cl. Thus, we consider the following induction hypothesis:
For all $i \leq r-1$ it holds for all actions $a \in \mathcal{A}ct'$ with $\mathsf{cl}(a) = i$ that $\mathsf{cl}(a) \leq \mathsf{cl}'(a)$.
Let $a \in \mathcal{A}ct'$ be an action with $\mathsf{cl}(a) = r$.

- Base case: $r = 0$.
  By definition no lower classification rank is possible. Thus, it follows $\mathsf{cl}(a) \leq \mathsf{cl}'(a)$.
- Inductive case: $r > 0$.
  1. Assume indirectly that $\mathsf{cl}'(a) < r$ holds.
  2. Consider the classification of a by Procedure 1. Action a is treated in one and only one iteration of the repeat-until loop in line 5. Thus, for later reference, let A denote the equivalence class with $a \in A$ and $CS_a \subseteq A$

the conflict set of a as determined by the algorithm in this iteration. We distinguish two cases how the algorithm computes the value of the classification rank of action a.
Case 1: The value is set in line 9. Then, there exist $b \in \mathcal{A}ct'$ and $c \in A$ such that

$$\mathsf{vioAfter}(\mathcal{K}, c) \sqsupset \mathsf{vioAfter}(\mathcal{K}, b) \text{ and } \mathsf{cl}(b) = r - 1. \tag{6}$$

Since actions a, c are in the same equivalence class A, it follows that $\mathsf{vioAfter}(\mathcal{K}, a) \sim \mathsf{vioAfter}(\mathcal{K}, c)$. Using elementary arguments on the definitions of $\sqsupset$ and $\sim$, we can show that together with (6) the latter implies $\mathsf{vioAfter}(\mathcal{K}, a) \sqsupset \mathsf{vioAfter}(\mathcal{K}, b)$. Due to the relation $\sqsupset$ and Principle I.1, it holds

$$\mathsf{cl}'(a) > \mathsf{cl}'(b). \tag{7}$$

Now, we apply the induction hypothesis on $\mathsf{cl}(b)$, since $\mathsf{cl}(b) = r - 1$ holds by (6) and obtain that $\mathsf{cl}'(b) \geq \mathsf{cl}(b)$. By assumption, it holds $r > \mathsf{cl}'(a)$ and thus $\mathsf{cl}'(b) \geq \mathsf{cl}(b) = r - 1 \geq \mathsf{cl}'(a)$. This contradicts that $\mathsf{cl}'(a) > \mathsf{cl}'(b)$ must hold by (7).
Case 2: If Case 1 does not apply, the value for the classification of a in line 17 is set in line 22. Thus, there exists $b \in A$ with $\mathsf{cl}(b) = r - 1$ and $b \in CS_b \subset A$ with $CS_b \neq CS_a$ such that there exists $c \in CS_a$ with b mvio c. More precisely, in the previous iteration of the for-loop in line 15 the conflict set CS_b must still be in $conflictSets$ as one reason why a has been not classified with $r - 1$, but then CS_b is removed from $conflictSets$ in line 21.
Since b, c are in the same equivalence class A, it holds $\mathsf{vioAfter}(\mathcal{K}, c) \sim \mathsf{vioAfter}(\mathcal{K}, b)$. Further, since $c \in CS_a$ and $b \in CS_b$ and $CS_a \neq CS_b$, by definition of $\mathsf{conflictSets}$, there do not exist actions $a_1, \ldots, a_n \in \mathcal{A}ct'$ such that $a_1 = c$, $a_n = b$ and a_i mvio a_{i+1} for all $i \in \{1, \ldots, n - 1\}$. Hence, the premises of Principle I.2 are satisfied which implies

$$\mathsf{cl}'(c) > \mathsf{cl}'(b). \tag{8}$$

Next, we argue that $a, c \in CS_a$ implies $\mathsf{cl}'(a) = \mathsf{cl}'(c)$. This follows from an inductive argument using the local mitigation requirement of Principle I.2 and the definition of CS_a. By (8), it holds $\mathsf{cl}'(a) > \mathsf{cl}'(b)$. Now, we apply the induction hypothesis on $\mathsf{cl}(b) = r - 1$ so that it follows $\mathsf{cl}'(b) \geq \mathsf{cl}(b)$. By assumption, it holds $r > \mathsf{cl}'(a)$ and thus $\mathsf{cl}'(b) \geq \mathsf{cl}(b) = r - 1 \geq \mathsf{cl}'(a)$. This contradicts that $\mathsf{cl}'(a) > \mathsf{cl}'(b)$ must hold. □

An AIF-Based Labeled Argumentation Framework

Maximiliano C.D. Budán[1,2,3], Mauro J. Gómez Lucero[2], and Guillermo Ricardo Simari[2]

[1] Consejo Nacional de Investigaciones Científicas y Técnicas (CONICET)
[2] Laboratorio de Investigación y Desarrollo en Inteligencia Artificial (LIDIA) - Universidad Nacional del Sur
[3] Departamento de Matemática - Universidad Nacional de Santiago del Estero
{mcdb,mjg,grs}@cs.uns.edu.ar

Abstract. Adding meta-level information to the arguments in the form of labels extends the representational capabilities of an argumentation formalism, expanding its capabilities. Labels allow the representation of different features: skill, reliability, strength, time availability, or any other that might be related to arguments; then, this information can be used to determine the strength of an argument and assist in the process of determining argument's acceptability.

We have developed a framework called Labeled Argumentation Framework (LAF) based in the Argument Interchange Format (AIF), integrating the handling of labels; thus, labels associated with arguments will be combined and propagated according to argument interactions, such as support, conflict, and aggregation. Through this process, we will establish argument acceptability, where the final labels propagated to the acceptable arguments provide additional acceptability information, such as degree of justification, or explanation, among others.

1 Introduction

Argumentation is a human-like reasoning process that follows a commonsense strategy to obtain support for claims; that is, the formalization of this process mimics how humans decide what to believe, particularly in the context of disagreement. In a general sense, argumentation can be associated with the interaction of arguments for and against claims or conclusions, with the ultimate purpose of determining which conclusions are acceptable in the context of a given knowledge base. The argumentation theories are applied in many areas such as legal reasoning [2], intelligent web search [11,10], recommender systems [11,6], autonomous agents and multi-agent systems [24,35], and many others [4,5,29,33].

The aim of this work is to introduce the consideration of meta-level information in the argumentative reasoning. The meta-information will take the form of labels attached to arguments to enhance the representational capabilities of the framework and increase the ability of modeling real-world situations. Furthermore, a reason for this extension is that, besides the all-important property

C. Beierle and C. Meghini (Eds.): FoIKS 2014, LNCS 8367, pp. 117–135, 2014.

of acceptability of an argument, there exist other features to take into account, for instance, the strength associated with an argument [3], reliability varying on time [8], possibilistic [13], among others. Labels can be defined to handle a set of features describing the distinguishing characteristics of an argument and then the interaction between arguments (such as support, aggregation, and conflict) can affect these labels appropriately. The introduction of labels will allow the representation of uncertainty, reliability, possibilistic, strength, or any other relevant feature of the arguments, providing a useful way of improving and refining the process of determining argument acceptability. The definition of an acceptability threshold is among the potential applications for the labels. This feature will help in determining if a given argument is strong enough to be accepted; moreover, labels will also help in the specification and manipulation of preferences associated with the arguments for determining which is more important in a specific domain [32,34,1].

In this paper we present a framework called *Labeled Argumentation Framework* (*LAF*) which combine the knowledge representation capabilities provided by the *Argument Interchange Format* (*AIF*) [14], and the treatment and management of labels proposed by an algebra of argumentation labels developed for this purpose. The labels associated with arguments will be combined and propagated according to argument interactions, such as support, aggregation, and conflict through the operations defined in an algebra associated with each of these interactions; in particular, aggregation has been studied in the form of argument accrual [27,36,20]. Through this process, we will establish argument acceptability, where the final labels propagated to the accepted arguments provide additional acceptability information, such as degree of justification, restrictions on justification, explanation, etc.

To facilitate the development of the paper, in Section 2 we will present an example to motivate and illustrate the objectives of our work. Then, in Section 3 we will introduce a formalism, an *Algebra of Argumentation Labels*, a particular abstract algebra for handling labels associated with the arguments. In Section 4, we will give a brief introduction to the *Argument Interchange Format* (*AIF*) containing the elements we need for our development. The main contribution of the paper is presented in Section 5 as the formalism *Labeled Argumentation Framework* (*LAF*) together with an example of application. Finally, in Section 6 we will discuss the related work associated with the central issue in this paper, and in Section 7 we will offer conclusions and propose future development of the ideas presented.

2 An Initial Example

In this work we aim to contribute to the successful integration of argumentation to different artificial intelligence applications, such as intelligent web search, knowledge management, natural language processing, among others. In this section we will introduce an example to motivate the usefulness of our formalization in the particular context of recommendation systems; more specifically, we choose a *movie recommender system* as an example where the formalism could help achieve a better performance.

Lets assume we want to develop a movie recommender system and to make it available on the web, *i.e.*, the system will give advice to users about possible movies to watch. Its behavior will be based on the particular preferences expressed by them on previously recommended movies, and it will integrate these preferences with previous feedback provided by other users of the system covering different aspects of the movies. Thus, the system should have a representation of user preferences and of her evaluations, together with a consideration of the influence of each aspect of a movie that receives an evaluation on the final recommendation for the movie. Our reasoning mechanism will be based on argumentation, and the recommendation for the movie will be obtained through a dispute over which recommendation to give that will decide the matter. For instance, to determine wether to recommend the film *"Oz: The Great And The Powerful"* to the user Brian, the system will consider the following arguments:

- A *Recommend the film, because the genre is adventure and Brian likes adventure films.*
- B *Recommend the film, because the film has a good rating.*
- C *Recommend the film, because the actors of the film are excellent, and they are the right actors for the roles they play in the film.*
- D *Do not recommend the film, because the script is bad.*
- E *The script of the film is bad, because it is not faithful to L. Frank Baum's original story.*
- F *The script of the film is good, because the story line is interesting.*
- G *Do not recommend the film, because the movie's soundtrack is poor.*

This example illustrates how the knowledge used to make recommendations can be naturally expressed as arguments, involving interactions such as support between the arguments (*e.g.*, E and D), aggregation in the form of different arguments for the same conclusion (*e.g.*, A and B), and conflicts represented as contradictory conclusions (*e.g.*, A and D).

Each aspect of a movie could be evaluated by the users, and could possibly influence the final recommendation with different strength, depending on the preferences of the particular user. This illustrates the need of representing meta-information associated with arguments, in this particular case a measure of strength modeling this additional knowledge.

In many applications of argumentation, particularly in recommender systems, it's more natural to analyze together arguments with the same conclusion than to appraise them individually. That is known as aggregation (or accrual), and it is based on the intuition that having more reasons (in the form of arguments) for a given conclusion makes such a conclusion more credible [27,36,20]. In the example, the three arguments A, B, and C, concluding that the movie should be recommended must be considered – and weighted – together against the arguments supporting the conclusion against to recommend it (D and G). Another reason to consider some form of meta-information that is related to the different recommendations that could be given is that, in general, several movies could be recommended, and the meta-information helps in sorting out which one should be selected.

Finally, as is common in argumentation theories, if two arguments X and Y are in conflict, and X is stronger than Y, then Y becomes defeated by X, or its attack is just disregarded, but X remains unaffected by Y attack. Some application domains need a more complex treatment of conflict evaluation, capturing the notion that X is in some way affected by the conflict brought about by the attack of Y. For instance, the strength of the recommendation assigned to a movie should not be the same in the case that is free of counter-arguments compared with the case when it is controversial. That is, we need to model the weakening of an undefeated argument reflecting the effect produced by its existing counter-arguments.

We will propose a general framework allowing the representation through labels of meta-information associated with arguments, providing the capability of defining acceptability by combining and propagating labels according to support, aggregation, and conflict interactions. We will instantiate the proposed formalization to model the recommendation example presented in this section.

3 Abstract Algebra and Labels

In formal sciences, the process of abstraction is used to focus our interest on what is relevant for a particular purpose; thus, by abstracting away details we obtain conceptual generality. Mathematicians have created theories of various structures that apply to many objects. For instance, mathematical systems based on sets benefit from the results obtained in set theory, *i.e.*, sets equipped with a single binary operation form a magma (also known as groupoid), results obtained studying this structure can be applied to any particular set in which a binary operation is defined, and all the set theoretic properties also apply to them.

Abstract algebra, evolving from earlier forms of arithmetic, reached its potential by this process of successive abstraction which has allowed to obtain systems increasingly more complex without loosing the mathematical purity and inherent beauty. In this sub-area of Mathematics algebraic structures such as groups, rings, fields, modules, vector spaces, and algebras are studied. The axiomatic nature of abstract algebra deals with systems which are based on sets whose elements are of unspecified type, together with certain operations that satisfy a prescribed lists of axioms. Next, we propose the introduction of an algebraic structure, called algebra of argumentation labels.

3.1 Algebra of Argumentation Labels

Here, we will present an algebrization of the representation of meta-level information through labels attached to arguments. This algebra will consist of a set of labels equipped with a collection of operators to be used to combine and propagate these labels according to the interactions of arguments of support, conflict and aggregation. Below, we formalize our Algebra of Argumentation Labels and show a particular instantiation of this algebra allowing to model strength measure of arguments.

Definition 1 (Algebra of Argumentation Labels). *An Algebra of Argumentation Labels is described as a 6-tuple* $\mathcal{A} = \langle \mathbb{E}, \uplus, \oplus, \ominus, \varnothing, \epsilon \rangle$ *where:*

- $\mathbb{E}$ *is a set of labels called* domain of argumentation labels,
- $\uplus : \mathbb{E} \times \mathbb{E} \rightarrow \mathbb{E}$, *called the* argumentation support operator,
- $\oplus : \mathbb{E} \times \mathbb{E} \rightarrow \mathbb{E}$, *called the* argumentation aggregation operator,
- $\ominus : \mathbb{E} \times \mathbb{E} \rightarrow \mathbb{E}$, *called the* argumentation conflict operator,
- *the support operator* $\uplus$ *and the aggregation operator* $\oplus$, *are commutative and associative, that is, if* $\alpha, \beta, \gamma \in \mathbb{E}$ *then:*
  $\alpha \uplus \beta = \beta \uplus \alpha$, *and* $\alpha \uplus (\beta \uplus \gamma) = (\alpha \uplus \beta) \uplus \gamma$,
  $\alpha \oplus \beta = \beta \oplus \alpha$, *and* $\alpha \oplus (\beta \oplus \gamma) = (\alpha \oplus \beta) \oplus \gamma$,
- $\varnothing$ *is the neutral (or identity) element for the argumentation support operator* $\uplus$, *that is, for all* $\alpha \in \mathbb{E}$ *then:*
  $\alpha \uplus \varnothing = \alpha$,
- ϵ *is the neutral (or identity) element for the argumentation aggregation operator* $\oplus$ *and for the conflict operator* $\ominus$, *that is, for all* $\alpha \in \mathbb{E}$ *then:*
  $\alpha \oplus \epsilon = \alpha$,
  $\alpha \ominus \epsilon = \alpha$.

We will drop the "argumentation" word when referring to the operators when no confusion can happen.

Note that $\mathbb{E}$, the carrier set of the algebra, is a set of labels associated with arguments. The *support* operator will be used to obtain the label associated with the conclusion of an inference from the labels associated with the premises. The *aggregation* operator will be used to obtain the label representing the collective strengthening of the reasons supporting the same conclusion, reflecting that a conclusion is more credible by having several reasons behind it. Finally, the *conflict* operator defines the label corresponding to a conclusion after considering the effect of its conflicts with other claims. Also note that the operators remain undefined for now, we will introduce the corresponding definitions later.

We will show how the labels allow to represent uncertainty, reliability, time availability, or any other feature concerning arguments, and provide assistance, by taking this information into account, in the process of determining argument acceptability. We will now present a core ontology for the *Argument Interchange Format* (AIF); then, based on this ontology, we will then develop the *Labeled Argumentation Framework* (LAF).

4 Argument Interchange Format (AIF)

The *Argument Interchange Format* (AIF) is a proposal for an abstract model for the representation and exchange of data between various argumentation tools and agent-based applications (full details can be found in [14]). The AIF core ontology models a set of argument-related concepts, which can be extended to capture a variety of argumentation formalisms and schemes, under the assumtion that argument entities can be represented as nodes in a directed graph described as an *argument network*. A particular node can also have a number of internal attributes, denoting things such as the author, textual details, certainty degree, acceptability status, etc.

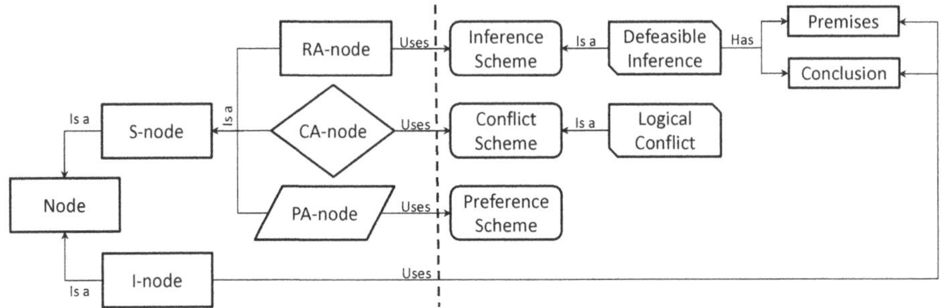

Fig. 1. AIF Core Ontology

In AIF two types of nodes are defined: *information* nodes (I-nodes) and *scheme* nodes (S-nodes), depicted with boxes and cans respectively in Figure 1. Information nodes are used to represent passive declarative information contained in an argument, such as a claim, premise, data, etc.; scheme nodes capture the application of schemes (*i.e.*, patterns of reasoning). Such schemes could represent domain-independent patterns of reasoning, with the appearance of deductive rules of inference, but also may represent non-deductive inference rules. The type of schemes that will be possible to represent will also include other relations, and the full set can be classified further into: *rule of inference* schemes, *conflict* schemes, and *preference* schemes. They respectively yield to three types of S-nodes: *rule application* nodes (RA-nodes), which denote applications of an inference rule or scheme; *conflict application* nodes (CA-nodes), which denote a specific conflict; and *preference application* nodes (PA-nodes), which denote specific preferences.

Particular nodes have different attributes such as title, creator, type (*e.g.*, decision, action, goal, belief), creation date, evaluation (or strength, or conditional evaluation table), acceptability, and polarity (*e.g.*, values such as *pro* or *con*); since these attributes correspond to specific applications will not be part of the core ontology. Some attributes correspond to the node itself, but others are derived. The latter type, derived attributes, of which acceptability is an example, may be computed from node-specic attributes. Nodes are used to build an AIF argument network, which is defined as follows.

Definition 2 (Argument Network [30]). *An AIF argument network is a digraph* $G = (N, E)$*, where:*

- $N = I \cup RA \cup CA \cup PA$ *is the set of nodes in* G*, where* I *is a set of I-Nodes,* RA *is a set of RA-Nodes,* CA *is a set of CA-Nodes, and* PA *is a set of PA-Nodes; and*
- $E \subseteq (N \times N) \setminus (I \times I)$*, is the set of the edges in* G*.*

Given an argument network representing argument-based concepts and relations, a node A is said to *support* another node B if and only if there is an edge running from A to B. The edges of the argument network are not necessarily marked, labelled, or be attached with semantic pointers. There are two types of edges: *Scheme edges* that start in S-nodes and they support conclusions following from the S-node, these conclusions may either be I-nodes or S-nodes; and, *Data edges* coming out of I-nodes and they must end in S-nodes, these edges supply data, or information, to scheme applications. Thus, there exist I-to-S edges (information or data supplying edges), S-to-I edges (conclusion edges), and S-to-S edges (warrant edges). Figure 2 presents a summary of the relations associated with the semantics of support (as proposed in [14]).

	to I-Node	to RA-node	to PA-node	to CA-node
from I-node		I-node data used in applying an inference	I-node data used in applying a preference	I-node data in conflict with information in node supported by CA-node
from RA-node	inferring a conclusion (claim)	inferring a conclusion in the form of an inference application	inferring a conclusion in the form of a preference application	inferring a conclusion in the form of a conflict denfition application
from PA-node	preference over data in I-node	preference over inference application in RA-node	meta-preferences: applying a preference over preference application in supported PA-node	preference application in supporting PA-node in conflict with preference application in PA-node supported by CA-node
from CA-node	incoming conflict to data in I-node	applying conflict definition to inference application in RA-node	applying conflict definition to preference application in PA-node	showing a conflict holds between a conflict definition and some other piece of information

Fig. 2. Relation Between Nodes in AIF

In this work, we will concentrate only in the relationships that are relevant to our formalization (these relationships highlighted in figure 2). Note that is not possible to have two I-nodes connected by an edge, the reason is that I-nodes need an explanation justifying the connection; that is, always will be a scheme, justification, inference, or rationale behind a relation between two or more I-nodes, and that will be represented by an S-node. Furthermore, I-nodes are unique in the sense that are the only type of node without incoming edges, since S-nodes relate two or more components: in RA-nodes, at least one antecedent should be used to support at least one conclusion; and in CA-nodes, at least one claim is in conflict with at least another claim.

Given an argument network, it is possible to identify arguments in it. A *simple argument* is be represented by linking a set of I-Nodes denoting the necessary premises of the argument to an I-Node denoting a conclusion or claim of it, making use of a particular RA-Node representing the inference. Formally, following [28]:

Definition 3 (Simple Argument). *Let $G = (N, E)$ be an AIF argument network with $N = I \cup RA \cup CA \cup PA$. A simple argument in G is a tuple (P, R, C) where $P \subseteq I$ represents the premises, $R \in RA$ the inference, and $C \in I$ the claim of the argument, such that for all $p \in P$, there exists $(p, R) \in E$ and there exists $(R, C) \in E$.*

The abstract AIF ontology, as presented here, is intended as a language for expressing arguments. In order to do anything meaningful with such arguments (*e.g.*, visualize, query, evaluate, or other similar actions), they must be expressed in a language more concrete to be able of being processed using additional tools and methods.

5 Labeled Argumentation Framework

Now we will introduce *Labeled Argumentation Frameworks* (LAF), combining the knowledge representation features provided by AIF, and the processing of meta-information using the algebra of argumentation labels. This framework will allow the representation of special characteristics of claims through labels which are combined and propagated according to the natural interactions occurring in the process of argumentation, such as support, conflict, and aggregation. We will use this propagation of labels to establish acceptability, and the final labels attached to the acceptable claims will provide additional acceptability information, such as restrictions on justification, explanation, etcetera.

Definition 4 (Labeled Argumentation Framework). *A Labeled Argumentation Framework (LAF) is a tuple represented as $\Phi = \langle \mathcal{L}, \mathcal{R}, \mathcal{K}, \mathcal{A}, \mathcal{F} \rangle$, where:*

- *$\mathcal{L}$ is a logical language used to represent claims. $\mathcal{L}$ is a set of expressions possibly involving the symbol "$\sim$" denoting strong negation* [1]*, such that there is no element in $\mathcal{L}$ involving a subexpression "$\sim\sim x$", and $\mathcal{L}$ is closed by complement with respect to "$\sim$".*[2]
- *$\mathcal{R}$ is a set of domain independent inference rules $R_1, R_2, \ldots, R_n$ defined in terms of $\mathcal{L}$, i.e., with premises and conclusion in $\mathcal{L}$.*
- *$\mathcal{K}$ is the knowledge base, a set of formulas of $\mathcal{L}$ describing the knowledge about the domain of discourse.*

[1] The term strong negation has its roots in Logic and refers to the concept of constructible falsity introduced by Nelson in [23] and later presented in the form of an axiomatic system by Vorob'ev in [37]. Also see [17].

[2] if $x \in \mathcal{L}$ then the complement of x is $\overline{x}$, and $\overline{x} \in \mathcal{L}$, where $\overline{x}$ is $\sim x$, and $\overline{\sim x}$ is x.

– *$\mathcal{A}$ is an algebra of argumentation labels (Def. 1).*
– *$\mathcal{F}$ is a function that assigns a label to each element of $\mathcal{K}$, i.e., $\mathcal{F} : \mathcal{K} \longrightarrow \mathbb{E}$.*

Associated with Φ we have the set of arguments Arg_Φ that can be built from $\mathcal{K}$ and $\mathcal{R}$.

Remark: To simplify the definition of conflict between claims, the occurrence of two or more consecutive "$\sim$" in $\mathcal{L}$ expressions is not allowed; this does not limit its expressive power or generality of the representation.

In this framework, we will use expressions in the language $\mathcal{L}$ to represent knowledge about a particular domain, obtaining a knowledge base $\mathcal{K}$ from which is possible to perform inferences through the specification of inference rules in $\mathcal{R}$. In LAF, inference rules represent domain-independent patterns of reasoning or inferences, such as deductive inference rules (*modus ponens, modus tollens, etc.*) or non-deductive or defeasible inference rules (*defeasible modus ponens, defeasible modus tollens, etc.*).

In the application of the algebra of argumentation labels on LAF, the support operator will be used to obtain the label associated with the conclusion of an inference from the set of labels associated with the premises, where this set of premises does not have a specific order. Thus, we define the support operator as a binary operation satisfying commutativity and associativity, and the justification for the aggregation operator is analogous to the support operator. Next we present an instantiation of LAF modeling the running example described in Section 2.

Example 1. *Let $\Phi = \langle \mathcal{L}, \mathcal{R}, \mathcal{K}, \mathcal{A}, \mathcal{F} \rangle$ be a LAF, where:*

– *$\mathcal{L}$ is a logical language that allows the construction of literals as described in Definition 4.*
– *$\mathcal{R}$= { dMP}, where the inference rule dMP is defined as follows:*

$$dMP: \ \frac{A_1, \ldots, A_n \qquad A \prec A_1, \ldots, A_n}{A} \ (\textit{Defeasible Modus Ponens})$$

In the rule $A \prec A_1, \ldots, A_n$, the first component A is a literal, called the head *of the rule, and the second component $A_1, \ldots, A_n$ is a finite non-empty set of literals called the* body *of the rule [31,18].*

– *$\mathcal{K}$ is the knowledge base shown below; in these rules, we use 'rec' for 'recommend'. In particular, the values attached to rules represent the strength of the connection between the antecedent and consequent of the rule.*
– *$\mathcal{A} = \langle \mathbb{E}, \uplus, \oplus, \ominus, \varnothing, \epsilon \rangle$ is an Algebra of Argumentation Labels, instantiated to represent and manipulate argument strengths, where strength of an argument means how relevant is it for the user,in the following way:*

$\mathbb{E} = \mathbb{N} \cup \{0\}$, represents the strength domain; $\epsilon = 0$ and $\varnothing = \infty$ are the neutral elements.

Let $\alpha, \beta \in \mathbb{E}$ be two labels, the operators over labels of support $\uplus$, aggregation $\oplus$ (both clearly commutative and associative), and conflict $\ominus$, are specified as follows:

$\alpha \uplus \beta = min(\alpha, \beta)$, i.e., the support operator reflects that an argument is as strong as its weakest support.

$\alpha \oplus \beta = \alpha + \beta$, i.e., the aggregation operator states that if we have more than one argument for a conclusion, its strength is the sum of the strengths of the arguments that support it.

$\alpha \ominus \beta = max(\alpha - \beta, 0)$, i.e., the conflict operator models that the strength of a conclusion is weakened by the strength of its counterargument.

- *$\mathcal{F}$ assigns strength values to each element of $\mathcal{K}$, this value is indicated between brackets and using a colon to separate it from said element.*

Below, after introducing the notion of argumentation graph, we will show some of the arguments in Arg_Φ (see the argumentation graph in Fig. 5 where arguments are circled with dotted lines).

$$\left\{\begin{array}{ll}
r_1 : rec(M) \prec likeGenres(M) : [14] & goodRating(oz) : [6] \\
r_2 : rec(M) \prec goodRating(M) : [9] & likeActors(oz) : [15] \\
r_3 : rec(M) \prec likeActors(M) : [12] & likeGenres(oz) : [10] \\
r_4 : \sim rec(M) \prec \sim goodScript(M) : [17] & poorStrack(oz) : [7] \\
r_5 : \sim goodScript(M) \prec notFaithful(M) : [17] & notFaithful(oz) : [15] \\
r_6 : \sim rec(M) \prec poorSTrack(M) : [6] & poorSTrack(oz) : [7] \\
r_7 : goodScript(M) \prec goodStory : [17] & goodStory(oz) : [4]
\end{array}\right\}$$

Next we introduce argumentation graphs, which will be used to represent the argumentative analysis derived from a *LAF*. Under our model of knowledge representation, we assume that in a given graph there are no two nodes which are labelled with the same sentence of $\mathcal{L}$, so we will use the labeling sentence to refer to the I-node in the graph.

Definition 5 (Argumentation graph). *Let $\Phi = \langle \mathcal{L}, \mathcal{R}, \mathcal{K}, \mathcal{A}, \mathcal{F} \rangle$ be a LAF and Arg_Φ the corresponding set of arguments. The argumentation graph G_Φ associated with Arg_Φ is an AIF digraph $G = (N, E)$, where N is the set of nodes, E is the set of the edges and the following conditions hold:*

- *each element of $\mathcal{K}$ is represented in N through an I-node.*
- *for each application of an inference rule in $\mathcal{R}$, there exists an RA-node $R \in N$ with premises $P_1, P_2, \ldots, P_n$ and conclusion C, and it holds that:*
  - *i) the premises $P_1, P_2, \ldots, P_n$ are all I-nodes in N,*
  - *ii) the conclusion C is an I-node in N, and*
  - *iii) every I-node in G is either a conclusion of an RA-node, or is an I-node that is in G because its content belongs to $\mathcal{K}$.*
- *if X and $\sim X$ are I-nodes in G, then there exists a CA-node in N such that connects the I-nodes for X and $\sim X$.*

– *for all $X \in N$, there is no path from X to X in G (i.e., G is an acyclic graph).*

Condition *iv)* forbids cycles. This appears as to be too restrictive, but it must be noted that RA-node and CA-node cycles are mostly generated by fallacious specifications.

From an argumentation graph, it is possible to identify arguments. A simple argument, as was defined in AIF (see Def. 3), can be represented by linking a set of I-nodes denoting premises to an I-node denoting a conclusion via a particular RA-node. Also, it is possible for two or more arguments to share a conclusion. This corresponds to the notion of argument accrual developed in [27,21,36,22], where the strength of the shared conclusion is the aggregation of the strengths of each individual argument supporting it.

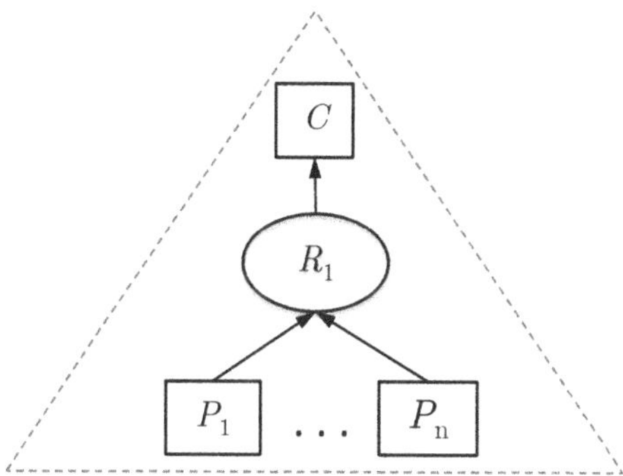

Fig. 3. Representation of a simple argument

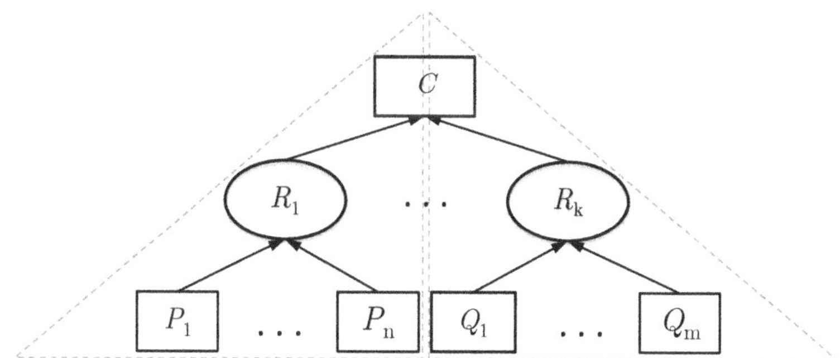

Fig. 4. Representation of argument accrual

Example 2. *From the set of arguments Arg_{Φ} obtained from the knowledge base $\mathcal{K}$ presented in the Example 1, we get the argumentation graph in Fig. 5, where arguments are shown circled with dotted lines, in the upper group there are three arguments for the literal rec(oz). In the group to the right there is an argument for* $\sim$*goodScript(oz) that feeds one of the premises for one of the arguments for* $\sim$*rec(oz) being the other one* r_4*, and there is another argument for the same claim with poorStrack(oz) and* r_6 *as premises. Finally, on the left there is an argument for the claim goodScript(oz). Notice the CA-nodes between complementary literals.*

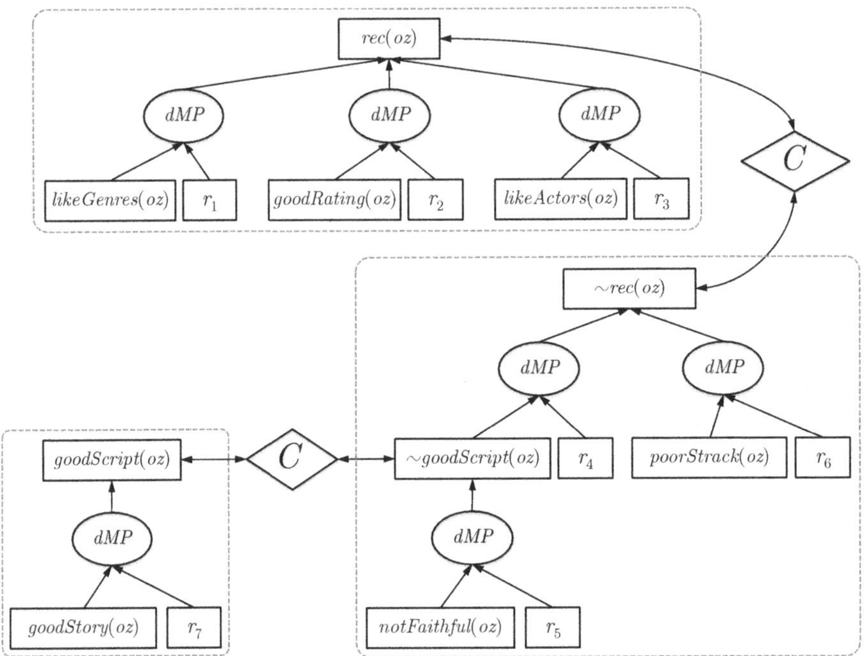

Fig. 5. Representation of an argumentation graph

Once we obtain a representation of Arg_{Φ} through the argumentative graph G_{Φ}, we will attach two labels to each I-node in G_{Φ} representing the aggregation and conflict values respectively. Note that, these labels are obtained through the operations defined in the algebra of argumentation labels which are applied considering the relationship between the knowledge pieces. The resulting graph is called *Labeled Argumentation Graph*, and the labeling process is captured in the following definition.

Definition 6 (Labeled argumentation graph). *Let* $\Phi = \langle \mathcal{L}, \mathcal{R}, \mathcal{K}, \mathcal{A}, \mathcal{F} \rangle$ *be a LAF, and* Arg_{Φ} *be the corresponding Argumentation Graph. A Labeled Argumentation Graph, denoted* Arg^{*}_{Φ} *is an argumentation graph where each I-node*

X has two labels (or attributes): μ^X that accounts for the aggregation of the reasons supporting the claim, and δ^X that holds the state of the claim after taking conflict into account. Thus, let X be an I-node, then:

- *X has no inputs, then X has a label that corresponds to it as an element of $\mathcal{K}$, thus it holds that $\mu^X = \delta^X = \mathcal{F}(X)$.*
- *X has inputs from the RA-nodes $R_1, \ldots, R_k$, where each R_i has premisses $X_1^{R_i}, \ldots, X_n^{R_i}$, then:*
$$\mu^X = \oplus_{i=1}^{k}(\uplus_{j=1}^{n} \delta^{X_j^{R_i}})$$
- *X has input from a CA-node C representing conflict with an I-node $\sim X$, then:*
$$\delta^X = \mu^X \ominus \mu^{\sim X}$$

Notice that δ^X represents the support of the I-node X after being affected by a conflict represented by a C-node.

Property 1. *Let $\Phi = \langle \mathcal{L}, \mathcal{R}, \mathcal{K}, \mathcal{A}, \mathcal{F} \rangle$ be a LAF, and Arg_Φ be the corresponding Argumentation Graph. The label μ^X associated with an I-node X is always greater than or equal to the label δ^X.*

Once the labeled argumentative graph is obtained, we are able to define the acceptability status for each I-node as follows.

Definition 7 (Acceptability status). *Let $\Phi = \langle \mathcal{L}, \mathcal{R}, \mathcal{K}, \mathcal{A}, \mathcal{F} \rangle$ be a LAF, and Arg_Φ be the corresponding Argumentation Graph. Each I-node X in Arg_Φ has assigned one of three possible acceptability status accordingly to their associated labels:*

- *Strictly Accepted iff $\mu^X = \delta^X$.*
- *Weakly Accepted iff $\mu^X \neq \delta^X$ and $\delta^X \neq \epsilon$.*
- *Rejected iff $\mu^X \neq \delta^X$ and $\delta^X = \epsilon$.*

The following property is self evident from the definition above.

Property 2. *The acceptability status for an I-node is unique in an Arg_Φ.*

Example 3. *Going back to Example 2, next we will calculate the labels for each claim (or I-node) of the argumentation graph, following Definition 6.*

$$\begin{aligned}
\mu^{rec(oz)} &= ((\delta^{likeGenres(oz)} \uplus \delta^{r_1}) \oplus (\delta^{goodRating(oz)} \uplus \delta^{r_2})) \oplus (\delta^{likeActors(oz)} \uplus \delta^{r_3}) \\
&= (min(10, 14) + min(6, 9)) + min(15, 12) = 28 \\
\mu^{\sim rec(oz)} &= min(11, 17) + min(7, 6) = 11 + 6 = 17 \\
\delta^{rec(oz)} &= \mu^{rec(oz)} \ominus \mu^{\sim rec(oz)} = max(28 - 17, 0) = 11 \\
\delta^{\sim rec(oz)} &= max(17 - 28, 0) = 0 \\
\mu^{\sim goodScript(oz)} &= min(15, 17) = 15
\end{aligned}$$

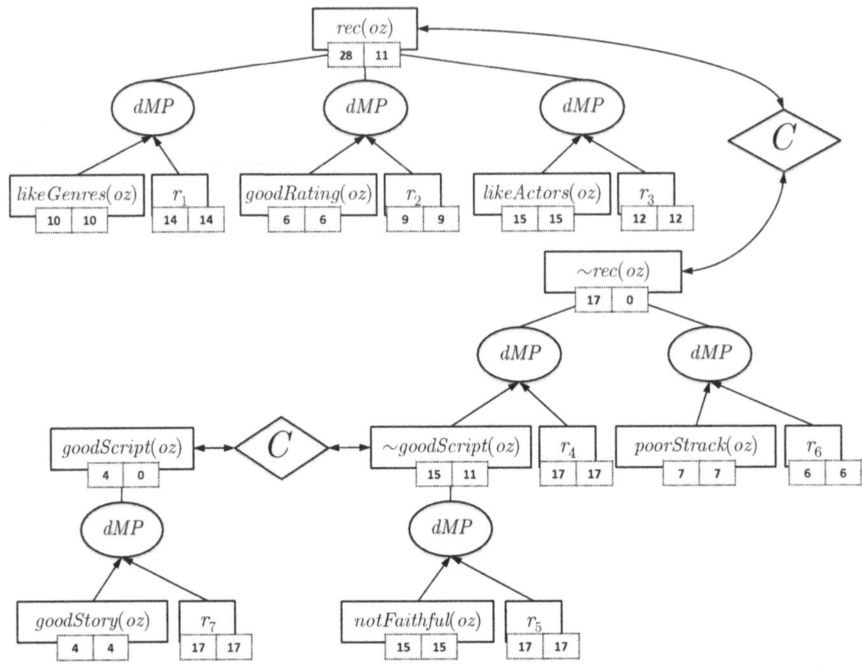

Fig. 6. A labeled argumentation graph

$$\mu^{goodScript(oz)} = min(4, 17) = 4$$
$$\delta^{\sim goodScript(oz)} = max(15 - 4, 0) = 11$$
$$\delta^{goodScript(oz)} = max(4 - 15, 0) = 0$$

We show the acceptability status for each node in the Labeled argumentation graph.

– $S = S^A \cup S^W$ *is the set of accepted claims, where*

$S^A = \{goodRating(oz), likeActors(oz), likeGenres(oz), poorSTrack(oz),$ $goodStory(oz), notFaitful(oz)\}$ *is the set of strictly accepted claims, because* $\mu^X = \delta^X$.

$S^W = \{rec(oz), {\sim}goodScript(oz)\}$ *is the set of weakly accepted claims, because* $\mu^X \neq \delta^X$ *and* $\delta^X \neq \epsilon$.

– $S^R = \{{\sim}rec(oz), goodScript(oz)\}$ *is the set of defeated claims, since* $\mu^X \neq \delta^X$ *and* $\delta^X = \epsilon$.

The final recommendation for the movie has a force of 11 *over* 28 *possible, being weakened by the existence of reasons for not recommending it. Thus, all the information was taken into account affecting the acceptability status of the arguments.*

6 Discussion and Related Work

Dov Gabbay's groundbreaking work on Labeled Deductive Systems [16,15], has provided a clear motivation for this work. The introduction of a flexible and rigorous formalism to tackle complex problems using logical frameworks that include labeled deduction capabilities has permitted to address research problems in areas such as temporal logics, database query languages, and defeasible reasoning systems. In labeled deduction, the formulæ are replaced by labeled formulæ, expressed as $L : \phi$, where L represents a label associated with the logical formula ϕ. Labels are used to carry additional information that enrich the representation language. The intuitions attached to labels may vary accordingly with the system that is necessary to model. The idea of structuring labels as an algebra was present from the very inception of labeled systems [16].

The full generality of Gabbay's proposal has been brought to focus on argumentation systems in [12]. In that work, the authors proposed a framework with the main purpose of formally characterizing and comparing different argument-based inference mechanisms through a unified framework; in particular, two non-monotonic inference operators are used to model argument construction and dialectical analysis, in the form of warrant. Labels were used in the framework to represent arguments and dialectical trees.

In common with the works mentioned above, our proposal also involves the use of labels and an algebra of argumentation labels. However, our intention is entirely different, our purpose is not to unify and formally compare different logics, but to *extend* the representational capabilities of argumentation frameworks by allowing them to represent additional domain specific information. Although it can be argued that, due to its extreme generality, Gabbays framework could also be instantiated in some way to achieve this purpose, we are proposing a concrete way in the context of the Argument Interchange Format, advancing in how to propagate labels in the specific case of argument interactions, such as aggregation, support, and conflict.

Cayrol and Lagasquie-Schiex in [9], described the argumentation process as a process which is divided into two steps: a valuation of the relative strength of the arguments and the selection of the most acceptable among them. They focused on defining a gradual valuation of arguments based on their interactions, and then established a graduality in the concept of acceptability of arguments. In their work, they do not consider argument structure, and the evaluations of the arguments are based on the interactions between their interaction directly applying it to them. In our work, we determine the valuation of arguments through their structures and the different interactions between them. We provide the ability of assigning more than one value to the arguments depending on the features associated with them we want to model. Finally, after analyzing all the interactions between arguments we obtain final values assigned to each argument; then, through these values theacceptability status (strictly accepted, weakly accepted and rejected) of the arguments is obtained.

T. J. M. Bench-Capon and J. L. Pollock have introduced systems that are currently have great influence over the research in argumentation, we will discuss them in turn. Bench-Capon [3] argues in his research that oftentimes it is impossible to conclusively demonstrate in the context of disagreement that either party is wrong, particularly in situations involving practical reasoning. The fundamental role of argument in such cases is to persuade rather than to prove, demonstrate, or refute. Bench-Capon argues that *"The point is that in many contexts the soundness of an argument is not the only consideration: arguments also have a force which derives from the value they advance or protect."* [3] He also cites Perelman [25], and the work on jurisprudence as a source of telling examples where values become important. Pollock [26] advances the idea that most semantics for defeasible reasoning ignore the fact that some arguments are better than others, thus supporting their conclusions more strongly. But once we acknowledge the fact that arguments can differ in strength and that conclusions can differ in their degree of justification, things become more complicated. In particular, he introduces the notion of *diminishers*, which are defeaters that cannot completely defeat their target, but instead lower the degree of justification of that argument. We have consider both contributions in the work presented here.

S. Kaci and L. van der Torre introduced in [19] a generalization of value-based argumentation theory, considering that an argument can promote multiple values, and that preference among these values or arguments can be further specified. In their work used the minimal and maximal specificity principle to define a preference relation. To calculate acceptability over a set of arguments they combine algorithms for reasoning about preferences with algorithms developed in argumentation theory.

Using the intuitions of these three research lines, we formalized an argumentative framework, additionally integrating AIF into the system. Labels provide the way to represent the characteristics of the arguments, completely generalizing the notion of value. he interaction between arguments can affect the labels they have associated, so that these changes can cause strengthening (through a form of accrual) and weakening (accomplishing a form of diminishing) of arguments. It is important to note that the characteristics or properties associated with an argument could vary over time and be affected by various characteristics that influence the real world; for instance, the reliability of a given source [8,7]. Using this framework, we established argument acceptability, where the final labels propagated to the accepted arguments provide additional acceptability information, such as degree of justification, restrictions on justification, explanation, and others.

7 Conclusions and Future Work

In argumentation applications, it is interesting to associate meta-information to arguments to increase the information available with the goal of determining their acceptability status. For instance, in an agent implementation, it would be beneficial to establish a degree of the success obtained by reaching a given objective; or, in the domain of recommender systems, it is interesting to provide recommendations together with an uncertainty measure or the reliability degree associated with it.

Our work has focused on the development of the framework called *Labeled Argumentation Framework* (LAF), which combine the knowledge representation capabilities provided by the *Argument Interchange Format* (*AIF*) and the treatment and management of labels by an algebra of argumentation labels developed with this purpose. Various relationships between arguments have associated operations defined on the algebra of argumentation labels, allowing to propagate meta-information in the argumentation graph. Through the argumentation graph labeling, it is possible to determine the acceptability of arguments, and the resulting meta-data associated with them providing extra information justifying their acceptability status.

We are studying the formal properties of the operations of the algebra of argumentation labels that we have defined here, and we will analyze the effect of these notions on the acceptability relation. We will develop an implementation of LAF instantiating it in the existing DeLP system [3] as a basis; the resulting implementation will be exercised in different domains requiring to model extra information associated with the arguments, taking as motivation studies and analysis of P-DeLP.

References

1. Amgoud, L., Cayrol, C.: A reasoning model based on the production of acceptable arguments. Annals of Mathematics and Artificial Intelligence 34(1-3), 197–215 (2002)
2. Amgoud, L., Prade, H.: Using arguments for making and explaining decisions. Artificial Intelligence 173(3-4), 413–436 (2009)
3. Bench-Capon, T.J.M.: Value-based argumentation frameworks. In: Benferhat, S., Giunchiglia, E. (eds.) NMR, pp. 443–454 (2002)
4. Bench-Capon, T.J.M., Dunne, P.E.: Argumentation in artificial intelligence. Artificial Intelligence 171(10-15), 619–641 (2007)
5. Besnard, P., Hunter, A.: Elements of Argumentation. MIT Press (2008)
6. Briguez, C.E., Budán, M.C., Deagustini, C.A.D., Maguitman, A.G., Capobianco, M., Simari, G.R.: Towards an argument-based music recommender system. In: COMMA, pp. 83–90 (2012)
7. Budán, M.C.D., Lucero, M.G., Chesñevar, C.I., Simari, G.R.: An approach to argumentation considering attacks through time. In: Hüllermeier, E., Link, S., Fober, T., Seeger, B. (eds.) SUM 2012. LNCS, vol. 7520, pp. 99–112. Springer, Heidelberg (2012)
8. Budán, M.C., Lucero, M.J.G., Chesñevar, C.I., Simari, G.R.: Modelling time and reliability in structured argumentation frameworks. In: Brewka, G., Eiter, T., McIlraith, S.A. (eds.) KR. AAAI Press (2012)
9. Cayrol, C., Lagasquie-Schiex, M.-C.: Graduality in argumentation. J. Artif. Intell. Res(JAIR) 23, 245–297 (2005)
10. Chesñevar, C.I., Maguitman, A.G., Simari, G.R.: Recommender system technologies based on argumentation 1. In: Emerging Artificial Intelligence Applications in Computer Engineering, pp. 50–73 (2007)

[3] See `http://lidia.cs.uns.edu.ar/delp`

11. Chesñevar, C.I., Maguitman, A.G., Simari, G.R.: A first approach to argument-based recommender systems based on defeasible logic programming. In: NMR, pp. 109–117 (2004)
12. Chesñevar, C.I., Simari, G.R.: Modelling inference in argumentation through labelled deduction: Formalization and logical properties. Logica Universalis 1(1), 93–124 (2007)
13. Chesñevar, C.I., Simari, G.R., Alsinet, T., Godo, L.: A logic programming framework for possibilistic argumentation with vague knowledge. In: Proceedings of the 20th Conference on Uncertainty in Artificial Intelligence, pp. 76–84. AUAI Press (2004)
14. Chesñevar, C.I., McGinnis, J., Modgil, S., Rahwan, I., Reed, C., Simari, G.R., South, M., Vreeswijk, G., Willmott, S.: Towards an argument interchange format. Knowledge Eng. Review 21(4), 293–316 (2006)
15. Gabbay, D.: Labelling Deductive Systems (vol. 1). Oxford Logic Guides, vol. 33. Oxford University Press (1996)
16. Gabbay, D.: Labelled deductive systems: a position paper. In: Oikkonen, J., Vaananen, J. (eds.) Proceedings of Logic Colloquium 1990. Lecture Notes in Logic, vol. 2, pp. 66–88. Springer-Verlag (1993)
17. García, A.J., Simari, G.R.: Strong and default negation in defeasible logic programming. In: 4th Dutch-German Workshop on Nonmonotonic Reasoning Techniques and Their Applications, Institute for Logic, Language and Computation. University of Amsterdam (March 1999)
18. García, A.J., Simari, G.R.: Defeasible logic programming: An argumentative approach. Theory Practice of Logic Programming 4(1), 95–138 (2004)
19. Kaci, S., van der Torre, L.: Preference-based argumentation: Arguments supporting multiple values. International Journal of Approximate Reasoning 48(3), 730–751 (2008)
20. Lucero, M.J.G., Chesñevar, C.I., Simari, G.R.: Modelling argument accrual with possibilistic uncertainty in a logic programming setting. Inf. Sci. 228, 1–25 (2013)
21. Gómez Lucero, M.J., Chesñevar, C.I., Simari, G.R.: Modelling argument accrual in possibilistic defeasible logic programming. In: Sossai, C., Chemello, G. (eds.) ECSQARU 2009. LNCS, vol. 5590, pp. 131–143. Springer, Heidelberg (2009)
22. Lucero, M.J.G., Chesñevar, C.I., Simari, G.R.: Modelling argument accrual with possibilistic uncertainty in a logic programming setting. Inf. Sci. 228, 1–25 (2013)
23. Nelson, D.: Constructible falsity. The Journal of Symbolic Logic 14(1), 16–26 (1949)
24. Pasquier, P., Hollands, R., Rahwan, I., Dignum, F., Sonenberg, L.: An empirical study of interest-based negotiation. Autonomous Agents and Multi-Agent Systems 22(2), 249–288 (2011)
25. Perelman, C.: Justice, Law and Argument, Reidel, Dordrecht, Holland. Synthese Library, vol. 142 (1980)
26. Pollock, J.L.: Defeasible reasoning and degrees of justification. Argument & Computation 1(1), 7–22 (2010)
27. Prakken, H.: A study of accrual of arguments, with applications to evidential reasoning. In: ICAIL 20 05: Proc. of the 10th Int. Conf. on Artificial Intelligence and Law, pp. 85–94. ACM, New York (2005)
28. Rahwan, I., Reed, C.: The argument interchange format. In: Rahwan, I., Simari, G.R. (eds.) Argumentation in Artificial Intelligence, pp. 383–402. Springer (2009)
29. Rahwan, I., Simari, G.R.: Argumentation in Artificial Intelligence. Springer Verlag (2009)
30. Rahwan, I., Zablith, F., Reed, C.: Laying the foundations for a world wide argument web. Artif. Intell. 171(10-15), 897–921 (2007)

31. Simari, G.R., Loui, R.P.: A Mathematical Treatment of Defeasible Reasoning and its Implementation. Artificial Intelligence 53, 125–157 (1992)
32. Simari, G.R., Loui, R.P.: A mathematical treatment of defeasible reasoning and its implementation. Artificial intelligence 53(2), 125–157 (1992)
33. Simari, G.R.: A brief overview of research in argumentation systems. In: Benferhat, S., Grant, J. (eds.) SUM 2011. LNCS, vol. 6929, pp. 81–95. Springer, Heidelberg (2011)
34. Stolzenburg, F., García, A.J., Chesñevar, C.I., Simari, G.R.: Computing generalized specificity. Journal of Applied Non-Classical Logics 13(1), 87–113 (2003)
35. van der Weide, T.L., Dignum, F., Meyer, J.-J.C., Prakken, H., Vreeswijk, G.A.W.: Multi-criteria argument selection in persuasion dialogues. In: AAMAS 2011, vol. 3, pp. 921–928 (2011)
36. Verheij, B.: Accrual of arguments in defeasible argumentation. In: Proceedings of the 2nd Dutch/German Workshop on Nonmonotonic Reasoning, pp. 217–224, Utrecht (1995)
37. Vorob'ev, N.N.: A constructive propositional calculus with strong negation. Doklady Akademii Nauk SSR 85, 465–468 (1952)

On the Semantics of Partially Ordered Bases

Claudette Cayrol, Didier Dubois, and Fayçal Touazi

IRIT, University of Toulouse, France
{ccayrol,dubois,faycal.touazi}@irit.fr

Abstract. This paper presents first results toward the extension of possibilistic logic when the total order on formulas is replaced by a partial preorder. Few works have dealt with this matter in the past but they include some by Halpern, and Benferhat *et al*. Here we focus on semantic aspects, namely the construction of a partial order on interpretations from a partial order on formulas and conversely. It requires the capability of inducing a partial order on subsets of a set from a partial order on its elements. The difficult point lies in the fact that equivalent definitions in the totally ordered case are no longer equivalent in the partially ordered one. We give arguments for selecting one approach extending comparative possibility and its preadditive refinement, pursuing some previous works by Halpern. It comes close to non-monotonic inference relations in the style of Kraus Lehmann and Magidor. We define an intuitively appealing notion of closure of a partially ordered belief base from a semantic standpoint, and show its limitations in terms of expressiveness, due to the fact that a partial ordering on subsets of a set cannot be expressed by means of a single partial order on the sets of elements. We also discuss several existing languages and syntactic inference techniques devised for reasoning from partially ordered belief bases in the light of this difficulty. The long term purpose is to find a proof method adapted to partially ordered formulas, liable of capturing a suitable notion of semantic closure.

1 Introduction

The basic concept of ordered knowledge base expressing the relative strength of formulas has been studied for more than twenty years in Artificial intelligence. To our knowledge this concept goes back to Rescher's work on plausible reasoning [1]. The idea of reasoning from formulas of various strengths is even older, since it goes back to antiquity with the texts of Theophrastus, a disciple of Aristotle, who claimed that the validity of a chain of reasoning is the validity of its weakest link. Possibilistic logic [2] is a typical example of logic exploiting a totally ordered base and implementing the weakest link principle. It is an extension of propositional logic, sound and complete with respect to a semantics in terms of possibility theory, where a set of models is replaced by a possibility distribution on the interpretations (which are then more or less plausible). It enables problems of inconsistency management [3], of revision [4] and of information fusion [5] to be handled in a natural way.

This simple approach has limitations in expressive power. We may go beyond it in several respects:

- extending the syntax to give a meaning to negations and disjunctions of weighted formulas, thus joining the syntactic framework of modal logic [6];

C. Beierle and C. Meghini (Eds.): FoIKS 2014, LNCS 8367, pp. 136–153, 2014.

- improving the treatment of the degrees attached to the formulas via a refinement of the induced possibility distribution, possibly by means of a partial order [3];
- making the approach more qualitative, by replacing the weights of certainty by the elements of a lattice, or by a partial order over a finite set of formulas [7].

This paper paves the way to the systematic study of the last two points, based on scattered existing works. Possibilistic logic exploits the equivalence between the deductive closure of a set of weighted formulas, and a possibility distribution on the interpretations. In case of logical inconsistency, it reasons with the formulas whose certainty level exceeds the global inconsistency degree, leaving out some highly uncertain formulas not concerned by inconsistency (this is called "the drowning effect"). Our idea is to preserve this kind of relation between semantics and syntax in the setting of weaker algebraic frameworks (with partial preorders), while proposing concepts of partially ordered closure accordingly. The following questions seem natural:

- Is it possible to represent a partially ordered set of formulas by a partially ordered set of models?
- Is it possible to represent a partially ordered set of models by a partially ordered set of formulas?
- Is it possible to define inference rules that account for such a semantics?

To address these questions, we first review how to go from a partial ordering on a set to a partial ordering on its subsets. This point, already reviewed by Halpern [8], is tricky because equivalent definitions in the case of a total order are no longer so in the partially ordered setting. Properties of partial orders among sets induced by partial order on elements are studied in detail. Then these results are applied to the definition of semantic inference from partially ordered knowledge bases. This definition poses the problem of representing the semantics of a partially ordered base in terms of a single partial order of its interpretations. We show that in general the partial order on formulas cannot be recovered from the partial order on interpretations it induces, contrary to the totally ordered case. Finally we briefly review existing proposals of syntactic inference that may be used to reason from partially ordered formulas, in the light of this limitation.

2 Comparing Sets of Totally Ordered Elements

Let $(S, \geq)$ be a totally ordered set and let A and B be subsets of S. To extend $\geq$ to 2^S, a natural idea is to compare A with B by means of logical quantifiers. So, four kinds of relations can be defined:

Definition 1.

- ***Unsafe dominance***: $A \succeq_u B$ *iff* $\exists a \in A, b \in B, a \geq b$
- ***Optimistic dominance***: $A \succeq_o B$ *iff* $\forall b \in B, \exists a \in A, a \geq b$
- ***Pessimistic dominance***: $A \succeq_p B$ *iff* $\forall a \in A, \exists b \in B, a \geq b$
- ***Safe dominance***: $A \succeq_s B$ *iff* $\forall a \in A, \forall b \in B, a \geq b$

Strict counterparts of these definitions, namely $\succ_u, \succ_o, \succ_p, \succ_s$ *can be similarly defined, replacing* $(S, \geq)$ *by its strict part* $(S, >)$

Note that relations $\succ_x$ are not the strict parts of $\succeq_x$. The four kinds of weak relations in this definition can be rewritten by comparing maximal (resp. minimal) elements. Denoting by $\max(A)$ (resp. $\min(A)$) any maximal (resp. minimal) element in A:

- **Unsafe dominance**: $A \succeq_u B$ iff $\max(A) \geq \min(B)$
- **Optimistic dominance**: $A \succeq_o B$ iff $\max(A) \geq \max(B)$
- **Pessimistic dominance**: $A \succeq_p B$ iff $\min(A) \geq \min(B)$
- **Safe dominance**: $A \succeq_s B$ iff $\min(A) \geq \max(B)$

Note that $A \succeq_u B$ iff $\neg(B \succ_s A)$. The strict safe dominance $\succ_s$ is a strict partial order that can compare disjoint sets only; and $\succeq_s$ is not even reflexive. The unsafe dominance is not transitive even if reflexive. The optimistic and pessimistic comparisons are total orders dual to each other in the following sense: $A \succeq_o B$ iff $B \succeq'_p A$ where $\geq'$ denotes the inverse of $\geq$ on S, defined by $a \geq' b$ iff $b \geq a$. It is interesting to highlight the point that the latter comparisons can be defined equivalently as:

- **Optimistic dominance**: $A \succeq_o B$ iff $\exists a \in A, \forall b \in B, a \geq b$;
- **Pessimistic dominance**: $A \succeq_p B$ iff $\exists b \in B, \forall a \in A, a \geq b$

These notions can be applied to the representation of uncertainty. Let S denote a set of states. Assume π a possibility distribution on S such that $\pi(s)$ is the plausibility degree that s is the real world. Let Π the associated possibility measure defined by $\Pi(A) = \max_{s \in A} \pi(s)$ and N the dual necessity measure defined by $N(A) = \min_{s \notin A} 1 - \pi(s) = 1 - \Pi(\overline{A})$ [2]. We have:

- $\Pi(A) \geq \Pi(B)$ iff $\max(A) \geq \max(B)$ (this is $A \succeq_o B$)
- $N(A) \geq N(B)$ iff $\max(\overline{B}) \geq \max(\overline{A})$ (this is $\overline{B} \succeq_o \overline{A}$)

So the optimistic comparison between A and B is a comparative possibility measure in the sense of Lewis [9] (see also [10]), and the optimistic comparison between complements $\overline{A}$ and $\overline{B}$, which expresses relative certainty, is related to epistemic entrenchment in revision theory [11]. In the uncertainty framework, safe dominance is never used as it is not representable by a monotonically increasing set function. On the other hand, the pessimistic ordering is monotonically decreasing with inclusion. In the following, we thus concentrate on the optimistic comparison $\succeq_o$.

The above definitions also apply to the representation of preferences. Then stating $A \succeq B$ accounts for an agent declaring that the truth of the proposition whose models form the set A is preferred to the truth of the proposition whose models form the set B. Interpreting such a statement requires the knowledge of the attitude of the agent, which leads to choosing between the four orderings considered above. The safe dominance is natural in this setting as a very conservative risk-free understanding of $A \succeq B$, akin to interval orderings [12]. Pessimistic and optimistic dominance are milder views, and both make sense, as explored by Benferhat *et al.* [13], Kaci and van den Torre [14,15], contrary to the case of representing the plausibility and certainty of formulas. However, preference modelling is not in the scope of this paper.

3 Properties of Relative Likelihood Relations Comparing Subsets

Let $\trianglerighteq$ be a reflexive relation that compares subsets A and B of S, and $\rhd$ its strict part. We enumerate different properties that may be satisfied by these relations in the scope of modeling relative uncertainty.

1. Compatibility with set-theoretic operations (inclusion, intersection, union)
   - **Compatibility with Inclusion (CI)** If $B \subseteq A$ then $A \trianglerighteq B$
   - **Orderliness (O)** If $A \rhd B$, $A \subseteq A'$, and $B' \subseteq B$, then $A' \rhd B'$
   - **Stability for Union (SU)** If $A \trianglerighteq B$ then $A \cup C \trianglerighteq B \cup C$
   - **Preadditivity (P)** If $A \cap (B \cup C) = \varnothing$ then ($B \trianglerighteq C$ iff $A \cup B \trianglerighteq A \cup C$)
   - **Self-duality (D)** $A \trianglerighteq B$ iff $\overline{B} \trianglerighteq \overline{A}$

   Note that (CI) is never satisfied by a non-reflexive relation while (O) makes sense for a reflexive relation too. (SU) does not make sense for an asymmetric relation (take $C = S$). Preadditivity and self-duality, like (O) make sense for both $\trianglerighteq$ and its strict part. All these properties have been studied in the case of total orders: (CI) and (O) are expected when $A \rhd B$ expresses a greater confidence in A than in B; (SU) characterizes possibility relations $\succeq_o$ (Lewis[9], Dubois[10]). Preadditivity and self-duality hold for probability measures [16], but also for the relation $A \rhd B$ iff $A \setminus B \succ_o B \setminus A$ [17].
2. Properties reflecting a qualitative point of view
   - **Qualitativeness (Q)** If $A \cup B \rhd C$ and $A \cup C \rhd B$, then $A \rhd B \cup C$
   - **Negligibility (N)** If $A \rhd B$ and $A \rhd C$, then $A \rhd B \cup C$

   (Q) is satisfied by strict parts of possibility relations ($\succ_o$, but not $\succeq_o$) and is found in non-monotonic logic. Negligibility also works for $\succeq_o$, it says that one cannot compensate for the low plausibility of a set by adding elements of low plausibility.
3. Properties concerning the deductive closure of partially ordered bases (see Friedman and Halpern[18], Dubois and Prade [19] and Halpern [8]):
   - **Conditional Closure by Implication (CCI)** If $A \subseteq B$ and $A \cap C \rhd \overline{A} \cap C$ then $B \cap C \rhd \overline{B} \cap C$
   - **Conditional Closure by Conjunction (CCC)** If $C \cap A \rhd C \cap \overline{A}$ and $C \cap B \rhd C \cap \overline{B}$ then $C \cap (A \cap B) \rhd C \cap \overline{A \cap B}$
   - **Left Disjunction (OR)** If $A \cap C \rhd A \cap \overline{C}$ and $B \cap C \rhd B \cap \overline{C}$ then $(A \cup B) \cap C \rhd (A \cup B) \cap \overline{C}$
   - **Cut (CUT)** If $A \cap B \rhd A \cap \overline{B}$ and $A \cap B \cap C \rhd A \cap B \cap \overline{C}$ then $A \cap C \rhd A \cap \overline{C}$
   - **Cautious Monotony (CM)**: If $A \cap B \rhd A \cap \overline{B}$ and $A \cap C \rhd A \cap \overline{C}$ then $A \cap B \cap C \rhd A \cap B \cap \overline{C}$

   These properties are rather intuitive when the relation $A \rhd \overline{A}$ is interpreted as "A is an accepted belief", and $A \cap C \rhd \overline{A} \cap C$ as "A is an accepted belief in the context C" [20,18,21]. They hold in the total order setting for the optimistic relation $\succ_o$, but they are not interesting to consider for reflexive relations (e.g. $\succeq_o$).

Proposition 1. *It is easy to see that, for any relation $\gg$:*

1. *(O) implies CCI.*
2. *If the relation $\gg$ is qualitative (Q) and orderly (O), then it satisfies Negligibility (N) and (CCC), and the converse of (SU): If $A \cup C \gg B \cup C$ then $A \gg B$.*

Proof:

1. Suppose $A \subseteq B$ and $A \cap C \gg \overline{A} \cap C$. We have $A \cap C \subseteq B \cap C$ and $\overline{B} \cap C \subseteq \overline{A} \cap C$. Hence from (O), $B \cap C \gg \overline{B} \cap C$.
2. Assume that $\gg$ satisfies (Q) and (O).
Suppose $A \gg B$ and $A \gg C$. Then, by (O), $A \cup B \gg C$ and $A \cup C \gg B$. Hence from (Q), $A \gg B \cup C$. So $\gg$ satisfies (N).
Suppose $C \cap A \gg C \cap \overline{A}$ and $C \cap B \gg C \cap \overline{B}$. Let $A' = A \cap B \cap C$, $B' = A \cap C \cap \overline{B}$, $C' = \overline{A} \cap B \cap C$ and $D' = C \cap \overline{B} \cap \overline{A}$. So, we have $A' \cup B' \gg C' \cup D'$ and $A' \cup C' \gg B' \cup D'$. Hence from (O), $A' \cup B' \cup D' \gg C'$, then from (Q) $A' \gg B' \cup C' \cup D'$. So, $A \cap B \cap C \trianglerighteq C \cap (\overline{A} \cup \overline{B}) = C \cap \overline{A \cap B}$. So $\gg$ satisfies (CCC).
Suppose $A \cup C \gg B \cup C$. Then by (O), $A \cup (C \cup B) \gg C$. Hence from (Q), $A \gg B \cup C$, then (O) again: $A \gg B$. So $\gg$ satisfies the converse of (SU).

4 Comparing Sets of Partially Ordered Elements

In this section, we start from a partially ordered set $(S, \geq)$ and we consider the construction of a relation $\trianglerighteq$ induced by $\geq$ for comparing subsets of S. In the scope of representing comparative belief and plausibility, the last section has shown that we can restrict to the optimistic comparison of sets. In the following, we focus on the generalization of optimistic dominance to the case of partially ordered sets. It has been noticed (see section 2) that there are two possible definitions of the optimistic dominance, that are equivalent in the total order setting. However they are no longer so in the partial order setting, as first noticed by Halpern [8]. As a consequence, in order to define a semantics for partially ordered logical bases, we have to study these different relations and to choose an appropriate one according to the properties they satisfy.

As usual, given a reflexive and transitive relation on S, denoted by $\geq$, $s' > s$ is an abbreviation for "($s' \geq s$) and not ($s \geq s'$)". The relation $>$ is the strict partial order determined by $\geq$. It is an irreflexive and transitive relation on S. $s' \sim s$ is an abbreviation for "($s' \geq s$) and ($s \geq s'$)". The relation $\sim$ is the equivalence relation determined by $\geq$. $s' \not\approx s$ is an abbreviation for "(neither ($s' \geq s$) nor ($s \geq s'$)". It is the incomparability relation determined by $\geq$. If this relation is empty, the relation $\geq$ is a total preorder. On the contrary, given a transitive and asymetric relation $>$, the relation $s' \not\approx s$ if and only if neither $s' > s$ nor $s > s'$ is its associated incomparability relation (while $s' \sim s$ reduces to the equality relation).

Let $(S, \geq)$ a partially ordered set, and $X \subseteq S$. $s \in X$ is *maximal* for $\geq$ in X if and only if we do not have $s' > s$ for any $s' \in X$. $M(X, \geq)$ ($M(X)$ for short) denotes the set of the maximal elements in X according to $\geq$.

The optimistic comparison between A and B is based on the comparison between $M(A)$ and $M(B)$. In the total order case, it can be defined in two ways, which are no longer equivalent in the partial case. We call them weak optimistic dominance and strong optimistic dominance in the following.

4.1 Weak Optimistic Dominance

Here again, various definitions can be proposed according to whether one starts from a strict order or not on S.

Definition 2 (Weak optimistic dominance).

1. *Weak optimistic strict dominance:*
   $A \succ_{wos} B$ *iff* $A \neq \varnothing$ *and* $\forall b \in B,\ \exists a \in A, a > b$.
2. *Weak optimistic loose dominance:* $A \succeq_{wol} B$ *iff* $\forall b \in B,\ \exists a \in A, a \geq b$.
3. *Strict order determined by* $\succeq_{wol}$: $A \succ_{wol} B$ *iff* $A \succeq_{wol} B$ *and* $\neg(B \succeq_{wol} A)$. *In other words,* $A \succ_{wol} B$ *iff* $\forall b \in B,\ \exists a \in A, a \geq b$ *and* $\exists a' \in A, \forall b \in B$, *either* $a' > b$ *or* $a' \not\sim b$.

These relations are respectively denoted by $\succ^s, \succeq^s$ and $\succ'$ by Halpern [8]. The relation $\succ_{wos}$ is a strict partial order (asymetric and transitive) on 2^S. We have always $A \succ_{wos} \varnothing$, except if A is empty. The relation $\succeq_{wol}$ is reflexive and transitive and such that $A \succeq_{wol} \varnothing$, but not $\varnothing \succeq_{wol} B$ except if B is empty. Finally, if $A \succ_{wos} B$ then $A \succ_{wol} B$. The converse is generally false except if $\geq$ is a complete order.

The following proposition shows that the weak optimistic dominance is appropriate for representing relative plausibility.

Proposition 2. *The weak optimistic strict dominance* $\succ_{wos}$ *is a strict partial order which satisfies Qualitativeness (Q), Orderliness (O), Left Disjunction (OR), (CUT) and (CM).*

The weak optimistic loose dominance $\succeq_{wol}$ *satisfies Compatibility with inclusion (CI), Orderliness (O), Negligibility (N), Stability for union (SU).*

The relation $\succ_{wol}$ *satisfies Orderliness (O) and Conditional Closure by Implication (CCI).*

Corollary 1. *The weak optimistic strict dominance* $\succ_{wos}$ *satisfies the converse of (SU), Negligibility (N), Conditional Closure by Implication (CCI) and Conditional Closure by Conjunction (CCC).*

Note that, as shown by Halpern [8], the relation $\succeq_{wol}$ is generally not qualitative and the relation $\succ_{wol}$ does not satisfy the property of Negligibility. Moreover, the relation $\succ_{wol}$ neither satisfies the property OR nor the property CUT as shown below:

Example 1. *Let* $S = \{a, b, c, d, e, f, g, h\}$ *be a partially ordered set with* $f \sim h,\ e \sim g,\ f > a,\ e > b,\ a > c$ *and* $b > d$.

OR: *Let* $A = \{a, c, e, g\}, B = \{b, d, f, h\}$ *and* $C = \{a, b, e, f\}$ *be three subsets of* S. *We have* $A \cap C \succ_{wol} A \cap \overline{C}$ *and* $B \cap C \succ_{wol} B \cap \overline{C}$ *but not* $(A \cup B) \cap C \succ_{wol} (A \cup B) \cap \overline{C}$.

CUT: *Let* $A = \{a, b, c, d, e, f, g, h\}, B = \{a, b, c, e, g, h\}$ *and* $C = \{a, b, d, g, h\}$ *be three subsets of* S.
We have $A \cap B \succ_{wol} A \cap \overline{B}$ *and* $A \cap B \cap C \succ_{wol} A \cap B \cap \overline{C}$ *but not* $A \cap C \succ_{wol} A \cap \overline{C}$.

It is clear that as a result, the relation $\succ_{wos}$ is the richest one to represent relative plausibility. But note that it has no non trivial associated equivalence relation (but for $A \sim_{wos} B$ if and only if $A = B$).

4.2 Strong Optimistic Dominance

The alternative approach, not considered by [8], consists in assuming, if $A \rhd B$, that one element in A dominates all elements in B. As before, various definitions can be proposed according to whether one uses a strict order or not on S.

Definition 3 (Strong optimistic dominance).

1. *Strong optimistic strict dominance:* $A \succ_{Sos} B$ *iff* $\exists a \in A,\ \forall b \in B,\ a > b$
2. *Strong optimistic loose dominance:* $A \succeq_{Sol} B$ *iff* $\exists a \in A,\ \forall b \in B,\ a \geq b$
3. *Strict order determined by* $\succeq_{Sol}$*:* $A \succ_{Sol} B$ *iff* $A \succeq_{Sol} B$ *and* $\neg(B \succeq_{Sol} A)$. *In other words,* $A \succ_{Sol} B$ *iff* $\exists a \in A,\ \forall b \in B,\ a \geq b$ *and* $\forall b \in B, \exists a \in A$, *either* $a > b$ *or* $a \not\sim b$.

Note that with the above definitions, if $A \neq \varnothing$, $A \succ_{Sos} \varnothing$ and never $\varnothing \succ_{Sos} B$. The relation $\succ_{Sos}$ is a strict partial order on 2^S. Finally, if $A \succ_{Sos} B$ then $A \succ_{Sol} B$. Obviously, the strong relations are stronger than the weak relations, namely: If $A \succ_{Sos} B$ then $A \succ_{wos} B$ and if $A \succeq_{Sol} B$ then $A \succeq_{wol} B$. The converse is true only if $\geq$ is a complete order on S. So, we also have: If $A \succ_{Sos} B$ then $A \succ_{wol} B$. However there is no entailment between the relations $\succ_{Sol}$ and $\succ_{wol}$ as shown by the following counterexamples:

Example 2. *Let* $S = \{a_1, a_2, b_1, b_2, b_3\}$ *with* $a_1 \sim b_1 > b_3$ *and* $a_2 > b_2$. *Then* $\{a_1, a_2\} \succ_{wol} \{b_1, b_2, b_3\}$, *but it is false that* $\{a_1, a_2\} \succ_{Sol} \{b_1, b_2, b_3\}$.

Example 3. *Let* $S = \{a_1, a_2, b_1, b_2, b_3\}$ *with* $a_1 \sim b_1 > b_3$, *and* $a_1 > b_2 > a_2$. *Then* $\{a_1, a_2\} \succ_{Sol} \{b_1, b_2, b_3\}$, *but it is false that* $\{a_1, a_2\} \succ_{wol} \{b_1, b_2, b_3\}$.

As indicated by Benferhat, Lagrue, Papini [22], the relation $\succ_{Sos}$ contains many incomparabilities, and $\succeq_{Sol}$ does not satisfy Compatibility with Inclusion. Indeed, if $A \subseteq B$, it is not obvious that there exists $b \in B$ such that $b \geq a, \forall a \in A$. In fact, $\succeq_{Sol}$ is thus not even reflexive, even if it is transitive. Finally, $A \succ_{Sos} B$ implies $A \succ_{Sol} B$. The converse is not true except when $\geq$ is a complete order on S. As for properties:

Proposition 3. *The strong optimistic strict dominance* $\succ_{Sos}$ *is a strict order satisfying Orderliness (O) and Cautious Monotony (CM)*

However it fails to satisfy Negligibility, Qualitativeness, CUT and Left Disjunction (OR), as shown by the following examples.

Example 4. *Let* $S = \{a_1, a_2, b, c\}$ *with* $a_1 > b$ *and* $a_2 > c$, *and the subsets* $A = \{a_1, a_2\}, B = \{b\}$ *and* $C = \{c\}$. *We have* $A \succ_{Sos} B$, $A \succ_{Sos} C$ *but we don't have* $A \succ_{Sos} (B \cup C)$. *So (N) is not satisfied.*

Due to Proposition 1, the relation $\succ_{Sos}$ fails to satisfy Qualitativeness as well.

Example 5. *Let* $S = \{a, b, c, d\}$ *with* $a > b$ *and* $c > d$, *and the subsets* $A = \{a, b, c, d\}$, $B = \{a, c, d\}$ *and* $C = \{a, c\}$. *We have* $A \cap B \succ_{Sos} A \cap \overline{B}$, $A \cap B \cap C \succ_{Sos} A \cap B \cap \overline{C}$ *but we do not have* $A \cap C \succ_{Sos} A \cap \overline{C}$. *So (CUT) is not satisfied.*

Lastly, suppose $\exists a \in A \cap C,\ \forall x \in A \cap \overline{C}\ a > x$, and $\exists b \in B \cap C,\ \forall x \in B \cap \overline{C}\ b > x$ then if a and b are not comparable, there may be no $c \in (A \cup B) \cap C$ that alone can dominate all elements in $(A \cup B) \cap \overline{C}$. So (OR) is not satisfied either.

So the weak optimistic dominance is a richer concept than the strong one.

4.3 Refinement of Partial Preorders Induced between Subsets

None of the relations presented in the above sections satisfies the property of Preadditivity, which considers that the common part of two sets should play no role in the comparison. A preadditive approach for comparing two sets A and B consists in eliminating the common part and then comparing $A \setminus B$ and $B \setminus A$. This is not a new idea (see [8] for a bibliography). In the following we consider the refinement of the weak optimistic dominance.

Definition 4 (Preadditive dominance).

- *strict preadditive dominance:* $A \succ_d^{wos} B$ *if and only if* $A \neq B$ *and* $A \setminus B \succ_{wos} B \setminus A$.
- *loose preadditive dominance:* $A \succeq_d^{wol} B$ *if and only if* $A \setminus B \succeq_{wol} B \setminus A$.
- *strict order from* $\succeq_d^{wol}$*:* $A \succ_d^{wol} B$ *if and only if* $A \succeq_d^{wol} B$ *and* $\neg(B \succeq_d^{wol} A)$.

These relations are respectively denoted by $\rhd_6, \rhd_4$ and $\rhd_5$ in Halpern [8]. The two first relations are thoroughly studied in [23] and [24]. They coincide with $\succ_{wos}$ and $\succeq_{wol}$ on disjoint sets. When S is totally ordered, the relations $\succ_d^{wol}$ and $\succ_d^{wos}$ coincide. Neither the relation $\succeq_d^{wol}$, nor its strict part are transitive, as indicated by the following counterexample.

Example 6. *Let* $S = \{a_1, a_2, a_3, b_1, b_2, c\}$ *with* $a_1 \sim a_2 \sim b_1$, $a_1 > c$ *and* $a_3 > b_2$. *Let* $A = \{a_1, a_2, a_3\}$, $B = \{b_1, b_2\}$, $C = \{a_1, a_2, c\}$. *We have* $A \succeq_d^{wol} B$ *but not* $B \succeq_d^{wol} A$, $B \succeq_d^{wol} C$ *but not* $C \succeq_d^{wol} B$ *but we don't have* $A \succeq_d^{wol} C$.

The relation $\succ_d^{wos}$ seems to be more appropriate due to the following properties:

Proposition 4. *The relation* $\succ_d^{wos}$ *is a strict partial order that satisfies:*

- *Strict compatibility with Inclusion (SCI): if* $A \subset B$ *then* $B \succ_d^{wos} A$.
- *Self-duality (D) and Preadditivity (P)*
- *a weak form of Negligibility: If* $A \cap B = A \cap C$ *then (If* $A \succ_d^{wos} B$ *and* $A \succ_d^{wos} C$ *then* $A \succ_d^{wos} (B \cup C)$*).*
- *a weak form of Qualitativeness: If* $A \cap B = A \cap C = B \cap C$ *then (If* $A \cup C \succ_d^{wos} B$ *and* $A \cup B \succ_d^{wos} C$ *then* $A \succ_d^{wos} (B \cup C)$*).*

Note that since $\succ_d^{wos}$ is equal to $\succ_{wos}$ on disjoint sets, it satisfies (CCI), (CCC), (OR), CUT and CM as well.

The next property relates the optimistic dominance to the preadditive dominance.

Proposition 5. $\succ_d^{wos}$ *refines* $\succ_{wos}$ *and its dual variant:*

- *If* $A \succ_{wos} B$ *then* $A \succ_d^{wos} B$.
- *If* $\overline{B} \succ_{wos} \overline{A}$ *then* $A \succ_d^{wos} B$.
- *If* $A \succ_d^{wos} B$ *then* $A \succeq_{wol} B$ *and* $\overline{B} \succeq_{wol} \overline{A}$.

The preadditive dominance based on the weak optimistic dominance is thus well-adapted to plausible reasoning with partially ordered knowledge bases. Note that the properties of Conditional Closure by Implication (CCI) and Conditional Closure by Conjunction (CCC) are essential to extract a deductively closed set of most plausible formulae.

4.4 From Weak Optimistic Dominance to a Partial Order on Elements

Halpern [8] studied the problem to know if a preorder on 2^S can be generated by a preorder on S. The only known result deals with total preorders: If $\trianglerighteq$ is a total preorder on 2^S that satisfies the properties of orderliness and qualitativeness, then there exists a total preorder $\geq$ on S such that $\trianglerighteq$ and $\succeq_{wol}$ coincide on 2^S (a similar result where one replaces these properties by stability for the union is already in [10], because in this case indeed a comparative possibility measure [9] is characterized by a complete preorder of possibility on S).

In the partial order case, if a strict order $\triangleright$ on 2^S is generated by a strict order $>$ on S, one must have $\{a\} \triangleright \{b\}$ whenever $a > b$. Conversely, suppose $\triangleright$ satisfies the properties of orderliness and negligibility and define the relation $a >_{\triangleright} b$ by $\{a\} \triangleright \{b\}$. Then, $A \succ_{wos} B$ means $\forall b \in B, \exists a \in A, a >_{\triangleright} b$ that is to say $\forall b \in B, \exists a \in A, \{a\} \triangleright \{b\}$. We have:

- If $A \succ_{wos} B$ then $A \triangleright B$.
- Conversely, if $A \triangleright B$, it is easy to prove that $\forall b \in B, A \triangleright \{b\}$. *But nothing proves that* $\exists a \in A$ *such that* $\{a\} \triangleright \{b\}$.

So, the situation of partial orders is strikingly different from the case of total orders. Even equipped with the properties of orderliness and negligibility, a partial order on subsets is generally NOT characterized by its restriction on singletons.

Another way to induce a partial order on 2^S from a partial order $>$ on S is to consider the partial order $>$ as a family of total orders $>^i$ extending (or compatible with) this partial order. Let A and B be two subsets of S, and let $\triangleright^i$ denote the ordering relation on 2^S induced by $>^i$. Then two methods for building a partial order on 2^S can classically be proposed [25]:

Cautious principle considering all the total orders on S compatible with $>$: $A \triangleright B$ iff $\forall i = 1, \ldots, n, A \triangleright^i B$

Bold principle considering at least one total order on S compatible with $>$: $A \triangleright B$ iff $\exists i, A \triangleright^i B$

It turns out that if we consider the family of total orders $>^i$ extending a partial order $>$ on S, the cautious principle enables the weak optimist dominance $\succ_{wos}$ to be recovered:

Proposition 6. *Let A, B two subsets of S. We have:*

$$A \succ_{wos} B \iff \forall i = 1..n \; A \succ_o^i B$$

As a consequence, a weak optimistic strict order on subsets is characterised by several total orderings on elements, not by a single partial order on elements. Given the properties satisfied by $\succ_{wos}$, this result clearly bridges the gap between the weak optimistic dominance and the partially ordered non-monotonic inference setting of Kraus, Lehmann and Magidor [20] interpreting the dominance $A \succ_{wos} B$ when $A \cap B = \varnothing$ as the default inference of A from $A \cup B$.

Example 7. *Let $(S, >) = \{a, b, c, d, e\}$ be the partially ordered set defined by $e > c > a > d$, $c > b > d$. Let $>^1, >^2$ be the two linear orders that extend the partial order $>$ defined by $e >^1 c >^1 b >^1 a >^1 d$ and $e >^2 c >^2 a >^2 b >^2 d$. Let $A = \{e, c\}$, $B = \{b, d\}$ and $C = \{a, d\}$:*

- $\forall i = 1, 2, \max(A) >^i \max(B)$, *and it holds that $A \succ_{wos} B$.*
- $\max(C) >^2 \max(B)$, $\max(B) >^1 \max(C)$. *Neither $C \succ_{wos} B$ nor $B \succ_{wos} C$.*

5 Representations of an Epistemic State

Let us first formalize the concept of epistemic state based on the notion of partial order, from a syntactic and semantic point of view. In the following, V will denote a set of propositional variables, $\mathcal{L}$ a propositional language on V, and $\mathcal{K}$ a finite base of formulas built on $\mathcal{L}$.

5.1 Syntactic Representation

From the syntactic point of view, we can view an epistemic state as a finite set of propositional formulas equipped with a partial preorder. Let $(\mathcal{K}, >)$ be a partially ordered base of formulas. If ϕ and ψ are two formulas of $\mathcal{K}$, $\psi > \phi$ is interpreted by "ψ is more likely than ϕ" (typically the first one is more certain or plausible as the second one). It can more generally be interpreted in terms of "priority". If $\psi > \phi$ is viewed as a constraint, the presence of the likelihood relation can be a cause of inconsistency. For instance, it seems irrational to assert $\phi > \psi$ when $\phi \models \psi$. It can be regarded as a semantic contradiction.

In the particular case where the preorder is total, there is a alternative representation by means of a stratified base $(\mathcal{K}_1, \cdots, \mathcal{K}_n)$ where all the elements of $\mathcal{K}_i$ are set at the same priority level, and those of $\mathcal{K}_i$ are strictly preferred to those of $\mathcal{K}_j$ if $i > j$. However, possibilistic logic [2] does not consider stratification as a strict ordering constraint. It interprets $\phi \in \mathcal{K}_i$ as assigning a *minimal* absolute level to ϕ, that may fail to be its final one, i.e. ϕ can end up at some level $j > i$ in the totally ordered deductive closure (which represents an epistemic entrenchment relation). Therefore, the stratification of the base is never an additional source of inconsistency. On the contrary, if when $\psi \in \mathcal{K}_i$ and $\phi \in \mathcal{K}_j, j > i$ is understood as a constraint $\psi > \phi$, it means that the stratified knowledge base is viewed as a fragment of a likelihood relation (epistemic entrenchment or necessity measure). The complexity of finding the deductive closure is higher in the last situation due to the possibility of a semantic contradiction between the likelihood relation at the syntactic level and logical entailment.

5.2 Semantic Representation

Let Ω be the set of interpretations of $\mathcal{L}$. At the semantic level, suppose that an epistemic state is modelled by a partial preorder on the interpretations of a propositional language, $(\Omega, \rhd)$. If ω and ω' represent two elements of Ω, the assertion $\omega' \rhd \omega$ is interpreted as ω' being more plausible than ω. In the knowledge representation literature, the main concern is often to extract the closed set of accepted beliefs $\mathcal{K}_{\rhd}$ (or belief set) associated with $(\Omega, \rhd)$. It is often defined as the deductively closed set of formulas whose models form the set $M(\Omega, \rhd)$ of most plausible models. Our aim is to go further and to define a deductive closure which is a partial order induced by $(\Omega, \rhd)$ on the language, $(\Omega, \rhd)$ being itself induced by a partially ordered base $(\mathcal{K}, >)$. The idea is to attach a semantics to $\phi > \psi$ in terms of a partial order on the interpretations, and then to build a partial order on $\mathcal{L}$ which is, as much as possible, in agreement with $(\mathcal{K}, >)$. The question is thus to go from $(\mathcal{K}, >)$ to $(\Omega, \rhd)$ and back, namely:

From $(\mathcal{K}, >)$ **to** $(\Omega, \rhd)$**:** Starting from a partially ordered base, the problem is to build a partial preorder on the set of interpretations of $\mathcal{K}$. A natural approach is to compare two interpretations ω and ω' by comparing subsets of formulas of $\mathcal{K}$ built from these interpretations.

A first proposal is to compare two interpretations ω and ω' by comparing the two subsets of formulas of $\mathcal{K}$ respectively *satisfied* by each of these interpretations. That is to say: ω' is more plausible than ω if for each formula ϕ satisfied by ω, there exists a formula preferred to ϕ and satisfied by ω'.

A dual proposal consists in comparing ω and ω' by comparing the two subsets of formulas of $\mathcal{K}$ respectively *falsified* by each of these interpretations. That is to say: ω' is more plausible than ω if for each formula ϕ' falsified by ω', there exists a formula falsified by ω preferred to ϕ'.

From $(\Omega, \rhd)$ **to** $(\mathcal{L}, \succ)$**:** Starting from a partial preorder on Ω, the problem is to build a partial preorder on the set of the formulas of the language $\mathcal{L}$. To this end, it is natural to compare two formulas ϕ and ϕ' by comparing subsets of interpretations built from these formulas. In the same way as above, a first proposal is to compare ϕ and ψ by comparing the sets of models of these formulas. One can alternatively compare ϕ and ψ by comparing their sets of counter-models, that is the models of $\neg\psi$ and $\neg\phi$.

In fact the choice between the two alternative approaches must be guided by the meaning of the relations on the families of sets. If $(\mathcal{K}, >)$ is interpreted in terms of relative certainty as in possibilistic logic, it is natural to compare the subsets of falsified formulas of $\mathcal{K}$ for assessing the relative plausibility of interpretations. Indeed, an interpretation ω is all the less plausible as it violates more certain propositions.

In the same way, starting from a plausibility relation on the interpretations $(\Omega, \rhd)$, we can express the idea of relative certainty $\phi \succ \psi$ on the language, by comparing sets of models of $\neg\psi$ and $\neg\phi$; for instance, in the case of a total order, a relation of comparative necessity, dual to comparative possibility, can be defined by $\phi \succ_N \psi$ iff $\neg\psi \succ_o \neg\phi$ [1].

This approach was thoroughly studied within the possibilistic framework for completely ordered bases [2], but much less often in the partially ordered case [22].

Some questions will arise naturally from this research program:

- Is the partial preorder $\succ$ built on $\mathcal{L}$ from $(\Omega, \rhd)$ compatible with $(\mathcal{K}, >)$? A strict meaning of compatibility would require that $>$ is preserved and refined. Note that

[1] Relation $\succ_o$ is introduced in Definition 1.

this is not the case in possibilistic logic if the lower bounds on certainty weights are not in conformity with classical deduction. Here again, it may happen that the relation $\succ$ on formulas induced from $\rhd$ does not preserve the original ordering $(\mathcal{K}, >)$, since the latter can be in conflict with semantic entailment, if supplied by some expert.

- Is the preorder built on Ω from $(\mathcal{K}, >)$ unique? The answer is almost obviously no, as it depends on how the ordering $>$ is understood in terms of a relation between sets of models of formulas appearing in K.
- Is it still possible to use a principle of minimal commitment in order to select a complete preorder on Ω in a non arbitrary way? In possibilistic logic [2], this is the principle of minimal specificity that yields the least informative possibility distribution on Ω (akin to the most compact ranking in system Z [26]).

The two transformations: from $(\mathcal{K}, >)$ to $(\Omega, \rhd)$ and from $(\Omega, \rhd)$ to $(\mathcal{L}, \succ)$ can be reduced to the problem of extending a partial order on a set S to a partial order on the set of the subsets of S, discussed in Section 4.

6 Optimistic Dominance on Partially Ordered Belief Bases

As in possibilistic logic, we assume that the relation $>$ expresses relative certainty, therefore we use the definitions based on falsified formulas. According to the previous sections, two approaches can be followed, using the weak optimistic dominance and its preadditive refinement. In the following, we consider both approaches consecutively. We do not consider the strong optimistic dominance as it allows to compare much less subsets, and we restrict here to strict dominance.

6.1 Weak Optimistic Dominance Semantics

Let $(\mathcal{K}, >)$ be a finite partially ordered set of formulas of the propositional language $\mathcal{L}$ build on V. $\mathcal{K}(\omega)$ (resp. $\overline{\mathcal{K}(\omega)}$) denotes the subset of formulas of $\mathcal{K}$ satisfied (resp. falsified) by the interpretation $\omega \in \Omega$. $[\phi]$ denotes the set of the models of ϕ, a subset of Ω.

Definition 5. *[From $(\mathcal{K}, >)$ to $(\Omega, \rhd)$] $\forall \omega, \omega' \in \Omega$, $\omega \rhd_{wos} \omega'$ iff $\overline{\mathcal{K}(\omega')} \succ_{wos} \overline{\mathcal{K}(\omega)}$*

In the spirit of possibilistic logic, it defines the dominance on interpretations in terms of the violation of the most certain formulas. But here these formulas may be incomparable.

Definition 6. *[From $(\Omega, \rhd)$ to $(\mathcal{L}, \succ_N)$] $\forall \phi, \psi \in \mathcal{L}$, $\phi \succ_N \psi$ iff $\overline{[\psi]} \rhd_{wos} \overline{[\phi]}$.*

In the case of a total order, it would define a necessity relation on the language. The partially ordered deductive closure of $(\mathcal{K}, >)$ is then defined by

$$\mathcal{C}(\mathcal{K}, >)_{\succ_N} = \{(\phi, \psi) \in \mathcal{L}^2 : \phi \succ_N \psi\}.$$

And we denote $(\phi, \psi) \in \mathcal{C}(\mathcal{K}, >)_{\succ_N}$ by $\mathcal{K} \models_{wos} \phi \succ_N \psi$. Besides, in agreement with [21], one may extract from $\mathcal{C}(\mathcal{K}, >)_{\succ_N}$ the set of accepted beliefs when ϕ is known to be true as $\mathcal{A}_\phi(\mathcal{K}, >)_{\succ_N} = \{\psi : (\phi \to \psi, \phi \to \neg\psi) \in \mathcal{C}(\mathcal{K}, >)_{\succ_N}\}$. Note that these are generic definitions that make sense for any variant of the optimistic strict order on $\mathcal{K}$.

Proposition 7.

- *the relation $\rhd_{wos}$ respects inclusion: If $\mathcal{K}(\omega) \subseteq \mathcal{K}(\omega')$ then $\omega \rhd_{wos} \omega'$ does not hold; and $\rhd_{wos}$ is orderly too.*
- *If the relation $>$ is the strict part of a total preorder, possibilistic logic is recovered (order "best out" in [3]).*
- *If ϕ is a logical consequence of ψ, it does not hold that $\psi \succ_N \phi$.*
- *$\succ_N$ verifies the converse of the stability for intersection: if $\phi \wedge \chi \succ_N \psi \wedge \chi$, then $\phi \succ_N \psi$.*

As a consequence, if the partial order on $\mathcal{K}$ violates the partial order induced by classical inference, $\succ_N$ will not refine it, but will correct it.

Example 8. $(\mathcal{K}, >) = \{x > x \wedge y\}$. *As usual, the four interpretations are denoted by* $xy, x\bar{y}, \bar{x}y, \bar{x}\bar{y}$.

Clearly $\overline{\mathcal{K}(xy)} = \varnothing, \overline{\mathcal{K}(x\bar{y})} = \{x \wedge y\}, \overline{\mathcal{K}(\bar{x}y)} = \overline{\mathcal{K}(xy)} = \mathcal{K}$. *Hence,* $xy \rhd_{wos} \{x\bar{y}, \bar{x}y, \bar{x}\bar{y}\}$ *and* $x\bar{y} \rhd_{wos} \{\bar{x}y, \bar{x}\bar{y}\}$. *Then it is easy to see that* $\mathcal{K} \models_{wos} x \succ_N y$ *(since* $[\neg y] = \{x\bar{y}, \bar{x}\bar{y}\} \rhd_{wos} [\neg x] = \{\bar{x}y, \bar{x}\bar{y}\}$ *but* $\mathcal{K} \not\models_{wos} y \succ_N x \wedge y$, *since it does not hold that* $[\neg x \vee \neg y] = \{\bar{x}y, x\bar{y}, \bar{x}\bar{y}\} \rhd_{wos} [\neg y] = \{x\bar{y}, \bar{x}\bar{y}\}$.

If (by mistake) we set $(\mathcal{K}', >) = \{x \wedge y > x\}$, *note that we still have that* $xy \rhd_{wos} \{x\bar{y}, \bar{x}y, \bar{x}\bar{y}\}$ *but not* $x\bar{y} \rhd_{wos} \{\bar{x}y, \bar{x}\bar{y}\}$. *Then*

$\mathcal{K}' \not\models_{wos} x \wedge y \succ_N x$, *that is, we correct this inconsistency via the semantics.*

However, the fact, pointed out in Section 2, that a partial order over a power set cannot be characterized by a single partial order on the set of elements may cause some available pieces of knowledge in $(\mathcal{K}, >)$ to be lost in $\mathcal{C}(\mathcal{K}, >)_{\succ_N}$, as shown thereafter.

Example 9. *Let* $(\mathcal{K}, >) = \{x, \neg x \vee y, x \wedge y, \neg x\}$ *be a partially ordered base, where* $>$ *is the strict partial order given as follows:* $\neg x \vee y > x \wedge y > \neg x$ *and* $x > \neg x$.
Let us apply the definitions 5 and 6:

- *From* $(\mathcal{K}, >)$ *to* $(\Omega, \rhd)$: *we obtain* $xy \rhd_{wos} \{\bar{x}y, x\bar{y}, \bar{x}\bar{y}\}$
- *From* $(\Omega, \rhd)$ *to* $(\mathcal{L}, \succ_N)$: *we obtain* $x \succ_N \neg x$, $x \wedge y \succ_N \neg x$ *and* $\neg x \vee y \succ_N \neg x$ *but not* $\neg x \vee y \succ_N x \wedge y$

We notice that, in the final order over formulas, $\neg x \vee y$ *and* $x \wedge y$ *become incomparable. The reason is that some information has been lost when going from* $(\mathcal{K}, >)$ *to* $(\Omega, \rhd)$. *Indeed, if the strict partial order* $>$ *of the base* $\mathcal{K}$ *is interpreted as the strict part* $\succ_N$ *of a necessity ordering, applying Definition 6 enables the following constraints to be obtained:*

- *Due to* $\neg x \vee y \succ_N x \wedge y$ *we must have* $\bar{x}y \rhd x\bar{y}$ *or* $\bar{x}\bar{y} \rhd x\bar{y}$
- *Due to* $x \wedge y \succ_N \neg x$ *we must have* $xy \rhd x\bar{y}$ *and* $xy \rhd \bar{x}y$ *or* $x\bar{y} \rhd \bar{x}y$ *and* $xy \rhd \bar{x}\bar{y}$ *or* $x\bar{y} \rhd \bar{x}\bar{y}$
- *Due to* $x \succ_N \neg x$ *we must have* $xy \rhd \bar{x}y$ *or* $x\bar{y} \rhd \bar{x}y$ *and* $xy \rhd \bar{x}\bar{y}$ *or* $x\bar{y} \rhd \bar{x}\bar{y}$

It is easy to see that these constraints imply that $xy \rhd \{\bar{x}y, x\bar{y}, \bar{x}\bar{y}\}$ *and* $(\bar{x}y \rhd x\bar{y}$ *or* $\bar{x}\bar{y} \rhd x\bar{y})$. *That is stronger than the partial order* $\rhd_{wos}$ *and not representable by a single partial order.*

One observes that the impossibility of representing the partial order $(\mathcal{K}, >)$ by a partial ordering on interpretations is the cause for losing the piece of information $\neg x \vee y \succ_N x \wedge y$. It suggests that the partially ordered deductive closure $\mathcal{C}(\mathcal{K}, >)_{\succ_N}$ is too weak to account for semantic entailment in partially ordered knowledge bases.

6.2 Preadditive Semantics

Under the preadditive semantics, the weak optimistic semantics in Definitions 5 and 6 is strengthened as follows:

Definition 7. *Let ω, ω' be two interpretations:*

- *From $(\mathcal{K}, >)$ to $(\Omega, \rhd)$: $\omega \rhd_d^{wos} \omega'$ iff $\mathcal{K}(\omega) \succ_d^{wos} \mathcal{K}(\omega')$.*
- *From $(\Omega, \rhd)$ to $(\mathcal{L}, \succ_d)$: $\phi \succ_d \psi$ iff $[\phi] \rhd_d^{wos} [\psi]$.*

The notion of semantic consequence and deductive closure are defined similarly, replacing $\succ_N$ by $\succ_d$. The following results hold:

Proposition 8. *It is clear that:*

- *the relation $\rhd_d^{wos}$ strictly respects inclusion: If $\mathcal{K}(\omega) \subset \mathcal{K}(\omega')$ then $\omega' \rhd_d^{wos} \omega$.*
- *If ϕ is a proper logical consequence of ψ, then $\phi \succ_d \psi$.*
- *if $\chi \wedge (\phi \vee \psi) = \bot$ then $\phi \succ_d \psi$ implies $\phi \wedge \chi \succ_d \psi$.*
- *$\phi \wedge \chi \succ_d \psi \wedge \chi$ implies $\phi \succ_d \psi$*

Example 9 (continued)

- *From $(\mathcal{K}, >)$ to $(\Omega, \rhd)$: $xy \rhd_d^{wos} \{\bar{x}y, x\bar{y}, \bar{x}\bar{y}\}$;*
- *From $(\Omega, \rhd)$ to $(\mathcal{L}, \succ_d)$:*
  - $x \succ_d x \wedge y \succ_d \neg x$
  - $\neg x \vee y \succ_d x \wedge y \succ_d \neg x$

We notice that the relation $\succ_d$ has preserved and extended *the initial strict partial order.*

However, the relation $\succ_d$ does not always preserve the initial strict partial order as shown thereafter.

Example 10. *Let $(\mathcal{K}, >) = \{x, \neg x \vee \neg y, x \wedge y, \neg x\}$ be a partially ordered base, where $>$ is the strict partial order given as follows: $\neg x \vee \neg y > x \wedge y > \neg x$ and $x > \neg x$.*

- *From $(\mathcal{K}, >)$ to $(\Omega, \rhd)$: we obtain $x\bar{y} \rhd_d^{wos} \{\bar{x}y, xy, \bar{x}\bar{y}\}$*
- *From $(\Omega, \rhd)$ to $(\mathcal{L}, \succ_d)$: we obtain $\neg x \vee \neg y \succ_d x \wedge y$, $\neg x \vee \neg y \succ_d \neg x$, $x \succ_d x \wedge y$ but not $x \wedge y \succ_d \neg x$.*

Finally, let us consider the particular case of flat bases, interpreted as containing formulas that are all equivalent or all incomparable. The former case corresponds to classical logic. Suppose formulas in $\mathcal{K}$ are either incomparable or equivalent (for no $\phi, \psi \in \mathcal{K}$ do we have $\phi > \psi$). Then the set of logical consequences is no longer flat. The induced orderings are as follows.

- From (flat) $\mathcal{K}$ to $(\Omega, \rhd)$: $\omega' \rhd_d^{wos} \omega$ iff $\mathcal{K}(\omega') \supset \mathcal{K}(w)$
- From $(\Omega, \rhd)$ to $(\mathcal{L}, \succ_d)$: $\phi \succ_d \psi$ iff $\forall \omega' \in [\psi] \setminus [\phi], \exists w \in [\phi] \setminus [\psi]$ such that $\mathcal{K}(\omega) \supset \mathcal{K}(\omega')$.

Thus it is easy to see that, for flat bases, $\phi \succ_d \psi$ if and only if ϕ is a proper logical consequence of ψ, which enriches the semantics of classical logic.

7 Towards Syntactic Inference with Partially Ordered Belief Bases

Once the semantics of partially ordered belief bases and their deductive closure are well-defined, the next step is to devise a syntactic inference relation $\vdash$ that enables to directly build the ordered deductive closure $\mathcal{C}(\mathcal{K}, >)_{\succ}$ from $(\mathcal{K}, >)$ (for a suitable choice of $\succ$) in agreement with the semantics, namely $(\mathcal{K}, >) \vdash \phi \succ \psi$, whenever $(\phi, \psi) \in \mathcal{C}(\mathcal{K}, >)_{\succ}$. It appears that this question has been little discussed in the partially ordered case, except by Halpern [8] and more recently by Benferhat and Prade [7], with very different approaches.

Several methods of inference from a partially ordered belief base have been proposed. Is is possible to:

1. map $(\mathcal{K}, >)$ to a partially ordered set $(L_{\mathcal{K}}, >)$ of absolute levels of certainty, and replace $(\mathcal{K}, >)$ by a possibilistic knowledge base B made of pairs $(\phi, \lambda), \lambda \in L_{\mathcal{K}}$, such that whenever $\phi > \psi$, $(\phi, \lambda), (\psi, \mu) \in B$ and $\lambda > \mu$. Then we can adapt the techniques of possibilistic logic to this setting.
2. consider a partial order as a family of total orders that extend it. So, a partially ordered base is seen as a set of (virtual) stratified bases.
3. reason directly with formulas $\phi > \psi$ in a suitable language.
4. reason in a classical way with consistent subsets of formulas extracted using the partial order.

The first approach was studied by Benferhat and Prade [7]. Let $(L_{\mathcal{K}}, >)$ be a finite ordered set associated with $\mathcal{K}$ by a homomorphism $\iota : \mathcal{K} \to L_{\mathcal{K}}$ such that $\phi \geq \psi \in \mathcal{K}$ iff $\iota(\phi) \geq \iota(\psi)$. Let us denote $\{\mu : \mu \geq \lambda\}$ by $\lambda^{\uparrow}$. The inequality $\lambda_1 \geq \lambda_2$ in $(L_{\mathcal{K}}, \geq)$ is encoded by $A_2 \vee \neg A_1$ with $A_i = \lambda_i^{\uparrow}$, and the pair (ϕ, λ) is encoded by $\neg A \vee \phi$, with $A = \lambda^{\uparrow}$. Then classical propositional deduction can be used.

Actually, there is a more direct way to apply possibilistic logic to the partially ordered case. It is well-known that in standard possibilistic logic $B \vdash (\phi, \lambda) \iff B_\lambda \vdash \phi$ where B_λ is the set of formulas with weights at least λ. In the partially ordered case, we could define, when $\psi \in \mathcal{K}$, $\mathcal{K} \vdash \phi \succ \psi$ by $\mathcal{K}_{\psi}^{>} \vdash \phi$ where $\mathcal{K}_{\psi}^{>} = \{\alpha \in \mathcal{K} : \alpha > \psi\}$ and likewise $\mathcal{K} \vdash \phi \succeq \psi$ by $\mathcal{K}_{\psi}^{\geq} \vdash \phi$ where $\mathcal{K}_{\psi}^{\geq} = \{\alpha \in \mathcal{K} : \psi \not> \alpha\}$. For instance if statements $\phi_i > \psi_i$ in $(K, >)$ are interpreted on the set 2^{Ω} by $\overline{[\psi_i]} \rhd \overline{[\phi_i]}$ where the relation $\rhd$ satisfies Negligibility and Orderliness, it does hold that $\overline{[\psi]} \rhd \overline{[\phi]}$ whenever $\mathcal{K}_{\psi}^{>} \vdash \phi$. A particular case occurs when $\rhd$ is $\succ_{wos}$.

However, as shown in Example 9, if the consequence $\phi \succ \psi$ is interpreted as $\phi \succ_N \psi$ using Definitions 5 and 6 (that is,via a partial order on interpretations derived from $(K, >)$) it may fail to hold that $\phi \succ_N \psi$ whenever $\mathcal{K}_{\psi}^{>} \vdash \phi$ (in Example 9, $\phi > \psi$ appears in $(K, >)$, and is absent from the semantic closure). This fact indicates a weakness in the semantics based on a partial order on interpretations, as opposed to a more complex semantics based on the partial ordering on subsets of interpretations reflecting $(K, >)$.

The second approach is described by Yahi *et al.* [25]. $(\mathcal{K}, >)$ is viewed as a set of possible stratifications of $\mathcal{K}$. So $\psi \geq \phi$ of $(\mathcal{K}, >)$ means that ψ is more certain than ϕ (in the sense of possibilistic logic) in all the stratified bases compatible with $(\mathcal{K}, >)$. Results in the previous section indicate the strong link between this view and the weak optimistic relation $\succ_{wos}$ (hence its dual $\succ_N$ and their preadditive refinement $\succ_d$). Note that in the first approach [7], the partial order on $L_{\mathcal{K}}$ is actually viewed as a set of possible total orders. The weights are symbolic in the sense of being partially unknown quantities on a totally ordered scale. To use this approach in practice may turn out to be difficult, because the set of total extensions of a given partial order may be large.

In the third approach, the key idea is to consider expressions of the form $\phi \succ \psi$ as the basic syntactic entities of the language encoding the preferences. It requires a higher order language for handling atomic propositions of the form $\phi \succ \psi$, their conjunctions, disjunctions and negations, with specific axioms for describing properties of the relation $\succ$. For instance, with the axiom: if $\psi \models \phi$ then $\neg(\psi \succ \phi)$, semantical contradictions will be found, thus enabling to repair the partially ordered base. This approach, which is the most natural one, goes back to Lewis [9] conditional logics (see also Hájek [27], p. 212) in the case of total preorders, for possibility theory). Halpern [8] has outlined such a logic to handle the relation $\succ_{wos}$. This is certainly the most general approach with the richest language. Especially it would readily allow for a semantics in terms of a partial order over the set of subsets of interpretations of the language, which would obviate difficulties pointed out by Examples 9 and 10 when we use a partial ordering on interpretations. However only a subset of the consequences $(K, >)$ from a set of $\phi \succ \psi$ statements will correspond to the (properly defined) semantic closure $\mathcal{C}(\mathcal{K}, >)$ (since, for instance, the latter does not contain disjunctions of such statements).

In the fourth approach [22], the partial order on $\mathcal{K}$ is just used to select preferred consistent subsets of formulas, and the deductive closure is a classical set of accepted beliefs. So, as pointed out in Benferhat and Yahi [28], the deductive closure of a partially ordered base $(\mathcal{K}, >)$ is just a deductively closed set (in the classical sense), obtained from preferred subbases. Then the inference $(\mathcal{K}, >) \vdash \phi$ is defined by: ϕ is consequence of all the preferred subsets of formulas. The notion of preference can be defined in various ways based on the partial order. This order between formulas is to some extent lost by the process of inference. In particular this kind of approach reduces to classical inference when $\mathcal{K}$ is classically consistent. By construction, this approach does not enable to deduce preferences between formulas, but essentially extracts accepted beliefs.

In the future, we plan to investigate whether or not the above syntactic inference schemes are sound (and if possible complete) with respect to our notion of partial-order-driven semantic closure. We have already noticed that a semantics based on a single partial ordering over interpretations may be problematic as seen in Example 9. This result motivates the use of a modal-like language with formulas of the form $\phi > \psi$, $\phi \geq \psi$, or $\phi \sim \psi$ with relational semantics on the powerset of the set of interpretations of the language where ϕ, ψ are expressed, whereby $\phi > \psi$ is viewed as a relation between $\overline{[\psi_i]}$ and $\overline{[\phi_i]}$, etc. Then the properties of the semantic relation can be used as inference rules at the syntactic level. However, one may wish to restrict the inference machinery to consequences of the form $\phi \succ \psi$ and $\phi \succeq \psi$.

8 Conclusion

The issue addressed in this paper concerns the extension of possibilistic logic when formulas weighted by certainty levels are replaced by a partial order on the belief base.

Defining proper semantics for such partially ordered bases requires the study of how to go from a partial order on elements to a partial order on subsets of such elements and conversely. Some preliminary results are offered in this paper. They indicate that many important concepts in the case of complete orders have several non equivalent definitions in the partial case. When going from a partial order on a set to a partial order on its subsets, it seems that the weak optimistic relation possesses the best properties. Moreover it seems that a straightforward adaptation of possibilistic logic to the partial order setting is not possible.

The question then becomes the one of finding the most natural understanding of a base partially ordered in terms of relative certainty. Our paper explains how to go from formulas to models and back, thus defining a semantic notion of deductive closure. We indicate that the expressive power of a partial order on the set of interpretations is limited, and one must stick to a partial order on its power set, or alternatively a set of total orders on the set of interpretations, to be on the safe side. Existing works proposing proof methods have been reviewed, but they all consider different points of view on the definition of inference in the partially ordered context, sometimes with unclear semantics. In contrast, our purpose is to eventually define a semantic closure that preserves and extends the partial order on $\mathcal{K}$ to the whole language, while correcting the initial assessment to make it comply with the classical deduction. Once this issue has been clarified, we have to choose an appropriate syntax, an axiomatization and a syntactic inference method. Some hints are provided above, but this is left for further research.

This work has potential applications for the revision and the fusion of beliefs, as well as preference modeling [29].

References

1. Rescher, N.: Plausible Reasoning. Van Gorcum, Amsterdam (1976)
2. Dubois, D., Lang, J., Prade, H.: Possibilistic logic. In: Gabbay, D., Hogger, C., Robinson, J., Nute, D. (eds.) Handbook of Logic in Artificial Intelligence and Logic Programming, vol. 3, pp. 439–513. Oxford University Press (1994)
3. Benferhat, S., Cayrol, C., Dubois, D., Lang, J., Prade, H.: Inconsistency management and prioritized syntax-based entailment. In: Bajcsy, R. (ed.) Proc. of the 13th IJCAI, pp. 640–645. Morgan-Kaufmann, Chambéry (1993)
4. Dubois, D., Prade, H.: Epistemic entrenchment and possibilistic logic. Artificial Intelligence 50, 223–239 (1991)
5. Benferhat, S., Dubois, D., Kaci, S., Prade, H.: Logique possibiliste et fusion d'informations. Technique et Science Informatique 22, 1035–1064 (2003)
6. Dubois, D., Prade, H., Schockaert, S.: Régles et métarègles en théorie des possibilités. Revue d'Intelligence Artificielle 26, 773–793 (2012)
7. Benferhat, S., Prade, H.: Encoding formulas with partially constrained weights in a possibilistic-like many-sorted propositional logic. In: Kaelbling, L.P., Saffiotti, A. (eds.) IJCAI, pp. 1281–1286. Professional Book Center (2005)
8. Halpern, J.Y.: Defining relative likelihood in partially-ordered preferential structures. Journal of Artificial intelligence Research 7, 1–24 (1997)
9. Lewis, D.: Counterfactuals and comparative possibility. Journal of Philosophical Logic 2, 418–446 (1973)

10. Dubois, D.: Belief structures, possibility theory and decomposable confidence measures on finite sets. Computers and Artificial Intelligence (Bratislava) 5, 403–416 (1986)
11. Alchourrón, C.E., Gärdenfors, P., Makinson, D.: On the logic of theory change: partial meet contraction and revision functions. Journal of Symbolic Logic 50, 510–530 (1985)
12. Fishburn, P.: Interval Orderings. Wiley, New-York (1987)
13. Benferhat, S., Dubois, D., Prade, H.: Towards a possibilistic logic handling of preferences. Appl. Intell. 14, 303–317 (2001)
14. Kaci, S., van der Torre, L.: Reasoning with various kinds of preferences: Logic, non-monotonicity and algorithms. Annals of Operations Research 163, 89–114 (2008)
15. Kaci, S.: Working With Preferences: Less Is More. Springer (2012)
16. Fishburn, P.C.: The axioms of subjective probability. Statistical Science 1, 335–358 (1986)
17. Dubois, D., Fargier, H., Prade, H.: Possibilistic likelihood relations. In: Proceedings of 7th International Conference on Information Processing and Management of Uncertainty in Knowledge-based Systems (IPMU 1998), Paris, Editions EDK, pp. 1196–1202 (1998)
18. Friedman, N., Halpern, J.Y.: Plausibility measures: A user's guide. In: Proc. of the Eleventh Annual Conference on Uncertainty in Artificial Intelligence, Montreal, Quebec, August 18-20, pp. 175–184 (1995)
19. Dubois, D., Prade, H.: Numerical representations of acceptance. In: Proc. of the Eleventh Annual Conference on Uncertainty in Artificial Intelligence, Montreal, Quebec, August 18-20, pp. 149–156 (1995)
20. Kraus, S., Lehmann, D., Magidor, M.: Nonmonotonic reasoning, preferential models and cumulative logics. Artificial Intelligence 44, 167–207 (1990)
21. Dubois, D., Fargier, H., Prade, H.: Ordinal and probabilistic representations of acceptance. J. Artif. Intell. Res. (JAIR) 22, 23–56 (2004)
22. Benferhat, S., Lagrue, S., Papini, O.: Reasoning with partially ordered information in a possibilistic logic framework. Fuzzy Sets and Systems 144, 25–41 (2004)
23. Cayrol, C., Royer, V., Saurel, C.: Management of preferences in assumption-based reasoning. In: Bouchon-Meunier, B., Valverde, L., Yager, R.R. (eds.) IPMU 1992. LNCS, vol. 682, pp. 13–22. Springer, Heidelberg (1993)
24. Geffner, H.: Default reasoning: Causal and Conditional Theories. MIT Press (1992)
25. Yahi, S., Benferhat, S., Lagrue, S., Sérayet, M., Papini, O.: A lexicographic inference for partially preordered belief bases. In: Brewka, G., Lang, J. (eds.) KR, pp. 507–517. AAAI Press (2008)
26. Pearl, J.: System Z: A natural ordering of defaults with tractable applications to default reasoning. In: Vardi, M. (ed.) Proc. of the 3rd Conference of Theoretical Aspects of Reasoning about Knowledge, pp. 121–135. Morgan-Kaufmann (1990)
27. Hájek, P.: The Metamathematics of Fuzzy Logics. Kluwer Academic (1998)
28. Benferhat, S., Yahi, S.: Etude comparative des relations d'inférence à partir de bases de croyance partiellement ordonnées. Revue d'Intelligence Artificielle 26, 39–61 (2012)
29. Dubois, D., Prade, H., Touazi, F.: Conditional preference nets and possibilistic logic. In: van der Gaag, L.C. (ed.) ECSQARU 2013. LNCS (LNAI), vol. 7958, pp. 181–193. Springer, Heidelberg (2013)

The Structure of Oppositions in Rough Set Theory and Formal Concept Analysis - Toward a New Bridge between the Two Settings

Davide Ciucci[1], Didier Dubois[2], and Henri Prade[2]

[1] DISCo, Università di Milano–Bicocca, viale Sarca 336 U14, 20126 Milano, Italy
ciucci@disco.unimib.it

[2] IRIT, Université Paul Sabatier, 118 route de Narbonne, 31062 Toulouse cedex 9, France
{dubois,prade}@irit.fr

Abstract. Rough set theory (RST) and formal concept analysis (FCA) are two formal settings in information management, which have found applications in learning and in data mining. Both rely on a binary relation. FCA starts with a formal context, which is a relation linking a set of objects with their properties. Besides, a rough set is a pair of lower and upper approximations of a set of objects induced by an indistinguishability relation; in the simplest case, this relation expresses that two objects are indistinguishable because their known properties are exactly the same. It has been recently noticed, with different concerns, that any binary relation on a Cartesian product of two possibly equal sets induces a cube of oppositions, which extends the classical Aristotelian square of oppositions structure, and has remarkable properties. Indeed, a relation applied to a given subset gives birth to four subsets, and to their complements, that can be organized into a cube. These four subsets are nothing but the usual image of the subset by the relation, together with similar expressions where the subset and / or the relation are replaced by their complements. The eight subsets corresponding to the vertices of the cube can receive remarkable interpretations, both in the RST and the FCA settings. One facet of the cube corresponds to the core of RST, while basic FCA operators are found on another facet. The proposed approach both provides an extended view of RST and FCA, and suggests a unified view of both of them.

Keywords: rough set, formal concept analysis, square of oppositions, possibility theory.

1 Introduction

Rough set theory (RST) [31,32,33,36,35,34] and formal concept analysis (FCA) [2,41,18,17] are two theoretical frameworks in information management which have been developed almost independently for thirty years, and which are of particular interest in learning and in data mining. Quite remarkably, both rely

C. Beierle and C. Meghini (Eds.): FoIKS 2014, LNCS 8367, pp. 154–173, 2014.

on a binary relation. In FCA, the basic building block is a relation that links a set of objects with a set of properties, called a formal context. A rough set is a pair of lower and upper approximations of a set of objects induced by an indistinguishability relation, objects being indistinguishable in particular when they have exactly the same known properties.

Besides, in a recent paper dealing with abstract argumentation [1], it has been noticed that in fact any binary relation is associated with a remarkable and rich structure, called cube of oppositions, which is closely related to the Aristotelian square of oppositions. The purpose of this paper is to take advantage of this cube for revisiting both RST and FCA in a unified manner. The expected benefit is twofold. On the one hand, it may provide an enriched view of each framework individually, on the other hand it should contribute to a better understanding of the relations and complementarities between the two frameworks.

The paper is organized as follows. In the next section, we present a detailed account of the structure of oppositions associated with a binary relation, which is represented by a cube laying bare four different squares of oppositions, each of which may be completed into hexagons in a meaningful way. This systematic study substantially extends preliminary remarks made in [10,1]. Then, Sections 3 and 4 respectively restate RST and FCA in the setting of this cube and its hexagons, providing an extended view of their classical frameworks. This leads to new results on the rough set cube and hexagons, which are then related to previous results on oppositions in rough sets [6]. Similarly this leads to a renewed presentation of results in FCA. Section 5, after surveying the various attempts in the literature at bridging or mixing RST and FCA in one way or another, suggests new directions of research which can benefit from the unified view presented in this paper, before concluding.

2 Structure of Oppositions Induced by a Binary Relation

Let us start with a refresher on the Aristotelian square of opposition [30]. The traditional square involves four logically related statements exhibiting universal or existential quantifications: it has been noticed that a statement (A) of the form "every x is p" is negated by the statement (O) "some x is not p", while a statement like (E) "no x is p" is clearly in even stronger opposition to the first statement (A). These three statements, together with the negation of the last one, namely (I) "some x is p", give birth to the Aristotelian square of opposition in terms of quantifiers $A : \forall x\ p(x)$, $E : \forall x\ \neg p(x)$, $I : \exists x\ p(x)$, $O : \exists x\ \neg p(x)$, pictured in Figure 1. Such a square is usually denoted by the letters A, I (affirmative half) and E, O (negative half). The names of the vertices come from a traditional Latin reading: **A**ff**I**rmo, n**E**g**O**). As can be seen, different relations hold between the vertices. Namely,

- (a) A and O are the negation of each other, as well as E and I;
- (b) A entails I, and E entails O (we assume that there are some x);
- (c) A and E cannot be true together, but may be false together;
- (d) I and O cannot be false together, but may be true together.

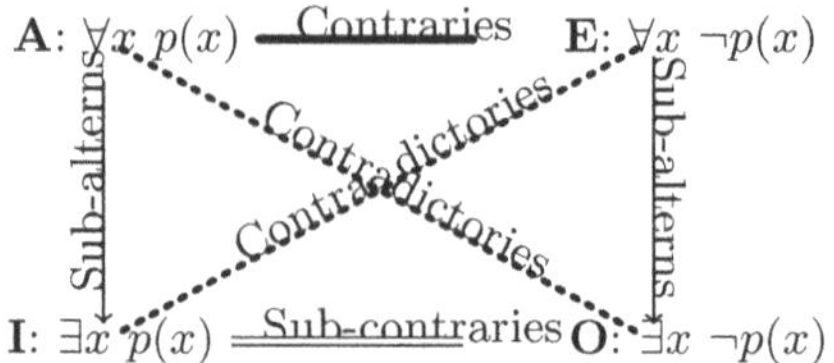

Fig. 1. Square of opposition

Another well-known instance of this square is in terms of the *necessary* ($\Box$) and *possible* ($\Diamond$) modalities, with the following reading $A : \Box p$, $E : \Box\neg p$, $I : \Diamond p$, $O : \Diamond\neg p$, where $\Diamond p =_{def} \neg\Box\neg p$ (with $p \neq \bot, \top$).

2.1 The Square of Relations

Let us now consider a binary relation R on a Cartesian product $X \times Y$ (one may have $Y = X$). We assume $R \neq \emptyset$. Let xR denote the set $\{y \in Y | (x, y) \in R\}$, and we write xRy when $(x, y) \in R$ holds, and $\neg(xRy)$ when $(x, y) \notin R$. Let R^t denote the transpose relation, defined by xR^ty if and only if yRx, and yR^t will be also denoted as $Ry = \{x \in X | (x, y) \in R\}$.

Moreover, we assume that $\forall x,\ xR \neq \emptyset$, which means that the relation R is *serial*, namely $\forall x, \exists y$ such that xRy; this is also referred to in the following as the X-*normalization condition*. In the same way R^t is also supposed to be serial, i.e., $\forall y,\ Ry \neq \emptyset$. We further assume that the complementary relation $\overline{R}$ ($x\overline{R}y$ iff $\neg(xRy)$), and its transpose are also serial, i.e. $\forall x,\ xR \neq Y$ and $\forall y,\ Ry \neq X$.

Let S be a subset of Y. The relation R and the subset S, also considering its complement $\overline{S}$, give birth to the two following subsets of X, namely the (left) images of S and $\overline{S}$ by R

$$R(S) = \{x \in X | \exists s \in S, xRs\} = \{x \in X |\ \ S \cap xR \neq \emptyset\} \tag{1}$$

$$R(\overline{S}) = \{x \in X | \exists s \in \overline{S}, xRs\}$$

and their complements

$$\overline{R(S)} = \{x \in X | \forall s \in S, \neg(xRs)\}$$

$$\overline{R(\overline{S})} = \{x \in X | \forall s \in \overline{S}, \neg(xRs)\} = \{x \in X |\ \ xR \subseteq S\} \tag{2}$$

The four subsets thus defined can be nicely organized into a square of opposition. See Figure 2. Indeed, it can be checked that the set counterparts of the relations existing between the logical statements of the traditional square of oppositions still hold here. Namely,

- (a) $\overline{R(\overline{S})}$ and $R(\overline{S})$ are complements of each other, as $\overline{R(S)}$ and $R(S)$; they correspond to the diagonals of the square;

- (b) $\overline{R(\overline{S})} \subseteq R(S)$, and $\overline{R(S)} \subseteq R(\overline{S})$,

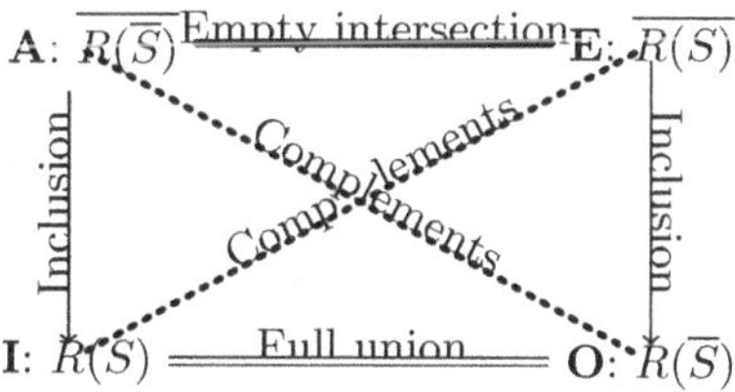

Fig. 2. Square of oppositions induced by a relation R and a subset S

thanks to the X-*normalization condition* $\forall x,\ xR \neq \emptyset$. These inclusions are represented by vertical arrows in Figure 2;

- (c) $\overline{R(\overline{S})} \cap \overline{R(S)} = \emptyset$ (this empty intersection corresponds to a thick line in Figure 2), and one may have $\overline{R(\overline{S})} \cup \overline{R(S)} \neq Y$;

- (d) $R(S) \cup R(\overline{S}) = X$ (this full union corresponds to a double thin line in Figure 2), and one may have $R(S) \cap R(\overline{S}) \neq \emptyset$.

Conditions (c)-(d) hold also thanks to the X-normalization of R.

Note that one may still have a modal logic reading of this square where R is viewed as an accessibility relation, and S as the set of models of a proposition.

2.2 The Cube of Relations

Let us also consider the complementary relation $\overline{R}$, namely $x\overline{R}y$ if and only if $\neg(xRy)$. We further assume that $\overline{R} \neq \emptyset$ (i.e., $R \neq X \times Y$). Moreover we have also assumed the X-normalization of $\overline{R}$, i.e. $\forall x, \exists y\ \neg(xRy)$. In the same way as previously, we get four other subsets of X from $\overline{R}$. Namely,

$$\overline{R}(\overline{S}) = \{x \in X | \exists s \in \overline{S}, \neg(xRs)\} = \{x \in X | S \cup xR \neq X\} \tag{3}$$

$$\overline{R}(S) = \{x \in X | \exists s \in S, \neg(xRs)\}$$

and their complements

$$\overline{\overline{R}(\overline{S})} = \{x \in X | \forall s \notin S, xRs\}$$

$$\overline{\overline{R}(S)} = \{x \in X | \forall s \in S, xRs\} = \{x \in X | S \subseteq xR\} \tag{4}$$

The eight subsets involving R and its complement can be organized into a cube of oppositions [7] (see Figure 3). Similar cubes have been recently exhibited as extending the traditional square of oppositions in terms of quantifiers [10], or in the particular setting of abstract argumentation (for the complement of the attack relation) [1]. As can be seen, the front facet of the cube in Figure 3 is nothing but the square in Figure 2, and the back facet is a similar square associated with $\overline{R}$. Neither the top and bottom facets, nor the side facets are squares of opposition in the above sense. Indeed, condition (a) is violated: Diagonals do not link complements in these squares. More precisely, in the top and bottom squares, diagonals change R into $\overline{R}$ and vice versa; there is no counterpart of condition (b), and either condition (c) holds for the pairs of subsets associated with vertices **A-E** and with vertices **a-e**, while condition (d) fails (top square), or conversely in the bottom square, condition (d) holds for the pairs of subsets associated with vertices **I-O** and with vertices **i-o** while condition (c) fails. For side facets, condition (b) clearly holds, while both conditions (c)-(d) fail. In side facets, vertices linked by diagonals are exchanged by changing R into $\overline{R}$ (and vice versa) and by applying the overall complementation. These diagonals express set inclusions: $\overline{R(S)} = \{x \in X | \forall s \in S, \neg(xRs)\} \subseteq \{x \in X | \exists s \in S, \neg(xRs)\} = \overline{R}(S)$. In the same way, we have $\overline{\overline{R}(\overline{S})} \subseteq R(\overline{S})$, $\overline{\overline{R}(S)} \subseteq R(S)$, and $\overline{R(\overline{S})} \subseteq \overline{R}(\overline{S})$, as pictured in Figure 3.

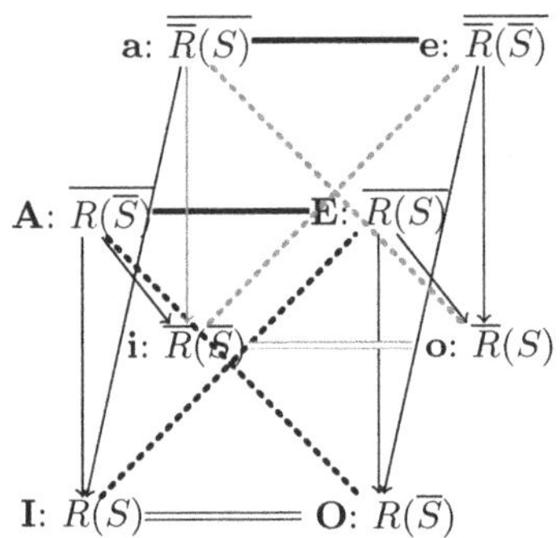

Fig. 3. Cube of oppositions induced by a relation R and a subset S

Moreover, the top and bottom facets exhibit other empty intersection relationships and full union relationships respectively. Indeed in the top facet, e.g. $\overline{\overline{R}(S)} \cap \overline{R(S)} = \emptyset$, since $\overline{\overline{R}(S)} = \{x \in X \mid S \subseteq xR\}$ and $\overline{R(S)} = \{x \in X \mid S \cap xR = \emptyset\}$. Similarly in the bottom facet, e.g. $\overline{R}(S) \cup R(S) = X$, since $\overline{R}(S) = \{x \in X | \exists s \in S, \neg(xRs)\}$ and $R(S) = \{x \in X | \exists s \in S, xRs\}$. This is pictured in Figure 4 (in order not to overload Figure 3).

Thus, while diagonals in front and back facets express complementations, they express inclusions in side facets, empty intersections in top facet, and full union in bottom facets.
It is important to keep in mind that the 4 subsets $\overline{R(\overline{S})}$, $\overline{R(S)}$, $\overline{\overline{R}(S)}$, and $\overline{\overline{R}(\overline{S})}$ (or their complements $R(\overline{S})$, $R(S)$, $\overline{R}(S)$, and $\overline{R}(\overline{S})$) constitute distinct pieces of information in the sense that one cannot be deduced from the others. Indeed the conditions $xR \subseteq S$, $S \cap xR = \emptyset$, $S \subseteq xR$, and $S \cup xR = X$ express the four possible inclusion relations of xR wrt S or $\overline{S}$ and define distinct subsets of X.

The 8 subsets corresponding to the vertices of the cube of oppositions can receive remarkable interpretations, both in the RST and FCA settings. As we shall see in Sections 3 and 4, the front facet of the cube corresponds to the core of RST, while basic FCA operators are on the left-hand side facet. However, before

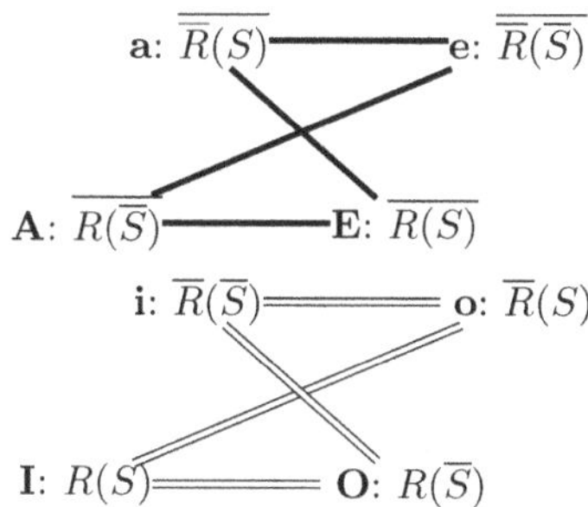

Fig. 4. Top and bottom facets of the cube of oppositions

moving to RST and FCA, it is interesting to complete the different squares corresponding to the facets of the cube into hexagons, as we are going to see.

2.3 From Squares to Hexagons

As proposed and advocated by Blanché [4,5], it is always possible to complete a classical square of opposition into a hexagon by adding the vertices $\mathbf{Y} =_{def} \mathbf{I} \wedge \mathbf{O}$, and $\mathbf{U} =_{def} \mathbf{A} \vee \mathbf{E}$. It fully exhibits the logical relations inside a structure of oppositions generated by the three mutually exclusive situations **A**, **E**, and **Y**, where two vertices linked by a diagonal are contradictories, **A** and **E** entail **U**, while **Y** entails both **I** and **O**. Moreover $\mathbf{I} = \mathbf{A} \vee \mathbf{Y}$ and $\mathbf{O} = \mathbf{E} \vee \mathbf{Y}$. Conversely, three mutually exclusive situations playing the roles of **A**, **E**, and **Y** always give birth to a hexagon [10], which is made of three squares of opposition: **AEOI**, **AYOU**, and **EYIU**, as in Figure 5. The interest of this hexagonal construct has been rediscovered and advocated again by Béziau [3] in the recent years in particular for solving delicate questions in paraconsistent logic modeling.

Applying this idea to the front facet of the cube of oppositions induced by a relation and a subset, we obtain the hexagon of Figure 5, associated with the tri-partition $\{\overline{R(\overline{S})}, \overline{R(S)}, R(S) \cap R(\overline{S})\}$. Note that indeed $R(\overline{S}) = \overline{R(S)} \cup (R(S) \cap R(\overline{S}))$ (since $R(\overline{S}) \supseteq \overline{R(S)}$). Similarly, $R(S) = \overline{R(\overline{S})} \cup (R(S) \cap R(\overline{S}))$. In Figure 5, arrows ($\rightarrow$) indicate set inclusions ($\subseteq$). A similar hexagon is associated with the back facet, changing R into $\overline{R}$.

Another type of hexagon can be associated with side facets. The one corresponding to the left-hand side facet is pictured in Figure 6. Now, not only the arrows of the sides of the hexagon correspond to set inclusions, but also the diagonals (oriented downwards). Indeed $\overline{R(\overline{S})} \subseteq \overline{R}(\overline{S})$ and $\overline{\overline{R}(S)} \subseteq R(S)$. Moreover, since $\overline{R(S)} \subseteq \overline{R}(S)$ and using the inclusions corresponding to the vertical edges of the cube, we get

$$\overline{R(\overline{S})} \cup \overline{\overline{R}(S)} \subseteq R(S) \cap \overline{R}(\overline{S}),$$

and for the right-hand side square, by De Morgan duality, we have

$$\overline{R(S)} \cup \overline{\overline{R}(\overline{S})} \subseteq R(\overline{S}) \cap \overline{R}(S).$$

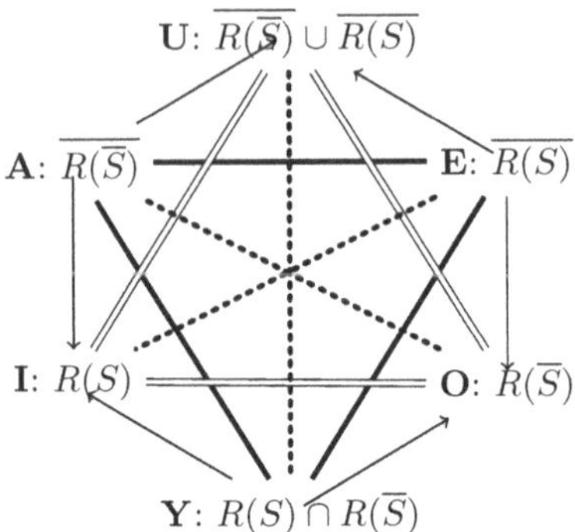

Fig. 5. Hexagon associated with the front facet of the cube

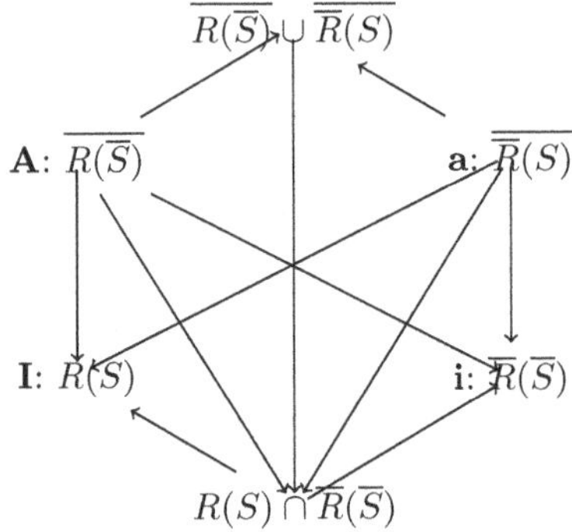

Fig. 6. Hexagon induced by the left-hand side square

One may wonder if one can build useful hexagons from the bottom and top squares of the cube. It is less clear. Indeed, if we consider the four subsets involved in the bottom square, namely $R(S)$, $R(\overline{S})$, $\overline{R}(S)$ and $\overline{R}(\overline{S})$ (the top square has their complements as vertices), they are weakly related through $R(S) \cup R(\overline{S}) = X$, $\overline{R}(S) \cup \overline{R}(\overline{S}) = X$, $R(S) \cup \overline{R}(S) = X$ and $R(\overline{S}) \cup \overline{R}(\overline{S}) = X$. Still $R(S) \cup \overline{R}(\overline{S})$ or $\overline{R}(S) \cup R(\overline{S})$ (or their complements in the top square, $\overline{R(S)} \cap \overline{\overline{R}(\overline{S})}$ or $\overline{\overline{R}(S)} \cap \overline{R(\overline{S})}$) are compound subsets that may make sense for some particular understanding of relation R. Note that similar combinations changing $\cap$ into $\cup$ and vice versa already appear in the hexagons associated with the side facets of the cube, while $R(S) \cap R(\overline{S})$ and $\overline{R}(S) \cap \overline{R}(\overline{S})$ are the **Y**-vertices of the hexagons associated with the front and back facets of the cube of oppositions. Besides, it would be also possible to complete the top facet into yet another type of hexagon by taking the complements of $\overline{\overline{R}(S)} \cup \overline{R(\overline{S})}$ or of $\overline{R(S)} \cup \overline{\overline{R}(\overline{S})}$, which clearly have empty intersections with the subsets attached to vertices **A** and **a** and to vertices **E** and **e** respectively. However, the intersection of the two resulting subsets, namely $\overline{R}(S) \cap R(\overline{S})$ and $R(S) \cap \overline{R}(\overline{S})$ is not necessarily empty. A dual construct could be proposed for the bottom facet.

In the cube of oppositions, three negations are at work, the usual outside one, and two inside ones respectively applying to the relation and to the subset - this gives birth to the eight vertices of the cube - while in the front and back squares (but also in the top and bottom squares) only two negations are at work. Besides, it is obvious that a similar cube can be built for the transpose relation R^t and a subset $T \subseteq X$, then inducing eight other remarkable subsets, now, in Y. This leads us to assume the X-normalizations of R^t and $\overline{R^t}$ $(= \overline{R}^t)$, which is nothing but the Y-normalizations of R and $\overline{R}$ $(\forall y, \exists t, tRy$, and $\forall y, \exists t', \neg(t'Ry))$, as already announced. We now apply this setting to rough sets.

3 The Cube in the Rough Set Terminology

Firstly defined by Pawlak [32], rough set theory is a set of tools to represent and manage information where the available knowledge cannot accurately describe reality. From an application standpoint it is used mainly in data mining, machine learning and pattern recognition [34].

At the basis of the theory, there is the impossibility to accurately give the intension of a concept knowing its extension. That is, given a set of objects S we cannot characterize it precisely with the available features (attributes) but, on the other hand, we can accurately define a pair of sets, the lower and upper approximations $L(S), U(S)$, which bound our set: $L(S) \subseteq S \subseteq U(S)$. The interpretation attached to the approximations is that the objects in the lower bound surely belong to S and the objects in the boundary region $U(S) \setminus L(S)$ possibly belong to S. As a consequence we have that the objects in the so-called exterior region $\overline{U(S)}$ do not belong to S for sure.

The starting point of the theory are *information tables* (or *information systems*) [31,36], which have been defined to represent knowledge about objects in terms of observables (attributes).

Definition 1. *An* information table *is a structure* $\mathcal{K}(X) = \langle X, A, val, F\rangle$ *where: the universe* X *is a non empty set of* objects*;* A *is a non empty set of* condition attributes*;* val *is the set of all* possible values *that can be observed for all attributes;* F *(called the* information map*) is a mapping* $F : X \times A \to val$ *which associates to any pair object–attribute, the value* $F(x,a) \in val$ *assumed by* a *for the object* x*.*

Given an information table, the *indiscernibility* relation with respect to a set of attributes $B \subseteq A$ is defined as

$$xR_By \quad \text{iff} \quad \forall a \in B,\ F(x,a) = F(y,a)$$

This relation is an equivalence one, which partitions X in equivalence classes xR_B. Due to a lack of knowledge we are not able to distinguish objects inside the granules, thus, it can happen that not all subsets of X can be precisely characterized in terms of the available attributes B. However, any set $S \subseteq X$ can be approximated by a *lower* and an *upper* approximation, respectively defined as:

$$L_B(S) = \{x : xR_B \subseteq S\} \tag{5a}$$
$$U_B(S) = \{x : xR_B \cap S \neq \emptyset\} \tag{5b}$$

The pair $(L_B(S), U_B(S))$ is called a *rough set*. Clearly, omitting subscript B, we have, $L(S) \subseteq S \subseteq U(S)$, which justifies the names lower/upper approximations. Moreover, the *boundary* is defined as the objects belonging to the upper but not to the lower: $Bnd(S) = U(S) \setminus L(S)$ and the *exterior* is the collection of objects not belonging to the upper: $E(S) = \overline{U(S)}$. The interpretation attached to these regions is that the objects in the lower approximation surely belong to S, the objects in the exterior surely do not belong to S and the objects in the boundary possibly belong to S.

Several generalizations of this standard approach are known, here we are interested in weakening the requirements on the relation. Indeed, in some situations it seems too demanding to ask for a total equality on the attributes and more natural to investigate the similarity of objects, for instance to have only a certain amount of attributes in common or to have equal values up to a fixed tolerance [39,38]. Thus, we are now considering a general binary relation $R \subseteq X \times X$ in place of the indiscernibility (equivalence) one. Instead of the equivalence classes, we have the *granules of information* $xR = \{y \in X : xRy\}$ and as a consequence, we no longer have a partition of the universe, but, in the general case, a partial covering, that is the granules can have non-empty intersection and some object can be outside all the granules. The lower and upper approximations are defined exactly as in Equations 5.

The normalization condition about the seriality of the relation R, nicely reflects in this framework. Indeed, we have that if the relation is serial then the covering is total and not partial (all objects belong to at least one granule) and also the following result [45]:

The relation R is serial iff $\forall S$, $L(S) \subseteq U(S)$

Given these definitions, a square of oppositions naturally arises from approximations:

- $R(S) = U_R(S)$ is the upper approximation of S wrt the relation R;
- $\overline{R(\overline{S})} = L_R(S)$ is the lower approximation of S wrt the relation R;
- $\overline{R(S)} = L_R(\overline{S}) = \overline{U_R(S)} = E(S)$ is the exterior region of S;
- $R(\overline{S}) = U_R(\overline{S}) = \overline{L_R(S)}$.

With respect to the cube of oppositions, it is the front face with the corners involving R in a positive way. To capture also the corners involving the negation of R, we have to consider other operators we can find in the rough set literature. Namely, the *sufficiency operator* (widely studied by Orlowska and Demri [14,28,15]), defined as:

$$[[S]]_R = \{x \in X | x\overline{R} \subseteq \overline{S}\};$$

and the dual operator $\ll S \gg_R = \{x \in X | x\overline{R} \cap \overline{S} \neq \emptyset\} = \{x \in X | S \cup xR \neq X\}$. So, we have

- $\overline{\overline{R}(S)} = [[S]] = \{x : S \subseteq xR\}$, that can be interpreted as the set of all x which are in relation to all $y \in S$;
- $\overline{R}(\overline{S}) = \ll S \gg$, that represents the set of objects which are not in relation to at least one object in $\overline{S}$;
- $\overline{\overline{R}(\overline{S})} = [[\overline{S}]]$ and $\overline{R}(S) = \ll \overline{S} \gg$.

Let us notice that in case of an equivalence relation the sufficiency operator $[[S]]$ is trivial, since it gives either the empty set or the set S itself if S is one of the equivalence classes. Dually, $\ll S \gg$ is either the universe or $\overline{S}$.

In Figure 7, all these rough set operators are put into the cube. The normalization condition on $\overline{R}$ requires that also $\overline{R}$ is serial, that is, also in the rough set cube, we have to require that xR is not the entire set of objects X.

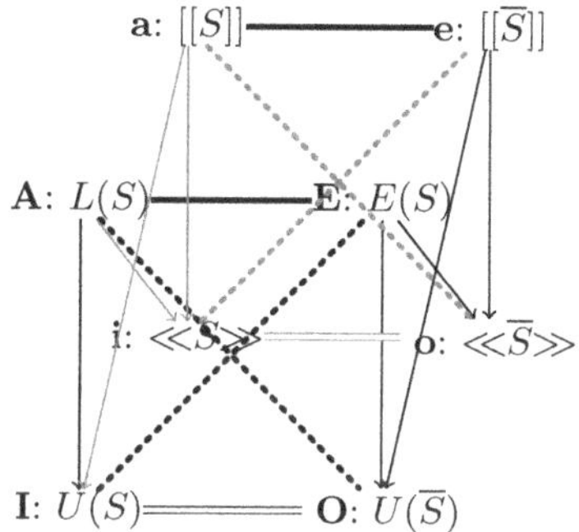

Fig. 7. Cube of oppositions induced by rough approximations

Remark 1. Often, the operators $[[S]]$ and $\ll S \gg$ are introduced together with a more general definition of information table. Indeed, it is considered that to each pair attribute-object we can associate more than one value, i.e. we have a many-valued table. For instance, if the attribute a is *color* with range $Val_a = \{white, green, red, blue, yellow, black\}$, an object can have both values $\{blue, yellow\} \subset Val_a$ whereas classically each object can assume only a single value in Val_a. In this way, we can define also several forms of generalized indiscernibility relation. For instance, we can ask that two objects are similar if $a(x) \cap a(y) \neq \emptyset$ (in the classical setting, it would be: $a(x) = a(y)$). Different relations of this kind are considered in [28] both of indistinguishability (based on R) and distinguishability (based on $\overline{R}$) type.

Let us note that the front square of the cube was already defined in [6], but considering only the equivalence relations and all the rest of the cube is a new organization of rough set operators. Indeed, a cube of opposition was also studied in [6] (see Figure 8) but it is of a different kind. In this last case, it is supposed that L and U are not dual, that is $L(\overline{S}) \neq \overline{U(S)}$. This is true in some generalized models such as *variable precision rough sets* (VPRS) [22]. VPRS are a generalization of Pawlak rough sets obtained by relaxing the notion of subset. Indeed, the lower and upper approximations are defined as $l_\alpha(H) = \{y \in X : \frac{|H \cap [y]|}{|[y]|} \geq 1 - \alpha\}$ and $u_\alpha(H) = \{y \in X : \frac{|H \cap [y]|}{|[y]|} > \alpha\}$. That is, we admit an error α in the subsethood relation $[y] \subseteq H$ and if $\alpha = 0$ we recover classical rough set approximations.

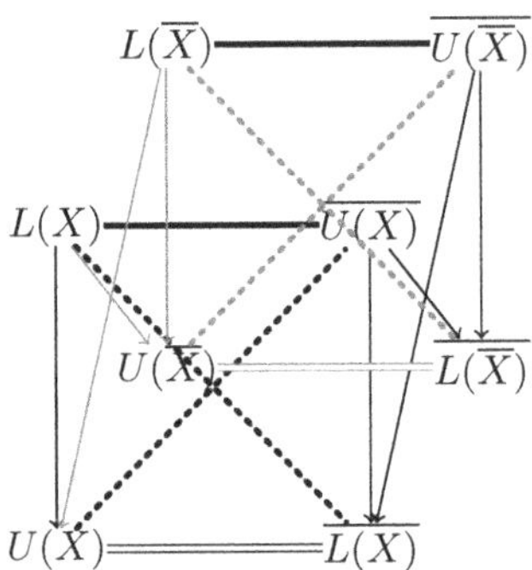

Fig. 8. Cube of opposition induced by generalized approximations

So, an open issue concerns the operators [[]] and ≪≫ in these generalized contexts. By applying them in *variable precision rough sets* we can expect to obtain another cube of $\overline{R}$ approximations and more interestingly, it should be investigated if this new setting has some practical application.

3.1 Hexagon

As we have seen, the front face contains the main operators in rough set theory. Also the hexagon, built in the standard way from the front square, is related to rough set operators. This hexagon has been built in [6] and it is reported in Figure 9.

So, on the top we have the set of objects on which we have a clear view: either they belong or not to the set S. On the contrary, on the bottom we have the boundary, that is the collection of *unknown* objects.

If we consider the back face, then $[[S]] \cup [[\overline{S}]]$ is the set of objects which are similar to all objects in the universe, whereas $\ll S \gg \cap \ll \overline{S} \gg$ is the set of objects which are different from at least one object in S and one object in $\overline{S}$.

Considering the face **AaIi** (the face **EeOo** is handled similarly), which corresponds to the FCA framework, from the top of the square, one may build

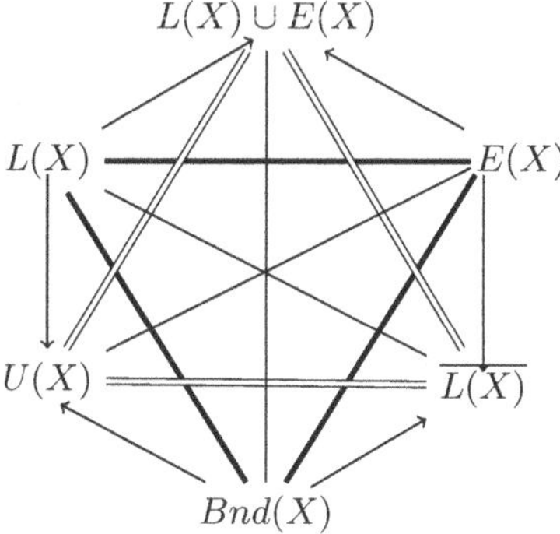

Fig. 9. Hexagon induced by Pawlak approximations

$$L(S) \cap [[S]] = \{x \in X | S \subseteq xR\} \cap \{x \in X | xR \subseteq S\} = \{x \in X | xR = S\}$$

That is, this intersection defines the set of objects which are in relation to all and only the objects in S. So, if R is an equivalence relation, it is either the empty set or the set S. On the contrary, in more general situations (when we have a covering, not a partition), it can be a non-empty subset of S.

With the bottom of the square, we may consider dually

$$U(S) \cup \ll S \gg = \{x \in X | xR \cap S \neq \emptyset\} \cup \{x \in X | xR \cup S \neq X\}$$

that corresponds to the set of objects that are in relation with at least one object in S or that are not in relation with at least one object in $\overline{S}$.

The counterpart of the hexagon of Fig. 6 makes also sense, and we have

$$L(S) \cup [[S]] \subseteq U(S) \cap \ll S \gg .$$

3.2 Other Sources of Oppositions

Let us notice that up to now we have considered oppositions arising from operators $L, U, [[]], \ll\gg$ definable in rough sets starting from a binary relation R. Other oppositions can be put forward based on other sources. First of all, relations. On the same set of data we can consider several relations besides the standard indiscernibility one, which can be in some kind of opposition among them. In [6], we have defined a classical square of oppositions based on four relations: equivalence (A), similarity (I), preclusivity (E), discernibility (O). Moreover, if we want to aggregate any two of this four relations (for instance, they can represent two agents point of view), we have 16 ways to do it and so we get a tetrahedron of oppositions.

Another source of opposition is given by attributes of an information table. Indeed, they can be characterized as useful or useless with respect to a classification task. More precisely, Yao in [47], defines a square of opposition classifying attributes as Core (A), Useful (I), NonUseful (E), NonCore (I). The Core contains the set of attributes which are in all the reducts[1], Useful attributes are those belonging to at least one reduct, NonUseful attributes are in none of the reducts and finally, NonCore attributes are not in at least one reduct.

4 The Cube in Formal Concept Analysis

In formal concept analysis [18], the relation R is defined between a set of objects X and a set of properties Y, and is called a *formal context*. It represents a data table describing objects in terms of their Boolean properties. In contrast with the data tables mentioned in the previous section on rough sets, we only consider binary attributes here, whose values correspond to the satisfaction or not of properties. As such, no particular constraint is assumed on R, except that it is serial. Indeed, let xR be the set of properties possessed by object x, and Ry is the set of objects having property y. Then, it is generally assumed in practice that $xR \neq \emptyset$, i.e. any object x should have at least one property in Y. It is also assumed that $xR \neq Y$, i.e., no object has all the properties in Y (i.e., $\overline{xR}$ cannot be empty). This is the bi-normalization of R assumed for avoiding existential import problems in the front and back facets of the cube of oppositions. Similarly, the bi-normalization of R^t means here that no property holds for all objects, or none object. In other words, the data table has no empty or full line and no empty or full column.

Given a set $S \subseteq Y$ of properties, four remarkable sets of objects can be defined in this setting (corresponding to equations (1)-(4)):

- $R^{\Pi}(S) = \{x \in X | xR \cap S \neq \emptyset\} = \cup_{y \in S} Ry$, which is the set of objects having at least one property in S;
- $R^{N}(S) = \{x \in X | xR \subseteq S\} = \cap_{y \notin S} \overline{Ry}$, which is the set of objects having no property outside S;
- $R^{\Delta}(S) = \{x \in X | xR \supseteq S\} = \cap_{y \in S} Ry$, which is the set of objects sharing all properties in S (they may have other ones).
- $R^{\nabla}(S) = \{x \in X | xR \cup S \neq Y\} = \cup_{y \notin S} \overline{Ry}$, which is the set of objects that are missing at least one property outside S.

With respect to the notations of the cube of oppositions in Figure 3, we have $R^{\Pi}(S) = R(S)$, $R^{N}(S) = \overline{R(\overline{S})}$, $R^{\Delta}(S) = \overline{\overline{R}(S)}$, $R^{\nabla}(S) = \overline{R}(\overline{S})$. They constitute, as already said, four distinct pieces of information. The names given here refer to the four possibility theory set functions Π, N, Δ, and ∇ [9] that are closely related to ideas of non-empty intersection, or of inclusion. The four formal concept analysis operators have been originally introduced in analogy with the four possibility theory set functions [8]. They correspond to the left side facet of the cube of oppositions. The full cube is recovered by introducing the complements, giving birth to the right side facet. Since $\overline{R^{\Pi}(S)} = R^{N}(\overline{S})$, and

[1] A reduct is a minimal subset of attributes which generates the same partition as the whole set.

$\overline{R^{\Delta}(S)} = R^{\nabla}(\overline{S})$, the classical square of oppositions **AEOI** is given by the four corners $R^N(S)$, $R^N(\overline{S})$, $R^{\Pi}(\overline{S})$, and $R^{\Pi}(S)$, whereas the square of oppositions **aeoi** on the back of the cube is given by $R^{\Delta}(S)$, $R^{\Delta}(\overline{S})$, $R^{\nabla}(\overline{S})$, and $R^{\nabla}(S)$. See Figure 10.

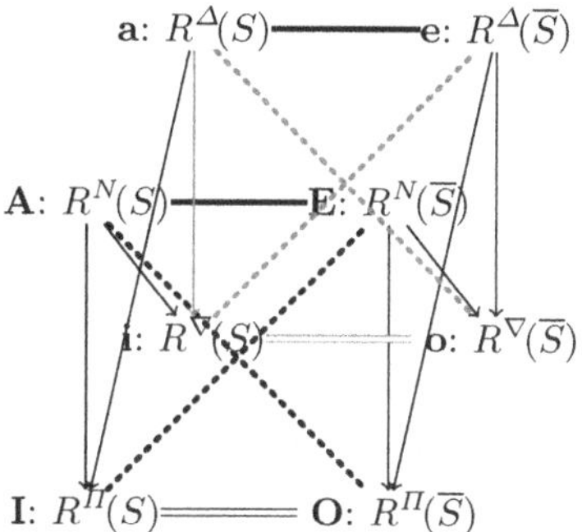

Fig. 10. Cube of oppositions in formal concept analysis

The counterpart of the hexagon of Figure 6 is given in Figure 11 where all edges are uni-directed, including the diagonal ones, and express inclusions. Indeed, as already directly established [8], under the bi-normalization hypothesis, the following inclusion relation holds: $R^N(S) \cup R^{\Delta}(S) \subseteq R^{\Pi}(S) \cap R^{\nabla}(S)$.

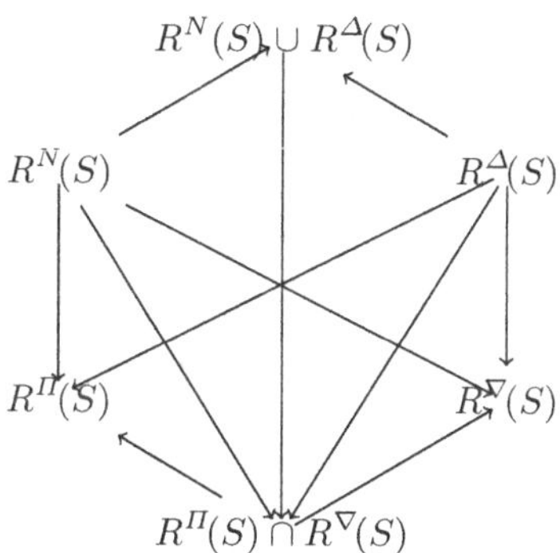

Fig. 11. Hexagon induced by the 4 operators underlying formal concept analysis

In fact, standard formal concept analysis [18] only exploits the third set function R^{Δ}. This function is enough for defining a *formal concept* as a pair made of its *extension* T and its *intension* S such that $R^{\Delta}(S) = T$ and $R^{t\Delta}(T) = S$, where $(T, S) \subseteq X \times Y$. Equivalently, a formal concept is a maximal pair (T, S) in the sense of set inclusion such that $T \times S \subseteq R$. Likewise, it has been recently established [11] that pairs (T, S) such that $R^N(T) = S$ and $R^{tN}(T) = S$ are characterizing *independent sub-contexts* which are such that $R \subseteq (T \times S) \cup (\overline{T} \times \overline{S})$. Indeed the two sub-contexts of R contained respectively in $T \times S$ and in $\overline{T} \times \overline{S}$ do not share then any object or property. The other connections $R^{\Theta}(S) = T$ and $R^{t\Lambda}(T) = S$ where $\Theta, \Lambda \in \{\Pi, N, \Delta, \nabla\}$ are worth considering. The cases $\Theta = \Lambda = \Pi$ and $\Theta = \Lambda = \nabla$ redefine the formal sub-contexts, and the formal concepts respectively [11], but the other mixed connections with $\Theta \neq \Lambda$ have still to be investigated systematically.

5 Towards Integrating RST and FCA

In FCA, one starts with an explicit relation between objects and properties, from which one defines formal concepts. This gives birth to a relation between objects (two objects are in relation if they belong together to at least one concept) and similarly to a relation between properties. RST starts with a relation between objects, but implicitly comes from an information table as in FCA linking objects and attribute values. The classical FCA setting identifies, in the information table, formal concepts from which association rules can be derived. RST, which is based on the idea of indiscernible sets of objects (note also that in the context of the intent properties, the objects in a formal concept are indiscernible as well), focuses on the ideas of reducts and core for identifying important attributes. Besides as we have shown, the structure of the cube of oppositions underlying both RST and FCA provides a richer view of the two frameworks. Thus, for instance, FCA extended with new operators has enabled us to identify independent sub-contexts inside an information table.

In the following, we both provide a synthesis of related works aiming at relating or hybridizing RST and FCA, and indicate various directions for further research taking advantages of the unified view we have introduced in this paper.

5.1 Related Works

Several authors have investigated the possibility of mixing the two theories or finding common points (see [25,48]), e.g. relating concept lattices and partitions [20], or the place of Galois connections in RST [43], or computing rough concepts [19]. In this section we intend to point out relationships of these works with our approach. Indeed, some of these works are linked to the operators provided by our cubes and to the ideas presented in the previous sections. At first, we consider the works that, more or less explicitly, put forward some of the relations between the two theories that we discussed before. Then, we will give some hint on the possibility to mix the two theories, this discussion will lead us to the next section about the possibility of integrating RST and FCA.

At first, let us note that Aristotelean oppositions are explicitly mentioned by Wille in [42]. With the aim of defining a concept logic, he generalizes the idea of formal concept and introduces two kinds of negation, one named *weak opposition* to model the idea of "contrary". Given a formal concept (T, S) its opposition is the pair $(R^{\Delta}(\overline{S}), R^{t\Delta}R^{\Delta}(\overline{S}))$. Clearly, the set $R^{\Delta}(\overline{S})$ is the contrary of the (standard) extent of $R^{\Delta}(S)$, as outlined in the FCA cube.

In several studies, it has been pointed out that the basic operators in the two theories are the four modal-like operators of the FCA square of oppositions and that the basic FCA operator is a sufficiency–like operator [12,13,48]. This fits well with our setting where it is possible to see that R^{Δ} (the sufficiency in FCA) and $[[]]$ (the sufficiency in RST) are on the same corner of the two cubes. We will better develop this issue in the following section. See also [44] for a modal logic reading of FCA and RST.

Also Pagliani and Chakraborty [29] consider three of the basic operators of the FCA square: $R^{N}, R^{\Pi}, R^{\Delta}$. Of course, they point out that R^{Δ} is the usual operator in FCA and moreover they show that in modified versions of FCA, $(R^{\Pi}(R^{tN}(X)), R^{tN}(X))$ is an "object oriented concept" [46] and $(R^{N}(R^{t\Pi}(X)), R^{t\Pi}(X))$ a "property oriented concept" [12]. Further, generalized upper and lower approximations are introduced on the set of objects as $U(X) = R^{N}(R^{t\Pi}(X))$ and $L(X) = R^{\Pi}(R^{tN}(X))$. Finally, the special case $X = Y$ is discussed, providing interpretation of the operators (similarly as we do in Section 3) and showing that in case of an equivalence relation, classical rough sets are obtained.

In the attempt at mixing the two theories, several authors considered rough approximations of concepts in the FCA settings [16,25,37,48]. The basic idea is (see for instance p.205 of [25]) that given any set of objects T, the lower and upper approximations are

$$L_{\mathcal{L}}(T) = R^{\Delta}(R^{t\Delta}(\bigcup\{X|(X,Y) \in \mathcal{L}, X \subseteq T\}) \tag{6a}$$

$$U_{\mathcal{L}}(T) = R^{\Delta}(R^{t\Delta}(\bigcup\{X|(X,Y) \in \mathcal{L}, T \subseteq X\}) \tag{6b}$$

where $\mathcal{L}$ is the concept lattice of all formal concepts. Clearly, definable sets (i.e., not rough) coincide with the extents of formal concepts.

A different approach to use RST ideas in FCA is given in [27]. Li builds a covering of the universe as the collection $R^{\Delta}(\{a\})$ for all attributes a. Then, he shows that one of the possible upper approximations on coverings (there are more than twenty ones available [49]) is equal to $R^{\Delta}(R^{t\Delta}(X))$ for all set of objects X. Once established this equality, rough sets ideas such as reducts are explored in the FCA setting.

In the other direction, that is from FCA to RST, Kang et al. [21] introduce a formal context from the indiscernibility relation of an information table and then re-construct rough sets tools (approximations, reducts, attribute dependencies) using formal concepts.

Finally, the handling of many-valued (or fuzzy) formal contexts has been a motivation for developing rough approximations of concepts [23,24,26].

5.2 Some Possible Directions for the Integration of RST and FCA

If we try to articulate the FCA cube with the RST one, the necessity of making explicit the role of attributes in the RST case is evident. In turn this requires to fix a relation. For the moment, let us consider the usual equivalence relation on attributes. As outlined in the previous section, the standard FCA operator R^{Δ} is a sufficiency operator and the corresponding corner of the RST cube is $[[T]]$. In FCA, $R^{\Delta}(S)$ is the set of objects sharing all properties in S. In RST, $[[T]]$ is the set of objects which are equivalent to all the objects $y \in T$. Using the attributes point of view: $x \in [[T]]$ iff $\forall y \in T, \forall a, F(x,a) = F(y,a)$. So, once fixed a formal concept (T,S) we have that $[[T]]_S = R^{\Delta}(S)$, where $[[]]_S$ means that the equivalence relation is computed only with respect to attributes in S.

If we do not fix a formal concept then it is not so straightforward to obtain all the formal concepts using rough set constructs. The problem is that the set of attributes used to compute $[[T]]_S$ depends on S which differs from concept to concept. An idea could be to collect all the equivalence classes generated by all subsets $S \subseteq A$, as well as all operators $[[T]]_S$ and the ones given by other corners of the cube. Among all these $[[T]]_S$ (or, it is the same, among all the equivalence classes) we have all the formal concepts. If among this (huge) collection of operators and equivalence classes, we could pick up exactly the formal concepts then we would have a new common framework for the two theories. In this direction it can be useful to explore the ideas presented in [40], where it is shown that the extents of any formal context are the definable sets on some approximation space.

A different approach in order to obtain formal concepts from rough constructs is to consider the covering made by extents of formal concepts and then consider on it the covering-based rough set lower and upper approximations (as already said more than twenty approximations of this kind are known). Then, we may wonder if any of these rough approximations coincides with the **A**, **I** corners of the FCA cube. Let us remark that Li's approach [27] previously described, is different from this idea, since based on a different covering.

Further, Li's scope is to use RST constructs such as reducts in FCA. Another issue is to explore the possibility of using the constructs in the FCA cube in order to express the basic notions of RST (besides approximations) such as reducts and cores. We may think that something more is needed, in order to define the notion of "carrying the same information" (at the basis of reducts), where "same information" may mean same equivalence classes, or same covering, or same formal concepts, etc...

The last line of investigation we would like to put forward, is related to ideas directly connected to possibility theory, in particular to a generalization of FCA named *approximate formal concept* outlined in [11]. Indeed, we may think of such approximate concept as a *rough* concept, which we want to approximate. Different solutions can be put at work. At first, we can suppose that definable (or exact) concepts coincide with the extent of formal concepts and that any other set of objects should be considered as rough. In this case, we can approximate a rough concept with a pair of formal concepts using equations 6.

On the other hand, given an approximate formal concept (T,S), we may think of it as representing an "approximate set" of objects which share an "approximate set" of properties. So it makes sense to ask which objects in T surely/

possibly have properties in S (and dually, which properties are surely/possibly shared by T). This means to try to define a pair of lower-upper approximation of the approximate formal concept. For instance, we could consider the pair $(R^{\Delta}(S), R^{\Delta,l}(S))$ where $R^{\Delta,k}(S)$ is a tolerant version of R^{Δ} admitting up to k errors. Of course, this framework should be further analyzed and put at work on concrete examples, to understand the potential of approximating formal concepts and of their approximations.

6 Conclusion

In this paper, we have shown that both RST and FCA share the same type of underlying structure, namely the one of a cube of oppositions. We have pointed out how having in mind this structure may lead to substantially enlarge the theoretical settings of both RST and FCA. Finally, this has helped us to provide an organized view of the related literature and to suggest new directions worth investigating. In the long range, it is expected, that such a structured view, which also includes possibility theory (and modal logic), may contribute to the foundations of a basic framework for information processing.

References

1. Amgoud, L., Prade, H.: A formal concept view of abstract argumentation. In: van der Gaag, L.C. (ed.) ECSQARU 2013. LNCS (LNAI), vol. 7958, pp. 181–193. Springer, Heidelberg (2013)
2. Barbut, M., Montjardet, B.: Ordre et classification: Algèbre et combinatoire, Hachette, vol. 2, ch. V (1970)
3. Béziau, J.-Y.: New light on the square of oppositions and its nameless corner. Logical Investigations 10, 218–233 (2003)
4. Blanché, R.: Sur l'opposition des concepts. Theoria 19, 89–130 (1953)
5. Blanché, R.: Structures intellectuelles. Essai sur l'organisation systématique des concepts, Vrin, Paris (1966)
6. Ciucci, D., Dubois, D., Prade, H.: Oppositions in rough set theory. In: Li, T., Nguyen, H.S., Wang, G., Grzymala-Busse, J., Janicki, R., Hassanien, A.E., Yu, H. (eds.) RSKT 2012. LNCS, vol. 7414, pp. 504–513. Springer, Heidelberg (2012)
7. Ciucci, D., Dubois, D., Prade, H.: Rough sets, formal concept analysis, and their structures of opposition (extended abstract). In: 4th Rough Set Theory Workshop, October 10, 1 p. Dalhousie University, Halifax (2013)
8. Dubois, D., Dupin de Saint-Cyr, F., Prade, H.: A possibility-theoretic view of formal concept analysis. Fundamenta Informaticae 75(1-4), 195–213 (2007)
9. Dubois, D., Prade, H.: Possibility theory: Qualitative and quantitative aspects, Quantified Representation of Uncertainty and Imprecision. In: Gabbay, D.M., Smets, P. (eds.) Handbook of Defeasible Reasoning and Uncertainty Management Systems, vol. 1, pp. 169–226. Kluwer Acad. Publ (1998)
10. Dubois, D., Prade, H.: From Blanché's hexagonal organization of concepts to formal concept analysis and possibility theory. Logica Univers. 6, 149–169 (2012)
11. Dubois, D., Prade, H.: Possibility theory and formal concept analysis: Characterizing independent sub-contexts. Fuzzy Sets and Systems 196, 4–16 (2012)

12. Düntsch, I., Gediga, G.: Modal-style operators in qualitative data analysis. In: Proc. IEEE Int. Conf. on Data Mining, pp. 155–162 (2002)
13. Düntsch, I., Gediga, G.: Approximation operators in qualitative data analysis. In: de Swart, H., Orłowska, E., Schmidt, G., Roubens, M. (eds.) TARSKI. LNCS, vol. 2929, pp. 214–230. Springer, Heidelberg (2003)
14. Düntsch, I., Orlowska, E.: Complementary relations: Reduction of decision rules and informational representability. In: Polkowski, L., Skowron, A. (eds.) Rough Sets in Knowledge Discovery 1, pp. 99–106. Physica–Verlag (1998)
15. Düntsch, I., Orlowska, E.: Beyond modalities: Sufficiency and mixed algebras. In: Orlowska, E., Szalas, A. (eds.) Relational Methods for Computer Science Applications, pp. 277–299. Physica–Verlag, Springer (2001)
16. Ganter, B., Kuznetsov, S.O.: Scale coarsening as feature selection. In: Medina, R., Obiedkov, S. (eds.) ICFCA 2008. LNCS (LNAI), vol. 4933, pp. 217–228. Springer, Heidelberg (2008)
17. Ganter, B., Stumme, G., Wille, R. (eds.): Formal Concept Analysis. LNCS (LNAI), vol. 3626. Springer, Heidelberg (2005)
18. Ganter, B., Wille, R.: Formal Concept Analysis: Mathematical Foundations. Springer–Verlag, Berlin (1999)
19. Ho, T.B.: Acquiring concept approximations in the framework of rough concept analysis. In: Kangassalo, H., Charrel, P.-J. (eds.) Preprint Proc. 7th Eur.-Jap. Conf. Informat. Modelling and Knowledge Bases, Toulouse, May 27-30, pp. 186–195 (1997)
20. Qi, J.-J., Wei, L., Li, Z.-Z.: A partitional view of concept lattice. In: Ślęzak, D., Wang, G., Szczuka, M.S., Düntsch, I., Yao, Y. (eds.) RSFDGrC 2005. LNCS (LNAI), vol. 3641, pp. 74–83. Springer, Heidelberg (2005)
21. Kang, X., Li, D., Wang, S., Qu, K.: Rough set model based on formal concept analysis. Information Sciences 222, 611–625 (2013)
22. Katzberg, J.D., Ziarko, W.: Variable precision extension of rough sets. Fundamenta Informaticae 27(2,3), 155–168 (1996)
23. Kent, R.E.: Rough concept analysis. In: Ziarko, W.P. (ed.) Rough Sets, Fuzzy Sets and Knowledge Discovery (RSKD), pp. 248–255. Springer (1994)
24. Kent, R.E.: Rough concept analysis. Fundamenta Informaticae 27, 169–181 (1996)
25. Kuznetsov, S.O., Poelmans, J.: Knowledge representation and processing with formal concept analysis. WIREs Data Mining Knowl. Discov. 3, 200–215 (2013)
26. Lai, H., Zhang, D.: Concept lattices of fuzzy contexts: Formal concept analysis vs. rough set theory. Int. J. Approx. Reasoning 50(5), 695–707 (2009)
27. Li, T.-J.: Knowledge reduction in formal contexts based on covering rough sets. In: Wen, P., Li, Y., Polkowski, L., Yao, Y., Tsumoto, S., Wang, G. (eds.) RSKT 2009. LNCS, vol. 5589, pp. 128–135. Springer, Heidelberg (2009)
28. Orlowska, E.: Introduction: What you always wanted to know about rough sets. In: Orlowska, E. (ed.) Incomplete Information: Rough Set Analysis, pp. 1–20. Springer, Heidelberg (1998)
29. Pagliani, P., Chakraborty, M.K.: Formal topology and information systems. Trans. on Rough Sets 6, 253–297 (2007)
30. Parsons, T.: The traditional square of opposition. In: Zalta, E.N. (ed.) The Stanford Encyclopedia of Philosophy, Fall 2008 edn. (2008)
31. Pawlak, Z.: Information systems - Theoretical foundations. Information Systems 6, 205–218 (1981)
32. Pawlak, Z.: Rough sets. Int. J. of Computer and Infor. Sci. 11, 341–356 (1982)
33. Pawlak, Z.: Rough Sets. Theoretical Aspects of Reasoning about Data. Kluwer (1991)

34. Pawlak, Z., Skowron, A.: Rough sets and Boolean reasoning. Information Sciences 177, 41–73 (2007)
35. Pawlak, Z., Skowron, A.: Rough sets: Some extensions. Information Sciences 177, 28–40 (2007)
36. Pawlak, Z., Skowron, A.: Rudiments of rough sets. Information Sciences 177, 3–27 (2007)
37. Saquer, J., Deogun, J.S.: Formal rough concept analysis. In: Zhong, N., Skowron, A., Ohsuga, S. (eds.) RSFDGrC 1999. LNCS (LNAI), vol. 1711, pp. 91–99. Springer, Heidelberg (1999)
38. Skowron, A., Komorowski, J., Pawlak, Z., Polkowski, L.: Rough sets perspective on data and knowledge. In: Handbook of Data Mining and Knowledge Discovery, pp. 134–149. Oxford University Press, Inc. (2002)
39. Slowinski, R., Vanderpooten, D.: A generalized definition of rough approximations based on similarity. IEEE Trans. Knowl. Data Eng. 12(2), 331–336 (2000)
40. Wasilewski, P.: Concept lattices vs. approximation spaces. In: Ślęzak, D., Wang, G., Szczuka, M., Düntsch, I., Yao, Y. (eds.) RSFDGrC 2005. LNCS (LNAI), vol. 3641, pp. 114–123. Springer, Heidelberg (2005)
41. Wille, R.: Restructuring lattice theory: An approach based on hierarchies of concepts. In: Rival, I., Reidel, D. (eds.) Ordered Sets, pp. 445–470, Dordrecht-Boston (1982)
42. Wille, R: Boolean concept logic, Conceptual Structures: Logical, Linguistic, and Computational Issues. In: Ganter, B., Mineau, G.W. (eds.) ICCS 2000. LNCS, vol. 1867, pp. 317–331. Springer, Heidelberg (2000)
43. Wolski, M.: Galois connections and data analysis. Fund. Infor. 60, 401–415 (2004)
44. Wolski, M.: Formal concept analysis and rough set theory from the perspective of finite topological approximations. Trans. Rough Sets III 3400, 230–243 (2005)
45. Yao, Y.Y.: Constructive and algebraic methods of the theory of rough sets. J. of Information Sciences 109, 21–47 (1998)
46. Yao, Y.: A comparative study of formal concept analysis and rough set theory in data analysis. In: Tsumoto, S., Słowiński, R., Komorowski, J., Grzymała-Busse, J.W. (eds.) RSCTC 2004. LNCS (LNAI), vol. 3066, pp. 59–68. Springer, Heidelberg (2004)
47. Yao, Y.Y.: Duality in rough set theory based on the square of opposition. Fundamenta Informaticae 127, 49–64 (2013)
48. Yao, Y., Chen, Y.: Rough set approximations in formal concept analysis. In: Peters, J.F., Skowron, A. (eds.) Transactions on Rough Sets V. LNCS, vol. 4100, pp. 285–305. Springer, Heidelberg (2006)
49. Yao, Y.Y., Yao, B.X.: Covering based rough set approximations. Information Sciences 200, 91–107 (2012)

Enriching Taxonomies of Place Types Using Flickr

Joaquín Derrac and Steven Schockaert

School of Computer Science & Informatics, Cardiff University,
5 The Parade, CF24 3AA, Cardiff, United Kingdom
{j.derrac,s.schockaert}@cs.cardiff.ac.uk

Abstract. Place types taxonomies tend to have a shallow structure, which limits their predictive value. Although existing place type taxonomies could in principle be refined, the result would inevitably be highly subjective and application-specific. Instead, in this paper, we propose a methodology to enrich place types taxonomies with a ternary betweenness relation derived from Flickr. In particular, we first construct a semantic space of place types by applying dimensionality reduction methods to tag co-occurrence data obtained from Flickr. Our hypothesis is that natural properties of place types should correspond to convex regions in this space. Specifically, knowing that places $P_1, ..., P_n$ have a given property, we could then induce that all places which are located in the convex hull of $\{P_1, ..., P_n\}$ in the semantic space are also likely to have this property. To avoid relying on computationally expensive convex hull algorithms, we propose to derive a ternary betweenness relation from the semantic space, and to approximate the convex hull at the symbolic level based on this relation. We present experimental results which support the usefulness of our approach.

1 Introduction

Taxonomies encode which categories exist in a given domain and how these categories are related. Often they are restricted to is-a relations (also called hyponym/hypernym relations or subsumption), although other relations may be considered as well. Taxonomies feature perhaps most prominently in biology, where their purpose is to group organisms with common characteristics. In information systems, taxonomies are often used to organise content. For example, online shops such as Amazon[1] use taxonomies of products to let users browse their site. Similarly, web sites may use taxonomies of music genres[2], movie genres[3] or place types[4] to allow for easier navigation. In this paper, we will focus on place types, although similar considerations apply to music and movie genres, research areas, apps, and many other domains.

[1] http://www.amazon.co.uk
[2] http://www.bbc.co.uk/music/genres
[3] http://dvd.netflix.com/AllGenresList
[4] http://aboutfoursquare.com/foursquare-categories/

C. Beierle and C. Meghini (Eds.): FoIKS 2014, LNCS 8367, pp. 174–192, 2014.

Whereas biological taxonomies are closely tied to evolution, taxonomies of place types merely reflect a perceived similarity between natural language labels. As a result, such taxonomies are usually application-specific. For example, in the Foursquare taxonomy, *bakery*, *ice cream shop* and *steakhouse* are all found in the same category, grouping venues related to *food*. Wordnet[5], on the other hand, classifies *bakery* as a hyponym of *shop* and *steakhouse* as a hyponym of *restaurant*, with *shop* and *restaurant* both being direct hyponyms of *building*. The vagueness of many natural language labels further complicates the problem of designing suitable taxonomies. For example, on Tripadvisor[6], ice cream shops tend to be listed under the category *restaurant*, while we may consider it more natural to consider ice cream shops under the category *shopping*.

When place type taxonomies are used for organising content, the aforementioned issues are inevitable: even though place types can be grouped in many meaningful ways, one particular hierarchy needs to be selected. However, one other important reason why taxonomies are important in biology is because they have predictive value. Taxonomies are used, for instance, to predict which species will become invasive [15] or which species are likely to be ecologically similar [16]. Similarly, place type taxonomies could potentially be valuable to support various forms of inductive reasoning. For example, consider a user who is looking for recommendations about places to visit on a day out with children. Current place recommendation system such as Foursquare or Yelp[7] do not support such queries. By analysing user reviews, however, we may discover that zoos, theme parks and beaches are suitable places for a day out with children. In principle, a sufficiently fine-grained place type taxonomy should then allow us to identify places which are taxonomically close to zoos, theme parks and beaches, and these places are likely to be suitable as well.

Current place type taxonomies are too shallow to support such predictive inferences. One solution would be to refine existing taxonomies, either manually or using automated methods [11,17,35]. However, this would only partially solve the problem, as the resulting taxonomy would still be application-dependent. For example, *beach* may be grouped with other sea-related places such as *harbour* and *oil platform*, with other coastal features such as *cape*, *cliff* and *polder*, or with other sand-related types of land cover such as *desert*. We therefore argue that to support inductive reasoning about place types, a richer structure is needed to model relatedness of place types, which can take account of the fact that place types can be categorised in many meaningful ways.

To derive information about the relatedness of place types, we propose the following methodology. First, from Flickr[8], a popular photo-sharing website, we derive a vector-space representation for all places of interest, for which we can exploit the fact that photos on Flickr often have a number of tags (i.e. short textual descriptions). Previous work [30,31] has already indicated that Flickr

[5] `http://wordnet.princeton.edu`
[6] `http://www.tripadvisor.co.uk`
[7] `http://www.yelp.com`
[8] `http://www.flickr.com`

tags can be successfully used for discovering places of a given type, suggesting that Flickr tags indeed have potential for modelling place types. Specifically, we first represent each place type as a vector based on its associated tags on Flickr and then use a dimensionality reduction method such as Singular Value Decomposition (SVD), MultiDimensional Scaling [5] or Isomap [28] to obtain a representation of each place type, either as a point or as a convex region in a lower-dimensional Euclidean space. Gärdenfors' theory of conceptual spaces [10] posits that natural properties tend to correspond to convex regions in some suitable metric space. Our assumption is that a similar property will hold for the representation we obtain from Flickr, i.e. we assume that this representation can be viewed as the approximation of a conceptual space. Knowing that place types $P_1, ..., P_n$ satisfy a given natural property (e.g. being suitable places for a day out with children), we could then induce that place types which are in the convex hull of the representations of $P_1, ...P_n$ are also likely to satisfy this property[9]. However, from an application point of view, this is still not fully satisfactory since it requires checking the convex hull membership of a potentially large number of place types, every time an inductive inference is made. As an alternative, we propose to extract a ternary betweenness relation from the vector-space representations of the place types, e.g. encoding that a *tapas bar* is between a *restaurant* and a *pub*. Note that this notion of betweenness at the same time corresponds to a geometric notion of betweenness in a Euclidean space and to the conceptual notion of having intermediate properties. In summary, the research questions we address in this paper are the following:

1. To what extent is Flickr a useful source for acquiring information about the relatedness of place types?
2. To what extent is a ternary betweenness relation useful to identify natural cateogories of place types?

The paper is structured as follows. After reviewing related work in the next section, Section 3 describes how the vector-space representations of the place types has been obtained. Subsequently, Section 4 discusses different ways of measuring betweenness. Section 5 reports the results of our experiments, after which we conclude.

2 Related Work

2.1 Vector-Space Models of Meaning

Vector-space models are widely used to represent the meaning of natural language concepts in fields such as information retrieval [22,6], natural language processing [19,20,9], cognitive science [14], and artificial intelligence [10,26].

[9] It should be noted that this notion of convex hull relates to the representations of the meanings of these place types in a conceptual space; it is unrelated to the geographic locations of instances of the place types.

Most approaches represent natural language terms as points (or vectors) in a Euclidean space. One notable exception is the work of Gärdenfors on conceptual spaces [10], where properties and concepts are represented using convex regions. While computationally more demanding, using regions instead of point has a number of advantages. First, it allows us to distinguish borderline instances of a category from more prototypical instances, by taking the view that instances which are closer to the center a region are more typical [10]. A second advantage of using regions to represent the meaning of natural language terms is that it makes it clear whether one concept subsumes another (e.g. every pizzeria is a restaurant), whether two concepts are mutually exclusive (e.g. no restaurant can also be a beach), or whether they are overlapping (e.g. some bars serve wine but not all, some establishments which serve wine are bars but not all). Region-based models have been shown to outperform point-based models in some natural language processing tasks [8], although point-based remain more popular.

In information retrieval, it is common to represent documents as vectors with one component for every term occurring in the corpus. In many other applications (as well as sometimes in information retrieval) some form of dimensionality reduction is used to obtain vectors whose components correspond to concepts. One of the most popular techniques, called latent semantic analysis (LSA [6]), uses singular value decomposition (SVD) for this purpose. Apart from many applications in natural language processing, LSA has proven useful for common sense reasoning. For example, [26] applies LSA on vector representations obtained from ConceptNet[10] to identify properties that concepts are likely to have, even if they are not explicitly mentioned in the knowledge base. A method inspired by LSA, called latent relational analysis, is used in [29] to discover analogous pairs of words such as cat:meow and dog:bark. Multi-dimensional scaling (MDS [5]) is another popular method for dimensionality reduction, which builds a vector-space representation from pairwise similarity judgements. It is among others used in cognitive science to interpret pairwise similarity judgements obtained from human assessors. It is also possible to reduce the dimensionality of document-term matrices by selecting the most informative terms, and omitting the remaining dimensions. We refer to [36] for a survey of such term selection methods. Finally, note that all of the aforementioned methods for dimensionality reduction correspond to a linear mapping from a high-dimensional space to a lower-dimensional space. Isomap [28] is an extension of MDS which essentially preprocesses the dissimilarity matrix used by MDS to obtain a non-linear mapping which is supposed to be a more faithful embedding of the original space. Another popular non-linear dimensionality reduction method is Locally Linear Embedding (LLE [21]).

2.2 Similarity Based Reasoning

Similarity based reasoning is a form of common sense reasoning which is based on the assumption that similar concepts tend to have similar properties [27,7].

[10] `http://conceptnet5.media.mit.edu`

Applying similarity based reasoning in practice first of all requires access to a similarity relation. This task of estimating the similarity of natural language terms is strongly related to the problem of dimensionality reduction. In fact, one of the main reasons why e.g. LSA is used in practice is to obtain better estimates of the similarity between terms. There are, however, two further problems with the principle of similarity based reasoning: similarity degrees are context-sensitive and the principle does not make clear how similar two concepts need to be to obtain meaningful conclusions.

It is common to model changes in context by rescaling the dimensions of a semantic space [13,10]. When an explicit representation of the context is available, for example as a set of terms, an appropriate similarity relation can thus be obtained. Another solution has been proposed in [24], which is based on the idea of replacing numerical similarity by qualitative spatial relations which remain invariant under linear transformations such as rescaling dimensions. One of the most important examples of such a qualitative spatial relation is betweenness. If *tapas bar* is between *restaurant* and *pub* in a semantic space, it will remain so regardless of how the dimensions of that space are rescaled. Given access to a ternary betweenness relation, the principle of similarity based reasoning could be replaced by the assumption that intermediate concepts have intermediate properties. The resulting form of commonsense reasoning is called interpolation, due to its similarities with numerical interpolation. In [23], a methodology was proposed to learn a betweenness relation (as well as analogical proportions) for music genres from the music recommendation web site last.fm[11], although the result was not formally evaluated. The methodology we use in this paper extends this previous work and we compare our method with the method from [23] in Section 5.3. Numerical similarity degrees can also be avoided by using a comparative similarity relation ("a is more similar to b than to c"), where needed [25]. However, in contrast to betweenness, comparative similarity is not invariant under linear transformations and is thus context-dependent. For example, whether *beach* is more similar to *zoo* than to *polder* depends on the application. Indeed, it is this context-dependent nature of comparative similarity that prevents us from constructing a single application-independent taxonomy of place types.

The k-nearest neighbours algorithm (k-NN [4]) can be seen as the counterpart of similarity based reasoning in machine learning. While it often works well in practice, its performance also depends on the availability of a similarity relation which is appropriate for the application context [34,1] . Some approaches have been studied which are similar in spirit to the idea of using betweenness in knowledge representation. These approaches are based on the idea that natural categories are convex regions, and include methods which represent training examples of a given class by their convex hull [33,18]. Analogical classifiers [3] avoid similarity degrees in a different way, based on the assumption that analogical changes in the features of an item to be classified should lead to analogical changes in their class labels. From a geometric point of view, the basic relation here is parallelism, which is also invariant under linear transformations. Yet an-

[11] `http://www.last.fm`

other way of avoiding similarity degrees is proposed in [12], where Markov logic is used to learn clusters of objects and clusters of properties such that membership of an object in a given cluster determines whether it satisfies the properties in a given cluster, with high probability. This essentially corresponds to learning a particular Boolean similarity relation where two objects or properties are either related (i.e. in the same cluster) or not. To deal with context-dependence, the approach from [12] learns multiple clusterings (i.e. multiple Boolean similarity relations). Hence *beach* and *zoo* could be in the same cluster in one of the clusterings but in different clusters in another. This is similar in spirit to our proposal of learning a betweenness relation instead of a hiearchical clustering of place types. However, while our method is unsupervised, the method from [12] requires all relevant properties to be known in advance.

3 Constructing a Semantic Space of Place Types

Our approach revolves on representing place types as either points or regions in a semantic space. In this section we detail how such spaces can be constructed using the meta-data of a large collection of photos.

3.1 Data Acquisition

The initial data set was constructed by analyzing the meta-data of a database of more than 105 million photos, which was introduced in [32]. This database was originally obtained using a publicly available API from Flickr [12]. Every photo in the database is associated with a set of tags, a pair of coordinates and information about the accuracy of these coordinates, among others. Photos with inaccurate locations (viz. those with an accuracy level lower than 12) and photos with too many (more than 100) or too few (less than 2) tags were discarded. The resulting photos are treated as lists of tags. Photos taken at the same place by the same user are combined and treated as a single list, containing the union of the tags associated with these photos. In this way, no tag is counted more than once per user/place pair, which is important to limit the extent to which a single user can influence the representation of a place. This resulted in more than 20 million tag lists. Two existing place type taxonomies [13] have been used as reference throughout the experiments:

- GeoNames[14] organises place types in 9 categories, encompassing both man-made features such as buildings or railroads and natural features such as mountains or forests.

[12] `http://www.flickr.com/services/api/`

[13] Other available place type taxonomies include those of Google Places API `https://developers.google.com/places/documentation/supported_types`, WordNet `http://wordnetweb.princeton.edu`, DBpedia `http://mappings.dbpedia.org/server/ontology/classes/` or Yago `http://www.mpi-inf.mpg.de/yago-naga/yago/downloads.html`

[14] `http://www.geonames.org/export/codes.html`

– Foursquare[15] also uses 9 top-level categories, but focuses mainly on urban man-made places such as restaurants, bars and shops. Although a few of these categories include sub-categories, the taxonomy is mostly flat, and we will only consider the top-level categories in this paper.

We associate each place type from these taxonomies with the corresponding set of photos, that is, the photos whose tags include the name of the place type. For composite names such as "football stadium", photos with the tags *football* and *stadium* were accepted, in addition to those including the concatenation of the whole name, *footballstadium*. Then, the place types with fewer than 1000 associated photos were discarded. This left the final version of the GeoNames taxonomy with 238 of the initial 667 place types, from 7 categories [16], whereas the final version of the Foursquare taxonomy included 354 of the initial 435 place types, from all 9 categories. The average number of (combined) photos that was thus associated with each place type is 20767. The total number of (combined) photos associated with any of the place types is 7350443 for the GeoNames taxonomy and 4943648 for the Foursquare taxonomy.

3.2 Point-Based and Region-Based Representations

We will consider two different ways to represent place types in a semantic space: as points and as convex regions. For the point-based representation, we encode each place type as a single vector, with one component for each tag in the collection. The corresponding component is the number of times that tag occurs in a photo associated with the place type. To reduce noise, the frequency of tags with fewer than 4 occurrences is set to 0. As a result of these steps, 354 and 238 vectors have been obtained for Foursquare and Geonames, respectively.

For the region-based representation, we will identify a set of vectors for each place type, from which a convex region can then be obtained (e.g. by taking the convex hull of the points). To this end, the photos of each place type are clustered based on the cosine similarity between their tag sets. We have used the K-Means clustering algorithm, establishing the number of clusters as $K = max(10, \mu \cdot \log_{10} n)$, where n is the number of photos of the place type and μ is an equalizing factor, established as $\mu = 5$ for GeoNames place types and $\mu = 10$ for Foursquare place types [17]. We have also considered clustering the photos based on their coordinates using mean-shift clustering, but because the initial results were not encouraging we do not consider this option in this paper. As a cleaning step, we merge clusters with fewer than 5 photos with their nearest cluster. Finally, every remaining cluster is encoded as a frequency vector, in the same way as for the point-based representation. In this way, the 238 GeoNames place types are represented as 4379 points (using 6-28 points per place type),

[15] `http://aboutfoursquare.com/foursquare-categories/`

[16] The categories A: country, state, region,... and P: city, village,... did not include any well-represented place type name.

[17] The values for μ are chosen so the average number of clusters per place type is approximately the same in both problems.

whereas the 354 Foursquare place types are described by 6593 points (using 5-56 points per place type). On average, about 18 points are used to represent each place type.

3.3 Dimensionality Reduction

Although in principle we could use the vector representations obtained in Section 3.2 to encode place types, such a representation has several disadvantages: many of the components in these vectors are correlated, the vectors tend to be sparse (i.e. tags which are relevant to a place type may have a frequency count of 0), and they do not take into account the fact that some tags are more relevant than others to encode the meaning of a place type. A common solution to this problem is to use a dimensionality reduction method. In this paper, we consider three such methods: SVD, MDS and Isomap.

The input to SVD is a matrix with a row for each considered vector. In this context, this means that we have one row for each place type (in the point-based representation) or one row for each cluster of places (in the region-based representation). SVDLIBC[18] was used to compute the singular value decomposition.

MDS requires a dissimilarity matrix between the points. Given that all the terms in the vectors (tag count) are positive, such dissimilarities can be obtained as 1 minus their cosine similarity. We have used the implementation of classical multidimensional scaling of the MDSJ java library[19].

Finally, the idea of Isomap is to preprocess the dissimilarity matrix and apply MDS on the resulting matrix. In this step, a fully connected weighted graph is constructed using as edge weights the values of the dissimilarity matrix. The graph is then pruned, removing those edges whose length (dissimilarity) is higher than a predefined threshold (the minimum value such that no region of the graph could be isolated). Missing edges are then recomputed as the shortest path between the nodes in the pruned graph (computed using Dijkstra's algorithm). Because the last step requires summing dissimilarities, instead of using the complement of the cosine similarity, we instead use the normalised angular distance:

$$d(\mathbf{a}, \mathbf{b}) = \frac{2 \cdot \arccos(\mathbf{a}, \mathbf{b})}{\pi}$$

The three methods allow to obtain an n-dimensional embedding of a set of high-dimensional data points. The initial high-dimensional points encode specific instances of the considered place types. Intuitively, by applying dimensionality reduction, the instances are generalised to obtain a better representation of the actual place type. The usefulness of these representations will depend on the chosen number of dimensions. Using too few dimensions limits the discriminative power of the representations. On the other hand, when using too many dimensions, the representations may not be sufficiently generalised.

Finally, to further clean the representations in the region based model, for each place type, outliers have been removed as follows. First, the medoid among

[18] `http://tedlab.mit.edu/~dr/SVDLIBC/`

[19] `http://www.inf.uni-konstanz.de/algo/software/mdsj/`

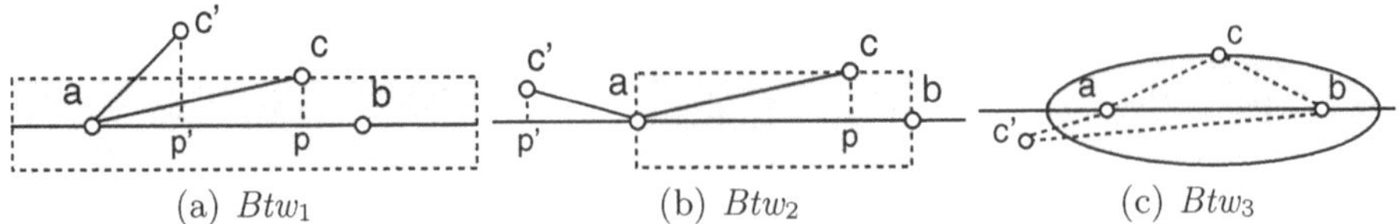

Fig. 1. Comparison of the three betweenness measures

the set of points corresponding to a given place type was determined as the vector m minimising the total distance $\sum_n d(n, m)$ to the other vectors associated with that place type. Then, of the points associated with that place type, the 10% that are furthest away from the medoid are interpreted as outliers and are discarded. Euclidean distance over the constructed spaces was used throughout this procedure.

4 Using betweenness

We propose to enrich taxonomies of place types with a ternary betweenness relation, which implicitly encodes different ways in which natural clusters of place types may be defined. The intuition is that a place type c is between place types a and b iff c has all the (important features) that a and b have in common. In what follows, we show how the semantic space representations from the previous section can be used to assign a score of how in-between place type c is with respect to place types a and b and we discuss in more detail how such betweenness relations could be used.

4.1 Measuring betweenness

Geometrically, a point c can only be between points a and b if a, b and c are collinear. In practice, the points corresponding to any three place types are unlikely to be exactly collinear, which means that we need a more flexible approach, measuring a degree of betweenness. The general idea is that a point c is considered between a and b to the degree that c is near the line segment connecting a and b. A first way to formalise this is as follows. Let p be the orthogonal projection of c on the line connecting a and b. We define:

$$Btw_1(a, c, b) = \|\overrightarrow{cp}\|$$

Note that higher scores correspond to weaker betweenness relations, and in particular that a score of 0 denotes perfect betweenness. This measure is illustrated in Figure 1(a). As the figure shows, the shorter the distance between c and p, the more c will be between a and b. Note that Btw_1 is actually quite naive as a measure of betweenness, as there is no guarantee that the point p is actually between a and b. Figure 1(b) shows an example of this situation. We define a second betweenness measure, in which we explicitly require the point p to be between a and b:

$$Btw_2(a,c,b) = \begin{cases} Btw_1(a,c,b) & \text{if } \cos\langle\overrightarrow{ab},\overrightarrow{ac}\rangle \geq 0 \text{ and } \cos\langle\overrightarrow{ba},\overrightarrow{bc}\rangle \geq 0 \\ +\infty & \text{otherwise} \end{cases} \tag{1}$$

based on the fact that $\cos\langle\overrightarrow{ab},\overrightarrow{ac}\rangle \geq 0$ and $\cos\langle\overrightarrow{ba},\overrightarrow{bc}\rangle \geq 0$ iff p lies on the line segment between a and b.

Our last betweenness measure is based on the rationale that c is exactly between a and b iff $\|\overrightarrow{ab}\| = \|\overrightarrow{ac}\| + \|\overrightarrow{bc}\|$:

$$Btw_3(a,c,b) = \frac{\|\overrightarrow{ab}\|}{\|\overrightarrow{ac}\| + \|\overrightarrow{bc}\|} \tag{2}$$

In contrast to the previous two meaures, higher values in the case of Btw_3 represent a stronger betweenness relation, with a score of 1 denoting perfect betweenness. This alternative definition (see Figure 1(c)) has the advantage that points near a or b will get some degree of betweenness, even if their projection p is not between a and b.

When using a point-based representation of place types, the aforementioned measures could be used as such. In the case of region-based representations, we have several options. Let P, Q and R be the sets of points corresponding to the place types $\mathcal{P}$, $\mathcal{Q}$ and $\mathcal{R}$. One option to assess the degree $btw(P,R,Q)$ to which $\mathcal{R}$ is between $\mathcal{P}$ and $\mathcal{Q}$, adopted in [23], is to use linear programming to check how many of the points in R are (perfectly) between a point in the convex hull of P and a point in the convex hull of Q. While this method works well in very low-dimensional spaces, we found that when using more than a few dimensions, often no points of Q are perfectly between points in the convex hulls of P and Q. An alternative would be to check how far the points in R are from the convex hull of $P \cup Q$, which can be done using quadratic programming, but this method is computationally expensive. Moreover, in preliminary experimental results, we found that the following method tends to outperform both of the aforementioned alternatives, despite being the cheapest in computational terms:

$$Btw_1^R(P,R,Q) = \frac{1}{|R|}\sum_{r\in R} \min_{p\in P}\min_{q\in Q} Btw_1(p,r,q)$$

By replacing Btw_1 by Btw_2 we obtain the measure Btw_2^R, and by replacing the minima by maxima and Btw_1 by Btw_3 we obtain the measure Btw_3^R.

4.2 Betweenness for Categorisation and Identification Problems

In principle, the sub-category structure in a place type taxonomy should encode hyponym/hypernym relationships, such as *italtian restaurant* is-a *restaurant*. In practice, there are at least two problems with this view of place type taxonomies. First, many categories are vague, and is-a hierarchies are ill-equipped to deal with this, e.g. it is not clear whether we should consider *tapas bar* is-a *restaurant* or *tapas bar* is-a *bar*. Second, in existing taxonomies, many categories merely group place types in thematic way, e.g. Foursquare has a category *food* which covers both restaurants and food-related shops. The problem with such thematic categories is that while they may be useful within the boundaries of a single application, they tend to be very context-sensitive.

To obtain a more generic conceptual model of place types, we propose to enrich an is-a hierarchy of place types with a graded betweenness relation, which can naturally handle vagueness due to its graded nature and implicitly encodes many alternative ways in which the considered set of place types can be clustered. This betweenness relation would also facilitate problems such as merging different taxonomies and detecting and repairing likely errors. Formally, we are interested in the following two tasks:

- Categorisation: Given a number of categories $C_1, ..., C_n$, each being represented as a set of place types, and a new place type p, decide to which of the categories C_i the place type p most likely belongs.
- Identification: Given a categegory C and a set of place types $p_1, ..., p_n$ which are known to belong to that category, decide which other place types are likely to belong to that category.

The categorisation problem can be tackled using the betweenness relation as follows: find the category C_i and the place types $q_1, q_2 \in C_i$ which maximise the degree to which p is between q_1 and q_2. To tackle the identification problem, we can consider all places q which are between places q_i and q_j from C to a sufficiently large extent.

5 Experimental Study

In this section, we experimentally evaluate the usefulness of the semantic space of place types we derived from Flickr, and the effectiveness of the proposed betweenness measures in solving the categorisation and identification problems. First, the experimental set-up is detailed (Section 5.1) and then the results are reported for both the point-based (Section 5.2) and region-based (Section 5.3) schemes. We conclude with a comparison to a human gold standard.

5.1 Experimental Set-Up

We compare the semantic spaces obtained using MDS, Isomap and SVD. Each time we let the number of dimensions vary between 2 and 50. For the categorisation problem, a 10-folds cross validation procedure was used. The result is evaluated using the classification accuracy, i.e. the percentage of all items which have been assigned to the correct category. For the identification problem, each category is considered individually and we report the average result we obtained over all categories. To evaluate the performance for a given category, we use a random sample of 25% of the places in that category as reference data. The task we consider is to rank the remaining places, which include 75% of the places in the considered category as well as all places belonging to the other categories. We then use Mean Average Precision (MAP) to measure the extent to which places from the considered category appear at the top of the ranking. MAP is computed as the mean of the average precision (AP) obtained for each of the categories, where AP is defined as:

$$AP = \frac{1}{n}\sum_{i=1}^{n} Prec@pos_i = \frac{1}{n}\sum_{i=1}^{n} \frac{i}{pos_i}$$

with $pos_1, ..., pos_n$ the positions in the ranking where the place types from the considered category occur and $Prec@k$ the precision of the first k elements.

For each configuration, we compare our betweenness-based methods with a nearest neighbour classifier (k-NN). Euclidean distance was considered as the dissimilarity measure. In experiments where $k > 1$, the final categorisation was obtained with a majority vote. For region-based representations, we also compare our results to the method described in [23], which we will refer to as CHRegion. This method evaluates the degree to which c is between a and b as the ratio of points of c which are between a point in the convex hull of a and a point in the convex hull of b.

In terms of computation time, the CHCRegion baseline is by far the most demanding method, requiring several hours or even days to complete a single run for problems of 30 to 50 dimensions. In contrast, all of the other methods complete a single run in less than a minute on a standard workstation.

5.2 Results Obtained: Point-Based Encoding

Table 1 shows the results obtained for the categorisation problem for the GeoNames and Foursquare taxonomies. The best result for each method is highlighted. The results show that the betweenness based approaches outperform 1-NN (only Btw_3 is outperformed by 1-NN, in some specific configurations). Somewhat surprisingly, there is no consistent difference between Btw_1 and Btw_2. For both GeoNames and Foursquare, MDS outperforms Isomap, which outperforms SVD. Finally, the best results obtained overall are 0.6134 for GeoNames (Btw_1, MDS, 30-D) and 0.7288 for Foursquare (Btw_1, MDS, 10-D).

While only the case $k = 1$ is considered in Table 1, similar results were obtained for other values of k, with $k = 1$ being an optimal or near-optimal choice in most cases. Figures 2(a) and 2(b) illustrate the result for different values of k obtained when using the best performing configuration of each method. For the betweeness approaches, a majority vote among the k-highest betweenness triples for each test place was used. From these figures, it can be seen that Btw_1 and Btw_2 for $k = 1$ outperform k-NN for any value of k. Note that while k-NN method performs considerably better for $k = 5$ than for $k = 1$ in the case of GeoNames, this result is very sensitive to the value of k and choosing an optimal k would therefore be difficult in practice.

Table 1. Classification accuracy for $k = 1$ for the point-based encoding

	GeoNames					Foursquare				
SVD	2-D	5-D	10-D	30-D	50-D	2-D	5-D	10-D	30-D	50-D
Btw_1	0.4034	0.4118	0.4454	**0.5168**	0.5000	0.2175	0.2881	0.4661	0.6102	**0.6186**
Btw_2	0.3908	0.4370	0.4328	**0.5210**	**0.5210**	0.2260	0.3531	0.4774	0.6045	**0.6243**
Btw_3	0.3908	0.4286	0.4328	**0.4874**	0.4454	0.2260	0.2260	0.2429	0.4859	**0.5113**
1-NN	0.2773	0.3235	0.3950	0.4412	**0.4748**	0.2119	0.3418	0.3277	0.5537	**0.5819**
MDS	2-D	5-D	10-D	30-D	50-D	2-D	5-D	10-D	30-D	50-D
Btw_1	0.3866	0.5000	0.5504	**0.6134**	0.5756	0.3729	0.5847	**0.7288**	0.6921	0.6780
Btw_2	0.4580	0.5210	0.5546	**0.6092**	0.5798	0.5141	0.6384	**0.7203**	0.7175	0.6723
Btw_3	0.4622	0.5420	0.5462	**0.5672**	0.5588	0.5000	0.5960	0.6667	**0.7006**	0.6780
1-NN	0.3571	0.5168	0.4790	0.5336	**0.5504**	0.4915	0.6130	**0.6780**	0.6695	0.6469
Isomap	2-D	5-D	10-D	30-D	50-D	2-D	5-D	10-D	30-D	50-D
Btw_1	0.4160	0.3908	0.4370	0.4496	**0.5000**	0.2542	0.4492	0.5904	**0.6186**	0.6102
Btw_2	0.4286	0.4538	0.4958	0.4412	**0.5000**	0.4350	0.5650	**0.6356**	0.6271	0.6299
Btw_3	0.4370	0.4370	0.4748	0.4706	**0.4748**	0.4294	0.5282	0.5424	**0.5593**	0.5141
1-NN	0.3950	0.3950	0.4160	0.4202	**0.4454**	0.4718	0.5791	0.5932	0.5847	**0.6045**

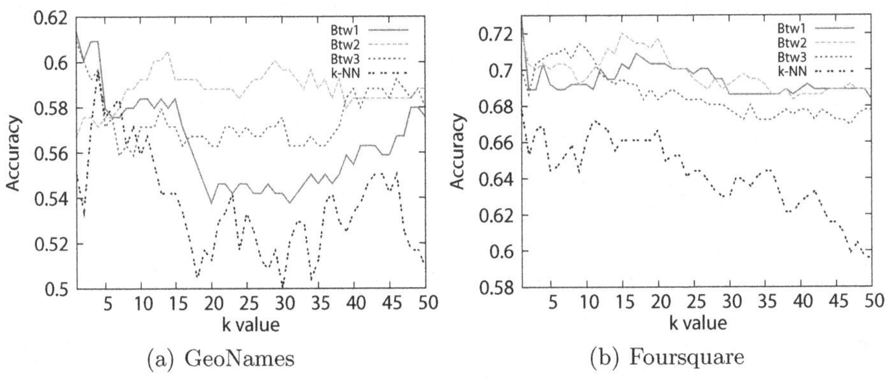

(a) GeoNames (b) Foursquare

Fig. 2. Influence of k on the classification accuracy for the point-based encoding

Table 2 shows the results obtained for the identification problem. The betweenness based approaches show a better performance than 1-NN in most cases, with Btw_2 performing slightly better than the other methods. MDS again outperforms the other dimensionality reduction methods. The best results found overall are 0.3137 (Btw_2, MDS, 50-D) for GeoNames and 0.5565 (Btw_2, MDS, 30-D) for Foursquare.

Table 2. Identification results (MAP) for the point-based encoding

	GeoNames					Foursquare				
SVD	2-D	5-D	10-D	30-D	50-D	2-D	5-D	10-D	30-D	50-D
Btw_1	**0.1445**	0.1370	0.1393	0.1405	0.1305	0.1083	0.1182	0.1512	**0.2099**	0.1846
Btw_2	**0.1634**	0.1532	0.1749	0.1525	0.1331	0.1331	0.1383	0.1555	**0.3030**	0.2244
Btw_3	0.1671	0.1575	0.1769	**0.1844**	0.1474	0.1287	0.1280	0.1613	**0.2497**	0.2156
1-NN	0.1184	**0.1513**	0.1346	0.1328	0.1262	0.1074	0.1193	0.1473	0.1921	**0.1989**
MDS	2-D	5-D	10-D	30-D	50-D	2-D	5-D	10-D	30-D	50-D
Btw_1	0.1419	0.2033	0.2568	0.2746	**0.3024**	0.1861	0.3812	0.4511	0.5199	**0.5257**
Btw_2	0.1558	0.2296	0.2623	0.2680	**0.3137**	0.3402	0.4880	0.5372	**0.5565**	0.5390
Btw_3	0.1625	0.2299	0.2599	0.2579	**0.3057**	0.3415	0.4613	0.4862	0.5094	**0.5149**
1-NN	0.1491	0.2008	0.2187	0.2525	**0.3073**	0.3218	0.3982	0.4452	0.5043	**0.5192**
Isomap	2-D	5-D	10-D	30-D	50-D	2-D	5-D	10-D	30-D	50-D
Btw_1	0.1528	0.1801	0.2485	0.2629	**0.2636**	0.1549	0.2613	**0.3547**	0.3457	0.3478
Btw_2	0.1607	0.2187	0.2160	0.2438	**0.2666**	0.2707	0.3947	**0.4449**	0.3723	0.3462
Btw_3	0.1510	0.1967	0.2432	0.2343	**0.2648**	0.2894	0.4136	**0.4160**	0.3575	0.3516
1-NN	0.1679	0.2020	0.2031	**0.2447**	0.2440	0.2947	**0.4121**	0.3743	0.3253	0.3325

Table 3. Classification accuracy for the region-based encoding

	GeoNames					Foursquare				
MDS	2-D	5-D	10-D	30-D	50-D	2-D	5-D	10-D	30-D	50-D
Btw_1^R	0.4370	0.5252	0.5630	0.5714	**0.5798**	0.3079	0.6017	0.6384	**0.6582**	0.6554
Btw_2^R	0.4538	0.5168	0.5588	**0.5798**	0.5714	0.4661	0.6102	0.6384	0.6582	**0.6723**
Btw_3^R	0.4244	0.5210	**0.5672**	0.5420	**0.5672**	0.4124	0.5565	0.6328	**0.6582**	0.6469
1-NN	0.3655	0.3908	0.5210	**0.5630**	0.4622	0.3950	0.4454	0.4706	0.4958	**0.5042**
CHRegion	**0.4748**	0.4538	0.4286	0.0084	0.0084	0.4407	**0.5198**	0.4407	0.0480	0.0480
Isomap	2-D	5-D	10-D	30-D	50-D	2-D	5-D	10-D	30-D	50-D
Btw_1^R	0.4580	0.5000	**0.5714**	0.5672	0.5588	0.2910	0.6045	0.6554	0.6864	**0.6921**
Btw_2^R	0.4244	0.5210	**0.5798**	0.5714	0.5588	0.5452	0.6356	0.6328	0.6723	**0.6949**
Btw_3^R	0.4496	0.5126	0.4622	0.5084	**0.5294**	0.5226	0.5367	0.5706	**0.5904**	0.5678
1-NN	0.3950	0.4454	0.4706	0.5042	**0.5210**	0.4774	0.5876	**0.6554**	0.6328	0.6497
CHRegion	0.4496	**0.4958**	0.1849	0.0084	0.0084	0.5028	**0.5056**	0.4972	0.0480	0.0480

5.3 Results Obtained: Region-Based Encoding

Table 3 shows the results obtained for the categorisation problem. Only MDS and Isomap were considered for these experiments, since they outperformed SVD for the point-based encoding and SVD is computationally more expensive. In Table 3 it can be observed that Btw_1^R and Btw_2^R clearly outperform the other methods, while Btw_3^R outperforms 1-NN in all cases, except when Isomap is used for the Foursquare taxonomy. Also, for the CHRegion method, it can be observed that, although its performance is competitive in very low dimensional

Table 4. Identification results (MAP) for the region-based encoding

	GeoNames					Foursquare				
MDS	2-D	5-D	10-D	30-D	50-D	2-D	5-D	10-D	30-D	50-D
Btw_1^R	0.1446	0.1768	0.1696	0.1740	**0.1899**	0.1562	0.2312	0.2521	0.2948	**0.3117**
Btw_2^R	0.1690	0.1770	0.1734	0.1768	**0.1897**	0.2073	0.2916	0.3042	**0.3138**	0.3079
Btw_3^R	0.1786	0.2166	0.2116	**0.2248**	0.2164	0.2154	0.3328	0.3756	**0.3799**	0.3697
1-NN	0.1477	0.1330	0.1473	0.1529	**0.1789**	0.1803	0.2710	**0.3023**	0.2847	0.2892
Isomap	2-D	5-D	10-D	30-D	50-D	2-D	5-D	10-D	30-D	50-D
Btw_1^R	0.1985	**0.3225**	0.3205	0.3080	0.3152	0.1898	0.4312	**0.4726**	0.4625	0.4426
Btw_2^R	0.2176	0.2648	0.3107	0.3072	**0.3148**	0.2960	0.4680	**0.4913**	0.4697	0.4463
Btw_3^R	0.2108	0.2440	0.2761	0.2662	**0.2985**	0.2939	0.4261	0.4573	**0.4676**	0.4439
1-NN	0.1554	0.2506	**0.2843**	0.2455	0.2421	0.2550	0.4477	0.4736	**0.4847**	0.4592

Table 5. Comparison with human-based categorisation

GeoNames			Foursquare		
Humans (average): 0.5716			Humans (average): 0.7667		
	Point-based	Region-based		Point-based	Region-based
Btw_1	0.5957	0.6383	Btw_1^R	0.7571	0.7714
Btw_2	0.5957	0.6170	Btw_2^R	0.7714	0.7571
Btw_3	0.5957	0.5957	Btw_3^R	0.7714	0.7429
k-NN	0.5957	0.5957	k-NN	0.7286	0.7000

spaces (up to 10-D), it quickly degrades if the dimensionality of the data is increased, the reason being that the method relies on finding points in the three regions which are in a perfect betweenness relation and such triples tend to be very rare in higher-dimensional spaces. Note that the best results for MDS are slightly inferior to the ones obtained with the point-based approaches, while the performance of Isomap is better.

Table 4 shows the results obtained for the identification problem. The betweenness based approaches outperform 1-NN in all cases, except when Isomap is used for the Foursquare taxonomy. Note that in contrast to the results discussed previously, when MDS is used, Btw_3^R outperforms both Btw_1^R and Btw_2^R. Isomap results are slightly better than those for the point-based approach, but MDS results are worse.

5.4 Comparison with a Human Gold Standard

While the accuracy scores in Tables 1 and 3 are relatively low at first glance, there are many place types in the GeoNames and Foursquare taxonomies whose categorisation is debatable. For example, in Foursquare, *piano bar* is in the same category as *aquarium* and *bowling alley* (category: *arts & entertainment*) but in a different category as *juice bar*, *café* and *gastropub* (category: *food*), which are in turn in a different category as *lounge*, *wine bar* or *pub* (category: *nightlife spot*). To enable a better interpretation of the results, we compared our methods with how humans perform on the same task. This experiment consisted of a single hold-out instance of the categorisation problem. Specifically, we asked three people to categorise 20% of the Foursquare place types, being given the remaining 80% of the instances of each of the categories (but not the names of the categories). Three different people were asked to categorise 20% of the Geonames place types. The same conditions were fixed for the betweenness-based methods and k-NN. Only the best configuration of each method was considered.

Table 5 shows the results of the comparison. For the GeoNames taxonomy, the betweenness methods and k-NN obtain a better accuracy than the humans, in both the point-based and region-based schemes. For the Foursquare taxonomy, Btw_2^R and Btw_3^R of the point-based scheme, and Btw_1^R of the region-based scheme outperform the humans, while k-NN performed slightly worse. The fact that the betweenness methods are capable of outperforming humans on this task is remarkable, and strongly supports our hypothesis that betweenness relations obtained from Flickr may indeed be a usuful addition to is-a hierarchies.

Cohen's kappa [2] was used to measure the agreement among the humans and between humans and the considered methods. In GeoNames, average human agreement was 0.3981 on average, whereas agreement among the four point-based methods was 0.7061 and agreement among the region-based methods was 0.6227. Also, the average agreement between all possible ways in which a human can be paired with one of the methods was 0.3720. In Foursquare, average agreement was 0.6457 among humans, 0.8249 for point-based methods, and 0.8313 for region-based methods. An average agreement of 0.6152 was found between humans and the methods. The higher values in the latter case suggest that the Foursquare taxonomy is more intuitive than the GeoNames taxonomy. Both point-based and region-based methods can be considered stable since they show more agreement than humans. It is notable that agreement among humans is only slightly higher than agreement between humans and the considered methods. This suggests that our methods could be useful to improve existing taxonomies, adapting them to a representation which is nearer to human perception.

6 Conclusions

We have proposed a method to obtain information about the conceptual betweenness of place types, where a place type r is said to be between place types p and q if r has all the properties which are shared by p and q. Specifically, we first induce a semantic space of place types by applying a dimensionality reduction method to frequency vectors obtained from Flickr, and then derive a betweenness relation from the resulting geometric representation. By augmenting classical place type taxonomies with a (graded) betweenness relation, a more flexbile and robust representation is obtained. This betweenness relation could be used to merge, repair, or extend existing taxonomies in an automated way. Moreover, it implicitly encodes many alternative ways in which place types could be grouped, which is useful in applications where taxonomies are used for their predictive value. Our experimental results show that using betweenness outperforms similarity-based methods such as k-NN on a number of place type classification tasks. Surprisingly, using betweenness was also shown to outperform humans (when they are not shown the names of the categories). Although we have specifically focused on place types in this paper, the idea of augmenting taxonomies with a betweenness relation is likely to be useful in other domains, such as movie genres, music genres or research areas. Furthermore, our method could easily be applied to improve or enrich multi-level hierarchical taxonomies, by repeatedly applying the same strategy at each of the levels of the taxonomy.

Acknowledgments. This work was supported by EPSRC grant EP/K021788/1. The authors would like to thank Olivier Van Laere for his help with obtaining the Flickr data.

References

1. Chen, Y., Garcia, E.K., Gupta, M.R., Rahimi, A., Cazzanti, L.: Similarity-based classification: Concepts and algorithms. Journal of Machine Learning Research 10, 747–776 (2009)
2. Cohen, J.: A coefficient of agreement for nominal scales. Educational and Psychological Measurement 20(1), 37–46 (1960)
3. Correa, W.F., Prade, H., Richard, G.: Trying to understand how analogical classifiers work. In: Hüllermeier, E., Link, S., Fober, T., Seeger, B. (eds.) SUM 2012. LNCS (LNAI), vol. 7520, pp. 582–589. Springer, Heidelberg (2012)
4. Cover, T., Hart, P.: Nearest neighbor pattern classification. IEEE Transactions on Information Theory 13(1), 21–27 (1967)
5. Cox, T.F., Cox, M.A.A., Cox, T.F.: Multidimensional Scaling. Chapman & Hall/CRC (2001)
6. Deerwester, S., Dumais, S.T., Furnas, G.W., Landauer, T.K., Harshman, R.: Indexing by latent semantic analysis. Journal of the American Society for Information Science 41(6), 391–407 (1990)
7. Dubois, D., Prade, H., Esteva, F., Garcia, P., Godo, L.: A logical approach to interpolation based on similarity relations. International Journal of Approximate Reasoning 17(1), 1–36 (1997)
8. Erk, K.: Representing words as regions in vector space. In: Proceedings of the Thirteenth Conference on Computational Natural Language Learning, pp. 57–65 (2009)
9. Gabrilovich, E., Markovitch, S.: Computing semantic relatedness using wikipedia-based explicit semantic analysis. In: Proceedings of the 20th International Joint Conference on Artificial Intelligence, vol. 6, pp. 1606–1611 (2007)
10. Gärdenfors, P.: Conceptual Spaces: The Geometry of Thought. MIT Press (2000)
11. Hearst, M.A.: Automatic acquisition of hyponyms from large text corpora. In: Proceedings of the 14th Conference on Computational Linguistics, pp. 539–545 (1992)

12. Kok, S., Domingos, P.: Statistical predicate invention. In: Proceedings of the 24th International Conference on Machine Learning, pp. 433–440 (2007)
13. Kozima, H., Ito, A.: Context-sensitive word distance by adaptive scaling of a semantic space. In: Mitkov, R., Nicolov, N. (eds.) Recent Advances in Natural Language Processing. Current Issues in Linguistic Theory, vol. 136, pp. 111–124. John Benjamins Publishing Company (1997)
14. Krumhansl, C.: Concerning the applicability of geometric models to similarity data: The interrelationship between similarity and spatial density. Psychological Review 5, 445–463 (1978)
15. Lockwood, J.: Predicting which species will become invasive: what's taxonomy got to do with it? In: Purvis, J.G.A., Brooks, T. (eds.) Phylogeny and Conservation, pp. 365–386. Cambridge University Press (2005)
16. Losos, J.B.: Phylogenetic niche conservatism, phylogenetic signal and the relationship between phylogenetic relatedness and ecological similarity among species. Ecology Letters 11(10), 995–1003 (2008)
17. Maedche, A., Pekar, V., Staab, S.: Ontology Learning Part One - On Discovering Taxonomic Relations from the Web, pp. 301–322. Springer (2002)
18. Nalbantov, G.I., Groenen, P.J., Bioch, J.C.: Nearest convex hull classification. Technical report, Erasmus School of Economics, ESE (2006)
19. Padó, S., Lapata, M.: Dependency-based construction of semantic space models. Computational Linguistics 33(2), 161–199 (2007)
20. Reisinger, J., Mooney, R.J.: Multi-prototype vector-space models of word meaning. In: Proceedings of the 2010 Annual Conference of the North American Chapter of the Association for Computational Linguistics, pp. 109–117 (2010)
21. Roweis, S.T., Saul, L.K.: Nonlinear dimensionality reduction by locally linear embedding. Science 290(5500), 2323–2326 (2000)
22. Salton, G., Wong, A., Yang, C.S.: A vector space model for automatic indexing. Communications of the ACM 18(11), 613–620 (1975)
23. Schockaert, S., Prade, H.: Interpolation and extrapolation in conceptual spaces: A case study in the music domain. In: Rudolph, S., Gutierrez, C. (eds.) RR 2011. LNCS, vol. 6902, pp. 217–231. Springer, Heidelberg (2011)
24. Schockaert, S., Prade, H.: Interpolative and extrapolative reasoning in propositional theories using qualitative knowledge about conceptual spaces. Artificial Intelligence 202, 86–131 (2013)
25. Sheremet, M., Tishkovsky, D., Wolter, F., Zakharyaschev, M.: A logic for concepts and similarity. Journal of Logic and Computation 17(3), 415–452 (2007)
26. Speer, R., Havasi, C., Lieberman, H.: Analogyspace: Reducing the dimensionality of common sense knowledge. In: Proceedings of the 23rd AAAI Conference on Artificial Intelligence, pp. 548–553 (2008)
27. Sun, R.: Robust reasoning: integrating rule-based and similarity-based reasoning. Artificial Intelligence 75(2), 241–295 (1995)
28. Tenenbaum, J.B., de Silva, V., Langford, J.C.: A global geometric framework for nonlinear dimensionality reduction. Science 290(5500), 2319–2323 (2000)
29. Turney, P.D.: Measuring semantic similarity by latent relational analysis. In: Proceedings of the 19th International Joint Conference on Artificial Intelligence, pp. 1136–1141 (2005)
30. Van Canneyt, S., Schockaert, S., Van Laere, O., Dhoedt, B.: Detecting places of interest using social media. In: Proceedings of the IEEE/WIC/ACM International Joint Conferences on Web Intelligence and Intelligent Agent Technology, pp. 447–451 (2012)

31. Van Canneyt, S., Van Laere, O., Schockaert, S., Dhoedt, B.: Using social media to find places of interest: A case study. In: Proceedings of the 1st ACM SIGSPATIAL International Workshop on Crowdsourced and Volunteered Geographic Information, pp. 2–8 (2012)
32. Van Laere, O., Schockaert, S., Dhoedt, B.: Ghent university at the 2011 placing task. In: Working Notes of the MediaEval Workshop. CEUR-WS.org (2011)
33. Vincent, P., Bengio, Y.: K-local hyperplane and convex distance nearest neighbor algorithms. In: Advances in Neural Information Processing Systems, pp. 985–992 (2001)
34. Weinberger, K.Q., Saul, L.K.: Distance metric learning for large margin nearest neighbor classification. Journal of Machine Learning Research 10, 207–244 (2009)
35. Wu, W., Li, H., Wang, H., Zhu, K.Q.: Probase: A probabilistic taxonomy for text understanding. In: Proceedings of the 2012 ACM SIGMOD International Conference on Management of Data, pp. 481–492 (2012)
36. Yang, Y., Pedersen, J.O.: A comparative study on feature selection in text categorization. In: Proceedings of the 14th International Conference on Machine Learning, pp. 412–420 (1997)

Hintikka-Style Semantic Games for Fuzzy Logics

Christian G. Fermüller*

Theory and Logic Group
Vienna University of Technology

Abstract. Various types of semantics games for deductive fuzzy logics, most prominently for Łukasiewicz logic, have been proposed in the literature. These games deviate from Hintikka's original game for evaluating classical first-order formulas by either introducing an explicit reference to a truth value from the unit interval at each game state (as in [4]) or by generalizing to multisets of formulas to be considered at any state (as, e.g., in [12,9,7,10]). We explore to which extent Hintikka's game theoretical semantics for classical logic can be generalized to a many-valued setting without sacrificing the simple structure of Hintikka's original game. We show that rules that instantiate a certain scheme abstracted from Hintikka's game do not lead to logics beyond the rather inexpressive, but widely applied Kleene-Zadeh logic, also known as 'weak Łukasiewicz logic' or even simply as 'fuzzy logic' [27]. To obtain stronger logics we consider propositional as well as quantifier rules that allow for random choices. We show how not only various extensions of Kleene-Zadeh logic, but also proper extensions Łukasiewicz logic arise in this manner.

1 Introduction

Fuzzy logics "in Zadeh's narrow sense" [34,15], i.e. truth functional logics with the real unit interval as set of truth values, nowadays come in many forms and varieties. (We refer to the *Handbook of Mathematical Fuzzy Logics* [3] for an overview.) From an application oriented point of view, but also with respect to foundational concerns, this fact imparts enhanced significance to the problem of deriving *specific* logics from underlying *semantic principles of reasoning*. Among the various models that have been proposed in this vein are Lawry's voting semantics [22], Paris's acceptability semantics [28], re-randomising semantics [21], and approximation semantics [2,29]. Of particular importance in our context is Robin Giles's attempt, already in the 1970s [12,13] to justify Łukasiewicz logic, one of the most fundamental formalizations of deductive fuzzy logic, with respect to a game that models reasoning about dispersive experiments. While Giles explicitly acknowledged the influence of Paul Lorenzen's pioneering work on dialogical foundations for constructive logic [23,24], he did not refer to Hintikka's game theoretic semantics [18,20]. However, with the benefit of hindsight, one can classify Giles's game for Łukasiewicz logic as a *semantic game*, i.e. a game for evaluating a formula with respect to a given interpretation, guided by rules

* Supported by FWF grant P25417-G15.

C. Beierle and C. Meghini (Eds.): FoIKS 2014, LNCS 8367, pp. 193–210, 2014.

for the stepwise reduction of logically complex formulas into their subformulas. While this renders Giles's game closer to Hintikka's than to Lorenzen's game, Giles deviates in some important ways from Hintikka's concept, as we will explain in Section 2. Semantic games for Łukasiewicz logic that, arguably, are closer in their mathematical form to Hintikka's semantic game for classical logic have been introduced by Cintula and Majer in [4]. However, also these latter games exhibit features that are hardly compatible with Hintikka's motivation for introducing game theoretic semantics [18,19] as foundational approach to logic and language. In particular, they entail an explicit reference to some (in general non-classical) truth value at every state of a game. The just presented state of affairs triggers a question that will guide the investigations of this paper: *To what extent can deductive fuzzy logics be modeled by games that remain close in their format, if not in spirit, to Hintikka's classic game theoretic semantics?*

The paper is organized as follows. In Section 2 we present (notational variants) of the mentioned semantic games by Hintikka, Cintula/Majer, and Giles in a manner that provides a basis for systematic comparison. In particular, we observe that so-called Kleene-Zadeh logic KZ, a frequently applied fragment of Łukasiewicz logic Ł, is characterized already by Hintikka's classic game if one generalizes the set of possible pay-off values from $\{0,1\}$ to the unit interval $[0,1]$. In Section 3 we introduce a fairly general scheme of rules that may be added to Hintikka's game in a many-valued setting and show that each connective specified by such a rule is already definable in logic KZ. Adapting an idea from [8,10], we then show in Sections 4 and 5 how one can go beyond KZ, while retaining essential features of Hintikka's original game format. In particular, we introduce in Section 4 a propositional 'random choice connective' π by a very simple rule. We show that this rule for π in combination with a rule for doubling the pay-off for the player who is currently in the role of the 'Proponent' leads to a proper extension of propositional Łukasiewicz logic. In Section 5 we indicate how, at the first-order level, various families of rules that involve a random selection of witnessing domain elements characterize corresponding families of fuzzy quantifiers. We conclude in Section 6 with a brief summary, followed by remarks on the relation between our 'randomized game semantics' and the 'equilibrium semantics' for IF-logic [25,32] arising from considering incomplete information in Hintikka's game.

2 Variants of Semantic Games

Let us start by reviewing Hintikka's classic semantic game [18,20]. There are two players, called *Myself* (or simply *I*) and *You*, here, who can both act either in the role of the *Proponent* **P** or of the *Opponent* **O**[1] of a given first-order formula F, augmented by a variable assignment θ. Initially *I* act as **P** and *You* act as **O**.

[1] Hintikka uses *Nature* and *Myself* as names for the players and *Verfier* and *Falisifer* for the two roles. To emphasize our interest in the connection to Giles's game we use Giles's names for the players and Lorenzen's corresponding role names throughout the paper.

My aim — or, more generally, **P**'s aim at any state of the game — is to show that the initial formula is true in a given interpretation $\mathcal{M}$ with respect to θ. The game proceeds according to the following rules. Note that these rules only refer to the roles and the outermost connective of the *current formula*, i.e. the formula, augmented by an assignment of domain elements to free variables, that is at stake at the given state of the game. Together with a *role distribution* of the players, this *augmented formula* fully determines any state of the game.

($R^{\mathcal{H}}_{\wedge}$) If the current formula is $(F \wedge G)[\theta]$ then **O** chooses whether the game continues with $F[\theta]$ or with $G[\theta]$.
($R^{\mathcal{H}}_{\vee}$) If the current formula is $(F \vee G)[\theta]$ then **P** chooses whether the game continues with $F[\theta]$ or with $G[\theta]$.
($R^{\mathcal{H}}_{\neg}$) If the current formula is $\neg F[\theta]$, the game continues with $F[\theta]$, except that the roles of the players are switched: the player who is currently acting as **P**, acts as **O** at the the next state, and vice versa for the current **O**.
($R^{\mathcal{H}}_{\forall}$) If the current formula is $(\forall x F(x))[\theta]$ then **O** chooses an element c of the domain of $\mathcal{M}$ and the game continues with $F(x)[\theta[c/x]]$[2].
($R^{\mathcal{H}}_{\exists}$) If the current formula is $\exists x F(x)[\theta]$ then **P** chooses an element c of the domain of $\mathcal{M}$ and the game continues with $F(x)[\theta[c/x]]$.

Except for ($R^{\mathcal{H}}_{\neg}$), the players' roles remain unchanged. The game ends when an atomic (augmented) formula $A[\theta]$ is hit. The player who is currently acting as **P** *wins* and the other player, acting as **O**, *loses* if A is true with respect to θ in the given model $\mathcal{M}$. We associate pay-off 1 with winning and pay-off 0 with losing. We also include the truth constants $\top$ and $\bot$, with their usual interpretation, among the atomic formulas. The game starting with formula F and assignment θ is called the *$\mathcal{H}$-game for $F[\theta]$ under $\mathcal{M}$*.

Theorem 1 (Hintikka). *A formula F is true in a (classical) interpretation $\mathcal{M}$ with respect to the initial variable assignment θ (in symbols: $v^{\theta}_{\mathcal{M}}(F) = 1$) iff I have a winning strategy in the $\mathcal{H}$-game for $F[\theta]$ under $\mathcal{M}$.*

Our aim is to generalize Hintikka's Theorem to deductive fuzzy logics. As already mentioned in the introductions, contemporary mathematical fuzzy logic offers a plethora of logical systems. Here we focus on (extensions of) a system simply called 'fuzzy logic', e.g., in the well known textbook [27]. Following [1], we prefer to call this logic *Kleene-Zadeh logic*, or KZ for short. KZ is mostly considered only at the propositional level, where its semantics is given by extending an assignment $\mathcal{M}$ of atomic formulas to truth values in $[0, 1]$ as follows:

$$\begin{aligned}
&v_{\mathcal{M}}(F \wedge G) = \min(v_{\mathcal{M}}(F), v_{\mathcal{M}}(G)),\\
&v_{\mathcal{M}}(F \vee G) = \max(v_{\mathcal{M}}(F), v_{\mathcal{M}}(G)),\\
&v_{\mathcal{M}}(\neg F) = 1 - v_{\mathcal{M}}(F),\\
&v_{\mathcal{M}}(\bot) = 0,\\
&v_{\mathcal{M}}(\top) = 1.
\end{aligned}$$

[2] $\theta[c/x]$ denotes the variable assignment that is like θ, except for assigning c to x.

At the first-order level an interpretation $\mathcal{M}$ includes a non-empty set D as domain. With respect to an assignment θ of domain elements to free variables, the semantics of the universal and the existential quantifier is given by

$$v^{\theta}_{\mathcal{M}}(\forall x F(x)) = \inf_{d\in D}(v^{\theta[d/x]}_{\mathcal{M}}(F(x))),$$
$$v^{\theta}_{\mathcal{M}}(\exists x F(x)) = \sup_{d\in D}(v^{\theta[d/x]}_{\mathcal{M}}(F(x))).$$

It is interesting to observe that neither the rules nor the notion of a state in an $\mathcal{H}$-game have to be changed in order to characterize logic KZ. We only have to generalize the possible pay-off values for the $\mathcal{H}$-game from $\{0,1\}$ to the unit interval $[0,1]$. More precisely, the pay-off for the player who is in the role of **P** when a game under $\mathcal{M}$ ends with the augmented atomic formula $A[\theta]$ is $v^{\theta}_{\mathcal{M}}(A)$.

If the pay-offs are modified as just indicated and correspond to the truth values of atomic formulas specified by a many-valued interpretation $\mathcal{M}$, we will speak of an $\mathcal{H}$-mv-game, where the *pay-offs match* $\mathcal{M}$. A slight complication arises for quantified formulas in $\mathcal{H}$-mv-games: there might be no element c in the domain of $\mathcal{M}$ such that $v^{\theta[c/x]}_{\mathcal{M}}(F(x)) = \inf_{d\in D}(v^{\theta[d/x]}_{\mathcal{M}}(F(x)))$ or no domain element e such that $v^{\theta[e/x]}_{\mathcal{M}}(F(x)) = \sup_{d\in D}(v^{\theta[d/x]}_{\mathcal{M}}(F(x)))$. A simple way to deal with this fact is to restrict attention to so-called witnessed models [17], where constants that witness all arising infima and suprema are assumed to exist. In other words: infima are minima and suprema are maxima in witnessed models. A more general solution refers to optimal payoffs up to some ϵ.

Definition 1. *Suppose that, for every $\epsilon > 0$, player $\boldsymbol{X}$ has a strategy that guarantees her a pay-off of at least $w - \epsilon$, while her opponent has a strategy that ensures that $\boldsymbol{X}$'s pay-off is at most $w + \epsilon$, then w is called the* value for **X** of the game.

This notion, which corresponds to that of an ϵ-equilibrium as known from game theory, allows us to state the following generalization of Theorem 1.

Theorem 2. *A formula F evaluates to $v^{\theta}_{\mathcal{M}}(F) = w$ in a KZ-interpretation $\mathcal{M}$ with respect to the variable assignment θ iff the $\mathcal{H}$-mv-game for $F[\theta]$ with pay-offs matching $\mathcal{M}$ has value w for Myself.*

A proof of Theorem 2 can (essentially[3]) be found in [10].

From the point of view of continuous t-norm based fuzzy logics, as popularized by Petr Hájek [15,16], Kleene-Zadeh logic KZ is unsatisfying: while min is a t-norm, it's indicated residuum, which corresponds to implication in Gödel-Dummett logic is not expressible. Indeed, defining implication by $F \supset G =_{def} \neg F \vee G$ (in analogy to classical logic) in KZ, entails that $F \to F$ is not valid, i.e. $v_{\mathcal{M}}(F \to F)$ is not true in all interpretations.[4] In fact, formulas that do not contain truth constants are never valid in KZ. The most important fuzzy logic extending KZ arguably is Łukasiewicz logic Ł. The language

[3] A variant of $\mathcal{H}$-games is used in [10] and KZ is called 'weak Łukasiewicz logic' there.

[4] We suppress the reference to a variable assignment θ when referring to propositional connectives.

of Ł extends that of KZ by implication $\rightarrow$, strong conjunction $\otimes$, and strong disjunction $\oplus$. The semantics of these connectives is given by

$$\begin{aligned} v_{\mathcal{M}}(F \rightarrow G) &= \min(1, 1 - v_{\mathcal{M}}(F) + v_{\mathcal{M}}(G)),\\ v_{\mathcal{M}}(F \otimes G) &= \max(0, v_{\mathcal{M}}(F) + v_{\mathcal{M}}(G) - 1),\\ v_{\mathcal{M}}(F \oplus G) &= \min(1, v_{\mathcal{M}}(F) + v_{\mathcal{M}}(G)). \end{aligned}$$

In fact all other propositional connectives could by defined in Ł, e.g., from $\rightarrow$ and $\perp$, or from $\otimes$ and $\neg$, alone. However, neither $\rightarrow$ nor $\otimes$ nor $\oplus$ can be defined in KZ.[5] The increased expressiveness of Ł over KZ is particularly prominent at the first-order level: while in KZ there are only trivially valid formulas (which involve the truth constants in an essential manner), the set of valid first-order formulas in Ł is not even recursively enumerable, due to a classic result of Scarpellini [31].

It seems to be impossible to characterize full Łukasiewicz logic Ł by trivial extensions of the $\mathcal{H}$-game, comparable to the shift from $\mathcal{H}$-games to $\mathcal{H}$-mv-games. Before investigating, in Sections 4 and 5, how one can nevertheless generalize the $\mathcal{H}$-game to extensions of KZ, including Ł, without changing the concept of a game state as solely determined by an (augmented) formula and a role distribution, we review two types of semantic games for Ł that deviate more radically from Hintikka's classic game theoretical semantics: explicit evaluation games, due to Cintula and Majer [4], and Giles's dialogue and betting game [12,13].

In [4] Cintula and Majer present a game for Ł that conceptually differs from the $\mathcal{H}$-mv-game by introducing an explicit reference to a value $\in [0,1]$ at every state of the game. They simply speak of an 'evaluation game'; but since all games considered in this paper are games for evaluating formulas with respect to a given interpretation, we prefer to speak of an *explicit evaluation game*, or $\mathcal{E}$-game for short. Like above, we call the players *Myself* (*I*) and *You*, and the roles **P** and **O**. In the initial state *I* am in the role of **P** and *You* are acting as **O**. In addition to the role distribution and the current (augmented) formula[6], also a *current value* $\in [0,1]$ is included in the specification of a game state. We will not need to refer to any details of $\mathcal{E}$-games, but present the rules for $\oplus$, $\otimes$, $\neg$, and $\exists$ here, to assist the comparison with other semantic games:

($R^{\mathcal{E}}_{\otimes}$) If the current formula is $(F \otimes G)[\theta]$ and the current value[7] is r then **P** chooses a value $\bar{r} \leq 1 - r$ and **O** chooses whether to continue the game with $F[\theta]$ and value $r + \bar{r}$ or with $G[\theta]$ and value $1 - \bar{r}$.

($R^{\mathcal{E}}_{\oplus}$) If the current formula is $(F \oplus G)[\theta]$ and the current value is r then **P** chooses $\bar{r} \leq 1 - r$ and **O** chooses whether to continue with $F[\theta]$ and value $\bar{r}$ or with $G[\theta]$ and value $r - \bar{r}$.

($R^{\mathcal{E}}_{\neg}$) If the current formula is $\neg F[\theta]$ and the current value is r, then **O** chooses $\bar{r}$, where $0 < \bar{r} \leq r$, and the game continues with $F[\theta]$ and value $(1 - r) + \bar{r}$ after a role switch.

---

[5] Therefore KZ is sometimes called the 'weak (fragment of) Łukasiewicz logic'.

[6] I.e., the current formula, now over the language of Ł, augmented by an assignment of domain elements to free variables.

[7] All values mentioned here have to be in $[0,1]$.

($R^{\mathcal{H}}_{\exists}$) If the current formula is $\exists x F(x)[\theta]$ and the current value is r then **O** chooses $\bar{r} > 0$ and **P** picks an element c of the domain of $\mathcal{M}$ and the game continues with $F(x)[\theta[c/x]]$ and value $r - \bar{r}$.

The rules for $\wedge$, $\vee$, $\forall$ are analogous to the corresponding rules for the $\mathcal{H}$-mv-game: the current value remains unchanged. Cintula and Mayer [4] do not specify a rule for implication. However such a rule can be synthesized from the other rules, given the definability of $\rightarrow$ from the other connectives. As soon as the game reaches an augmented atomic formula $A[\theta]$ the game under interpretation $\mathcal{M}$ ends and the player in the current role of **P** wins (and the opposing player loses) if $v^{\theta}_{\mathcal{M}}(A) \geq r$. Otherwise the current **O** wins and the current **P** loses. Compared to Theorems 1 and 2, the adequateness theorem for the $\mathcal{E}$-game shows a somewhat less direct correspondence to the standard semantics of Ł.

Theorem 3 (Cintula/Mayer). *I have a winning strategy in the $\mathcal{E}$-game under $\mathcal{M}$ starting with $F[\theta]$ and value r iff $v^{\theta}_{\mathcal{M}}(F) \geq r$.*

A game based interpretation of Ł that arguably deviates even more radically from $\mathcal{H}$-games than $\mathcal{E}$-games was presented by Giles already in the 1970s [12,13]. In fact Giles did not refer to Hintikka, but rather to the dialogue games suggested by Lorenzen [23,24] as a foundation for constructive reasoning. Initially Giles proposed his game as a model of logical reasoning within theories of physics; but later he motivated the game explicitly as an attempt to provide "tangible meaning" for fuzzy logic [14]. We briefly review the essential features of Giles's game, in a variant called $\mathcal{G}$-game, that facilitates comparison with the other semantic games mentioned in this paper. Again we use *Myself* (*I*) and *You* as names for the players, and refer to the roles **P** and **O**. Unlike in $\mathcal{H}$-, $\mathcal{H}$-mv- or $\mathcal{E}$-games, a game state contains more that one current formula, in general. More precisely a state of a $\mathcal{G}$-game is given by

$$[F_1[\theta_1], \ldots, F_m[\theta_m] \mid G_1[\theta'_1], \ldots, G_n[\theta'_n]],$$

where $\{F_1[\theta_1], \ldots, F_m[\theta_m]\}$ is the *multiset* of augmented formulas currently asserted by *You*, called *your tenet*, and $\{G_1[\theta'_1], \ldots, G_n[\theta'_n]\}$ is the multiset of augmented formulas currently asserted by *Myself*, called *my tenet*. At any given state an occurrence of a non-atomic augmented formula $H[\theta]$ is picked arbitrarily and distinguished as *current formula*.[8] If $H[\theta]$ is in my tenet then *I* am acting as **P** and *You* are acting as **O**. Otherwise, i.e. if $H[\theta]$ is in your tenet, *I* am **O** and *You* are **P**. States that only contain atomic formulas are called *final*. At non-final states the game proceeds according to the following rules:

($R^{\mathcal{G}}_{\wedge}$) If the current formula is $(F \wedge G)[\theta]$ then the game continues in a state where the indicated occurrence of $(F \wedge G)[\theta]$ in **P**'s tenet is replaced by either $F[\theta]$ or by $G[\theta]$, according to **O**'s choice.

[8] It turns out that the powers of the players of a $\mathcal{G}$-game are not depended on the manner in which the current formula is picked at any state. Still, a more formal presentation of $\mathcal{G}$-games will employ the concepts of a regulation and of so-called internal states in formalizing state transitions. We refer to [7] for details.

($R_{\vee}^{\mathcal{G}}$) If the current formula is $(F \vee G)[\theta]$ then the game continues in a state where the indicated occurrence of $(F \vee G)[\theta]$ in **P**'s tenet is replaced by either $F[\theta]$ or by $G[\theta]$, according to **P**'s choice.

($R_{\to}^{\mathcal{G}}$) If the current formula is $(F \to G)[\theta]$ then the indicated occurrence of $(F \to G)[\theta]$ is removed from **P**'s tenet and **O** chooses whether to continue the game at the resulting state or whether to add $F[\theta]$ to **O**'s tenet and $G[\theta]$ to **P**'s tenet before continuing the game.

($R_{\forall}^{\mathcal{G}}$) If the current formula is $(\forall x F(x))[\theta]$ then **O** chooses an element c of the domain of $\mathcal{M}$ and the game continues in a state where the indicated occurrence of $(\forall x F(x))[\theta]$ in **P**'s tenet is replaced by $F(x)[\theta[c/x]]$.

($R_{\exists}^{\mathcal{G}}$) If the current formula is $(\exists x F(x))[\theta]$ then **P** chooses an element c of the domain of $\mathcal{M}$ and the game continues in a state where the indicated occurrence of $(\exists x F(x))[\theta]$ in **P**'s tenet is replaced by $F(x)[\theta[c/x]]$.

No rule for negation is needed if $\neg F$ is defined as $F \to \bot$. Likewise, rules for strong conjunction $\otimes$ and $\oplus$ can either be dispensed with by treating these connectives as defined from the other connectives or by introducing corresponding rules. (See [5,7] for a presentation of rules for strong conjunction.) If no non-atomic formula is left to pick as current formula, the game has reached a final state

$$[A_1[\theta_1], \ldots, A_m[\theta_m] \mid B_1[\theta'_1], \ldots, B_n[\theta'_n]],$$

where the $A_i[\theta_i]$ and $B_i[\theta'_i]$ are atomic augmented formulas. With respect to an interpretation $\mathcal{M}$ (i.e, an assignment of truth values to all atomic augmented formulas) the pay-off for *Myself* at this state is defined as

$$m - n + 1 + \sum_{1 \le i \le n} v_{\mathcal{M}}^{\theta}(B_i) - \sum_{1 \le i \le m} v_{\mathcal{M}}^{\theta}(A_i).$$

(Empty sums are identified with 0.) These pay-off values are said to *match* $\mathcal{M}$.

Just like for $\mathcal{H}$-mv-games, we need to take into account that suprema and infima are in general not witnessed by domain elements. Note that Definition 1 does not refer to any particular game. We may therefore apply the notion of the *value of a game* to $\mathcal{G}$-games as well. A $\mathcal{G}$-game where *my* tenet at the initial state consists of a single augmented formula occurrence $F[\theta]$, while *your* tenet is empty, is called a $\mathcal{G}$-game for $F[\theta]$. This allows us to express the adequateness of $\mathcal{G}$-games for Łukasiewicz logic in direct analogy to Theorem 2.

Theorem 4 (essentially Giles[9]). *A formula F evaluates to $v_{\mathcal{M}}^{\theta}(F) = w$ in a Ł-interpretation $\mathcal{M}$ with respect to the variable assignment θ iff the $\mathcal{G}$-game for $F[\theta]$ with pay-offs matching $\mathcal{M}$ has value w for Myself.*

At this point readers familiar with the original presentation of the game in [12,13] might be tempted to protest that we have skipped Giles's interesting

[9] Giles [12,13] only sketched a proof for the language without strong conjunction. For a detailed proof of the propositional case, where the game includes a rule for strong conjunction, we refer to [7].

story about betting money on the results of dispersive experiments associated with atomic assertions. Indeed, Giles proposes to assign an experiment E_A to each atomic formula A[10]. While each trial of an experiment yields either "yes" or "no" as its result, successive trials of the same experiment may lead to different results. However for each experiment E_A there is a known probability $\langle A \rangle$ that the result of a trial of E_A is negative. Experiment $E_\bot$ always yields a negative result; therefore $\langle \bot \rangle = 1$. Similarly $\langle \top \rangle = 0$. For each occurrence ('assertion') of an atomic formula in a player's final tenet, the corresponding experiment is run and the player has to pay one unit of money (say 1€) to the other player if the result is negative. Therefore Giles calls $\langle A \rangle$ the *risk* associated with A. For the final state $[A_1, \ldots, A_m \mid B_1, \ldots, B_n]$ the expected total amount of money that I have to pay to *You* (my total risk) is readily calculated to equal

$$\Big(\sum_{1 \le i \le m} \langle A_i \rangle - \sum_{1 \le i \le n} \langle B_i \rangle\Big)€.$$

Note that the total risk at final states translates into the pay-off specified above for $\mathcal{G}$-games via $v^\theta_\mathcal{M}(A) = 1 - \langle A \rangle$. To sum up: Giles's interpretation of truth values as inverted risk values associated with bets on dispersive experiments is totally independent from the semantic game for the stepwise reduction of complex formulas to atomic sub-formulas. In principle, one can interpret the pay-off values also for the $\mathcal{H}$-mv-game as inverted risk values and speak of bets on dispersive experiments at final states also there. The only (technically inconsequential) difference to the original presentation is that one implicitly talks about *expected* pay-off (inverted *expected* loss of money), rather than of certain pay-off when the betting scenario is used to interpret truth values.

Table 1 provides a summary of the general structure of the games reviewed in this section (where 'formula' means 'augmented formula' in the first-order case).

Table 1. Comparison of some semantic games

game	state determined by	pay-offs
$\mathcal{H}$-game	single formula + role distribution	bivalent
$\mathcal{H}$-mv-game	single formula + role distribution	many-valued
$\mathcal{E}$-game	single formula + role distribution + value	many-valued
$\mathcal{G}$-game	two multisets of formulas	many-valued

3 Generalized Propositional Rules for the $\mathcal{H}$-mv-game

At a first glimpse the possibilities for extending $\mathcal{H}$-mv-games to logics more expressive than KZ look very limited if, in contrast to $\mathcal{E}$-games and $\mathcal{G}$-games, we insist on *Hintikka's principle* that a state of the game is fully determined by a

[10] Giles ignores variable assignments, but stipulates that there is a constant symbol for every domain element. Thus only closed formulas need to be considered.

formula[11] and a distribution of the two roles (**P** and **O**) to the two players. One can come up with a more general concept of propositional game rules, related to those described in [6] for connectives defined by arbitrary finite deterministic and non-deterministic matrices. In order to facilitate a concise specification of all rules of that type, we introduce the following technical notion.

Definition 2. *An* n-selection *is a non-empty subset* S *of* $\{1, \ldots, n\}$*, where each element of* S *may additionally be marked by a* switch sign.

A game rule for an n-ary connective $\diamond$ in a *generalized* $\mathcal{H}$*-mv-game* is specified by a non-empty set $\{S_1, \ldots, S_m\}$ of n-selections. According to this concept, a round in a generalized $\mathcal{H}$-mv-game consists of two phases. The scheme for the corresponding game rule specified by $\{S_1, \ldots, S_m\}$ is as follows:

(Phase 1): If the current formula is $\diamond(F_1, \ldots, F_n)$ then **O** chooses an n-selection S_i from $\{S_1, \ldots, S_m\}$.

(Phase 2): **P** chooses an element $j \in S_i$. The game continues with formula F_j, where the roles of the players are switched if j is marked by a switch sign.

Remark 1. A variant of this scheme arises by letting **P** choose the n-selection S_i in phase 1 and **O** choose $j \in S_i$ in phase 2. But note that playing the game for $\diamond(F_1, \ldots, F_n)$ according to that role inverted scheme is equivalent to playing the game for $\neg \diamond (\neg F_1, \ldots, \neg F_n)$ using the exhibited scheme.

Remark 2. The rules $R^{\mathcal{H}}_{\wedge}$, $R^{\mathcal{H}}_{\vee}$, and $R^{\mathcal{H}}_{\neg}$ can be understood as instances of the above scheme:

- $R^{\mathcal{H}}_{\wedge}$ is specified by $\{\{1\}, \{2\}\}$,
- $R^{\mathcal{H}}_{\vee}$ is specified by $\{\{1, 2\}\}$, and
- $R^{\mathcal{H}}_{\neg}$ is specified by $\{\{1^*\}\}$, where the asterisk is used as switch mark.

Theorem 5. *In a generalized* $\mathcal{H}$*-mv-game, each rule of the type described above corresponds to a connective that is definable in logic* KZ.

Proof. The argument for the adequateness of all semantic games considered in this paper proceeds by backward induction on the game tree.

For (generalized) $\mathcal{H}$-mv-games the base case is trivial: by definition **P** receives pay-off $v_{\mathcal{M}}(A)$ and **O** receives pay-off $1 - v_{\mathcal{M}}(A)$ if the game ends with the atomic formula A.

For the inductive case assume that the current formula is $\diamond(F_1, \ldots, F_n)$ and that the rule for $\diamond$ is specified by the set $\{S_1, \ldots, S_m\}$ of n-selections, where $S_i = \{j(i, 1), \ldots, j(i, k(i))\}$ for $1 \leq i \leq m$ and $1 \leq k(i) \leq n$. Remember that the elements of S_i are numbers $\in \{1, \ldots, n\}$, possibly marked by a switch sign. For sake of clarity let us first assume that there are no switch signs, i.e. no role switches occur. Let us say that a player **X** *can force pay-off* w if **X** has a strategy that guarantees her a pay-off $\geq w$ at the end of the game. By the

[11] Since we focus on the propositional level, we will drop all explicit reference to variable assignments in Sections 3 and 4. However all statements remain valid if one replaces 'formula' by 'formula augmented by a variable assignment' throughout these sections.

induction hypothesis, **P** can force pay-off $v_{\mathcal{M}}(G)$ for herself and **O** can force pay-off pay-off $1 - v_{\mathcal{M}}(G)$ for himself if G is among $\{F_1, \ldots, F_n\}$ and does indeed occur at a successor state to the current one; in other words, if $G = F_{j(i,\ell)}$ for some $i \in \{1, \ldots, m\}$ and $\ell \in \{1, \ldots, k(i)\}$. Since **O** chooses the n-selection S_i, while **P** chooses an index number in S_i, **P** can force pay-off

$$\min_{1 \leq i \leq m} \max_{1 \leq \ell \leq k(i)} v_{\mathcal{M}}(F_{j(i,\ell)})$$

at the current state, while **O** can force pay-off

$$\max_{1 \leq i \leq m} \min_{1 \leq \ell \leq k(i)} (1 - v_{\mathcal{M}}(F_{j(i,\ell)})) = 1 - \min_{1 \leq i \leq m} \max_{1 \leq \ell \leq k(i)} v_{\mathcal{M}}(F_{j(i,\ell)}).$$

If both players play optimally these pay-off values are actually achieved. Therefore the upper expression corresponds to the truth function for $\diamond$. Both expressions have to be modified by uniformly substituting $1 - v_{\mathcal{M}}(F_{j(i,\ell)})$ for $v_{\mathcal{M}}(F_{j(i,\ell)})$ whenever $j(i,\ell)$ is marked by a switch sign in S_1 for $1 \leq i \leq m$ and $1 \leq k(i) \leq n$.

To infer that the connective $\diamond$ is definable in logic KZ it suffices to observe that its truth function, described above, can be composed from the functions $\lambda x(1-x)$, $\lambda x, y \min(x,y)$, and $\lambda x, y \max(x,y)$. But these functions are the truth functions for $\neg$, $\wedge$, and $\vee$, respectively, in KZ. □

4 Random Choice Connectives

In Section 2, following Giles, we have introduced the idea of expected pay-offs in a randomized setting. However, Giles applied this idea only to the interpretation of *atomic* formulas. For the interpretation of logical connectives and quantifiers in any of the semantic games mentioned in Section 2 it does not matter whether the players seek to maximize expected or certain pay-off or, equivalently, try to minimize either expected or certain payments to the opposing player. In [8,10] we have shown that considering random choices of witnessing constants in quantifier rules for *Giles-style* games, allows one to model certain (semi-)fuzzy quantifiers that properly extend first-order Łukasiewicz logic. In this section we want to explore the consequences of introducing random choices in rules for propositional connectives context of *Hintikka-style* games.

The results of Section 3 show that, in order to go beyond logic KZ with Hintikka-style games, a new variant of rules has to be introduced. As already indicated, a particularly simple type of new rules, that does not entail any change in the structure of game states, arises from randomization. So far we have only considered rules where either **P** or **O** chooses the sub-formula of the current formula to continue the game with. In game theory one often introduces *Nature* as a special kind of additional player, who does not care what the next state looks like, when it is her time to move and therefore is modeled by a uniformly random choice between all moves available to *Nature* at that state. As we will see below, introducing *Nature* leads to increased expressive power of semantic games. In fact, to keep the presentation of the games simple, we prefer to leave the role of *Nature* only implicit and just speak of random choices, without attributing them officially to a third player. The most basic rule of the indicated type refers to a new propositional connective π and can be formulated as follows.[12]

($R_\pi^{\mathcal{R}}$) If the current formula is $(F \pi G)$ then a uniformly random choice determines whether the game continues with F or with G.

Remark 3. Note that no role switch is involved in the above rule: the player acting as **P** remains in this role at the succeeding state; likewise for **O**.

We call the $\mathcal{H}$-mv-game augmented by rule ($R_\pi^{\mathcal{R}}$) the *(basic) $\mathcal{R}$-game* . We claim that the new rule gives raise to the following truth function, to be added to the semantics of logic KZ:

$$v_{\mathcal{M}}(F \pi G) = (v_{\mathcal{M}}(F) + v_{\mathcal{M}}(G))/2.$$

KZ(π) denotes the logic arising from KZ by adding π. To assist a concise formulation of the adequateness claim for the $\mathcal{R}$-game we have to adapt Definition 1 by replacing 'pay-off' with 'expected pay-off'. In fact, since we restrict attention to the propositional level here, we can use the following simpler definition.

Definition 3. *If player **X** has a strategy that leads to an expected pay-off for her of at least w, while her opponent has a strategy that ensures that **X**'s expected pay-off is at most w, then w is called the* expected value for **X** *of the game.*

Theorem 6. *A propositional formula F evaluates to $v_{\mathcal{M}}(F) = w$ in a KZ(π)-interpretation $\mathcal{M}$ iff the basic $\mathcal{R}$-game for F with pay-offs matching $\mathcal{M}$ has expected value w for Myself.*

Proof. Taking into account that $v_{\mathcal{M}}(F)$ coincides with the value of the $\mathcal{H}$-mv-game matching $\mathcal{M}$ if F does not contain the new connective π, we only have to add the case for a current formula of the form $G \pi H$ to the usual backward induction argument. However, because of the random choice involved in rule ($R_\pi^{\mathcal{R}}$), it is now her *expected* pay-off that **P** seeks to maximize and **O** seeks to minimize.

Suppose the current formula is $G \pi H$. By the induction hypothesis, at the successor state σ_G with current formula G (the player who is currently) **P** can force[13] an expected pay-off $v_{\mathcal{M}}(G)$ for herself, while **O** can force an expected pay-off $1 - v_{\mathcal{M}}(G)$ for himself. Therefore the expected value for **P** for the game starting in σ_G is $v_{\mathcal{M}}(G)$ for **P**. The same holds for H instead of G. Since the choice between the two successor states σ_G and σ_H is uniformly random, we conclude that the expected value for **P** for the game starting with $G \pi H$ is the average of $v_{\mathcal{M}}(F)$ and $v_{\mathcal{M}}(G)$, i.e. $(v_{\mathcal{M}}(F)+v_{\mathcal{M}}(G))/2$. The theorem thus follows from the fact that *I* (*Myself*) am the initial **P** in the $\mathcal{R}$-game for F. □

Since the function $\lambda x, y(x+y)/2$ cannot be composed solely from the functions $\lambda x(1-x)$, $\lambda x, y \min(x, y)$, $\lambda x, y \max(x, y)$ and the values 0 and 1, we can make the following observation.

[12] A similar rule is considered in [33] in the context of partial logic.

[13] We re-use the terminology introduced in the proof of Theorem 5, but applied to *expected* pay-offs here.

Proposition 1. *The connective* π *is not definable in logic* KZ.

But also the following stronger fact holds.

Proposition 2. *The connective* π *is not definable in Łukasiewicz logic* Ł.

Proof. By McNaughton's Theorem [26] a function $f : [0,1]^n \to [0,1]$ corresponds to a formula of propositional Łukasiewicz logic iff f is piecewise linear, where every linear piece has integer coefficients. But clearly the coefficient of $(x+y)/2$ is not an integer. □

Remark 4. We may also observe that, in contrast to Ł, not only $\overline{0.5} =_{def} \bot\pi\top$, but in fact every rational number in $[0,1]$ with a finite (terminating) expansion in the binary number system is definable as truth constant in logic $\mathsf{KZ}(\pi)$.

Conversely to Proposition 2 we also have the following.

Proposition 3. *None of the connectives* $\otimes$, $\oplus$, $\to$ *of* Ł *can be defined in* $\mathsf{KZ}(\pi)$.

Proof. Let Ψ denote the set of all interpretations $\mathcal{M}$, where $0 < v_{\mathcal{M}}(A) < 1$ for all propositional variables A. The following claim can be straightforwardly checked by induction.

For every formula F of $\mathsf{KZ}(\pi)$ one of the following holds:
(1) $0 < v_{\mathcal{M}}(F) < 1$ for all $\mathcal{M} \in \Psi$, or
(2) $v_{\mathcal{M}}(F) = 1$ for all $\mathcal{M} \in \Psi$, or
(3) $v_{\mathcal{M}}(F) = 0$ for all $\mathcal{M} \in \Psi$.

Clearly this claim does not hold for $A \otimes B$, $A \oplus B$, and $A \to B$. Therefore the connectives $\otimes$, $\oplus$, $\to$ cannot be defined in $\mathsf{KZ}(\pi)$. □

In light of the above propositions, the question arises whether one can come up with further game rules, that, like $(R^{\mathcal{R}}_{\pi})$, do not sacrifice what we above called *Hintikka's principle*, i.e., the principle that game state is determined solely by a formula and a role distribution. An obvious way to generalize rule $(R^{\mathcal{R}}_{\pi})$ is to allow for a (potentially) biased random choice:

$(R^{\mathcal{R}}_{\pi^p})$ If the current formula is $(F\pi^p G)$ then the game continues with F with probability p, but continues with G with probability $1-p$.

Clearly, π coincides with $\pi^{0.5}$. But for other values of p we obtain a new connective. However, it is straightforward to check that Proposition 3 also holds if replace π by π^p for any $p \in [0,1]$.

Interestingly, there is a fairly simple game based way to obtain a logic that properly extends Łukasiewicz logic by introducing a unary connective D that signals that the pay-off values for **P** is to be doubled (capped to 1, as usual) at the end of the game.

$(R^{\mathcal{R}}_{\mathrm{D}})$ If the current formula is DF then the game continues with F, but with the following changes at the final state. The pay-off, say x, for **P** is changed to $\max(1, 2x)$, while the the pay-off $1-x$ for **O** is changed to $1-\max(1, 2x)$.

Remark 5. Instead of explicitly capping the modified pay-off for **P** to 1 one may equivalently give **O** the opportunity to either continue that game with doubled pay-off for **P** (and inverse pay-off for **O** herself) or to simply end the game at that point with pay-off 1 for **P** and pay-off 0 for **O** herself.

Let us use $\mathsf{KZ}(\mathrm{D})$ for the logic obtained from KZ by adding the connective D with the following truth function to KZ:

$$v_{\mathcal{M}}(\mathrm{D}F) = \min(1, 2 \cdot v_{\mathcal{M}}(F)).$$

Moreover, we use $\mathsf{KZ}(\pi, \mathrm{D})$ to denote the extension of KZ with both π and D and call the $\mathcal{R}$-game augmented by rule $(R^{\mathcal{R}}_{\mathrm{D}})$ the D-*extended* $\mathcal{R}$-*game* .

Theorem 7. *A propositional formula F evaluates to $v_{\mathcal{M}}(F) = w$ in a $\mathsf{KZ}(\pi, \mathrm{D})$-interpretation $\mathcal{M}$ iff the* D*-extended $\mathcal{R}$-game for F with pay-offs matching $\mathcal{M}$ has expected value w for Myself.*

Proof. The proof of Theorem 6 is readily extended to the present one by considering the additional inductive case of $\mathrm{D}G$ as current formula. By the induction hypothesis, the expected value for **P** of the game for G (under the same interpretation $\mathcal{M}$) is $v_{\mathcal{M}}(G)$. Therefore rule $(R^{\mathcal{R}}_{\mathrm{D}})$ entails that the expected value for **P** of the game for $\mathrm{D}G$ is $\max(1, 2 \cdot v_{\mathcal{M}}(G))$. □

Given Proposition 3 and Theorem 7 the following simple observation is of some significance.

Proposition 4. *The connectives $\otimes$, $\oplus$ and $\to$ of Ł are definable in $\mathsf{KZ}(\pi, D)$.*

Proof. It is straightforward to check that the following definitions in $\mathsf{KZ}(\pi, D)$ match the corresponding truth functions for Ł: $G \oplus F =_{def} \mathrm{D}(G \pi F)$, $G \otimes F =_{def} \neg \mathrm{D}(\neg G \pi \neg F)$, $G \to F =_{def} \mathrm{D}(\neg G \pi F)$. □

Remark 6. Note that Proposition 4 jointly with Theorem 7 entails that one can provide game semantics for (an extension of) Łukasiewicz without dropping "Hintikka's principle" as done in $\mathcal{E}$-games and in $\mathcal{G}$-games.

Remark 7. The definitions mentioned in the proof of Proposition 4 give rise to corresponding additional rules for the D-extended $\mathcal{R}$-game. E.g., for strong disjunction we obtain:

$(R^{\mathcal{R}}_{\oplus})$ If the current formula is $G \oplus F$ then a random choice determines whether to continue the game with F or with G. But in any case the pay-off for **P** is doubled (capped to 1), while the pay-off for **O** remains inverse to that for **P**.

By further involving role switches similar rules for strong conjunction and for implications are readily obtained.

It remains to be seen whether these rules can assist in arguing for the plausibility of the corresponding connective in intended application scenarios. But in any case, it is clear that, compared to the sole specification of truth functions, the game interpretation provides an additional handle for assessing the adequateness of the Łukasiewicz connectives for formalizing reasoning with graded notions and vague propositions.

Like $(R_{\pi}^{\mathcal{R}})$, also rule $(R_{\mathrm{D}}^{\mathcal{R}})$ can be generalized in an obvious manner:

$(R_{\mathrm{M}_c}^{\mathcal{R}})$ If the current formula is $\mathrm{M}_c F$ then the game continues with F, but with the following changes at the final state. The pay-off, say x, for **P** is changed to $\max(1, c{\cdot}x)$, while the the pay-off $1-x$ for **O** is changed to $1-\max(1, c{\cdot}x)$.

Adding further instances of π^p and M_c to $\mathsf{KZ}(\pi, \mathrm{D})$ leads to more expressive logics, related to Rational Łukasiewicz Logic and to divisible MV-algebras [11].[14]

5 Random Witnesses for Quantifiers

The idea of allowing for random choices of witnessing elements in quantifier rules — beyond **O**'s choice of a witness for a universally quantified statement and **P**'s choice of a witness for an existentially quantified statement — has been introduced in [8,10]. But the rules there refer to Giles's game, not the $\mathcal{H}$-game; consequently a game state may comprise more than one formula. Moreover, attention has been restricted to so-called semi-fuzzy quantifiers in a two-tiered language variant of Łukasiewicz logic, where the predicates in the scope of such a quantifier are crisp. Here, we lift that restriction and moreover retain *Hintikka's principle* of game states as being determined by a single (augmented) formula and a current role distribution.

In picking a witness element randomly, we may in principle refer to any given distribution over the domain. However, as convincingly argued, e.g., in [30], the meaning of a quantifiers should remain invariant under isomorphism, i.e., under permutations of domain elements, if that quantifier is to be conceived as a *logical* particle. This principle entails that the random choice of witnessing elements has to refer to the *uniform* distribution over the domain. However, as is well known, only *finite domains* admit uniform distributions. The restriction to finite domains is moreover well justified by the intended applications that model linguistic phenomena connected to gradedness and vagueness. As a welcome side effect of this restriction, we may drop the more involved notion of a value of a game as arising from approximations of pay-offs (Definition 1) and define the value of a game as in Definition 3, i.e., without involving ϵ.

The rule for the simplest quantifier (denoted by Π) that involves a random witness element is as follows.

$(R_{\Pi}^{\mathcal{R}})$ If the current formula is $(\Pi x F(x))[\theta]$ then an element c from the (finite) domain of $\mathcal{M}$ is chosen randomly and the game continues with $F(x)[\theta[c/x]]$.

In analogy to case of $(R_{\pi}^{\mathcal{R}})$ a truth function for Π can be extracted from this rule (where $|D|$ is the cardinality of the domain D of $\mathcal{M}$):

$$v_{\mathcal{M}}^{\theta}(\Pi x F(x)) = \sum_{c \in D} \frac{v_{\mathcal{M}}^{\theta[c/x]}(F(x))}{|D|}$$

[14] As pointed out by a referee, the following observation by Hájek is relevant here: If one adds the truth constant 0.5 to Ł then *all* rational numbers are expressible. Therefore $\mathsf{KZ}(\pi, \mathrm{D})$ extends not only Ł, but also Rational Pavelka Logic, where all rationals truth constants are added to Ł (see [15]). On the other hand, neither (e.g.) $\pi^{1/3}$ or M_3 seem to be expressible in $\mathsf{KZ}(\pi, \mathrm{D})$.

By $\mathsf{KZ}(\Pi)$ we refer to the logic KZ augmented by the quantifier Π. The proof of the corresponding adequateness statement is analogous to that of Theorem 6 and is therefore left to the reader.

Theorem 8. *A formula F evaluates to $v_{\mathcal{M}}(F) = w$ in a $\mathsf{KZ}(\Pi)$-interpretation $\mathcal{M}$ iff the $\mathcal{R}$-game for F, extended by rule $(R^{\mathcal{R}}_{\Pi})$, with pay-offs matching $\mathcal{M}$ has expected value w for Myself.*

Remark 8. Like the propositional connective π, the quantifier Π can be seen as an 'averaging operator', that provides explicit access to the (uniform) average of the values of the sub-formulas or instances of a formula $F\pi G$ or $\Pi x F(x)$, respectively.

Remark 9. Obviously one may extend not just KZ, but also the extensions of KZ discussed in Section 3 with the random choice quantifier Π. This leads to first-order logics that are strictly more expressive than Łukasiewicz logic Ł.

In [10] it is demonstrated how random choices of witness elements allow for the introduction of different (infinite) families of semi-fuzzy quantifiers that are intended to address the problem to justify particular fuzzy models of informal quantifier expressions like 'few', 'many', or 'about half'. As already mentioned above, the corresponding quantifier rules in [10] (like those in [8]) employ Giles's concept of referring to multisets of formulas asserted by **P** and by **O**, respectively, at any given state of the game. However, even without sacrificing *Hintikka's principle* by moving to $\mathcal{G}$-games or to $\mathcal{E}$-games, one can come up with new quantifier rules. For example, one may introduce a family of quantifiers $\widehat{\Pi}^n$ by the following parameterized game rule:

$(R^{\mathcal{R}}_{\widehat{\Pi}^n})$ If the current formula is $(\widehat{\Pi}^n x F(x))[\theta]$ then n elements $c_1, \ldots, c_n$ from the domain of $\mathcal{M}$ are chosen randomly. **P** then chooses some $c \in \{c_1, \ldots, c_n\}$ and the game continues with $F(x)[\theta[c/x]]$.

A dual family of quantifiers is obtained by replacing **P** by **O** in rule $(R^{\mathcal{R}}_{\widehat{\Pi}^n})$.Yet another type of quantifiers arises by the following rule:

$(R^{\mathcal{R}}_{\widetilde{\Pi}^n})$ If the current formula is $(\widetilde{\Pi}^x F(x))[\theta]$ then an element c_1 from the domain of $\mathcal{M}$ is chosen randomly. **P** decides whether to continue the game with $F(x)[\theta[c_1/x]]$ or to ask for a further randomly chosen element c_2. This procedure is iterated until an element c_i, where $1 < i \leq n$ is accepted by **P**. (c_n has to be accepted if none of the earlier random elements was accepted.) The game then continues with $F(x)[\theta[c_i/x]]$.

Again, variants of this rule are obtained by replacing **P** with **O** in $(R^{\mathcal{R}}_{\widetilde{\Pi}^n})$, possibly only for certain $i \in \{1, \ldots, n\}$.

Truth functions corresponding to the above rules are readily computed by applying elementary principles of probability theory. We will not work out these examples here and leave the systematic investigation of logics arising from enriching the $\mathcal{H}$-mv-game or the $\mathcal{R}$-game in the indicated manner to future work.

6 Conclusion and Future Research

We began our investigations by observing (in Section 2) that Hintikka's well known game semantics for classical first-order logic (here referred to as $\mathcal{H}$-game) can be straightforwardly generalized to the $\mathcal{H}$-mv-game, where the pay-off values are taken from the unit interval $[0,1]$ instead of just $\{0,1\}$. Following [1], we call the resulting basic fuzzy logic Kleene-Zadeh logic KZ. At least two alternative types of semantic games, called $\mathcal{E}$-game and $\mathcal{G}$-game here, can be found in the literature (see, e.g., [4,12,5,9,7]). These games provide alternative semantics for Łukasiewicz logic Ł, which is considerably more expressive than KZ. Both, the $\mathcal{E}$-game and the $\mathcal{G}$-game, deviate quite drastically from the $\mathcal{H}$-mv-game (and therefore also from the $\mathcal{H}$-game) in their underlying concept of a game state. In this paper, we have explored the power of semantic games that adhere of 'Hintikka's principle', by which we mean the principle that each state of a game is determined by a single formula (possibly augmented by a variable assignment) and a role distribution (telling us who of the two players is currently acting as Proponent **P** and who is currently acting as Opponent **O**). In Section 3 we have shown that adding rules that instantiate a fairly general scheme of possible rules to the $\mathcal{H}$-game does not give rise to logics that are more expressible than KZ. However introducing random choices in game rules, either as an alternative or in addition to choices made by **P** or by **O**, leads to various proper extensions of KZ, as we have seen in Section 4 for propositional logics and in Section 5 for the first-order level. In particular, the combination of the basic random choice connective π with a unary connective that signals doubling of pay-offs for **P** (capped to 1) allowed us to characterize a logic, in which all connectives of Ł are definable. A more complete and systematic exploration of the rich landscape of new connectives and quantifiers that can be defined for 'randomized' $\mathcal{H}$-mv-games is an obvious topic for future research.

A further open question is to what extend Hintikka-style games can be formulated for fuzzy logics that, unlike Łukasiewicz logic, do not extend logic KZ. In particular the two other fundamental t-norm based logics, Gödel logic and Product logic are obvious candidates for corresponding investigations. Another important direction for further research concerns the relation to proof theory. In [7] a direct correspondence between the logical rules of a hypersequent system for propositional Łukasiewicz logic and the rules of the $\mathcal{G}$-game has been established. In principle, a similar correspondence between game rules and logical rules should also hold for other games and certain analytic proof systems.

We conclude with a brief remark on the relation between our 'randomized game semantics' and 'equilibrium semantics' for $\mathcal{H}$-games with imperfect information. We have only considered games of perfect information in this paper: the players always know all previous moves and thus have full knowledge of the current state of the game. However, the full power of Hintikka's game semantics arises from admitting that players may not be aware of all previous moves. This leads to *Independence Friendly* logic (IF-logic), where occurrences of quantifiers and connectives in a formula may be 'slashed' with respect to other such occurrences to indicate that the moves in the game that refer to those slashed occurrences are unknown to the current proponent. E.g., the formula $F = (G \vee_{/\{\wedge\}} H) \wedge (H \vee_{/\{\wedge\}} G)$ refers to an $\mathcal{H}$-game, where the choice by **P** of either the conjunct $G \vee_{/\{\wedge\}} H$ or $H \vee_{/\{\wedge\}} G$ is unknown to **O** when **O** has to choose either the right or the left disjunct of the remaining current formula. In [25] and in [32], Sandu and his colleagues present so-called equilibrium semantics for IF-logic, where mixed strategies for $\mathcal{H}$-games with incomplete information induce intermediate expected pay-off values in $[0,1]$, even if each atomic formula is evaluated to either 0 or 1. It is readily checked that the corresponding value for the above formula F is $(v_{\mathcal{M}}(G) + v_{\mathcal{M}}(H))/2$, where $v_{\mathcal{M}}(G)$ and $v_{\mathcal{M}}(H)$ are the values for G and H, respectively. In other words, we can simulate the effect of the random choice that induces our new connective π by the IF-formula F, and vice versa: π simulates effects of imperfect knowledge in games with classical pay-offs. Clearly, the connections between equilibrium semantics and (extended) $\mathcal{R}$-games deserves to be explored in more detail in future work.

Finally, we suggest that the results of this paper — in addition to the earlier results of Giles [12,13,14], Cintula/Majer [4], as well as Fermüller and co-authors [5,7,8,10] — may serve as a basis for discussing to what extent and in which manner the game semantic approach to fuzzy logic addresses the important challenge of deriving truth functions for fuzzy connectives and quantifiers from basic semantic principles and thus to guide the fuzzy modeler's task in many application scenarios.

References

1. Aguzzoli, S., Gerla, B., Marra, V.: Algebras of fuzzy sets in logics based on continuous triangular norms. In: Sossai, C., Chemello, G. (eds.) ECSQARU 2009. LNCS, vol. 5590, pp. 875–886. Springer, Heidelberg (2009)
2. Bennett, A.D.C., Paris, J.B., Vencovska, A.: A new criterion for comparing fuzzy logics for uncertain reasoning. Journal of Logic, Language and Information 9(1), 31–63 (2000)
3. Cintula, P., Hájek, P., Noguera, C. (eds.): Handbook of Mathematical Fuzzy Logic. College Publications (2011)
4. Cintula, P., Majer, O.: Towards evaluation games for fuzzy logics. In: Majer, O., Pietarinen, A.-V., Tulenheimo, T. (eds.) Games: Unifying Logic, Language, and Philosophy, pp. 117–138. Springer (2009)
5. Fermüller, C.G.: Revisiting Giles's game. In: Majer, O., Pietarinen, A.-V., Tulenheimo, T. (eds.) Games: Unifying Logic, Language, and Philosophy, Logic, Epistemology, and the Unity of Science, pp. 209–227. Springer (2009)
6. Fermüller, C.G.: On matrices, Nmatrices and games. Journal of Logic and Computation (2013) (page to appear)
7. Fermüller, C.G., Metcalfe, G.: Giles's game and the proof theory of Łukasiewicz logic. Studia Logica 92(1), 27–61 (2009)
8. Fermüller, C.G., Roschger, C.: Randomized game semantics for semi-fuzzy quantifiers. In: Greco, S., Bouchon-Meunier, B., Coletti, G., Fedrizzi, M., Matarazzo, B., Yager, R.R. (eds.) IPMU 2012, Part IV. CCIS, vol. 300, pp. 632–641. Springer, Heidelberg (2012)
9. Fermüller, C.G., Roschger, C.: From games to truth functions: A generalization of Giles's game. Studia Logica (2013) (to appear)

10. Fermüller, C.G., Roschger, C.: Randomized game semantics for semi-fuzzy quantifiers. Logic Journal of the IGPL (to appear)
11. Gerla, B.: Rational Łukasiewicz logic and DMV-algebras. Neural Networks World 11, 579–584 (2001)
12. Giles, R.: A non-classical logic for physics. Studia Logica 33(4), 397–415 (1974)
13. Giles, R.: A non-classical logic for physics. In: Wojcicki, R., Malinkowski, G. (eds.) Selected Papers on Łukasiewicz Sentential Calculi, pp. 13–51. Polish Academy of Sciences (1977)
14. Giles, R.: Semantics for fuzzy reasoning. International Journal of Man-Machine Studies 17(4), 401–415 (1982)
15. Hájek, P.: Metamathematics of Fuzzy Logic. Kluwer Academic Publishers (2001)
16. Hájek, P.: Why fuzzy logic? In: Jacquette, D. (ed.) Blackwell Companion to Philosophical Logic, pp. 596–606. Wiley (2002)
17. Hájek, P.: On witnessed models in fuzzy logic. Mathematical Logic Quarterly 53(1), 66–77 (2007)
18. Hintikka, J.: Language-games for quantifiers. In: Rescher, N. (ed.) Studies in Logical Theory, pp. 46–72. Blackwell, Oxford (1968); Reprinted in [19]
19. Hintikka, J.: Logic, language-games and information: Kantian themes in the philosophy of logic. Clarendon Press Oxford (1973)
20. Hintikka, J., Sandu, G.: Game-theoretical semantics. In: Handbook of Logic and Language. Elsevier (2010)
21. Hisdal, E.: Are grades of membership probabilities? Fuzzy Sets and Systems 25(3), 325–348 (1988)
22. Lawry, J.: A voting mechanism for fuzzy logic. International Journal of Approximate Reasoning 19(3-4), 315–333 (1998)
23. Lorenzen, P.: Logik und Agon. In: Atti Congr. Internaz. di Filosofia, Venezia, Settembre 12-18, vol. IV, Sansoni (1960)
24. Lorenzen, P.: Dialogspiele als semantische Grundlage von Logikkalkülen. Archiv Für Mathemathische Logik und Grundlagenforschung 11, 32–55, 73–100 (1968)
25. Mann, A.L., Sandu, G., Sevenster, M.: Independence-friendly logic: A game-theoretic approach. Cambridge University Press (2011)
26. McNaughton, R.: A theorem about infinite-valued sentential logic. Journal of Symbolic Logic 16(1), 1–13 (1951)
27. Nguyễn, H.T., Walker, E.A.: A first course in fuzzy logic. CRC Press (2006)
28. Paris, J.B.: A semantics for fuzzy logic. Soft Computing 1(3), 143–147 (1997)
29. Paris, J.B.: Semantics for fuzzy logic supporting truth functionality. In: Novák, V., Perfilieva, I. (eds.) Discovering the World with Fuzzy Logic, pp. 82–104. Physica-Verlag (2000)
30. Peters, S., Westerståhl, D.: Quantifiers in language and logic. Oxford University Press, USA (2006)
31. Scarpellini, B.: Die Nichtaxiomatisierbarkeit des unendlichwertigen Prädikatenkalküls von Łukasiewicz. Journal of Symbolic Logic 27(2), 159–170 (1962)
32. Sevenster, M., Sandu, G.: Equilibrium semantics of languages of imperfect information. Annals of Pure and Applied Logic 161(5), 618–631 (2010)
33. Wen, X., Ju, S.: Semantic games with chance moves revisited: from IF logic to partial logic. Synthese 190(9), 1605–1620 (2013)
34. Zadeh, L.A.: Fuzzy logic. IEEE: Computer 21(4), 83–93 (1988)

A Finite Axiomatization of Conditional Independence and Inclusion Dependencies*

Miika Hannula and Juha Kontinen

University of Helsinki, Department of Mathematics and Statistics, P.O. Box 68, 00014 Helsinki, Finland
{miika.hannula,juha.kontinen}@helsinki.fi

Abstract. We present a complete finite axiomatization of the unrestricted implication problem for inclusion and conditional independence atoms in the context of dependence logic. For databases, our result implies a finite axiomatization of the unrestricted implication problem for inclusion, functional, and embedded multivalued dependencies in the unirelational case.

1 Introduction

We formulate a finite axiomatization of the implication problem for inclusion and conditional independence atoms (dependencies) in the dependence logic context. The input of this problem is given by a finite set $\Sigma \cup \{\phi\}$ consisting of conditional independence atoms and inclusion atoms, and the question to decide is whether the following logical consequence holds

$$\Sigma \models \phi. \tag{1}$$

Independence logic [1] and inclusion logic [2] are recent variants of dependence logic the semantics of which are defined over sets of assigments (teams) rather than a single assignment as in first-order logic. By viewing a team X with domain $\{x_1, \dots, x_k\}$ as a relation schema $X[\{x_1, \dots, x_k\}]$, our results provide a finite axiomatization for the unrestricted implication problem of inclusion, functional, and embedded multivalued database dependencies over $X[\{x_1, \dots, x_k\}]$.

Dependence logic [3] extends first-order logic by dependence atomic formulas

$$=(x_1, \dots, x_n) \tag{2}$$

the meaning of which is that the value of x_n is functionally determined by the values of $x_1, \dots, x_{n-1}$. Independence logic replaces the dependence atoms by independence atoms

$$\boldsymbol{y} \perp_{\boldsymbol{x}} \boldsymbol{z},$$

the intuitive meaning of which is that, with respect to any fixed value of $\boldsymbol{x}$, the variables $\boldsymbol{y}$ are totally independent of the variables $\boldsymbol{z}$. Furthermore, inclusion logic is based on inclusion atoms of the form

$$\boldsymbol{x} \subseteq \boldsymbol{y},$$

* The authors were supported by grant 264917 of the Academy of Finland.

C. Beierle and C. Meghini (Eds.): FoIKS 2014, LNCS 8367, pp. 211–229, 2014.

with the meaning that all the values of $\boldsymbol{x}$ appear also as values for $\boldsymbol{y}$. By viewing a team X of assignments with domain $\{x_1, \ldots, x_k\}$ as a relation schema $X[\{x_1, \ldots, x_k\}]$, the atoms $=(\boldsymbol{x})$, $\boldsymbol{x} \subseteq \boldsymbol{y}$, and $\boldsymbol{y} \perp_{\boldsymbol{x}} \boldsymbol{z}$ correspond to functional, inclusion, and embedded multivalued database dependencies. Furthermore, the atom $=(x_1, \ldots, x_n)$ can be alternatively expressed as

$$x_n \perp_{x_1 \ldots x_{n-1}} x_n,$$

hence our results for independence atoms cover also the case where dependence atoms are present.

The team semantics of dependence logic is a very flexible logical framework in which various notions of dependence and independence can be formalized. Dependence logic and its variants have turned out to be applicable in various areas. For example, Väänänen and Abramsky have recently axiomatized and formally proved Arrow's Theorem from social choice theory and, certain No-Go theorems from the foundations of quantum mechanics in the context of independence logic [4]. Also, the pure independence atom $\boldsymbol{y} \perp \boldsymbol{z}$ and its axioms has various concrete interpretations such as independence $X \perp\!\!\!\perp Y$ between two sets of random variables [5], and independence in vector spaces and algebraically closed fields [6].

Dependence logic is equi-expressive with existential second-order logic (ESO). Furthermore, the set of valid formulas of dependence logic has the same complexity as that of full second-order logic, hence it is not possible to give a complete axiomatization of dependence logic [3]. However, by restricting attention to syntactic fragments [7,8,9] or by modifying the semantics [10] complete axiomatizations have recently been obtained. The axiomatization presented in this article is based on the classical characterization of logical implication between dependencies in terms of the *Chase* procedure [11]. The novelty in our approach is the use of the so-called *Lax* team semantics of independence logic to simulate the chase on the logical level using only inclusion and independence atoms and existential quantification.

In database theory, the implication problems of various types of database dependencies have been extensively studied starting from Armstrong's axiomatization for functional dependencies [12]. Inclusion dependencies were axiomatized in [13], and an axiomatization for pure independence atoms is also known (see [14,5,15]). On the other hand, the implication problem of embedded multivalued dependencies, and of inclusion dependencies and functional dependencies together, are known to be undecidable [16,17,18]. Still, the unrestricted implication problem of inclusion and functional dependencies has been finitely axiomatized in [19] using a so-called *Attribute Introduction Rule* that allows new attribute names representing derived attributes to be introduced into deductions. These new attributes can be thought of as implicitly existentially quantified. Our *Inclusion Introduction Rule* is essentially equivalent to the Attribute Introduction Rule of [19]. It is also worth noting that the chase procedure has been used to axiomatize the unrestricted implication problem of various classes of dependencies, e.g., *Template Dependencies* [20], and *Typed Dependencies* [21]. Finally we note that the role of inclusion atom in our axiomatization has some similarities to the axiomatization of the class of *Algebraic Dependencies* [22].

2 Preliminaries

In this section we define team semantics and introduce dependence, independence and inclusion atoms. The version of team semantics presented here is the Lax one, originally introduced in [2], which will turn out to be valuable for our purposes due to its interpretation of existential quantification.

2.1 Team Semantics

The semantics is formulated using sets of assignments called teams instead of single assignments. Let $\mathcal{M}$ be a model with domain M. An *assignment* s of $\mathcal{M}$ is a finite mapping from a set of variables into M. A *team* X over $\mathcal{M}$ with domain $\mathrm{Dom}(X) = V$ is a set of assignments from V to M. For a subset W of V, we write $X \upharpoonright W$ for the team obtained by restricting all the assignments of X to the variables in W.

If s is an assignment, x a variable, and $a \in A$, then $s[a/x]$ denotes the assignment (with domain $\mathrm{Dom}(s) \cup \{x\}$) that agrees with s everywhere except that it maps x to a. For an assignment s, and a tuple of variables $\boldsymbol{x} = (x_1, ..., x_n)$, we sometimes denote the tuple $(s(x_1), ..., s(x_n))$ by $s(\boldsymbol{x})$. For a formula ϕ, $\mathrm{Var}(\phi)$ and $\mathrm{Fr}(\phi)$ denote the sets of variables that appear in ϕ and appear free in ϕ, respectively. For a finite set of formulas $\Sigma = \{\phi_1, \ldots, \phi_n\}$, we write $\mathrm{Var}(\Sigma)$ for $\mathrm{Var}(\phi_1) \cup \ldots \cup \mathrm{Var}(\phi_n)$, and define $\mathrm{Fr}(\Sigma)$ analogously. When using set operations $\boldsymbol{x} \cup \boldsymbol{y}$ and $\boldsymbol{x} \setminus \boldsymbol{y}$ for sequences of variables $\boldsymbol{x}$ and $\boldsymbol{y}$, then these sequences are interpreted as the sets of elements of these sequences.

Team semantics is defined for first-order logic formulas as follows:

Definition 1 (Team Semantics). *Let $\mathcal{M}$ be a model and let X be any team over it. Then*

- *If ϕ is a first-order atomic or negated atomic formula, then $\mathcal{M} \models_X \phi$ if and only if for all $s \in X$, $\mathcal{M} \models_s \phi$ (in Tarski semantics).*
- *$\mathcal{M} \models_X \psi \vee \theta$ if and only if there are Y and Z such that $X = Y \cup Z$ and $\mathcal{M} \models_Y \psi$ and $\mathcal{M} \models_Z \theta$.*
- *$\mathcal{M} \models_X \psi \wedge \theta$ if and only if $\mathcal{M} \models_X \psi$ and $\mathcal{M} \models_X \theta$.*
- *$\mathcal{M} \models_X \exists v \psi$ if and only if there is a function $F : X \to \mathcal{P}(M) \backslash \{\emptyset\}$ such that $\mathcal{M} \models_{X[F/v]} \psi$, where $X[F/v] = \{s[m/v] : s \in X, m \in F(s)\}$.*
- *$\mathcal{M} \models_X \forall v \psi$ if and only if $\mathcal{M} \models_{X[M/v]} \psi$, where $X[M/v] = \{s[m/v] : s \in X, m \in M\}$.*

The following lemma is an immediate consequence of Definition 1.

Lemma 1. *Let $\mathcal{M}$ be a model, X a team and $\exists x_1 \ldots \exists x_n \phi$ a formula in team semantics setting where $x_1, \ldots, x_n$ is a sequence of variables. Then*

$\mathcal{M} \models_X \exists x_1 \ldots \exists x_n \phi$ iff for some function $F : X \to \mathcal{P}(M^n) \backslash \{\emptyset\}, \mathcal{M} \models_{X[F/x_1 \ldots x_n]} \phi$

where $X[F/x_1 \ldots x_n] := \{s[a_1/x_1] \ldots [a_n/x_n] \mid (a_1, \ldots, a_n) \in F(s)\}$.

If $\mathcal{M} \models_X \phi$, then we say that X *satisfies* ϕ in $\mathcal{M}$. If ϕ is a sentence (i.e. a formula with no free variables), then we say that ϕ is *true* in $\mathcal{M}$, and write $\mathcal{M} \models \phi$, if $\mathcal{M} \models_{\{\emptyset\}} \phi$ where $\{\emptyset\}$ is the team consisting of the empty assignment. Note that $\{\emptyset\}$ is different from the *empty team* $\emptyset$ containing no assignments.

In the team semantics setting, formula ψ is a *logical consequence* of ϕ, written $\phi \Rightarrow \psi$, if for all models $\mathcal{M}$ and teams X, with $\mathrm{Fr}(\phi) \cup \mathrm{Fr}(\psi) \subseteq \mathrm{Dom}(X)$,

$$\mathcal{M} \models_X \phi \Rightarrow \mathcal{M} \models_X \psi.$$

Formulas ϕ and ψ are said to be *logically equivalent* if $\phi \Rightarrow \psi$ and $\psi \Rightarrow \phi$. Logics $\mathcal{L}$ and $\mathcal{L}'$ are said to be equivalent, $\mathcal{L} = \mathcal{L}'$, if every $\mathcal{L}$-sentence ϕ is equivalent to some $\mathcal{L}'$-sentence ψ, and vice versa.

2.2 Dependencies in Team Semantics

Dependence, independence and inclusion atoms are given the following semantics.

Definition 2. *Let $\boldsymbol{x}$ be a tuple of variables and y a variable. Then* $=(\boldsymbol{x}, y)$ *is a* dependence atom *with the semantic rule*

– $\mathcal{M} \models_X =(\boldsymbol{x}, y)$ *if and only if for any $s, s' \in X$ with $s(\boldsymbol{x}) = s'(\boldsymbol{x})$, $s(y) = s'(y)$.*

Let $\boldsymbol{x}$, $\boldsymbol{y}$ and $\boldsymbol{z}$ be tuples of variables. Then $\boldsymbol{y} \perp_{\boldsymbol{x}} \boldsymbol{z}$ is a conditional independence atom *with the semantic rule*

– $\mathcal{M} \models_X \boldsymbol{y} \perp_{\boldsymbol{x}} \boldsymbol{z}$ *if and only if for any $s, s' \in X$ with $s(\boldsymbol{x}) = s'(\boldsymbol{x})$ there is a $s'' \in X$ such that $s''(\boldsymbol{x}) = s(\boldsymbol{x})$, $s''(\boldsymbol{y}) = s(\boldsymbol{y})$ and $s''(\boldsymbol{z}) = s'(\boldsymbol{z})$.*

Furthermore, we will write $\boldsymbol{x} \perp \boldsymbol{y}$ as a shorthand for $\boldsymbol{x} \perp_{\emptyset} \boldsymbol{y}$, and call it a pure independence atom.

Let $\boldsymbol{x}$ and $\boldsymbol{y}$ be two tuples of variables of the same length. Then $\boldsymbol{x} \subseteq \boldsymbol{y}$ is an inclusion atom *with the semantic rule*

– $\mathcal{M} \models_X \boldsymbol{x} \subseteq \boldsymbol{y}$ *if and only if for any $s \in X$ there is a $s' \in X$ such that $s(\boldsymbol{x}) = s'(\boldsymbol{y})$.*

Note that in the definition of an inclusion atom $\boldsymbol{x} \subseteq \boldsymbol{y}$, the tuples $\boldsymbol{x}$ and $\boldsymbol{y}$ may both have repetitions. Also in the definition of a conditional independence atom $\boldsymbol{y} \perp_{\boldsymbol{x}} \boldsymbol{z}$, the tuples $\boldsymbol{x}$, $\boldsymbol{y}$ and $\boldsymbol{z}$ are not necessarily pairwise disjoint. Thus any dependence atom $=(\boldsymbol{x}, y)$ can be expressed as a conditional independence atom $y \perp_{\boldsymbol{x}} y$. Also any conditional independence atom $\boldsymbol{y} \perp_{\boldsymbol{x}} \boldsymbol{z}$ can be expressed as a conjunction of dependence atoms and a conditional independence atom $\boldsymbol{y}^* \perp_{\boldsymbol{x}} \boldsymbol{z}^*$ where $\boldsymbol{x}$, $\boldsymbol{y}^*$ and $\boldsymbol{z}^*$ are pairwise disjoint. For disjoint tuples $\boldsymbol{x}$, $\boldsymbol{y}$ and $\boldsymbol{z}$, independence atom $\boldsymbol{y} \perp_{\boldsymbol{x}} \boldsymbol{z}$ corresponds to the embedded multivalued dependency $\boldsymbol{x} \twoheadrightarrow \boldsymbol{y}|\boldsymbol{z}$. Hence the class of conditional independence atoms corresponds to the class of functional dependencies and embedded multivalued dependencies in database theory.

Proposition 1 ([23]). *Let* $\boldsymbol{y} \perp_{\boldsymbol{x}} \boldsymbol{z}$ *be a conditional independence atom where* $\boldsymbol{x}$*,* $\boldsymbol{y}$ *and* $\boldsymbol{z}$ *are tuples of variables. If* $\boldsymbol{y}^*$ *lists the variables in* $\boldsymbol{y} - \boldsymbol{x} \cup \boldsymbol{z}$*,* $\boldsymbol{z}^*$ *lists the variables in* $\boldsymbol{z} - \boldsymbol{x} \cup \boldsymbol{y}$*, and* $\boldsymbol{u}$ *lists the variables in* $\boldsymbol{y} \cap \boldsymbol{z} - \boldsymbol{x}$*, then*

$$\mathcal{M} \models_X \boldsymbol{y} \perp_{\boldsymbol{x}} \boldsymbol{z} \Leftrightarrow \mathcal{M} \models_X \boldsymbol{y}^* \perp_{\boldsymbol{x}} \boldsymbol{z}^* \wedge \bigwedge_{u \in \boldsymbol{u}} =(\boldsymbol{x}, u).$$

The extension of first-order logic by dependence atoms, conditional independence atoms and inclusion atoms is called *dependence logic* (FO$(=(\ldots))$), *independence logic* (FO$(\perp_c)$) and *inclusion logic* (FO$(\subseteq)$), respectively. The fragment of independence logic containing only pure independence atoms is called *pure independence logic*, written FO$(\perp)$. For a collection of atoms $\mathcal{C} \subseteq \{=(\ldots), \perp_c, \subseteq\}$, we will write FO$(\mathcal{C})$ (omitting the set parenthesis of $\mathcal{C}$) for first-order logic with these atoms.

We end this section with a list of properties of these logics.

Proposition 2. *For* $\mathcal{C} = \{=(\ldots), \perp_c, \subseteq\}$*, the following hold.*

1. *(Empty Team Property) For all models* $\mathcal{M}$ *and formulas* $\phi \in$ FO$(\mathcal{C})$

$$\mathcal{M} \models_{\emptyset} \phi.$$

2. *(Locality [2]) If* $\phi \in$ FO$(\mathcal{C})$ *is such that* $Fr(\phi) \subseteq V$*, then for all models* $\mathcal{M}$ *and teams* X*,*

$$\mathcal{M} \models_X \phi \Leftrightarrow \mathcal{M} \models_{X \upharpoonright V} \phi.$$

3. *[2] An inclusion atom* $\boldsymbol{x} \subseteq \boldsymbol{y}$ *is logically equivalent to the pure independence logic formula*

$$\forall v_1 v_2 \boldsymbol{z}((\boldsymbol{z} \neq \boldsymbol{x} \wedge \boldsymbol{z} \neq \boldsymbol{x}) \vee (v_1 \neq v_2 \wedge \boldsymbol{z} \neq \boldsymbol{y}) \vee ((v_1 = v_2 \vee \boldsymbol{z} = \boldsymbol{y}) \wedge \boldsymbol{z} \perp v_1 v_2))$$

where v_1*,* v_2 *and* $\boldsymbol{z}$ *are new variables.*

4. *[24] Any independence logic formula is logically equivalent to some pure independence logic formula.*
5. *[3,1] Any dependence (or independence) logic sentence* ϕ *is logically equivalent to some existential second-order sentence* ϕ^**, and vice versa.*
6. *[25] Any inclusion logic sentence* ϕ *is logically equivalent to some positive greatest fixpoint logic sentence* ϕ^**, and vice versa.*

3 Deduction System

In this section we present a sound and complete axiomatization for the implication problem of inclusion and independence atoms. The implication problem is given by a finite set $\Sigma \cup \{\phi\}$ consisting of conditional independence and inclusion atoms, and the question is to decide whether $\Sigma \models \phi$.

Definition 3. *In addition to the usual introduction and elimination rules for conjunction, we adopt the following rules for conditional independence and inclusion atoms. Note that in Identity Rule and Start Axiom, the new variables should be thought of as implicitly existentially quantified.*

1. *Reflexivity:*
$$\boldsymbol{x} \subseteq \boldsymbol{x}.$$
2. *Projection and Permutation:*
$$\text{if } x_1 \dots x_n \subseteq y_1 \dots y_n, \text{ then } x_{i_1} \dots x_{i_k} \subseteq y_{i_1} \dots y_{i_k},$$
for each sequence $i_1, \dots, i_k$ *of integers from* $\{1, \dots, n\}$.
3. *Transitivity:*
$$\text{if } \boldsymbol{x} \subseteq \boldsymbol{y} \wedge \boldsymbol{y} \subseteq \boldsymbol{z}, \text{ then } \boldsymbol{x} \subseteq \boldsymbol{y}.$$
4. *Identity Rule:*
$$\text{if } ab \subseteq cc \wedge \phi, \text{ then } \phi',$$
where ϕ' *is obtained from* ϕ *by replacing any number of occurrences of* a *by* b.
5. *Inclusion Introduction:*
$$\text{if } \boldsymbol{a} \subseteq \boldsymbol{b}, \text{ then } \boldsymbol{a}x \subseteq \boldsymbol{b}c,$$
where x *is a* new *variable.*
6. *Start Axiom:*
$$\boldsymbol{ac} \subseteq \boldsymbol{ax} \wedge \boldsymbol{b} \perp_{\boldsymbol{a}} \boldsymbol{x} \wedge \boldsymbol{ax} \subseteq \boldsymbol{ac}$$
where $\boldsymbol{x}$ *is a sequence of pairwise distinct* new *variables.*
7. *Chase Rule:*
$$\text{if } \boldsymbol{y} \perp_{\boldsymbol{x}} \boldsymbol{z} \wedge \boldsymbol{ab} \subseteq \boldsymbol{xy} \wedge \boldsymbol{ac} \subseteq \boldsymbol{xz}, \text{ then } \boldsymbol{abc} \subseteq \boldsymbol{xyz}.$$
8. *Final Rule:*
$$\text{if } \boldsymbol{ac} \subseteq \boldsymbol{ax} \wedge \boldsymbol{b} \perp_{\boldsymbol{a}} \boldsymbol{x} \wedge \boldsymbol{abx} \subseteq \boldsymbol{abc}, \text{ then } \boldsymbol{b} \perp_{\boldsymbol{a}} \boldsymbol{c}.$$

In an application of Inclusion Introduction, the variable x is called the new variable of the deduction step. Similarly, in an application of Start Axiom, the variables of $\boldsymbol{x}$ are called the new variables of the deduction step. A deduction from Σ is a sequence of formulas $(\phi_1, \dots, \phi_n)$ such that:

1. Each ϕ_i is either an element of Σ, an instance of Reflexivity or Start Axiom, or follows from one or more formulas of $\Sigma \cup \{\phi_1, \dots, \phi_{i-1}\}$ by one of the rules presented above.
2. If ϕ_i is an instance of Start Axiom (or follows from $\Sigma \cup \{\phi_1, \dots, \phi_{i-1}\}$ by Inclusion Introduction), then the new variables of $\boldsymbol{x}$ (or the new variable x) must not appear in $\Sigma \cup \{\phi_1, \dots, \phi_{i-1}\}$.

We say that ϕ is provable from Σ, written $\Sigma \vdash \phi$, if there is a deduction $(\phi_1, \dots, \phi_n)$ from Σ with $\phi = \phi_n$ and such that no variables in ϕ are new in $\phi_1, \dots, \phi_n$.

4 Soundness

First we prove the soundness of these axioms.

Lemma 2. *Let* $(\phi_1, \ldots, \phi_n)$ *be a deduction from* Σ*, and let* $\boldsymbol{y}$ *list all the new variables of the deduction steps. Let* $\mathcal{M}$ *and* X *be such that* $\mathcal{M} \models_X \Sigma$ *and* $Var(\Sigma_n) \setminus \boldsymbol{y} \subseteq Dom(X)$ *where* $\Sigma_n := \Sigma \cup \{\phi_1, \ldots, \phi_n\}$*. Then*

$$\mathcal{M} \models_X \exists \boldsymbol{y} \bigwedge \Sigma_n.$$

Proof. We show the claim by induction on n. So assume that the claim holds for any deduction of length n. We prove that the claim holds for deductions of length $n+1$ also. Let $(\phi_1, \ldots, \phi_{n+1})$ be a deduction from Σ, and let $\boldsymbol{y}$ and $\boldsymbol{z}$ list all the new variables of the deduction steps $\phi_1, \ldots, \phi_n$ and ϕ_{n+1}, respectively. Note that ϕ_{n+1} might not contain any new variables in which case $\boldsymbol{z}$ is empty. Assume that $\mathcal{M} \models_X \Sigma$ for some $\mathcal{M}$ and X, where $\mathrm{Var}(\Sigma_{n+1}) \setminus \boldsymbol{yz} \subseteq \mathrm{Dom}(X)$. By Proposition 2.2 we may assume that $\mathrm{Var}(\Sigma_{n+1}) \setminus \boldsymbol{yz} = \mathrm{Dom}(X)$. We need to show that

$$\mathcal{M} \models_X \exists \boldsymbol{y} \exists \boldsymbol{z} \bigwedge \Sigma_{n+1}.$$

By the induction assumption,

$$\mathcal{M} \models_X \exists \boldsymbol{y} \bigwedge \Sigma_n$$

when by Lemma 1 there is a function $F : X \to \mathcal{P}(M^{|\boldsymbol{y}|}) \setminus \{\emptyset\}$ such that

$$\mathcal{M} \models_{X'} \bigwedge \Sigma_n \tag{3}$$

where $X' := X[F/\boldsymbol{y}]$. It suffices to show that

$$\mathcal{M} \models_{X'} \exists \boldsymbol{z} \bigwedge \Sigma_{n+1}.$$

If ϕ_{n+1} is an instance of Start Axiom, or follows from Σ_n by Inclusion Introduction, then it suffices to find a $G : X' \to \mathcal{P}(M^{|\boldsymbol{z}|}) \setminus \{\emptyset\}$, such that $\mathcal{M} \models_{X'[G/\boldsymbol{z}]} \phi_{n+1}$ (note that in the first case this is due to Lemma 1). For this note that no variable of $\boldsymbol{z}$ is in $\mathrm{Var}(\Sigma_n)$, and hence by Proposition 2.2 $\mathcal{M} \models_{X'[G/\boldsymbol{z}]} \Sigma_n$ follows from (3). Otherwise, if $\boldsymbol{z}$ is empty, then it suffices to show that $\mathcal{M} \models_{X'} \phi_{n+1}$.

The cases where ϕ_{n+1} is an instance of Reflexivity, or follows from Σ_n by a conjunction rule, Projection and Permutation, Transitivity or Identity are straightforward. We prove the claim in the cases where one of the last four rules is applied.

- Inclusion Introduction: Then ϕ_{n+1} is of the form $\boldsymbol{a}x \subseteq \boldsymbol{b}c$ where $\boldsymbol{a} \subseteq \boldsymbol{b}$ is in Σ_n. Let $s \in X'$. Since $\mathcal{M} \models_{X'} \boldsymbol{a} \subseteq \boldsymbol{b}$ there is a $s' \in X'$ such that $s(\boldsymbol{a}) = s'(\boldsymbol{b})$. We let $G(s) = \{s'(c)\}$. Since $x \notin \mathrm{Dom}(X')$ we conclude that $\mathcal{M} \models_{X'[G/x]} \boldsymbol{a}x \subseteq \boldsymbol{b}c$.

- Start Axiom: Then ϕ_{n+1} is of the form $\boldsymbol{ac} \subseteq \boldsymbol{ax} \wedge \boldsymbol{b} \perp_{\boldsymbol{a}} \boldsymbol{x} \wedge \boldsymbol{ax} \subseteq \boldsymbol{ac}$. We define $G : X' \to \mathcal{P}(M^{|\boldsymbol{x}|}) \setminus \{\emptyset\}$ as follows:

$$G(s) = \{s'(\boldsymbol{c}) \mid s' \in X', s'(\boldsymbol{a}) = s(\boldsymbol{a})\}.$$

  Again, since $\boldsymbol{x}$ does not list any of the variables in $\mathrm{Dom}(X')$, it is straightforward to show that

$$\mathcal{M} \models_{X'[G/\boldsymbol{x}]} \boldsymbol{ac} \subseteq \boldsymbol{ax} \wedge \boldsymbol{b} \perp_{\boldsymbol{a}} \boldsymbol{x} \wedge \boldsymbol{ax} \subseteq \boldsymbol{ac}.$$

- Chase Rule: Then ϕ_{n+1} is of the form $\boldsymbol{abc} \subseteq \boldsymbol{xyz}$ where

$$\boldsymbol{y} \perp_{\boldsymbol{x}} \boldsymbol{z} \wedge \boldsymbol{ab} \subseteq \boldsymbol{xy} \wedge \boldsymbol{ac} \subseteq \boldsymbol{xz} \in \Sigma_n.$$

  Let $s \in X'$. Since $\mathcal{M} \models_{X'} \boldsymbol{ab} \subseteq \boldsymbol{xy} \wedge \boldsymbol{ac} \subseteq \boldsymbol{xz}$ there are $s', s'' \in X'$ such that $s'(\boldsymbol{xy}) = s(\boldsymbol{ab})$ and $s''(\boldsymbol{xz}) = s(\boldsymbol{ac})$. Since $s'(\boldsymbol{x}) = s''(\boldsymbol{x})$ and $\mathcal{M} \models_{X'} \boldsymbol{y} \perp_{\boldsymbol{x}} \boldsymbol{z}$, there is a $s_0 \in X'$ such that $s_0(\boldsymbol{xyz}) = s(\boldsymbol{abc})$ which shows the claim.
- Final Rule: Then ϕ_{n+1} is of the form $\boldsymbol{b} \perp_{\boldsymbol{a}} \boldsymbol{c}$ where

$$\boldsymbol{ac} \subseteq \boldsymbol{ax} \wedge \boldsymbol{b} \perp_{\boldsymbol{a}} \boldsymbol{x} \wedge \boldsymbol{abx} \subseteq \boldsymbol{abc} \in \Sigma_n.$$

  Let $s, s' \in X'$ be such that $s(\boldsymbol{a}) = s'(\boldsymbol{a})$. Since $\mathcal{M} \models_{X'} \boldsymbol{ac} \subseteq \boldsymbol{ax}$ there is a $s_0 \in X'$ such that $s'(\boldsymbol{ac}) = s_0(\boldsymbol{ax})$. Since $\mathcal{M} \models_{X'} \boldsymbol{b} \perp_{\boldsymbol{a}} \boldsymbol{x}$ and $s(\boldsymbol{a}) = s_0(\boldsymbol{a})$ there is a $s_1 \in X'$ such that $s_1(\boldsymbol{abx}) = s(\boldsymbol{ab})s_0(\boldsymbol{x})$. And since $\mathcal{M} \models_{X'} \boldsymbol{abx} \subseteq \boldsymbol{abc}$ there is a $s'' \in X'$ such that $s''(\boldsymbol{abc}) = s_1(\boldsymbol{abx})$. Then $s''(\boldsymbol{abc}) = s(\boldsymbol{ab})s'(\boldsymbol{c})$ which shows the claim and concludes the proof. □

This gives us the following soundness theorem.

Theorem 1. *Let $\Sigma \cup \{\phi\}$ be a finite set of conditional independence and inclusion atoms. Then $\Sigma \models \phi$ if $\Sigma \vdash \phi$.*

Proof. Assume that $\Sigma \vdash \phi$. Then there is a deduction $(\phi_1, \ldots, \phi_n)$ from Σ such that $\phi = \phi_n$ and no variables in ϕ are new in $\phi_1, \ldots, \phi_n$. Let $\mathcal{M}$ and X be such that $\mathrm{Var}(\Sigma \cup \{\phi\}) \subseteq \mathrm{Dom}(X)$ and $\mathcal{M} \models_X \Sigma$. We need to show that $\mathcal{M} \models_X \phi$. Let $\boldsymbol{y}$ list all the new variables in $\phi_1, \ldots, \phi_n$, and let $\boldsymbol{z}$ list all the variables in $\mathrm{Var}(\Sigma_n) \setminus \boldsymbol{y}$ which are not in $\mathrm{Dom}(X)$. We first let $X' := X[\boldsymbol{0}/\boldsymbol{z}]$ for some dummy sequence $\boldsymbol{0}$ when by Theorem 2.2, $\mathcal{M} \models_{X'} \Sigma$. Then by Theorem 2, $\mathcal{M} \models_{X'} \exists \boldsymbol{y} \bigwedge \Sigma_n$ implying there exists a $F : X' \to \mathcal{P}(M^{|\boldsymbol{y}|}) \setminus \{\emptyset\}$ such that $\mathcal{M} \models_{X''} \phi$, for $X'' := X'[F/\boldsymbol{y}]$. Since $X'' = X[\boldsymbol{0}/\boldsymbol{z}][F/\boldsymbol{y}]$ and no variables of $\boldsymbol{y}$ or $\boldsymbol{z}$ appear in ϕ, we conclude by Theorem 2.2 that $\mathcal{M} \models_X \phi$. □

5 Completeness

In this section we will prove that the set of axioms and rules presented in Definition 3 is complete with respect to the implication problem for conditional independence and inclusion atoms. For this purpose we introduce a graph characterization for the implication problem in Sect. 5.1. This characterization is based on the classical characterization of the implication problem for various database dependencies using the chase procedure [11]. The completeness proof is presented in Sect. 5.2. Also, in this section we will write $X \models \phi$ instead of $\mathcal{M} \models_X \phi$, since we will only deal with atoms, and the satisfaction of an atom depends only on the team X.

5.1 Graph Characterization

We will consider graphs consisting of vertices and edges labeled by (possibly multiple) pairs of variables. The informal meaning is that a vertice will correspond to an assignment of a team, and an edge between s and s', labeled by uw, will express that $s(u) = s'(w)$. The graphical representation of the chase procedure is adapted from [26].

Definition 4. *Let $G = (V, E)$ be a graph where E consists of directed labeled edges $(u, w)_{ab}$ where ab is a pair of variables, and for every pair (u, w) of vertices there can be several ab such that $(u, w)_{ab} \in E$. Then we say that u and w are ab-connected, written $u \sim_{ab} w$, if $u = w$ and $a = b$, or if there are vertices $v_0, \ldots, v_n$ and variables $x_0, \ldots, x_n$ such that*

$$(u, v_0)_{ax_0}, (v_0, v_1)_{x_0x_1}, \ldots, (v_{n-1}, v_n)_{x_{n-1}x_n}, (v_n, w)_{x_nb} \in E^*$$

where $E^* := E \cup \{(w, u)_{ba} \mid (u, w)_{ab} \in E\}$.

Next we define a graph $G_{\Sigma,\phi}$ in the style of Definition 4 for a set $\Sigma \cup \{\phi\}$ of conditional independence and inclusion atoms.

Definition 5. *Let $\Sigma \cup \{\phi\}$ be a finite set of conditional independence and inclusion atoms. We let $G_{\Sigma,\phi} := (\bigcup_{n\in\mathbb{N}} V_n, \bigcup_{n\in\mathbb{N}} E_n)$ where $G_n = (V_n, E_n)$ is defined as follows:*

- *If ϕ is $\boldsymbol{b} \perp_{\boldsymbol{a}} \boldsymbol{c}$, then $V_0 := \{v^+, v^-\}$ and $E_0 := \{(v^+, v^-)_{aa} \mid a \in \boldsymbol{a}\}$. If ϕ is $\boldsymbol{a} \subseteq \boldsymbol{b}$, then $V_0 := \{v\}$ and $E_0 := \emptyset$.*
- *Assume that G_n is defined. Then for every $v \in V_n$ and $x_1 \ldots x_k \subseteq y_1 \ldots y_k \in \Sigma$ we introduce a new vertex v_{new} and new edges $(v, v_{\text{new}})_{x_iy_i}$, for $1 \leq i \leq k$. Also for every $u, w \in V_n$, $u \neq w$, and $\boldsymbol{y} \perp_{\boldsymbol{x}} \boldsymbol{z} \in \Sigma$ where $u \sim_{xx} w$, for $x \in \boldsymbol{x}$, we introduce a new vertex v_{new} and new edges $(u, v_{\text{new}})_{yy}$, $(w, v_{\text{new}})_{zz}$, for $y \in \boldsymbol{xy}$ and $z \in \boldsymbol{xz}$. We let V_{n+1} and E_{n+1} be obtained by adding these new vertices and edges to the sets V_n and E_n.*

Note that $G_{\Sigma,\phi} = G_0$ if $\Sigma = \emptyset$.

The construction of $G_{\Sigma,\phi}$ can be illustrated through an example. Suppose $\phi = b \perp_a c$ and $\Sigma = \{c \perp_a d, c \perp_b c, ab \subseteq bc\}$. Then, at level 0 of the construction of $G_{\Sigma,\phi}$, we have two nodes v^+ and v^- and an edge between them labeled by the pair aa.

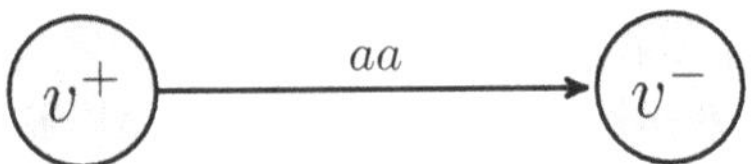

At level 1, four new nodes $v_1, \ldots, v_4$ and the corresponding edges are introduced: v_1 and v_2 for $c \perp_a d$, and v_3 and v_4 for $ab \subseteq bc$. The dashed node v_5 is an example of a new node introduced at level 2, due to $c \perp_b c \in \Sigma$ and $v_3 \sim_{bb} v_4$.

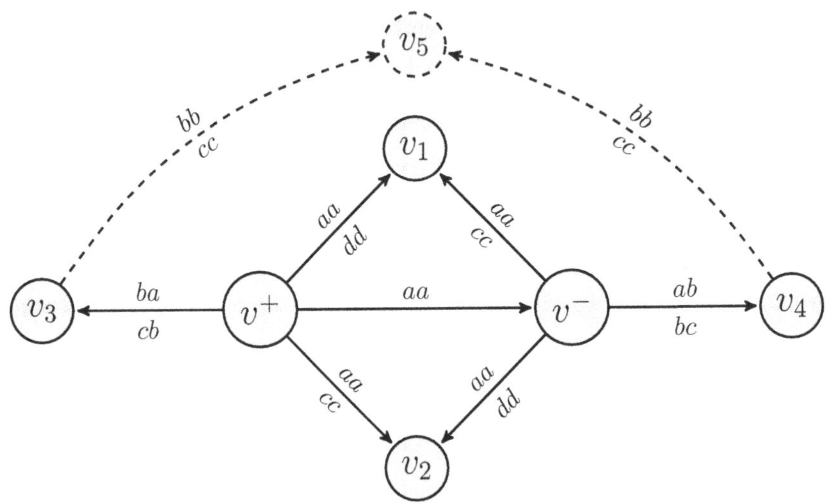

We will next show in detail how $G_{\Sigma,\phi}$ yields a characterization of the implication problem $\Sigma \models \phi$.

Theorem 2. *Let $\Sigma \cup \{\phi\}$ be a finite set of conditional independence and inclusion atoms.*

1. *If ϕ is $a_1 \ldots a_k \subseteq b_1 \ldots b_k$, then $\Sigma \models \phi \Leftrightarrow \exists w \in V_{\Sigma,\phi}(v \sim_{a_i b_i} w$ for all $1 \leq i \leq k)$.*
2. *If ϕ is $\boldsymbol{b} \perp_{\boldsymbol{a}} \boldsymbol{c}$, then $\Sigma \models \phi \Leftrightarrow \exists v \in V_{\Sigma,\phi}(v^+ \sim_{bb} v$ and $v^- \sim_{cc} v$ for all $b \in \boldsymbol{ab}$ and $c \in \boldsymbol{ac})$.*

Proof. We deal with cases 1 and 2 simultaneously. First we will show the direction from right to left. So assume that the right-hand side assumption holds. We show that $\Sigma \models \phi$. Let X be a team such that $X \models \Sigma$. We show that $X \models \phi$. For this, let $s, s' \in X$ be such that $s(\boldsymbol{a}) = s'(\boldsymbol{a})$. If ϕ is $\boldsymbol{b} \perp_{\boldsymbol{a}} \boldsymbol{c}$, then we need to find a s'' such that $s''(\boldsymbol{abc}) = s(\boldsymbol{ab})s'(\boldsymbol{c})$. If ϕ is $a_1 \ldots a_k \subseteq b_1 \ldots b_k$, then we need to find a s'' such that $s(a_1 \ldots a_k) = s''(b_1 \ldots b_k)$. We will now define inductively, for each natural number n, a function $f_n : V_n \to X$ such that $f_n(u)(x) = f_n(w)(y)$ if $(u, w)_{xy} \in E_n$. This will suffice for the claim as we will later show.

- Assume that $n = 0$.
  1. If ϕ is $a_1 \dots a_k \subseteq b_1 \dots b_k$, then $V_0 = \{v\}$ and $E_0 = \emptyset$, and we let $f_0(v) := s$.
  2. If ϕ is $\boldsymbol{b} \perp_{\boldsymbol{a}} \boldsymbol{c}$, then $V_0 = \{v^+, v^-\}$ and $E_0 = \{(v^+, v^-)_{aa} \mid a \in \boldsymbol{a}\}$. We let $f_0(v^+) := s$ and $f_0(v^-) := s'$. Then $f(v^+)(a) = f(v^-)(a)$, for $a \in \boldsymbol{a}$, as wanted.
- Assume that $n = m+1$, and that f_m is defined so that $f_m(u)(x) = f_m(w)(y)$ if $(u, w)_{xy} \in E_m$. We let $f_{m+1}(u) = f_m(u)$, for $u \in V_m$. Assume that $v_{\text{new}} \in V_{m+1} \setminus V_m$ and that there are $u \in V_m$ and $x_1 \dots x_l \subseteq y_1 \dots y_l \in \Sigma$ such that $(u, v_{\text{new}})_{x_i y_i} \in E_{m+1} \setminus E_m$, for $1 \leq i \leq l$. Since $X \models x_1 \dots x_l \subseteq y_1 \dots y_l$, there is a $s_0 \in X$ such that $f_{m+1}(u)(x_i) = s_0(y_i)$, for $1 \leq i \leq l$. We let $f_{m+1}(v_{\text{new}}) := s_0$ when $f_{m+1}(u)(x_i) = f_{m+1}(v_{\text{new}})(y_i)$, for $1 \leq i \leq l$, as wanted.

  Assume then that $v_{\text{new}} \in V_{m+1} \setminus V_m$ and that there are $u, w \in V_m$, $u \neq w$, and $\boldsymbol{y} \perp_{\boldsymbol{x}} \boldsymbol{z} \in \Sigma$ such that $(u, v_{\text{new}})_{yy}, (w, v_{\text{new}})_{zz} \in E_{m+1} \setminus E_m$, for $y \in \boldsymbol{xy}$ and $z \in \boldsymbol{xz}$. Then $u \sim_{xx} w$ in G_m, for $x \in \boldsymbol{x}$. This means that there are vertices $v_0, \dots, v_n$ and variables $x_0, \dots, x_n$, for $x \in \boldsymbol{x}$, such that

$$(u, v_0)_{xx_0}, (v_0, v_1)_{x_0 x_1}, \dots, (v_{n-1}, v_n)_{x_{n-1} x_n}, (v_n, w)_{x_n x} \in E_m^*,$$

  where $E_m^* := E_m \cup \{(w, u)_{ba} \mid (u, w)_{ab} \in E_m\}$. By the induction assumption then

$$f_m(u)(x) = f_m(v_0)(x_0) = \dots = f_m(v_n)(x_n) = f_m(w)(x).$$

  Hence, since $X \models \boldsymbol{y} \perp_{\boldsymbol{x}} \boldsymbol{z}$, there is a s_0 such that $s_0(\boldsymbol{xyz}) = f_m(u)(\boldsymbol{xy}) f_m(w)(\boldsymbol{z})$. We let $f_{m+1}(v_{\text{new}}) := s_0$ and conclude that $f_{m+1}(u)(y) = f_{m+1}(v_{\text{new}})(y)$ and $f_{m+1}(w)(z) = f_{m+1}(v_{\text{new}})(z)$, for $y \in \boldsymbol{xy}$ and $z \in \boldsymbol{xz}$. This concludes the construction.

Now, in case 2 there is a $v \in V_{\Sigma,\phi}$ such that $v^+ \sim_{bb} v$ and $v^- \sim_{cc} v$ for all $b \in \boldsymbol{ab}$ and $c \in \boldsymbol{ac}$. Let n be such that each path witnessing this is in G_n. We want to show that choosing s'' as $f_n(v)$, $s''(\boldsymbol{abc}) = s(\boldsymbol{ab}) s'(\boldsymbol{c})$. Recall that $s = f_n(v^+)$ and $s' = f_n(v^-)$. First, let $b \in \boldsymbol{ab}$. The case where $v = v^+$ is trivial, so assume that $v \neq v^+$ in which case there are vertices $v_0, \dots, v_n$ and variables $x_0, \dots, x_n$ such that

$$(v^+, v_0)_{bx_0}, (v_0, v_1)_{x_0 x_1}, \dots, (v_{n-1}, v_n)_{x_{n-1} x_n}, (v_n, v)_{x_n b} \in E_n^*$$

when by the construction, $f_n(v^+)(b) = f_n(v)(b)$. Analogously $f_n(v^-)(c) = f_n(v)(c)$, for $c \in \boldsymbol{c}$, which concludes this case.

In case 1, s'' is found analogously. This concludes the proof of the direction from right to left.

For the other direction, assume that the right-hand side assumption fails in $G_{\Sigma,\phi}$. Again, we deal with both cases simultaneously. We will now construct a team X such that $X \models \Sigma$ and $X \not\models \phi$. We let $X := \{s_u \mid u \in V_{\Sigma,\phi}\}$ where each $s_u : \text{Var}(\Sigma \cup \{\phi\}) \to \mathcal{P}(V_{\Sigma,\phi})^{|\text{Var}(\Sigma \cup \{\phi\})|}$ is defined as follows:

$$s_u(x) := \prod_{y \in \text{Var}(\Sigma \cup \{\phi\})} \{w \in V_{\Sigma,\phi} \mid u \sim_{xy} w\}.$$

We claim that $s_u(x) = s_w(y) \Leftrightarrow u \sim_{xy} w$. Indeed, assume that $u \sim_{xy} w$. If now v is in the set with the index z of the product $s_u(x)$, then $u \sim_{xz} v$. Since $w \sim_{yx} u$, we have that $w \sim_{yz} v$. Thus v is in the set with the index z of the product $s_w(y)$. Hence by symmetry we conclude that $s_u(x) = s_w(y)$. For the other direction assume that $s_u(x) = s_w(y)$. Then consider the set with the index y of the product $s_w(y)$. Since $w \sim_{yy} w$ by the definition, the vertex w is in this set, and thus by the assumption it is in the set with the index y of the product $s_u(x)$. It follows by the definition that $u \sim_{xy} w$ which shows the claim.

Next we will show that $X \models \Sigma$. So assume that $\boldsymbol{y} \perp_{\boldsymbol{x}} \boldsymbol{z} \in \Sigma$ and that $s_u, s_w \in X$ are such that $s_u(\boldsymbol{x}) = s_w(\boldsymbol{x})$. We need to find a $s_v \in X$ such that $s_v(\boldsymbol{xyz}) = s_u(\boldsymbol{xy})s_w(\boldsymbol{z})$. Since $u \sim_{xx} w$, for $x \in \boldsymbol{x}$, there is a $v \in G_{\Sigma,\phi}$ such that $(u,v)_{yy}, (w,v)_{zz} \in E_{\Sigma,\phi}$, for $y \in \boldsymbol{xy}$ and $z \in \boldsymbol{xz}$. Then $s_u(\boldsymbol{xy}) = s_v(\boldsymbol{xy})$ and $s_w(\boldsymbol{xz}) = s_v(\boldsymbol{xz})$, as wanted. In case $x_1 \dots x_l \subseteq y_1 \dots y_l \in \Sigma$, $X \models x_1 \dots x_l \subseteq y_1 \dots y_l$ is shown analogously.

It suffices to show that $X \not\models \phi$. Assume first that ϕ is $\boldsymbol{b} \perp_{\boldsymbol{a}} \boldsymbol{c}$. Then $s_{v^+}(\boldsymbol{a}) = s_{v^-}(\boldsymbol{a})$, but by the assumption there is no $v \in V_{\Sigma,\phi}$ such that $v^+ \sim_{bb} v$ and $v^- \sim_{cc} v$ for all $b \in \boldsymbol{ab}$ and $c \in \boldsymbol{ac}$. Hence there is no $s_v \in X$ such that $s_v(\boldsymbol{ab}) = s_{v^+}(\boldsymbol{ab})$ and $s_v(\boldsymbol{ac}) = s_{v^-}(\boldsymbol{ac})$ when $X \not\models \boldsymbol{b} \perp_{\boldsymbol{a}} \boldsymbol{c}$. In case ϕ is $a_1 \dots a_k \subseteq b_1 \dots b_k$, $X \not\models \phi$ is shown analogously. □

Let us now see how to use this theorem with our concrete example (see the paragraph after Definition 5). First we notice that v_5 witnesses $v^+ \sim_{bb} v^-$. Also $v^+ \sim_{aa} v^-$ since $(v^+, v^-)_{aa} \in E_{\Sigma,\phi}$, and $v^- \sim_{xx} v^-$ for any x by the definition. Therefore, choosing v as v^-, we obtain $\Sigma \models b \perp_a c$ by the previous theorem.

5.2 Completeness Proof

We are now ready to prove the completeness. Let us first define some notation needed in the proof. We will write $x = y$ for syntactical identity, $x \equiv y$ for an atom of the form $xy \subseteq zz$ implying the identity of x and y, and $\boldsymbol{x} \equiv \boldsymbol{y}$ for an conjunction the form $\bigwedge_{i \leq |\boldsymbol{x}|} \mathrm{pr}_i(\boldsymbol{x}) \equiv \mathrm{pr}_i(\boldsymbol{y})$. Let $\boldsymbol{x}$ be a sequence listing $\mathrm{Var}(\Sigma \cup \{\phi\})$. If $\boldsymbol{x}_v$ is a vector of length $|\boldsymbol{x}|$ (representing vertex v of the graph $G_{\Sigma,\phi}$), and $\boldsymbol{a} = (x_{i_1}, \dots, x_{i_l})$ is a sequence of variables from $\boldsymbol{x}$, then we write $\boldsymbol{a}_v$ for

$$(\mathrm{pr}_{i_1}(\boldsymbol{x}_v), \dots, \mathrm{pr}_{i_l}(\boldsymbol{x}_v)).$$

Also, for a deduction d from Σ, we write $\Sigma \vdash^d \psi$ if ψ appears as a proof step in d. Note that then new variables of the proof steps are allowed to appear in ψ.

We will next prove the completeness by using the following lemma (which will be proved later). Recall that (V_n, E_n) refers to the nth level of the construction of $G_{\Sigma,\phi}$.

Lemma 3. *Let n be a natural number, $\Sigma \cup \{\phi\}$ a finite set of conditional independence and inclusion atoms, and $\boldsymbol{x}$ a sequence listing $Var(\Sigma \cup \{\phi\})$. Then there is a deduction $d = (\phi_1, \dots, \phi_N)$ from Σ such that for each $u \in V_n$, there is a sequence $\boldsymbol{x}_u$ of length $|\boldsymbol{x}|$ (and possibly with repetitions) such that $\Sigma \vdash^d \boldsymbol{x}_u \subseteq \boldsymbol{x}$, and for each $(u,w)_{x_i x_j} \in E_n^*$, $\Sigma \vdash^d pr_i(\boldsymbol{x}_u) \equiv pr_j(\boldsymbol{x}_w)$. Moreover,*

- *if ϕ is of the form $\boldsymbol{a} \subseteq \boldsymbol{b}$, then $\phi_1 = \boldsymbol{x}_v \subseteq \boldsymbol{x}$ (obtained by Reflexivity), for $\boldsymbol{x}_v$ defined as $\boldsymbol{x}$,*
- *if ϕ is of the form $\boldsymbol{b} \perp_{\boldsymbol{a}} \boldsymbol{c}$, then $\phi_1 = \boldsymbol{ac} \subseteq \boldsymbol{ac}^* \wedge \boldsymbol{b} \perp_{\boldsymbol{a}} \boldsymbol{c}^* \wedge \boldsymbol{ac}^* \subseteq \boldsymbol{ac}$ (obtained by Start Axiom), for $\boldsymbol{a}_{v^+}\boldsymbol{b}_{v^+}\boldsymbol{c}_{v^-} = \boldsymbol{abc}^*$.*

Theorem 3. *Let $\Sigma \cup \{\phi\}$ be a finite set of conditional independence and inclusion atoms. Then $\Sigma \vdash \phi$ if $\Sigma \models \phi$.*

Proof. Let Σ and ϕ be such that $\Sigma \models \phi$. We will show that $\Sigma \vdash \phi$.
We have two cases: either

1. ϕ is $x_{i_1} \dots x_{i_m} \subseteq x_{j_1} \dots x_{j_m}$ and, by Theorem 2, there is a $w \in V_{\Sigma,\phi}$ such that $v \sim_{x_{i_k} x_{j_k}} w$ for all $1 \leq k \leq m$, or
2. ϕ is $\boldsymbol{b} \perp_{\boldsymbol{a}} \boldsymbol{c}$ and, by Theorem 2, there is a $v \in V_{\Sigma,\phi}$ such that $v^+ \sim_{x_i x_i} v$ and $v^- \sim_{x_j x_j} v$ for all $x_i \in \boldsymbol{ab}$ and $x_j \in \boldsymbol{ac}$.

Assume now first that ϕ is $\boldsymbol{a} \subseteq \boldsymbol{b}$ where $\boldsymbol{a} := x_{i_1} \dots x_{i_m}$ and $\boldsymbol{b} := x_{j_1} \dots x_{j_m}$. Then there is a $w \in V_{\Sigma,\phi}$ such that $v \sim_{x_{i_k} x_{j_k}} w$, for $1 \leq k \leq m$. Let n be such that all the witnessing paths are in G_n, and let $d = (\phi_1, \dots, \phi_N)$ be a deduction from Σ obtained by Lemma 3, for $\Sigma \cup \{\phi\}$, n and $\boldsymbol{x}$ listing $\mathrm{Var}(\Sigma \cup \{\phi\})$. For $\Sigma \vdash \phi$, it now suffices to show that $\Sigma \cup \{\phi_1, \dots, \phi_N\} \vdash \phi$ since, by Lemma 3, the variables that appear in ϕ appear already in ϕ_1 (as not new) and therefore cannot appear as new in any step of $(\phi_1, \dots, \phi_N)$.

Let first $1 \leq k \leq m$. We show that from $\Sigma \cup \{\phi_1, \dots, \phi_N\}$ we may derive

$$\mathrm{pr}_{i_k}(\boldsymbol{x}_v) \equiv \mathrm{pr}_{j_k}(\boldsymbol{x}_w). \tag{4}$$

If $w = v$ and $i_k = j_k$, then (4) is obtained by Reflexivity. If $w \neq v$ or $i_k \neq j_k$, then there are vertices $v_0, \dots, v_p \in V_n$ and variables $x_{l_0}, \dots, x_{l_p}$ such that

$$(v, v_0)_{x_{i_k} x_{l_0}}, (v_0, v_1)_{x_{l_0} x_{l_1}}, \dots, (v_{p-1}, v_p)_{x_{l_{p-1}} x_{l_p}}, (v_p, w)_{x_{l_p} x_{j_k}} \in E_n^*.$$

Then by Lemma 3,

$$\Sigma \vdash^d \mathrm{pr}_{i_k}(\boldsymbol{x}_v) \equiv \mathrm{pr}_{l_0}(\boldsymbol{x}_{v_0}) \wedge \dots \wedge \mathrm{pr}_{l_p}(\boldsymbol{x}_{v_p}) \equiv \mathrm{pr}_{j_k}(\boldsymbol{x}_w) \tag{5}$$

from which we obtain $\mathrm{pr}_{i_k}(\boldsymbol{x}_v) \equiv \mathrm{pr}_{j_k}(\boldsymbol{x}_w)$ by Identity Rule. Hence, we may now derive

$$\boldsymbol{a}_v \equiv \boldsymbol{b}_w. \tag{6}$$

Since $\Sigma \vdash^d \boldsymbol{x}_w \subseteq \boldsymbol{x}$ by Lemma 3, then by Permutation and Projection we obtain

$$\boldsymbol{b}_w \subseteq \boldsymbol{b}. \tag{7}$$

Note that by Lemma 3, $\boldsymbol{x}_v = \boldsymbol{x}$ when $\boldsymbol{a}_v = \boldsymbol{a}$. Thus we obtain $\boldsymbol{a} \subseteq \boldsymbol{b}$ from (6) and (7) using repeatedly Identity Rule. Since none of the steps above introduce any new variables, we get $\Sigma \cup \{\phi_1, \dots, \phi_N\} \vdash \phi$ which concludes case 1.

Assume then that ϕ is $\boldsymbol{b} \perp_{\boldsymbol{a}} \boldsymbol{c}$ when there is a $v \in V_{\Sigma,\phi}$ such that $v^+ \sim_{x_i x_i} v$ and $v^- \sim_{x_j x_j} v$ for all $x_i \in \boldsymbol{ab}$ and $x_j \in \boldsymbol{ac}$. Analogously to the previous case, by Lemma 3, we obtain a deduction $d = (\phi_1, \dots, \phi_N)$ from Σ for which

$$\Sigma \vdash^d \boldsymbol{x}_v \subseteq \boldsymbol{x} \tag{8}$$

and

$$\Sigma \vdash^d \boldsymbol{a}_v\boldsymbol{b}_v \equiv \boldsymbol{a}_{v^+}\boldsymbol{b}_{v^+} \wedge \boldsymbol{a}_v\boldsymbol{c}_v \equiv \boldsymbol{a}_{v^-}\boldsymbol{c}_{v^-}. \tag{9}$$

Again, for $\Sigma \vdash \phi$, it suffices to show that $\Sigma \cup \{\phi_1, \ldots, \phi_N\} \vdash \phi$. By Projection and Permutation we first deduce

$$\boldsymbol{a}_v\boldsymbol{b}_v\boldsymbol{c}_v \subseteq \boldsymbol{abc} \tag{10}$$

from (8), and using repeatedly Projection and Permutation and Identity Rule we get

$$\boldsymbol{a}_{v^+}\boldsymbol{b}_{v^+}\boldsymbol{c}_{v^-} \subseteq \boldsymbol{abc} \tag{11}$$

from (9) and (10). Note that by Lemma 3, $\boldsymbol{a}_{v^+}\boldsymbol{b}_{v^+}\boldsymbol{c}_{v^-} = \boldsymbol{abc}^*$ and $\Sigma \vdash^d \boldsymbol{ac} \subseteq \boldsymbol{ac}^* \wedge \boldsymbol{b} \perp_{\boldsymbol{a}} \boldsymbol{c}^*$. Therefore we can derive $\boldsymbol{b} \perp_{\boldsymbol{a}} \boldsymbol{c}$ with one application of Final Rule. Since none of the steps above introduce any new variables, we have $\Sigma \cup \{\phi_1, \ldots, \phi_N\} \vdash \phi$ which concludes case 2 and the proof. □

We are left to prove Lemma 3.

Proof (Lemma 3). Let n be a natural number, $\Sigma \cup \{\phi\}$ a finite set of conditional independence and inclusion atoms, and $\boldsymbol{x}$ a sequence listing $\mathrm{Var}(\Sigma \cup \{\phi\})$. We show the claim by induction on n. Note that at each step n it suffices to consider only edges $(u, w)_{x_i x_j} \in E_n$, since for $(w, u)_{x_j x_i} \in E_n^*$, $\mathrm{pr}_j(\boldsymbol{x}_w) \equiv \mathrm{pr}_i(\boldsymbol{x}_u)$ can be deduced from $\mathrm{pr}_i(\boldsymbol{x}_u) \equiv \mathrm{pr}_j(\boldsymbol{x}_w)$ (using Reflexivity for $\mathrm{pr}_i(\boldsymbol{x}_u)\mathrm{pr}_i(\boldsymbol{x}_u)$ and then Identity Rule).

- Assume that $n = 0$. We show in two cases how to construct a deduction d from Σ such that it meets the requirements of Lemma 3.
  1. Assume that ϕ is $\boldsymbol{a} \subseteq \boldsymbol{b}$ when $V_0 := \{v\}$ and $E_0 := \emptyset$. Then we let $\boldsymbol{x}_v := \boldsymbol{x}$ in which case we can derive $\boldsymbol{x}_v \subseteq \boldsymbol{x}$ as a first step by Reflexivity.
  2. Assume that ϕ is $\boldsymbol{b} \perp_{\boldsymbol{a}} \boldsymbol{c}$ when $V_0 := \{v^+, v^-\}$ and $E_0 := \{(v^+, v^-)_{x_i x_i} \mid x_i \in \boldsymbol{a}\}$. As a first step we use Start Axiom to obtain

$$\boldsymbol{ac} \subseteq \boldsymbol{ac}^* \wedge \boldsymbol{b} \perp_{\boldsymbol{a}} \boldsymbol{c}^* \wedge \boldsymbol{ac}^* \subseteq \boldsymbol{ac} \tag{12}$$

where $\boldsymbol{c}^*$ is a sequence of pairwise distinct new variables. Then using Inclusion Introduction and Projection and Permutation we may deduce

$$\boldsymbol{ab}^*\boldsymbol{c}^*\boldsymbol{d}^* \subseteq \boldsymbol{abcd} \tag{13}$$

from $\boldsymbol{ac}^* \subseteq \boldsymbol{ac}$ where $\boldsymbol{d}$ lists $\boldsymbol{x} \setminus \boldsymbol{abc}$ and $\boldsymbol{b}^*\boldsymbol{c}^*\boldsymbol{d}^*$ is a sequence of pairwise distinct new variables. By Projection and Permutation and Identity Rule we may assume that $\boldsymbol{ab}^*\boldsymbol{c}^*\boldsymbol{d}^*$ has repetitions exactly where $\boldsymbol{abcd}$ has. Therefore we can list the variables of $\boldsymbol{ab}^*\boldsymbol{c}^*\boldsymbol{d}^*$ in a sequence $\boldsymbol{x}_{v^-}$ of length $|\boldsymbol{x}|$ where

$$\boldsymbol{ab}^*\boldsymbol{c}^*\boldsymbol{d}^* = (\mathrm{pr}_{i_1}(\boldsymbol{x}_{v^-}), \ldots, \mathrm{pr}_{i_l}(\boldsymbol{x}_{v^-})),$$

for $\boldsymbol{abcd} = (x_{i_1}, \ldots, x_{i_l})$. Then $\boldsymbol{a}_{v^-}\boldsymbol{b}_{v^-}\boldsymbol{c}_{v^-}\boldsymbol{d}_{v^-} = \boldsymbol{ab}^*\boldsymbol{c}^*\boldsymbol{d}^*$, and we can derive $\boldsymbol{x}_{v^-} \subseteq \boldsymbol{x}$ from (13) by Projection and Permutation. We also let $\boldsymbol{x}_{v^+} := \boldsymbol{x}$ when $\boldsymbol{x}_{v^+} \subseteq \boldsymbol{x}$ is derivable by Reflexivity and $\boldsymbol{a}_{v^+}\boldsymbol{b}_{v^+}\boldsymbol{c}_{v^-} = \boldsymbol{abc}^*$. Moreover, $\boldsymbol{a}_{v^+} \equiv \boldsymbol{a}_{v^-}$ is derivable by Reflexivity because $\boldsymbol{a}_{v^+} = \boldsymbol{a}_{v^-}$. This concludes the case $n = 0$.

– Assume that $n = m + 1$. Then by the induction assumption, there is a deduction d such that for each $u \in V_m$ there is a sequence $\boldsymbol{x}_u$ such that $\Sigma \vdash^d \boldsymbol{x}_u \subseteq \boldsymbol{x}$, and for each $(u, w)_{x_i x_j} \in E_m$ also $\Sigma \vdash^d \mathrm{pr}_i(\boldsymbol{x}_u) \equiv \mathrm{pr}_j(\boldsymbol{x}_w)$. Assume that $v_{\mathrm{new}} \in V_{m+1} \setminus V_m$ is such that there are $u \in V_m$ and $x_{i_1} \ldots x_{i_l} \subseteq x_{j_i} \ldots x_{j_l} \in \Sigma$ for which we have added new edges $(u, v_{\mathrm{new}})_{x_{i_k} x_{j_k}}$ to V_{m+1}, for $1 \leq k \leq l$. We will introduce a sequence $\boldsymbol{x}_{v_{\mathrm{new}}}$ and show how to extend d to a deduction d^* such that $\Sigma \vdash^{d^*} \boldsymbol{x}_{v_{\mathrm{new}}} \subseteq \boldsymbol{x}$ and $\Sigma \vdash^{d^*} \mathrm{pr}_{i_k}(\boldsymbol{x}_u) \equiv \mathrm{pr}_{j_k}(\boldsymbol{x}_{v_{\mathrm{new}}})$, for $1 \leq k \leq l$.
By Projection and Permutation we deduce first

$$\mathrm{pr}_{i_1}(\boldsymbol{x}_u) \ldots \mathrm{pr}_{i_l}(\boldsymbol{x}_u) \subseteq x_{i_1} \ldots x_{i_l} \tag{14}$$

from $\boldsymbol{x}_u \subseteq \boldsymbol{x}$. Then we obtain

$$\mathrm{pr}_{i_1}(\boldsymbol{x}_u) \ldots \mathrm{pr}_{i_l}(\boldsymbol{x}_u) \subseteq x_{j_i} \ldots x_{j_l} \tag{15}$$

from (14) and the assumption $x_{i_1} \ldots x_{i_l} \subseteq x_{j_i} \ldots x_{j_l}$ by Transitivity.
Then by Reflexivity we may deduce $\mathrm{pr}_{i_1}(\boldsymbol{x}_u) \subseteq \mathrm{pr}_{i_1}(\boldsymbol{x}_u)$ from which we derive by Inclusion Introduction

$$\mathrm{pr}_{i_1}(\boldsymbol{x}_u) y_1 \subseteq \mathrm{pr}_{i_1}(\boldsymbol{x}_u) \mathrm{pr}_{i_1}(\boldsymbol{x}_u) \tag{16}$$

where y_1 is a new variable. Then from (15) and (16) we derive by Identity Rule

$$y_1 \mathrm{pr}_{i_2}(\boldsymbol{x}_u) \ldots \mathrm{pr}_{i_l}(\boldsymbol{x}_u) \subseteq x_{j_1} \ldots x_{j_l}. \tag{17}$$

Iterating this procedure l times leads us to a formula

$$\bigwedge_{1 \leq k \leq l} \mathrm{pr}_{i_k}(\boldsymbol{x}_u) \equiv y_k \wedge y_1 \ldots y_l \subseteq x_{j_1} \ldots x_{j_l} \tag{18}$$

where $y_1, \ldots, y_l$ are pairwise distinct new variables. Let $x_{j_{l+1}}, \ldots, x_{j_{l'}}$ list $\boldsymbol{x} \setminus \{x_{j_1}, \ldots, x_{j_l}\}$. Repeating Inclusion Introduction for the inclusion atom in (18) gives us a formula

$$y_1 \ldots y_{l'} \subseteq x_{j_1} \ldots x_{j_{l'}} \tag{19}$$

where $y_{l+1}, \ldots, y_{l'}$ are pairwise distinct new variables. Let $\boldsymbol{y}$ now denote the sequence $y_1 \ldots y_{l'}$ when

$$\bigwedge_{1 \leq k \leq l} \mathrm{pr}_{i_k}(\boldsymbol{x}_u) \equiv \mathrm{pr}_k(\boldsymbol{y}) \wedge \boldsymbol{y} \subseteq x_{j_1} \ldots x_{j_{l'}} \tag{20}$$

is the formula obtained from (18) by replacing its inclusion atom with (19). By Projection and Permutation and Identity Rule we may assume that $\mathrm{pr}_k(\boldsymbol{y}) = \mathrm{pr}_{k'}(\boldsymbol{y})$ if and only if $j_k = j_{k'}$, for $1 \leq k \leq l'$. Analogously to the case $n = 0$, we can then order the variables of $\boldsymbol{y}$ as a sequence $\boldsymbol{x}_{v_{\mathrm{new}}}$ of length $|\boldsymbol{x}|$ such that $\mathrm{pr}_{j_k}(\boldsymbol{x}_{v_{\mathrm{new}}}) = \mathrm{pr}_k(\boldsymbol{y})$, for $1 \leq k \leq l'$. Then

$$\bigwedge_{1 \leq k \leq l} \mathrm{pr}_{i_k}(\boldsymbol{x}_u) \equiv \mathrm{pr}_{j_k}(\boldsymbol{x}_{v_{\mathrm{new}}}) \wedge \mathrm{pr}_{j_1}(\boldsymbol{x}_{v_{\mathrm{new}}}) \ldots \mathrm{pr}_{j_{l'}}(\boldsymbol{x}_{v_{\mathrm{new}}}) \subseteq x_{j_1} \ldots x_{j_{l'}} \tag{21}$$

is the formula (20). By Projection and Permutation we can now deduce $\boldsymbol{x}_{v_{\text{new}}} \subseteq \boldsymbol{x}$ from the inclusion atom in (21). Hence $\boldsymbol{x}_{v_{\text{new}}}$ is such that $\boldsymbol{x}_{v_{\text{new}}} \subseteq \boldsymbol{x}$ and $\text{pr}_{i_k}(\boldsymbol{x}_u) \equiv \text{pr}_{j_k}(\boldsymbol{x}_{v_{\text{new}}})$ can be derived, for $1 \leq k \leq l$. This concludes the case for inclusion.

Assume then that $v_{\text{new}} \in V_{m+1} \setminus V_m$ is such that there are $u, w \in V_m$, $u \neq w$, and $\boldsymbol{q} \perp_{\boldsymbol{p}} \boldsymbol{r} \in \Sigma$ for which we have added new edges $(u, v_{\text{new}})_{x_i x_i}, (w, v_{\text{new}})_{x_j x_j}$ to V_{m+1}, for $x_i \in \boldsymbol{pq}$ and $x_j \in \boldsymbol{pr}$. We will introduce a sequence $\boldsymbol{x}_{v_{\text{new}}}$ and show how to extend d to a deduction d^* such that $\Sigma \vdash^{d^*} \boldsymbol{x}_{v_{\text{new}}} \subseteq \boldsymbol{x}$, and $\Sigma \vdash^{d^*} \text{pr}_i(\boldsymbol{x}_u) \equiv \text{pr}_i(x_{v_{\text{new}}})$ and $\Sigma \vdash^{d^*} \text{pr}_j(\boldsymbol{x}_w) \equiv \text{pr}_j(x_{v_{\text{new}}})$, for $x_i \in \boldsymbol{pq}$ and $x_j \in \boldsymbol{pr}$. The latter means that

$$\Sigma \vdash^{d^*} \boldsymbol{p}_u \boldsymbol{q}_u \equiv \boldsymbol{p}_{v_{\text{new}}} \boldsymbol{q}_{v_{\text{new}}} \wedge \boldsymbol{p}_w \boldsymbol{r}_w \equiv \boldsymbol{p}_{v_{\text{new}}} \boldsymbol{r}_{v_{\text{new}}}.$$

First of all, we know that $u \sim_{x_k x_k} w$ in G_m for all $x_k \in \boldsymbol{p}$. Thus there are vertices $v_0, \ldots, v_n \in V_m$ and variables $x_{i_0}, \ldots, x_{i_n}$ such that

$$(u, v_0)_{x_k x_{i_0}}, (v_0, v_1)_{x_{i_0} x_{i_1}}, \ldots, (v_{n-1}, v_n)_{x_{i_{n-1}} x_{i_n}}, (v_n, w)_{x_{i_n} x_k} \in E_m^*.$$

Hence by the induction assumption and Identity Rule, there are $\boldsymbol{x}_u$ and $\boldsymbol{x}_w$ such that $\Sigma \vdash^d \boldsymbol{x}_u \subseteq \boldsymbol{x}$ and $\Sigma \vdash^d \boldsymbol{x}_w \subseteq \boldsymbol{x}$, and $\Sigma \vdash^d \text{pr}_k(\boldsymbol{x}_u) \equiv \text{pr}_k(\boldsymbol{x}_w)$, for $x_k \in \boldsymbol{p}$. In other words,

$$\Sigma \vdash^d \boldsymbol{p}_u \equiv \boldsymbol{p}_w. \tag{22}$$

By Projection and Permutation we first derive

$$\boldsymbol{p}_u \boldsymbol{q}_u \subseteq \boldsymbol{pq} \tag{23}$$

and

$$\boldsymbol{p}_w \boldsymbol{r}_w \subseteq \boldsymbol{pr} \tag{24}$$

from $\boldsymbol{x}_u \subseteq \boldsymbol{x}$ and $\boldsymbol{x}_w \subseteq \boldsymbol{x}$, respectively. Then we derive

$$\boldsymbol{p}_u \boldsymbol{r}_w \subseteq \boldsymbol{pr} \tag{25}$$

from $\boldsymbol{p}_u \equiv \boldsymbol{p}_w$ and (24) by Identity Rule. By Chase Rule we then derive

$$\boldsymbol{p}_u \boldsymbol{q}_u \boldsymbol{r}_w \subseteq \boldsymbol{pqr} \tag{26}$$

from the assumption $\boldsymbol{q} \perp_{\boldsymbol{p}} \boldsymbol{r}$, (23) and (25). Now it can be the case that $x_i \in \boldsymbol{pq}$ and $x_i \in \boldsymbol{r}$, but $\text{pr}_i(\boldsymbol{x}_u) \neq \text{pr}_i(\boldsymbol{x}_w)$. Then we can derive

$$\text{pr}_i(\boldsymbol{x}_u)\text{pr}_i(\boldsymbol{x}_w) \subseteq x_i x_i \tag{27}$$

from (26) by Projection and Permutation, and

$$\boldsymbol{p}_u \boldsymbol{q}_u \boldsymbol{r}_w (\text{pr}_i(\boldsymbol{x}_u)/\text{pr}_i(\boldsymbol{x}_w)) \subseteq \boldsymbol{pqr} \tag{28}$$

from (27) and (26) by Identity Rule. Let now $\boldsymbol{r}^*$ be obtained from $\boldsymbol{r}_w$ by replacing, for each $x_i \in \boldsymbol{pq} \cap \boldsymbol{r}$, the variable $\text{pr}_i(\boldsymbol{x}_w)$ with $\text{pr}_i(\boldsymbol{x}_u)$. Iterating the previous derivation gives us then

$$\boldsymbol{r}^* \equiv \boldsymbol{r}_w \wedge \boldsymbol{p}_u \boldsymbol{q}_u \boldsymbol{r}^* \subseteq \boldsymbol{pqr}. \tag{29}$$

Let $\boldsymbol{s}$ list the variables in $\boldsymbol{x} \setminus \boldsymbol{pqr}$. From the inclusion atom in (29) we derive by Inclusion Introduction

$$\boldsymbol{p}_u\boldsymbol{q}_u\boldsymbol{r}^*\boldsymbol{s}^* \subseteq \boldsymbol{pqrs} \tag{30}$$

where $\boldsymbol{s}^*$ is a sequence of pairwise distinct new variables. Then $\boldsymbol{p}_u\boldsymbol{q}_u\boldsymbol{r}^*\boldsymbol{s}^*$ has repetitions at least where $\boldsymbol{pqrs}$ has, and hence we can define $\boldsymbol{x}_{v_{\text{new}}}$ as the sequence of length $|\boldsymbol{x}|$ where

$$\boldsymbol{p}_u\boldsymbol{q}_u\boldsymbol{r}^*\boldsymbol{s}^* = (\text{pr}_{i_1}(\boldsymbol{x}_{v_{\text{new}}}), \ldots, \text{pr}_{i_l}(\boldsymbol{x}_{v_{\text{new}}})), \tag{31}$$

for $\boldsymbol{pqrs} = (x_{i_1}, \ldots, x_{i_l})$. Then $\boldsymbol{p}_{v_{\text{new}}}\boldsymbol{q}_{v_{\text{new}}}\boldsymbol{r}_{v_{\text{new}}}\boldsymbol{s}_{v_{\text{new}}} = \boldsymbol{p}_u\boldsymbol{q}_u\boldsymbol{r}^*\boldsymbol{s}^*$, and we can thus derive

$$\boldsymbol{x}_{v_{\text{new}}} \subseteq \boldsymbol{x} \tag{32}$$

from (30) by Projection and Permutation. Moreover,

$$\boldsymbol{p}_{v_{\text{new}}}\boldsymbol{q}_{v_{\text{new}}} \equiv \boldsymbol{p}_u\boldsymbol{q}_u \tag{33}$$

can be derived by Reflexivity, and

$$\boldsymbol{p}_{v_{\text{new}}}\boldsymbol{r}_{v_{\text{new}}} \equiv \boldsymbol{p}_w\boldsymbol{r}_w \tag{34}$$

is derivable since (34) is the conjunction of $\boldsymbol{p}_u \equiv \boldsymbol{p}_w$ in (22) and $\boldsymbol{r}^* \equiv \boldsymbol{r}_w$ in (29). Hence, for $\boldsymbol{x}_{v_{\text{new}}}$ we can derive

$$\boldsymbol{x}_{v_{\text{new}}} \subseteq \boldsymbol{x} \wedge \boldsymbol{p}_{v_{\text{new}}}\boldsymbol{q}_{v_{\text{new}}} \equiv \boldsymbol{p}_u\boldsymbol{q}_u \wedge \boldsymbol{p}_{v_{\text{new}}}\boldsymbol{r}_{v_{\text{new}}} \equiv \boldsymbol{p}_w\boldsymbol{r}_w$$

which concludes the case $n = m + 1$ and the proof. □

By Theorem 1 and Theorem 3 we now have the following.

Corollary 1. *Let $\Sigma \cup \{\phi\}$ be a finite set of conditional independence and inclusion atoms. Then $\Sigma \vdash \phi$ if and only if $\Sigma \models \phi$.*

The following example shows how to deduce $b \perp_a c \vdash c \perp_a b$ and $b \perp_a cd \vdash b \perp_a c$.

Example 1.

- $b \perp_a c \vdash c \perp_a b$:
  1. $ab \subseteq ab' \wedge c \perp_a b' \wedge ab' \subseteq ab$ (Start Axiom)
  2. $ac \subseteq ac$ (Reflexivity)
  3. $b \perp_a c \wedge ab' \subseteq ab \wedge ac \subseteq ac \vdash ab'c \subseteq abc$ (Chase Rule)
  4. $ab'c \subseteq abc \vdash acb' \subseteq acb$ (Projection and Permutation)
  5. $ab \subseteq ab' \wedge c \perp_a b' \wedge acb' \subseteq acb \vdash c \perp_a b$ (Final Rule)
- $b \perp_a cd \vdash b \perp_a c$:
  1. $ac \subseteq ac' \wedge b \perp_a c' \wedge ac' \subseteq ac$ (Start Axiom)
  2. $ac'd' \subseteq acd$ (Inclusion Introduction)
  3. $ab \subseteq ab$ (Reflexivity)
  4. $b \perp_a cd \wedge ab \subseteq ab \wedge ac'd' \subseteq acd \vdash abc'd' \subseteq abcd$ (Chase Rule)
  5. $abc' \subseteq abc$ (Projection and Permutation)
  6. $ac \subseteq ac' \wedge b \perp_a c' \wedge abc' \subseteq abc \vdash b \perp_a c$ (Final Rule)

Our results show that for any consequence $\boldsymbol{b} \perp_{\boldsymbol{a}} \boldsymbol{c}$ of Σ there is a deduction starting with an application of Start Axiom and ending with an application of Final Rule.

References

1. Grädel, E., Väänänen, J.: Dependence and independence. Studia Logica 101(2), 399–410 (2013)
2. Galliani, P.: Inclusion and exclusion dependencies in team semantics: On some logics of imperfect information. Annals of Pure and Applied Logic 163(1), 68–84 (2012)
3. Väänänen, J.: Dependence Logic. Cambridge University Press (2007)
4. Abramsky, S., Väänänen, J.: Dependence logic, social choice and quantum physics (in preparation)
5. Geiger, D., Paz, A., Pearl, J.: Axioms and algorithms for inferences involving probabilistic independence. Information and Computation 91(1), 128–141 (1991)
6. Paolini, G., Väänänen, J.: Dependence Logic in Pregeometries and ω-Stable Theories. ArXiv e-prints abs/1310.7719 (2013)
7. Väänänen, J., Yang, F.: Propositional dependence logic (2013) (in preparation)
8. Hannula, M.: Axiomatizing first-order consequences in independence logic. CoRR abs/1304.4164 (2013)
9. Kontinen, J., Väänänen, J.A.: Axiomatizing first-order consequences in dependence logic. Ann. Pure Appl. Logic 164(11), 1101–1117 (2013)
10. Galliani, P.: General models and entailment semantics for independence logic. Notre Dame Journal of Formal Logic 54(2), 253–275 (2013)
11. Maier, D., Mendelzon, A.O., Sagiv, Y.: Testing implications of data dependencies. ACM Trans. Database Syst. 4(4), 455–469 (1979)
12. Armstrong, W.W.: Dependency Structures of Data Base Relationships. In: Proc. of IFIP World Computer Congress, pp. 580–583 (1974)
13. Casanova, M.A., Fagin, R., Papadimitriou, C.H.: Inclusion dependencies and their interaction with functional dependencies. In: Proceedings of the 1st ACM SIGACT-SIGMOD Symposium on Principles of Database Systems, PODS 1982, pp. 171–176. ACM, New York (1982)
14. Paredaens, J.: The interaction of integrity constraints in an information system. J. Comput. Syst. Sci. 20(3), 310–329 (1980)
15. Kontinen, J., Link, S., Väänänen, J.: Independence in database relations. In: Libkin, L., Kohlenbach, U., de Queiroz, R. (eds.) WoLLIC 2013. LNCS, vol. 8071, pp. 179–193. Springer, Heidelberg (2013)
16. Herrmann, C.: On the undecidability of implications between embedded multivalued database dependencies. Information and Computation 122(2), 221–235 (1995)
17. Herrmann, C.: Corrigendum to "on the undecidability of implications between embedded multivalued database dependencies". Inf. Comput. 122, 221–235 (1995); Inf. Comput. 204(12), 1847–1851 (2006)
18. Chandra, A.K., Vardi, M.Y.: The implication problem for functional and inclusion dependencies is undecidable. SIAM Journal on Computing 14(3), 671–677 (1985)
19. Mitchell, J.C.: The implication problem for functional and inclusion dependencies. Information and Control 56(3), 154–173 (1983)
20. Sadri, F., Ullman, J.D.: Template dependencies: A large class of dependencies in relational databases and its complete axiomatization. J. ACM 29(2), 363–372 (1982)
21. Beeri, C., Vardi, M.Y.: Formal systems for tuple and equality generating dependencies. SIAM J. Comput. 13(1), 76–98 (1984)
22. Yannakakis, M., Papadimitriou, C.H.: Algebraic dependencies. J. Comput. Syst. Sci. 25(1), 2–41 (1982)

23. Galliani, P., Hannula, M., Kontinen, J.: Hierarchies in independence logic. In: Rocca, S.R.D. (ed.) Computer Science Logic 2013 (CSL 2013). Leibniz International Proceeding3s in Informatics (LIPIcs), vol. 23, pp. 263–280. Dagstuhl, Germany, Schloss Dagstuhl–Leibniz-Zentrum fuer Informatik (2013)
24. Galliani, P., Väänänen, J.A.: On dependence logic. In: Baltag, A., Smets, S. (eds.) Johan van Benthem on Logical and Informational Dynamics. Springer (to appear)
25. Galliani, P., Hella, L.: Inclusion Logic and Fixed Point Logic. In: Rocca, S.R.D. (ed.) Computer Science Logic (CSL 2013), Dagstuhl, Germany. Leibniz International Proceedings in Informatics (LIPIcs), vol. 23, pp. 281–295. Schloss Dagstuhl–Leibniz-Zentrum fuer Informatik (2013)
26. Naumov, P., Nicholls, B.: R.E. axiomatization of conditional independence. In: Proceedings of the 14th Conference on Theoretical Aspects of Rationality and Knowledge (TARK 2013), pp. 131–137 (2013)

Guard Independence and Constraint-Preserving Snapshot Isolation

Stephen J. Hegner

Umeå University, Department of Computing Science
SE-901 87 Umeå, Sweden
hegner@cs.umu.se
http://www.cs.umu.se/~hegner

Abstract. A method for detecting potential violations of integrity constraints of concurrent transactions running under snapshot isolation (SI) is presented. In contrast to methods for ensuring full serializability under snapshot isolation, violations of integrity constraints may be detected by examining certain read-write interaction of only two transactions at a time. The method, called *constraint-preserving snapshot isolation (CPSI)*, thus provides greater isolation than ordinary SI in that results do not violate any integrity constraints, while requiring substantially less overhead, and involving fewer false positives, than typical for enhancements to SI which guarantee full serializable isolation.

1 Introduction

Over the course of the past few decades, *snapshot isolation (SI)* has become one of the preferred modes of transaction isolation for concurrency control in database-management systems (DBMSs). In SI, each transaction operates on its own private copy of the database (its *snapshot*). To implement commit for concurrent transactions, the results of these individual snapshots must be integrated. If there is a write conflict; that is, if more than one transaction writes the same data object, then only one transaction is allowed to commit. The others must abort if they are not naturally terminated in some other way.

On the one hand, SI avoids many of the update anomalies associated with policies such as read uncommitted (RU) and read (latest) committed (RC), such as dirty and nonrepeatable reads, respectively [6, p. 61]. On the other hand, with the now widespread use of multiversion concurrency control (MVCC), it admits very efficient implementation, avoiding many of the performance bottlenecks associated with lock-based *rigorous two-phase locking (rigorous 2PL)* [4], more commonly called *strong strict two-phase locking (SS2PL)* nowadays. Nevertheless, it does allow certain types of undesirable behavior which do not occur under view serialization, such as read and write skew [2].

Because true view serializability [14, Sec. 2.4] is the gold standard for isolation of transactions, there has been substantial recent interest in extending SI to achieve such true serializablity, the idea being to achieve the desirable

C. Beierle and C. Meghini (Eds.): FoIKS 2014, LNCS 8367, pp. 230–249, 2014.

properties of true serializablity while exploiting the efficiency of SI. As a consequence, *serializable SI* (SSI), has been developed [8,5]. In stark contrast to SS2PL, SSI is an optimistic approach. On top of standard SI, it looks for *dangerous structures*, which are sequences of two consecutive read-write edges of concurrent transactions in the multiversion conflict graph for the transactions. If such a structure is found, one of the participating transactions is required to terminate without committing its results. The existence of a dangerous structure in the conflict graph is a necessary condition for the nonserializablity of a set of concurrent transactions under SI, but it is not a sufficient one. Thus, the SSI strategy is subject to false positives. To illustrate, let $n \geq 2$ be a natural number and let $\mathbf{E}_0$ be a database schema which includes n integer-valued data objects $d_0, d_1, \ldots, d_{n-1}$. Let τ_i be the transaction which replaces the value of d_i with the current value of $d_{(i+1) \bmod n}$; i.e., which executes $d_i \leftarrow d_{(i+1) \bmod n}$. Running the set $\mathbf{T}_0 = \{\tau_i \mid 0 \leq i < n\}$ of transactions concurrently under snapshot isolation results in a permutation of the values of the d_i's, with the new value of d_i being the old value of $d_{(i+1) \bmod n}$, since each transaction sees the old values of the d_i's in its snapshot. However, no serial schedule of $\mathbf{T}_0$ can produce this permutation result. Indeed, if τ_i is run first and commits before any other transaction begins, then the old value of d_i will be overwritten before $\tau_{(i+1) \bmod n}$ is able to read it. Thus, $\mathbf{T}_0$ is not view serializable. Formally, there is a read-write dependency [8, Def. 2.2] (or *antidependency* [1, 4.4.2]) from $\tau_{i \bmod n}$ to $\tau_{(i+1) \bmod n}$ for data object $d_{(i+1) \bmod n}$; these dependencies are represented using the *multiversion serialization graph* or *multiversion conflict graph* as illustrated in Fig. 1. As argued above, (and in general since this graph contains a cycle [1, Sec. 5.3]), no view serialization is possible. However, if any transaction is removed from $\mathbf{T}_0$, the remaining set is serializable. Indeed, if τ_i is removed, then execution in the serial order $\tau_{i+1}\tau_{i+2}\ldots\tau_{n-1}\tau_0\tau_1\ldots\tau_{i-1}$ is equivalent to concurrent execution under SI. Thus, for any natural number n, there is a set $\mathbf{T}$ of n transactions whose execution under SI is not equivalent to any serial schedule, but execution of any proper subset of $\mathbf{T}$ under SI is equivalent to a serial execution. In other words, to determine whether a set of n transactions run under SI is view serializable, a test involving all n transactions must be performed. Since a dangerous structure of SSI [5] involves at most three transactions, the approach must necessarily involve false positives. Nevertheless, benchmarks reported in [5] are impressive, but of course the transaction mix must be taken into account. Recently, PostgreSQL, as of version 9.1, became the first widely used DBMS to put SSI into practice, employing a variant for the implementation of its serializable isolation level [15].

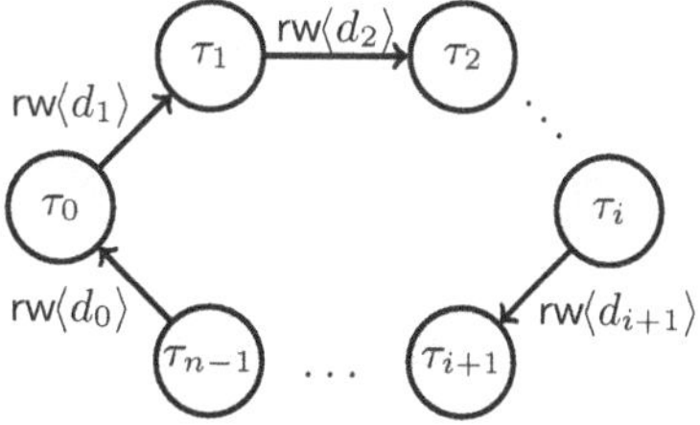

Fig. 1. An SI rw-conflict cycle of length n

In *Precisely SSI (PSSI)* [16], the entire multiversion conflict graph is constructed. This avoids virtually all false positives, but may involve a high overhead for transaction mixes involving long cycles, although reported benchmarks are favorable.

Despite the impressive performance statistics in the benchmark results, and the recent use in a widely used DBMS, it must be acknowledged that SSI and PSSI are not appropriate for all application domains. In particular, in any setting which involves long-running and interactive transactions, a policy for enforcing isolation which requires frequent aborts and/or waits is highly undesirable. Interactive business processes are one such domain, and it is in particular the context of cooperative transactions within that setting [13,11] which motivated the work of this paper. The central notion is an augmentation of SI, named *constraint-preserving SI (CPSI)*, which ensures that all integrity constraints will be satisfied. It is strictly weaker than SSI, in that nonserializable behavior which does not result in constraint violation is not ruled out. On the other hand, CPSI involves only a relatively simple check of a property of the conflict graph which, at least under one definition, is both necessary and sufficient to guarantee constraint satisfaction; that is, it does not produce any false positives. To illustrate, let $\mathbf{E}_1$ be identical to $\mathbf{E}_0$, save that the constraint $\varphi_1 = \sum_{i=0}^{n-1} d_i > 100 \cdot n$ is enforced. In concrete terms, think of each d_i as the balance in a bank account, and the constraint requiring that the average of the balances must exceed 100. Let $\mathbf{T}_1 = \{\tau'_i \mid 0 \leq i \leq n-1\}$ be the set of transactions on $\mathbf{E}_1$ with τ'_i defined by $d_i \leftarrow d_i - 1$. This is a generalization of the write-skew example of [2]. In order to determine whether the update to d_i will preserve the constraint φ_1, it is necessary for τ'_i to read *every* other d'_j. This means that there is a read-write conflict between *any* two τ_{j_1}, τ_{j_2}, as illustrated in Fig. 2. As a specific example, let M_{10} be the database which has $d_0 = 101$ and $d_j = 100$ for $1 \leq j \leq n-1$. Then any single transaction from $\mathbf{E}_1$, run in isolation, preserves φ_1, while the concurrent execution under SI of any two distinct members of $\mathbf{E}_1$ does not. The main result of this paper is that this two-element characterization holds in general; if a set of transactions running under SI results in a constraint violation, then there is a two-element subset which has this property when run on some legal database. This may in fact be further refined if transaction reads are separated into those required to verify constraints (called *guard reads*) and those required for other reasons. If the multiversion conflict graph is free of two-element cycles consisting of read-write edges when only writes and guard reads are considered, then the transactions under consideration cannot cause a constraint violation when run concurrently under SI, regardless of the initial database to which the transactions are applied.

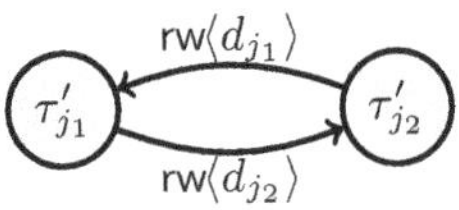

Fig. 2. An SI rw-conflict cycle of length 2

Although it may seem unacceptable to allow nonserializable results, this is in fact done all the time. For reasons of efficiency, lack of true serializability is routinely accepted with lower levels of isolation, such as RU and RC. On the other hand, results which violate integrity constraints, even those expressed via triggers or within application programs, are almost never acceptable. Separating the two, and providing checks for full serializability only when necessary, provides an avenue for much more efficient support for long-running and interactive transactions.

2 Schemata, Views, and Updates

Although the ideas surrounding database schemata, views, and updates which are used in this paper should already be familiar to the reader, the specific notation and conventions which are used nevertheless need to be spelled out carefully. While similar in many aspects to those used in previous papers, such as [9], [10], and [12], the framework employed here also differs in substantial ways. In particular, database schemata, while still being modelled by their sets of states, are characterized by both their overall state sets (which need not satisfy the integrity constraints) and their legal states (which must satisfy the integrity constraints). Furthermore, for the order and lattice structure of views, the syntactic congruence (on all states, ignoring the integrity constraints) rather than the semantic congruence (on just the legal states, taking into account equivalences implied by the integrity constraints), is employed. Therefore, while a notation consistent with these earlier works has been used wherever possible, it seems best to provide a self-contained presentation, with an acknowledgment that much has been drawn from those previous works.

For concepts related to order and lattices, the reader is referred to [7] for further clarification of notions utilized in this paper.

Summary 2.1 (Database schemata and views). A *database schema* $\mathbf{D}$ is characterized by two sets: $\mathsf{DB}(\mathbf{D})$ is the collection of all databases (or *states*), while $\mathsf{LDB}(\mathbf{D})$ is the subset of $\mathsf{DB}(\mathbf{D})$ consisting of just the *legal* databases (or *legal states*); i.e., those which satisfy the integrity constraints of the schema. In describing examples, it may be useful to identify explicitly a set $\mathsf{Constr}(\mathbf{D})$ of *constraints*, with $\mathsf{LDB}(\mathbf{D})$ consisting of precisely those members of $\mathsf{DB}(\mathbf{D})$ which satisfy the elements of $\mathsf{Constr}(\mathbf{D})$. However, such an explicit representation of constraints is not essential to the theory.

Given database schemata $\mathbf{D}_1$ and $\mathbf{D}_2$, a *database morphism* $f : \mathbf{D}_1 \to \mathbf{D}_2$ is a function $f : \mathsf{DB}(\mathbf{D}_1) \to \mathsf{DB}(\mathbf{D}_2)$. The morphism f is *semantic* if for every $M \in \mathsf{LDB}(\mathbf{D}_1)$, $f(M) \in \mathsf{LDB}(\mathbf{D}_2)$; i.e., if it maps legal databases to legal databases. The morphism f is said to be *fully surjective* if it is semantic, the function $f : \mathsf{DB}(\mathbf{D}_1) \to \mathsf{DB}(\mathbf{D}_2)$ is surjective, and for every $M_2 \in \mathsf{LDB}(\mathbf{D}_2)$, there is an $M_1 \in \mathsf{LDB}(\mathbf{D}_1)$ with $f(M_1) = M_2$. In other words, it is fully surjective if it is surjective as a function on all databases and also when it is restricted to just the legal databases of both $\mathbf{D}_1$ and $\mathbf{D}_2$.

A *view* over the database schema $\mathbf{D}$ is a pair $\Gamma = (\mathbf{V}, \gamma)$ in which $\mathbf{V}$ is a database schema with $\gamma : \mathbf{D} \to \mathbf{V}$ a fully surjective morphism. The set of all views on $\mathbf{D}$ is denoted $\mathsf{Views}(\mathbf{D})$. Full surjectivity is a natural property. By its very nature, the states (resp. legal states) of a view are determined by the states (resp. legal states) of the main schema. The notation for states of schemata extends naturally to views. Given a view $\Gamma = (\mathbf{V}, \gamma)$, $\mathsf{DB}(\Gamma)$ and $\mathsf{LDB}(\Gamma)$ are alternate notation for $\mathsf{DB}(\mathbf{V})$ and $\mathsf{LDB}(\mathbf{V})$ respectively.

The *syntactic congruence* of Γ is $\mathsf{SynCongr}(\Gamma) = \{(M_1, M_2) \in \mathsf{DB}(\mathbf{D}) \times \mathsf{DB}(\mathbf{D}) \mid \gamma(M_1) = \gamma(M_2)\}$. The *syntactic preorder* $\preceq_{\mathbf{D}}$ on $\mathsf{Views}(\mathbf{D})$ is defined by $\Gamma_1 \preceq_{\mathbf{D}} \Gamma_2$ iff $\mathsf{SynCongr}(\Gamma_2) \subseteq \mathsf{SynCongr}(\Gamma_1)$. The intuitive idea behind this order is that if $\Gamma_1 \preceq_{\mathbf{D}} \Gamma_2$, then Γ_2 preserves at least as much information about the state of the main schema $\mathbf{D}$ as does Γ_1.

For $\Gamma_1 = (\mathbf{V}_1, \gamma_1)$ and $\Gamma_2 = (\mathbf{V}_2, \gamma_2)$, with $\Gamma_2 \preceq_{\mathbf{D}} \Gamma_1$, Γ_2 may be regarded as a view of Γ_1. More precisely, define the relative morphism $\lambda\langle\Gamma_1, \Gamma_2\rangle : \mathbf{V}_1 \to \mathbf{V}_2$ to be the unique function $\lambda\langle\Gamma_1, \Gamma_2\rangle : \mathsf{DB}(\mathbf{V}_1) \to \mathsf{DB}(\mathbf{V}_2)$ which satisfies $\lambda\langle\Gamma_1, \Gamma_2\rangle \circ \gamma_1 = \gamma_2$. See [9, Def. 2.3] for a elaboration of this concept.

The *identity view* $\mathbf{1}_{\mathbf{D}}$ of $\mathbf{D}$ has schema is $\mathbf{D}$ and morphism the identity $\mathbf{D} \to \mathbf{D}$. Similarly, a *zero view* $\mathbf{0}_{\mathbf{D}}$ of $\mathbf{D}$ has a schema which has only one database, with the view whose schema is a one element set, with the view morphism sending every element of $\mathsf{DB}(\mathbf{D})$ to the unique element of that set. It is immediate that for any $\Gamma \in \mathsf{Views}(\mathbf{D})$, $\mathbf{0}_{\mathbf{D}} \preceq_{\mathbf{D}} \Gamma \preceq_{\mathbf{D}} \mathbf{1}_{\mathbf{D}}$.

Summary 2.2 (Updates). An update on $\mathbf{D}$ is a pair $\langle M_1, M_2\rangle \in \mathsf{LDB}(\mathbf{D}) \times \mathsf{LDB}(\mathbf{D})$. M_1 is the current or old state before the update, and M_2 is the new state afterwards. Note that updates always transform legal states to legal states. The set of all updates on $\mathbf{D}$ is denoted $\mathsf{Updates}(\mathbf{D})$.

Updates are often identified by name; therefore, it is convenient to have a shorthand for its components. To this end, if $u \in \mathsf{Updates}(\mathbf{D})$, then write $u^{(1)}$ and $u^{(2)}$ for the values of the state before and after the update, respectively; i.e., $u = \langle u^{(1)}, u^{(2)}\rangle$. The *composition* $u_1 \circ u_2$ of two updates $u_1, u_2 \in \mathsf{Updates}(\mathbf{D})$ is just their composition in the sense of mathematical relations. More precisely, $u_1 \circ u_2 = \{(M_1, M_3) \mid (\exists M_2 \in \mathsf{LDB}(\mathbf{D}))((M_1, M_2) \in u_1 \wedge (M_2, M_3) \in u_2)\}$.

It will also prove useful to be able to select just those updates which apply to a specific state. For $N \in \mathsf{LDB}(\mathbf{D})$, define $\mathbf{u}_{|N} = \{u \in \mathbf{u} \mid u^{(1)} = N\}$.

It will sometimes be necessary to map updates from one view to a second, smaller one. Recall the relative morphism $\lambda\langle\Gamma_1, \Gamma_2\rangle : \mathbf{V}_1 \to \mathbf{V}_2$ defined in Summary 2.1 above. Then, for $u \in \mathsf{Updates}(\Gamma_1)$, define $\lambda\langle\Gamma_1, \Gamma_2\rangle(u) = \langle\lambda\langle\Gamma_1, \Gamma_2\rangle(u^{(1)}), \lambda\langle\Gamma_1, \Gamma_2\rangle(u^{(2)})\rangle$, and for $\mathbf{u} \subseteq \mathsf{Updates}(\Gamma_1)$, define $\lambda\langle\Gamma_1, \Gamma_2\rangle(\mathbf{u}) = \{\lambda\langle\Gamma_1, \Gamma_2\rangle(u) \mid u \in \mathbf{u}\}$. The set $\mathbf{u} \subseteq \mathsf{Updates}(\Gamma_1)$ is *a partial identity* on Γ_2 if $\lambda\langle\Gamma_1, \Gamma_2\rangle(\mathbf{u})$ is a subset of the identity relation on $\mathsf{LDB}(\mathbf{V}_2)$. In this case, it is said that $\mathbf{u}$ *holds* Γ *constant.* The single update $u \in \mathsf{Updates}(\mathbf{V}_1)$ is *a partial identity* on Γ_2 if $\{u\}$ has this property.

Finally the notational conventions which permit view names to be used in lieu of their components is extended to updates. Specifically, for $\Gamma = (\mathbf{V}, \gamma)$, $\mathsf{Updates}(\Gamma)$ will be used as an alternate notation for $\mathsf{Updates}(\mathbf{V})$.

Definition 2.3 (Algebras of updateable views). In an investigation of the interaction of transactions, it is central to be able to model their read-write interaction; for example, to express succinctly that transaction T_1 does or does not write some data object which T_2 reads. Transactions typically read and write compound data objects (e.g., sets of rows or tuples in the relational context) rather than just single primitive objects (e.g., single rows or tuples). In order to express such interaction, it is convenient to model compound data objects as being built up from simple ones in a systematic way. The natural mathematical structure for such a model is the Boolean algebra [7, p. 94]. It is assumed that the reader is familiar with that notion; only notation will be recalled here. In the Boolean algebra $\mathbf{L} = (L, \vee, \wedge, ^{-}, \top, \bot)$, L is the underlying set, $\vee$ and $\wedge$ are the join and meet operators, respectively, $\top$ and $\bot$ are the identity and zero elements, respectively, and $^{-}$ is the complement operator, which is written as an overbar; i.e., the complement of x is $\overline{x}$. The join operation induces a partial order via $a \leq b$ iff $a \vee b = b$, called the *underlying partial order*. The order with equality excluded is denoted $<$. An *atom* $a \in L$ is a minimal element which is greater than $\bot$; i.e., $\bot < a$ and for no $b \in L$ is it the case that $\bot < b < a$. The set of all atoms of $\mathbf{L}$ is denoted $\mathsf{Atoms}_{\mathbf{L}}$. If $\mathbf{L}$ is finite, then every $a \in L$ has a unique representation as the join of atoms. In this case, define the *basis* of a to be $\mathsf{Basis}_{\mathbf{L}}\langle a \rangle = \{x \in \mathsf{Atoms}_{\mathbf{L}} \mid x \leq a\}$ [7, 5.5].

Now, given a database schema $\mathbf{D}$, an *algebra of updateable views* over $\mathbf{D}$ is a finite Boolean algebra $\boldsymbol{\mathcal{V}} = (\mathcal{V}, \vee, \wedge, ^{-}, \top, \bot)$ with $\mathcal{V} \subseteq \mathsf{Views}(\mathbf{D})$, whose underling partial order is the restriction of $\preceq_{\mathbf{D}}$ to $\mathcal{V}$. This requirement on the order structure has an important consequence. Given a view $\Gamma \in \mathcal{V}$, there is a natural correspondence between $\mathsf{DB}(\Gamma)$ and $\{\mathsf{DB}(\Gamma') \mid \Gamma' \in \mathsf{Basis}_{\mathcal{V}}\langle \Gamma \rangle\}$. Specifically, each $M \in \mathsf{DB}(\Gamma)$ has a unique representation as the set $\{\lambda\langle \Gamma, \Gamma' \rangle(M) \mid \Gamma' \in \mathsf{Basis}_{D}\langle \Gamma \rangle\}$. There is a bit of notational shorthand, based upon this observation, which will prove useful. If $\Gamma_1 = (\mathbf{V}_1, \gamma_1), \Gamma_2 = (\mathbf{V}_2, \gamma_2) \in \mathcal{V}$, with $\Gamma_1 \wedge \Gamma_2 = \bot$, and if $M_1 \in \mathsf{DB}(\Gamma_1)$ and $M_2 \in \mathsf{DB}(\Gamma_2)$, then $M_1 \vee M_2 \in \mathsf{DB}(\Gamma_1 \vee \Gamma_2)$ denotes the unique state with the representation $\{\lambda\langle \Gamma, \Gamma' \rangle(M) \mid \Gamma' \in \mathsf{Basis}_{D}\langle \Gamma_1 \rangle \cup \mathsf{Basis}_{D}\langle \Gamma_2 \rangle\}$.

The least element of $\boldsymbol{\mathcal{V}}$ is formally a zero view $\mathbf{0_D}$, but will frequently be written as $\bot$.

Example 2.4 (The algebra of updateable views of $\mathbf{E}_0$ and $\mathbf{E}_1$). For the schema $\mathbf{E}_0$ introduced in Sec. 1, let Let $\Omega_{d_i} = (\mathbf{W}_{d_i}, \omega_{d_i})$ be the view which retains just d_i, discarding all d_j for $j \neq i$. Thus, $\mathbf{W}_{d_i}$ contains just the data object d_i, while the view morphism $\omega_{d_i} : \mathbf{E}_0 \to \mathbf{W}_{d_i}$ retains just d_i from $D_{\mathbf{E}_0} = \{d_j \mid 0 \leq j \leq n-1\}$. $\{\Omega_{d_j} \mid 0 \leq j \leq n-1\}$ forms the set $\mathsf{Atoms}_{\boldsymbol{\mathcal{V}}_{\mathbf{E}_0}}$ of atoms of the algebra $\boldsymbol{\mathcal{V}}_{\mathbf{E}_0} = (\mathcal{V}_{\mathbf{E}_0}, \vee, \wedge, ^{-}, \top, \bot)$ of updateable views associated with $\mathbf{E}_0$. Each element of $\mathcal{V}_{\mathbf{E}_0}$ is of the form $\bigvee_{j \in S} \Omega_{d_j}$ for some $S \subseteq \{0, 1, \ldots, n-1\}$. In other words, the members of $\mathcal{V}_{\mathbf{E}_0}$ are in bijective correspondence with subsets of $D_{\mathbf{E}_0}$. The zero view corresponds to the empty set $\emptyset$, while the identity view corresponds to the entire set $D_{\mathbf{E}_0}$. Join and meet correspond to union and intersection on $\{d_j \mid 0 \leq j \leq n-1\}$, respectively. In short, $\boldsymbol{\mathcal{V}}_{\mathbf{E}_0} = (\mathcal{V}_{\mathbf{E}_0}, \vee, \wedge, ^{-}, \top, \bot)$ is isomorphic to the power-set algebra [7, 4.18(1)] of $D_{\mathbf{E}_0}$. The algebra for $\mathbf{E}_1$ is identical.

Definition 2.5 (The algebra of σ-views of a relational schema). It is instructive to illustrate the ideas of Definition 2.3 within a framework which recaptures common usage. Let $\mathbf{D}$ be a relational schema, and suppose that *row-level granularity* in relational DBMSs, in which the smallest grain of data access for a transaction is a single row (or tuple or atom), is employed. Let $\mathsf{GrAtoms}\langle \mathbf{D} \rangle$ denote the set of all ground atoms of $\mathbf{D}$ (tuples not involving variables, the values for columns/attributes are domain values only). In practice, $\mathsf{GrAtoms}\langle \mathbf{D} \rangle$ is always a finite set; this finiteness restriction is assumed to hold here as well. Further, assume that tuples are tagged with the relations in which they occur, so that it is not necessary to identify relations explicitly in selections. For $t \in \mathsf{GrAtoms}\langle \mathbf{D} \rangle$, define $\Sigma_t = (\Upsilon_t, \sigma_t)$ to be the view which selects t from the appropriate relation and discards everything else. $\mathsf{DB}(\Upsilon_t) = \{\emptyset, \{t\}\}$; that is, there are only two states to the view schema, one representing that t is present in the main schema, and the other that it is not. On Σ_t, the only update operations which are possible are to delete t, to insert t, and to do nothing. The set of all such selections on a single ground tuple, $\Sigma\text{-}\mathsf{GrAtoms}\langle \mathbf{D} \rangle = \{\Sigma_t \mid t \in \mathsf{GrAtoms}\langle \mathbf{D} \rangle\}$, forms the set of atoms for the lattice of data objects.

Extending this to compound objects, if $S = \{t_1, t_2, \ldots, t_n\} \subseteq \mathsf{GrAtoms}\langle \mathbf{D} \rangle$, then $\Sigma_S = (\Upsilon_S, \sigma_S)$ denotes the selection on all of S. The logical operation connecting the tuples is disjunction; all of the t_i's are selected. A longer but more descriptive representation might be $\Sigma_{\{t_1,t_2,\ldots,t_n\}} = (\Upsilon_{\{t_1,t_2,\ldots,t_n\}}, \sigma_{t_1 \vee t_2 \vee \ldots \vee t_n})$. The point here is that the view Σ_S is the join of a unique set of primitive selections, namely, $\{\Sigma_{t_i} \mid 1 \leq i \leq n\}$. As a degenerate but nevertheless very useful case, note also that $\Sigma_\emptyset$ is a zero view with $\emptyset$ as the only view state; no updates are possible on it.

Define $\Sigma\text{-}\mathsf{GrAtoms}\langle \mathbf{D} \rangle = \{\Sigma_t \mid t \in \mathsf{GrAtoms}\langle \mathbf{D} \rangle\}$, the set of all selections on a single ground tuple, and define $\Sigma\text{-}\mathsf{Views}\langle \mathbf{D} \rangle = \{\Sigma_S \mid S \subseteq \Sigma\text{-}\mathsf{GrAtoms}\langle \mathbf{D} \rangle\}$. Then $\Sigma\text{-}\mathsf{GrAtoms}\langle \mathbf{D} \rangle$ is the set of atoms of the lattice of updateable views whose elements are $\Sigma\text{-}\mathsf{Views}\langle \mathbf{D} \rangle$. Join, meet, and complement are given by union, intersection, and complement on the underlying sets of tuples: $\Sigma_{S_1} \vee \Sigma_{S_2} = \Sigma_{S_1 \cup S_2}$, $\Sigma_{S_1} \wedge \Sigma_{S_2} = \Sigma_{S_1 \cap S_2}$, and the complement $\overline{\Sigma_S}$ of Σ_S is $\Sigma_{\mathsf{GrAtoms}\langle \mathbf{D} \rangle \setminus S}$.

Notation 2.6 (Notation for views). Views appear frequently in that which follows, and it is convenient to have a uniform convention for identifying constituent parts. If Γ is the name of a view, possibly with subscripts or other annotations, then $\mathbf{V}$ and γ will be used to denote its schema and morphism, respectively with the same annotations. Thus, for the views Γ, Γ', Γ_1, and $\overline{\Gamma_1}$, the full expansions are assumed to be $\Gamma = (\mathbf{V}, \gamma)$, $\Gamma' = (\mathbf{V}', \gamma')$, $\Gamma_1 = (\mathbf{V}_1, \gamma_1)$, and $\overline{\Gamma_1} = (\overline{\mathbf{V}_1}, \overline{\gamma_1})$.

Abstract views, as used in definitions and theorems, will always use the (possibly annotated) $\Gamma = (\mathbf{V}, \gamma)$ notation. For specific examples, an alternate notation, using $\Omega = (\mathbf{W}, \omega)$, also with annotations, will be used. In this way, example views are always clearly distinguished from abstract ones. The same conventions apply; for example, the full expansion of $\overline{\Omega_1'}$ is $\overline{\Omega_1'} = (\overline{\mathbf{W}_1'}, \overline{\omega_1'})$.

Notation 2.7 (Some mathematical shorthand). It will often be necessary to assert that a partial function f is defined on an argument x. The shorthand $f(x)\!\downarrow$ will be used in this regard. In order to avoid the need to state independently that a function is defined on an argument, a statement such as $f(x)\!\downarrow\in Y$ will be used to indicate that both $f(x)\!\downarrow$ and $f(x) \in Y$. Similarly, $f(x)\!\uparrow$ denotes that $f(x)$ is undefined on x.

$\mathbb{N}$ denotes the set $\{0, 1, 2, \ldots\}$ of natural numbers. For $i, j \in \mathbb{N}$, $[i, j]$ denotes the set $\{i, i+1, \ldots, j\}$ of natural numbers between i and j inclusive, while $[j, \text{-}]$

denotes the set of all natural numbers which are greater than or equal to i. $\mathbb{Z}$ denotes the set of all integers, positive, negative, and zero.

$\mathsf{Card}(X)$ denotes the cardinality of the set X.

3 Snapshot Isolation

In this section, an overview of snapshot isolation is presented and the concurrency issues surrounding it are summarized. It is assumed that the reader has a basic understanding of transactions, and in particular serializability, as is presented in [18], [3], and Chapters 14-15 of the textbook [17].

Summary 3.1 (The transaction model of snapshot isolation). Before presenting the theory, it is appropriate to sketch the model of snapshot isolation (hereafter *SI*) which is used. A *transaction* performs read and write operations on data objects. Each transaction has a start time, as well as an end time at which its writes are *committed* to the database. Two transactions are *concurrent* if the start time of one lies between the start and end times of the other.

In SI, the transaction T always operates on a private copy of the database, called the *snapshot*, taken at the start time of the transaction. While it is running, it does not see updates performed by other transactions, and other transactions do not see its updates. When T finishes, its updates must be committed to the global database. Such commits are governed by the *first-committer wins (FCW)* rule. If any other transaction T' which is concurrent with T, and which has already committed has written a data object Γ which T has also written, then T is not allowed to commit.

Summary 3.2 (Variations of SI in practice). In practice, things are not quite as simple as sketched in Summary 3.1 for at least two reasons. First of all, the rule for conflict resolution which is used in practice is most often that which is called *first updater wins (FUW)*. With FUW, if some other concurrent transaction T' writes a data object Γ which T later is to write, then T is blocked from continuing to operate, even on its private copy, until T' commits (in which case T is aborted) or T' aborts (in which case T is allowed to continue). While FCW and FUW differ in implementation, they are identical in the definition of a conflict; namely, that concurrent transactions may not both write the same data object. They furthermore produce identical results when there is no write conflict. Thus, for a study of conflict and constraint violation, FUW may be used in lieu of FCW with no loss of generality. It will be used here because it admits a simpler conceptual model of a transaction which does not involve the order or points in time at which the transaction performs internal operations on its private copy.

A second reason why the FCW model is somewhat idealistic is that in most implementations of SQL, primary-key and uniqueness constraints are enforced immediately, and unless checking is declared to be `deferrable` and then `deferred`, foreign-key constraints will also be enforced immediately. In this work, which is of a more foundational nature, this "implementation detail" will be ignored. In any case, for the purposes of this work, integrity constraints include not only the usual database dependencies (e.g., key and foreign-key dependencies), but also rules specified in triggers and possibly even application programs. The latter two are often of central importance in business processes.

4 Constraint Preservation and Its Characterization

Notation 4.1 (Notational conventions). Throughout this section, unless stated specifically to the contrary, take $\mathbf{D}$ to be a database schema and $\boldsymbol{\mathcal{V}} = (\mathcal{V}, \vee, \wedge, {}^{-}, \top, \bot)$ a (finite Boolean) algebra of updateable views, as described in Summary 2.1 and Definition 2.3, respectively.

Definition 4.2 (Updateable objects). It is useful to combine an updateable view and the updates which may be applied to it into one package. Formally, an *updateable object* over $\boldsymbol{\mathcal{V}}$ is a pair $\langle \Gamma, \mathbf{u} \rangle$ with $\Gamma \in \mathcal{V}$ and $\mathbf{u} \subseteq \mathsf{Updates}(\Gamma)$. The set of all updateable objects over $\boldsymbol{\mathcal{V}}$ is denoted $\mathsf{UpdObj}(\boldsymbol{\mathcal{V}})$.

Call the updateable object $\langle \Gamma, \mathbf{u} \rangle$ *functional* if for any two $u_1, u_2 \in \mathbf{u}$, $u_1^{(1)} = u_2^{(1)}$ implies $u_1^{(2)} = u_2^{(2)}$. Thus, if $\langle \Gamma, \mathbf{u} \rangle$ is functional, there is at most one applicable update in $\mathbf{u}$ for each legal state of the associated view. The set of all functional updateable objects over $\boldsymbol{\mathcal{V}}$ is denoted $\mathsf{FUpdObj}(\boldsymbol{\mathcal{V}})$.

The *write view* of $\langle \Gamma, \mathbf{u} \rangle$ is the largest view $\Gamma^{(w)} \in \mathcal{V}$ with $\Gamma^{(w)} \preceq_{\mathbf{D}} \Gamma$ and the property that for some $u \in \mathbf{u}$, $\lambda\langle \Gamma, \Gamma^{(w)} \rangle(u^{(1)}) \neq \lambda\langle \Gamma, \Gamma^{(w)} \rangle(u^{(2)})$. The *read-only view* of $\langle \Gamma, \mathbf{u} \rangle$ is the largest view $\Gamma^{(r)} \in \mathcal{V}$ with $\Gamma^{(r)} \preceq_{\mathbf{D}} \Gamma$ and the property that for all $u \in \mathbf{u}$, $\lambda\langle \Gamma, \Gamma^{(r)} \rangle(u^{(1)}) = \lambda\langle \Gamma, \Gamma^{(r)} \rangle(u^{(2)})$. In other words, $\Gamma^{(r)}$ is the largest subview on which $\mathbf{u}$ is a partial identity. Clearly $\{\Gamma^{(r)}, \Gamma^{(w)}\}$ forms a decomposition of Γ into disjoint components; i.e., $\Gamma^{(r)} \wedge \Gamma^{(w)} = \bot$ and $\Gamma^{(r)} \vee \Gamma^{(w)} = \Gamma$.

For $\Gamma \in \mathcal{V}$ with $\Gamma' \preceq_{\mathbf{D}} \Gamma$, define the *projection* of $\langle \Gamma, \mathbf{u} \rangle$ onto Γ' to be the updateable object $\mathsf{Proj}_{\Gamma'}\langle\langle \Gamma, \mathbf{u} \rangle\rangle = \langle \Gamma', \lambda\langle \Gamma, \Gamma' \rangle(\mathbf{u}) \rangle$. In general, such a projection will not be functional, even if $\langle \Gamma, \mathbf{u} \rangle$ is. The projection $\mathsf{Proj}_{\Gamma^{(w)}}\langle\langle \Gamma, \mathbf{u} \rangle\rangle$ is called the *write object* of $\langle \Gamma, \mathbf{u} \rangle$.

The restriction of the write object of a functional updateable object to a single database $M \in \mathsf{LDB}(\mathbf{D})$ is, however, always functional, since there is at most one update in $\mathbf{u}$ which is applicable to M. Formally, define $\mathsf{Proj}_{\langle \Gamma' | M \rangle}\langle \Gamma, \mathbf{u} \rangle = \langle \Gamma', \lambda\langle \Gamma, \Gamma' \rangle(\mathbf{u}_{|M}) \rangle$.

Recall that $\mathbf{0_D}$ is the zero view on $\mathbf{D}$, and let $\mathbf{u}_{\mathbf{0_D}}$ denote the set containing just one element, the unique (identity) update $u_{\mathbf{0_D}}$ on the singleton set $\mathsf{LDB}(\mathbf{0_D})$. $\langle \mathbf{0_D}, \mathbf{u}_{\mathbf{0_D}} \rangle$, the unique updateable object of the zero view $\mathbf{0_D}$, is called the *zero object*. By itself, this updateable object is uninteresting, since no nontrivial updates are possible, but it will prove to be useful as a tool to assert succinctly that two views are disjoint (by asserting that their meet in the algebra $\boldsymbol{\mathcal{V}}$ of updateable views is the zero view).

Examples 4.3 (Read views and write views). The decomposition of the view of an updateable object into its write view and its read view is central, and deserves a closer look. To begin, consider again the schema $\mathbf{E}_0$ and the set $\mathbf{T}_0$ of transactions introduced in Sec. 1, with the corresponding algebra of updateable views described in Example 2.4. For the update family defined by τ_i; i.e., by $d_i \leftarrow d_{(i+1) \bmod n}$, the associated view is $\Omega_{d_i} \vee \Omega_{d_{(i+1) \bmod n}}$, with Ω_{d_i} the write view and $\Omega_{d_{(i+1) \bmod n}}$ the read-only view. The family of updates itself is $\mathbf{\upsilon}_{d_i} = \{\langle (n_1, n_2), (n_2, n_2) \rangle \mid n_1, n_2 \in \mathbb{Z}\}$, with a pair (n_1, n_2) representing the values for $(d_i, d_{(i+1) \bmod n})$. The updateable object is thus $\langle \Omega_{d_i} \vee \Omega_{d_{(i+1) \bmod n}}, \mathbf{\upsilon}_{d_i} \rangle$.

This simple introductory example does not cover all aspects of the framework. For a more comprehensive examination of the ideas, let $\mathbf{E}_2$ be a database schema which, for a fixed $q \in [3, \text{-}]$, includes three sets of data objects $\{x_1, x_2, \ldots, x_q\}$, $\{y_1, y_2, \ldots, y_q\}$, and $\{z_1, z_2, \ldots, z_q\}$, governed by the constraints $x_i + y_i \geq 500$ for each $i \in [1, q]$. Assume that each data object takes integer values, and in concordance with Definition 2.3, regard each x_i (resp. y_i) as a view Ω_{x_i} (resp. Ω_{y_i}).

Fix $i \in [1, q]$, and let $\mathbf{\upsilon}_{x_i}$ be the update set on Ω_{x_i} defined by $x_i \leftarrow x_i - 100$; more precisely, $\mathbf{\upsilon}_{x_i} = \{\langle n, n-100 \rangle \mid n \in \mathbb{Z}\}$. Define $\mathbf{\upsilon}_{y_i}$ on Ω_{y_i} similarly by $y_i \leftarrow y_{i-1} - 100$. The pairs $\langle \Omega_{x_i}, \mathbf{\upsilon}_{x_i} \rangle$ and $\langle \Omega_{y_i}, \mathbf{\upsilon}_{y_i} \rangle$ then form the associated updateable objects. In these simple cases, $\Omega_{x_i}^{(w)} = \Omega_{x_i}$ and $\Omega_{y_i}^{(w)} = \Omega_{y_i}$, with $\Omega_{x_i}^{(r)} = \Omega_{y_i}^{(r)} = \bot$.

Next, consider the update operation $x_i \leftarrow x_i - z_i$ on the view $\Omega_{x_i} \vee \Omega_{z_i}$. The set of associated updates is $\mathbf{\upsilon}_{x_i z_i} = \{\langle (n_1, n_2), (n_1 - n_2, n_2) \rangle \mid n_1, n_2 \in \mathbb{Z}\}$, with n_1 and n_2 representing the values of x_i and y_i, respectively. Here $(\Omega_{x_i} \vee \Omega_{z_1})^{(w)} = \Omega_{x_i}$ and $(\Omega_{x_i} \vee \Omega_{z_1})^{(r)} = \Omega_{z_i}$. The projection $\mathsf{Proj}_{\Omega_{x_i}} \langle\langle \Omega_{x_i} \vee \Omega_{z_i}, \mathbf{\upsilon}_{x_i z_i} \rangle\rangle = \mathbb{Z} \times \mathbb{Z}$, since any update is possible on x_i by choosing the appropriate value for z_i. More interesting is the projection based upon a particular state of the main schema. Let $M_{22} \in \mathsf{LDB}(\mathbf{E}_2)$ be any legal state with $x_i = 300$, $y_i = 300$, and $z_i = 100$. Then $\mathsf{Proj}_{\langle \Omega_{x_i} | M_{22} \rangle} \langle\langle \Omega_{x_i} \vee \Omega_{z_i}, \mathbf{\upsilon}_{x_i z_i} \rangle\rangle = \{\langle n, n-100 \rangle \mid n \in \mathbb{Z}\}$, which is exactly $\mathbf{\upsilon}_{x_i}$. The difference is in how they are realized. With $\langle \Omega_{x_i}, \mathbf{\upsilon}_{x_i} \rangle$, the parameter 100 is fixed in the update object itself, while in $\mathsf{Proj}_{\langle \Omega_{x_i} | M_{22} \rangle} \langle\langle \Omega_{x_i} \vee \Omega_{z_i}, \mathbf{\upsilon}_{x_i z_i} \rangle\rangle$, the parameter z_i is bound to 100 after reading the value of z_i from the state M_{22}. This distinction will prove to be critical, since the parameter z_i is used only internally, by the transaction, to determine which update it is to apply. From the point of view of constraint satisfaction, it does not matter how that parameter was obtained; only the update itself matters.

A similar construction applies for the update $y_i \leftarrow y_i - z_i$ on $\Omega_{y_i} \vee \Omega_{z_i}$.

Definition 4.4 (Lifting for an updateable data object). A central feature of updates performed by transactions is that they are localized. If u is an update to be performed on data object Γ, then the changes are made only to the state of the data object Γ; the states of atomic data objects not included in Γ remain fixed. The extension of a set $\mathbf{u}$ of updates to a larger environment, with the environment which is not part of Γ held constant, is called a *lifting*. Formally, let $\langle \Gamma, \mathbf{u} \rangle \in \mathsf{FUpdObj}(\mathcal{V})$, and let $\Gamma' \in \mathcal{V}$ with $\Gamma \wedge \Gamma' = \bot$. Define the *lifting* of $\langle \Gamma, \mathbf{u} \rangle$ to $\Gamma \vee \Gamma'$ (*with constant* Γ') as

$$\begin{aligned} \mathsf{Lift}_{\Gamma \vee \Gamma'} \langle\langle \Gamma, \mathbf{u} \rangle\rangle = \{ & (M, u^{(2)} \vee \lambda \langle \Gamma \vee \Gamma', \Gamma' \rangle (M)) \mid \\ & (M \in \mathsf{LDB}(\Gamma \vee \Gamma')) \wedge (u \in \mathbf{u}) \wedge \lambda \langle \Gamma \vee \Gamma', \Gamma \rangle (M) = u^{(1)} \} \end{aligned}$$

In parsing this definition, recall the notation shorthand for joining states of disjoint views introduced near the end of Definition 2.3.

Thus, $\mathsf{Lift}_{\Gamma\vee\Gamma'}\langle\langle\Gamma,\mathbf{u}\rangle\rangle$ is a relation on $\mathsf{LDB}(\Gamma\vee\Gamma')\times\mathsf{DB}(\Gamma\vee\Gamma')$. Since $\langle\Gamma,\mathbf{u}\rangle$ is functional, $\mathsf{Lift}_{\Gamma\vee\Gamma'}\langle\langle\Gamma,\mathbf{u}\rangle\rangle$ will be a function in the sense that for any $P\in\mathsf{LDB}(\Gamma\vee\Gamma')$, there is at most one $P'\in\mathsf{DB}(\Gamma\vee\Gamma')$ with $(P,P')\in\mathsf{Lift}_{\Gamma\vee\Gamma'}\langle\langle\Gamma,\mathbf{u}\rangle\rangle$. If such a P exists, it will be denoted $\mathsf{Lift}_{\Gamma\vee\Gamma'}\langle\langle\Gamma,\mathbf{u}\rangle\rangle(P)$. In other words, $\mathsf{Lift}_{\Gamma\vee\Gamma'}\langle\langle\Gamma,\mathbf{u}\rangle\rangle$ may be regard as a partial function on $\mathsf{LDB}(\Gamma\vee\Gamma')$.

Define $\mathsf{Compat}_{\Gamma\vee\Gamma'}\langle\langle\Gamma,\mathbf{u}\rangle\rangle$ to be the set of all legal states of $\Gamma\vee\Gamma'$ which are sent to legal states when lifted to $\Gamma\vee\Gamma'$ using $\langle\Gamma,\mathbf{u}\rangle$. Formally, $\mathsf{Compat}_{\Gamma\vee\Gamma'}\langle\langle\Gamma,\mathbf{u}\rangle\rangle = \{M\in\mathsf{LDB}(\Gamma\vee\Gamma') \mid \mathsf{Lift}_{\Gamma\vee\Gamma'}\langle\langle\Gamma,\mathbf{u}\rangle\rangle(M)\downarrow \in\mathsf{LDB}(\Gamma\vee\Gamma')\}$. This lifting is said to be *legal* (or *allowed*) for $P\in\mathsf{LDB}(\Gamma\vee\Gamma')$ if $P\in\mathsf{Compat}_{\Gamma\vee\Gamma'}\langle\langle\Gamma,\mathbf{u}\rangle\rangle$; otherwise it is *illegal* (or *disallowed*).

A special case occurs when $\Gamma'=\overline{\Gamma}$; i.e., for liftings to the entire main schema. Recalling that $\mathbf{1_D}$ is the identity view on $\mathbf{D}$, define $\mathsf{Lift}_{\mathbf{D}}\langle\langle\Gamma,\mathbf{u}\rangle\rangle$ to be $\mathsf{Lift}_{\mathbf{1_D}}\langle\langle\Gamma,\mathbf{u}\rangle\rangle$, and define $\mathsf{Compat}_{\mathbf{D}}\langle\langle\Gamma,\mathbf{u}\rangle\rangle$ to be $\mathsf{Compat}_{\mathbf{1_D}}\langle\langle\Gamma,\mathbf{u}\rangle\rangle$.

Liftings to $\mathbf{D}$ will be used to model the internal operation of database transactions, as described in Definition 4.6 below. The transaction must do something when the lifting is undefined or disallowed. The most useful solution is to have it perform the identity update; that is, to do nothing. Formally, for $\langle\Gamma,\mathbf{u}\rangle\in\mathsf{UpdObj}(\boldsymbol{\mathcal{V}})$ and $M\in\mathsf{LDB}(\mathbf{D})$, define

$$\begin{aligned}\mathsf{Lift}^{+}_{\Gamma\vee\Gamma'}\langle\langle\Gamma,\mathbf{u}\rangle\rangle &= (\mathsf{Lift}_{\Gamma\vee\Gamma'}\langle\langle\Gamma,\mathbf{u}\rangle\rangle\cap\mathsf{LDB}(\Gamma\vee\Gamma')\times\mathsf{LDB}(\Gamma\vee\Gamma'))\\ &\cup\{(M,M)\mid \mathsf{Lift}_{\Gamma\vee\Gamma'}\langle\langle\Gamma,\mathbf{u}\rangle\rangle(M)\downarrow\notin\mathsf{LDB}(\Gamma\vee\Gamma')\}\\ &\cup\{(M,M)\mid \mathsf{Lift}_{\Gamma\vee\Gamma'}\langle\langle\Gamma,\mathbf{u}\rangle\rangle(M)\uparrow\}\end{aligned}$$

and define $\mathsf{Lift}^{+}_{\mathbf{D}}\langle\langle\Gamma,\mathbf{u}\rangle\rangle$ to be $\mathsf{Lift}^{+}_{\mathbf{1_D}}\langle\langle\Gamma,\mathbf{u}\rangle\rangle$.

Examples 4.5 (Lifting). As a simple example of a lifting, return to the context of $\mathbf{E}_0$, as presented in Sec. 1, Example 2.4, and Examples 4.3. Let $i\in[0,n-1]$, let $i'=(i+1)\bmod n$, and let $J\subseteq[0,n-1]$ with Ω' denoting $\bigvee_{j\in J}\Omega_{d_j}$ and Ω'' denoting $\Omega_{d_i}\vee\Omega_{d_{i'}}$. Represent an $N\in\mathsf{DB}(\Omega''\vee\Omega')$ as a tuple indexed by $\{i,i'\}\cup J$ in which the element indexed by j is the value of d_j. Then $\mathsf{Lift}_{\Omega''\vee\Omega'}\langle\langle\Omega'',\mathbf{v}_{d_i}\rangle\rangle$ consists of those pairs of $(\{i,i'\}\cup J)$-indexed tuples (N,N') for which $\pi_i(N')=\pi_{i'}(N)$, and which agree on all indices other than i. In other words, $\mathbf{v}_{d_i}$ is extended to component views of the form Ω_{d_j} for $j\in J\setminus\{i,i'\}$ as the identity update.

To illustrate the interaction of lifting and constraints, in the context $\mathbf{E}_2$ of Examples 4.3, $\mathsf{Lift}_{\Omega_{x_i}\vee\Omega_{y_i}}\langle\langle\Omega_{x_i},\mathbf{v}_{x_i}\rangle\rangle = \{\langle(n_1,n_2),(n_1-100,n_2)\rangle \mid (n_1,n_2)\in\mathbb{Z}\times\mathbb{Z}\}$, with the tuple (n_1,n_2) representing the values (x_i,y_i). The lifting is allowed if $n_1+n_2-100\geq 500$, and disallowed otherwise.

Definition 4.6 (Black-box transactions and snapshot isolation). In a black-box model, the internal operations are hidden. Rather, just the interaction with the environment is modelled. A particularly simple version of a black-box model is appropriate for a theoretical study of snapshot isolation (SI). Indeed, under FCW, as described in Summary 3.1, the internal sequence of read and write operations of which a transaction is composed is not of interest. Rather, it is only the writes which are to be committed which are of relevance for modelling violations of integrity constraints. Thus, it is appropriate to regard a transaction under SI as a single update on an input database, taking that database as input at the beginning of the transaction and delivering an updated version at its end, a simplification which retains all necessary features for modelling conflicts.

More precisely, a *black-box transaction* T over $\mathcal{V}$ is represented by a pair $\langle \Gamma_T, \mathbf{u}_T \rangle \in \mathsf{FUpdObj}(\boldsymbol{\mathcal{V}})$. For an input state $M \in \mathsf{LDB}(\mathbf{D})$, the output state is $\mathsf{Lift}^{+}_{\mathbf{D}}\langle\langle \Gamma_T, \mathbf{u}_T \rangle\rangle(M)$. This defines a total operation on $\mathsf{LDB}(\mathbf{D})$; for any input state M, $\mathsf{Lift}^{+}_{\mathbf{D}}\langle\langle \Gamma_T, \mathbf{u}_T \rangle\rangle(M) \in \mathsf{LDB}(\mathbf{D})$ as well. The set of all black-box transactions over $\boldsymbol{\mathcal{V}}$ is denoted $\mathsf{BBTrans}_{\boldsymbol{\mathcal{V}}}$.

The notation $\langle \Gamma_T, u_T \rangle$ will be used throughout the rest of this paper to denote the update object which underlies the transaction T. No confusion should result, because transaction names will always take the form of T or τ, possibly with a prime and/or subscript. Thus, for example, the update object associated with T'_i is $\langle \Gamma_{T'_i}, \mathbf{u}_{T'_i} \rangle$. On the other hand, update objects not associated with a transaction will never use subscripts involving T or τ.

Definition 4.7 (Schedules of transactions under SI). The usual model of execution for a transaction T employs a start time $t_{\mathsf{Start}}\langle T \rangle$ and an end time $t_{\mathsf{End}}\langle T \rangle$. Concurrency properties are then defined in terms of these parameters. Specifically, two transactions T_1 and T_2 run *serially* if $t_{\mathsf{End}}\langle T_1 \rangle < t_{\mathsf{Start}}\langle T_2 \rangle$ or $t_{\mathsf{End}}\langle T_2 \rangle < t_{\mathsf{Start}}\langle T_1 \rangle$, and they run *concurrently* otherwise. In the theory presented here, the end time of a transaction is its commit time. As explained in Definition 4.4, a transaction which fails for some reason is modelled as executing the identity update.

The actual times do not matter; rather, it is only their ordering relative to each other which is of interest in terms of behavior. To this end, rather than working with explicit timestamps, an order-based representation will be employed. Let $\mathbf{T}$ be a finite subset of $\mathsf{BBTrans}_{\boldsymbol{\mathcal{V}}}$. Define $\mathsf{SCSet}\langle \mathbf{T} \rangle = \{T^s \mid T \in \mathbf{T}\} \cup \{T^c \mid T \in \mathbf{T}\}$, in which T^s and T^c represent the relative start and commit times of transaction T, respectively. A *SI-schedule* on $\mathbf{T}$ is given by a partial order $<_{\mathrm{T}}$ on $\mathsf{SCSet}\langle T \rangle$ with the property that for each $T \in \mathbf{T}$, $T^s < T^c$. It is important to understand that T^s and T^c are just symbols; the representation is only for the relative times; no numerical values are specified. In translating from a representation with explicit timestamps, $T_1^s <_{\mathrm{T}} T_2^s$ iff $t_{\mathsf{Start}}\langle T_1 \rangle < t_{\mathsf{Start}}\langle T_2 \rangle$, $T_1^c <_{\mathrm{T}} T_2^c$ iff $t_{\mathsf{End}}\langle T_1 \rangle < t_{\mathsf{End}}\langle T_2 \rangle$, $T_1^s <_{\mathrm{T}} T_2^c$ iff $t_{\mathsf{Start}}\langle T_1 \rangle < t_{\mathsf{End}}\langle T_2 \rangle$, and $T_1^c <_{\mathrm{T}} T_2^s$ iff $t_{\mathsf{End}}\langle T_1 \rangle < t_{\mathsf{Start}}\langle T_2 \rangle$.

For any $T \in \mathbf{T}$, $\mathsf{CSPred}_{<_{\mathrm{T}}}\langle T \rangle$ denotes the last transaction to commit before T starts, when it exists. Thus, $(\mathsf{CSPred}_{<_{\mathrm{T}}}\langle T \rangle)^c <_{\mathrm{T}} T^s$ and for no $T' \in \mathbf{T}$ is it the case that $(\mathsf{CSPred}_{<_{\mathrm{T}}}\langle T \rangle)^c <_{\mathrm{T}} T'^c <_{\mathrm{T}} T^c$.

Similarly, $\mathsf{CCPred}_{<_{\mathrm{T}}}\langle T \rangle$ denotes the last $T' \in \mathbf{T}$ which commits before T does, when it exists. Note that both $\mathsf{CSPred}_{<_{\mathrm{T}}}\langle - \rangle$ and $\mathsf{CCPred}_{<_{\mathrm{T}}}\langle - \rangle$ are partial functions, since some transactions will not have the required predecessors.

For $T_1, T_2 \in \mathbf{T}$, T_1 *serially precedes* T_2 if $T_1^c <_{\mathrm{T}} T_2^s$. If neither T_1 serially precedes T_2 nor T_2 serially precedes T_1, then T_1 and T_2 *execute concurrently* and and $\{T_1, T_2\}$ is said to form a *concurrent pair*.

A subset $\mathbf{S} \subseteq T$ is said to be *nonoverlapping* if for any $T_1, T_2 \in \mathbf{S}$, $\Gamma_1^{(w)} \wedge \Gamma_2^{(w)} = \bot$. In other words, the write views do not overlap. The schedule $<_{\mathbf{T}}$ is *nonoverlapping* if every concurrent pair $\{T_1, T_2\}$ is nonoverlapping. In the rest of this paper, schedules will always be taken to be nonoverlapping.

Definition 4.8 (Semantics of SI-schedules). In order to be able to model the interaction of transactions and to characterize constraint-preserving properties, it is necessary to have a formal model of the semantics of an SI-schedule; that is, to have a way of representing the overall behavior of the execution of a schedule of transactions, given the semantics of each individual transaction as described in Definition 4.6.

Let $\mathbf{T}$ be a finite subset of $\mathsf{BBTrans}_{\mathcal{V}}$ and let $<_{\mathbf{T}}$ be an SI-schedule for $\mathbf{T}$. For the execution of $<_{\mathbf{T}}$, three states in $\mathsf{LDB}(\mathbf{D})$ are defined for each transaction $T \in \mathbf{T}^+$ and each initial state $M \in \mathsf{LDB}(\mathbf{D})$ for the entire schedule:

$\mathsf{InitSnap}_{\langle <_{\mathbf{T}}:M\rangle}\langle T\rangle$: The initial state which transaction T reads at the beginning of its execution. In other words, it is the initial snapshot of T.

$\mathsf{BeforeCmt}_{\langle <_{\mathbf{T}}:M\rangle}\langle T\rangle$: The state of the database immediately before T commits.

$\mathsf{AfterCmt}_{\langle <_{\mathbf{T}}:M\rangle}\langle T\rangle$: The state of the database immediately after T commits.

For each initial state $M \in \mathsf{LDB}(\mathbf{D})$, The semantics are defined in a formal way as follows:

$$\mathsf{InitSnap}_{\langle <_{\mathbf{T}}:M\rangle}\langle T\rangle = \begin{cases} \mathsf{AfterCmt}_{\langle <_{\mathbf{T}}:M\rangle}\langle \mathsf{CSPred}_{<_{\mathbf{T}}}\langle T\rangle\rangle & \text{if } \mathsf{CSPred}_{<_{\mathbf{T}}}\langle T\rangle\downarrow \\ M & \text{otherwise} \end{cases}$$

$$\mathsf{BeforeCmt}_{\langle <_{\mathbf{T}}:M\rangle}\langle T\rangle = \begin{cases} \mathsf{AfterCmt}_{\langle <_{\mathbf{T}}:M\rangle}\langle \mathsf{CCPred}_{<_{\mathbf{T}}}\langle T\rangle\rangle & \text{if } \mathsf{CSPred}_{<_{\mathbf{T}}}\langle T\rangle\downarrow \\ M & \text{otherwise} \end{cases}$$

$$\mathsf{AfterCmt}_{\langle <_{\mathbf{T}}:M\rangle}\langle T\rangle = \mathsf{Lift}_{\mathbf{D}}\langle \mathsf{Proj}_{\langle \Gamma^{(w)} | \mathsf{InitSnap}_{\langle <_{\mathbf{T}}:M\rangle}\langle T\rangle\rangle}\langle\langle \Gamma, \mathbf{u}\rangle\rangle\rangle(\mathsf{BeforeCmt}_{\langle <_{\mathbf{T}}:M\rangle}\langle T\rangle)$$

Less formally, for an initial state M, $\mathsf{InitSnap}_{\langle <_{\mathbf{T}}:M\rangle}\langle T\rangle$ is the state of the global database just after the last commit operation which occurs before T starts, or the initial state M in the case that no such commit operation has occurred. $\mathsf{BeforeCmt}_{\langle <_{\mathbf{T}}:M\rangle}\langle T\rangle$ is the state of the global database just after the last commit operation which occurs before the commit operation of T. Finally, $\mathsf{AfterCmt}_{\langle <_{\mathbf{T}}:M\rangle}\langle T\rangle$ is the result of lifting, to $\mathsf{BeforeCmt}_{\langle <_{\mathbf{T}}:M\rangle}\langle T\rangle$, the projection of the update operation of T onto its write view.

It is important to note that it is only the update to the write view $\Gamma_T^{(w)}$, and not the entire update $\mathbf{u}_T$ to Γ_T, which is lifted upon commit. This is critical because the read view $\Gamma_T^{(r)}$ may have been updated by another concurrent transaction. For example, in the context of Examples 4.3, let $\tau_{x_i z_i}$ be the transaction whose update object is $\langle \Omega_{x_1} \vee \Omega_{z_i}, \mathbf{\upsilon}_{x_1 z_i}\rangle$. It is quite possible that another, concurrent transaction could write z_i after $\tau_{x_i z_i}$ begins but before it commits. In that case, lifting the entire update $\mathbf{\upsilon}_{x_i z_i}$ would not produce the correct result, since the transaction $\tau_{x_i z_i}$ does not change the value z_i, and should not restore its value to that when the transaction began. The correct approach, as defined above, is to lift only the projection of $\mathbf{v}_{x_i z_i}$ onto its write view, for the database state which $\tau_{x_i z_i}$ acquired for its snapshot. This uses the value of z_i at the beginning of $\tau_{x_i z_i}$, as required, to compute the changes on the write view defined by the update, and then commits only those changes. The value of z_i should not be written as part of the final update of $\tau_{x_i z_i}$.

Returning to the general context, call $<_{\mathbf{T}}$ *constraint preserving* if for every $M \in \mathsf{LDB}(\mathbf{D})$ and every $T \in \mathbf{T}$, $\mathsf{AfterCmt}_{\langle <_{\mathbf{T}} : M\rangle}\langle T\rangle \in \mathsf{LDB}(\mathbf{D})$. The goal is to show how to ensure that $<_{\mathbf{T}}$ has this property.

Definition 4.9 (Write-commuting pairs). The key abstract property to be used in guaranteeing constraint-preserving schedules is write commutativity, Roughly, two transactions T_1 and T_2 form a write-commuting pair if whenever they each may be executed concurrently on a given database state, then they may be executed serially as well, in either order. However, it is only the updates on their respective write views, and not the entire update of each transactions, which must obey this commutativity constraint. Thus, each transaction may overwrite the read-only view of the other with the two still forming a write-commuting pair.

Formally, call $\{T_1, T_2\} \subseteq \mathsf{BBTrans}_{\mathcal{V}}$ a *write-commuting pair* if it is nonoverlapping and for any $M \in \mathsf{LDB}(\mathbf{D})$ and any $u_1 \in \mathsf{Proj}_{\langle \Gamma_1^{(w)}|M\rangle}\langle\langle \Gamma_1, \mathbf{u}_1\rangle\rangle$ and $u_2 \in \mathsf{Proj}_{\langle \Gamma_2^{(w)}|M\rangle}\langle\langle \Gamma_2, \mathbf{u}_2\rangle\rangle$, with both $\mathsf{Lift}_{\mathbf{D}}\langle\langle \Gamma_{T_1}^{(w)}, \{u_1\}\rangle\rangle(M)\!\downarrow\, \in \mathsf{LDB}(\mathbf{D})$ and $\mathsf{Lift}_{\mathbf{D}}\langle\langle \Gamma_{T_2}^{(w)}, \{u_2\}\rangle\rangle(M)\!\downarrow\, \in \mathsf{LDB}(\mathbf{D})$, it is the case that both $\mathsf{Lift}_{\mathbf{D}}\langle\langle \Gamma_{T_1}^{(w)}, \{u_1\}\rangle\rangle \circ \mathsf{Lift}_{\mathbf{D}}\langle\langle \Gamma_{T_2}^{(w)}, \{u_2\}\rangle\rangle(M)\!\downarrow\in \mathsf{LDB}(\mathbf{D})$ and $\mathsf{Lift}_{\mathbf{D}}\langle\langle \Gamma_{T_2}^{(w)}, \{u_2\}\rangle\rangle \circ \mathsf{Lift}_{\mathbf{D}}\langle\langle \Gamma_{T_1}^{(w)}, \{u_1\}\rangle\rangle(M)\!\downarrow\in \mathsf{LDB}(\mathbf{D})$.

A large class of write-commuting pairs will be identified in Proposition 4.19 below. For now, the key property to observe, which justifies the name, is that such pairs produce the same result in either order of composition, with the consequence (Theorem 4.11) that SI schedules are constraint preserving.

Observation 4.10 (Write-commuting pairs produce the same result in either order). *If $\{T_1, T_2\}$ is a write-commuting pair, $M \in \mathsf{LDB}(\mathbf{D})$, and $u_1 \in \mathsf{Proj}_{\langle \Gamma_1^{(w)}|M\rangle}\langle\langle \Gamma_1, \mathbf{u}_1\rangle\rangle$, $u_2 \in \mathsf{Proj}_{\langle \Gamma_2^{(w)}|M\rangle}\langle\langle \Gamma_2, \mathbf{u}_2\rangle\rangle$ with both of the liftings $\mathsf{Lift}_{\mathbf{D}}\langle\langle \Gamma_{T_1}^{(w)}, \{u_1\}\rangle\rangle(M)\!\downarrow\in \mathsf{LDB}(\mathbf{D})$ and $\mathsf{Lift}_{\mathbf{D}}\langle\langle \Gamma_{T_2}^{(w)}, \{u_2\}\rangle\rangle(M)\!\downarrow\in \mathsf{LDB}(\mathbf{D})$, then*

$$\mathsf{Lift}_{\mathbf{D}}\langle\langle \Gamma_{T_1}^{(w)}, \{u_1\}\rangle\rangle \circ \mathsf{Lift}_{\mathbf{D}}\langle\langle \Gamma_{T_2}^{(w)}, \{u_2\}\rangle\rangle(M) = \mathsf{Lift}_{\mathbf{D}}\langle\langle \Gamma_{T_2}^{(w)}, \{u_2\}\rangle\rangle \circ \mathsf{Lift}_{\mathbf{D}}\langle\langle \Gamma_{T_1}^{(w)}, \{u_1\}\rangle\rangle(M).$$

Proof. This is immediate, since the updates are nonoverlapping. As long as both are defined, they must be the same. □

Theorem 4.11 (Write-commuting concurrent pairs guarantee constraint-preserving SI-schedules). *Let $\mathbf{T}$ be a finite subset of $\mathsf{BBTrans}_{\mathcal{V}}$, and let $<_{\mathbf{T}}$ be an SI-schedule for $\mathbf{T}$. If every concurrent pair of $<_{\mathbf{T}}$ is write commuting, then $<_{\mathbf{T}}$ is constraint preserving.*

Proof. The proof is by induction on the size of $\mathbf{T}$. For zero or one transaction, the result is immediate. For the inductive step, let $n \in \mathbb{N}$ and assume that the result is true whenever $\mathsf{Card}(\mathbf{T}) \leq n$. Then let $\mathsf{Card}(\mathbf{T}) = n + 1$ (with $n \geq 1$), and let T_n and T_{n+1} be the nth and $n + 1^{\text{st}}$ transactions to commit in $<_{\mathbf{T}}$, respectively (that is, the penultimate and last transactions to commit). If $n \geq 2$, i.e., if $n + 1 \geq 3$, let T_{n-1} be the transaction which commits just before T_n. Let $M \in \mathsf{LDB}(\mathbf{D})$ be the initial state for the schedule.

If T_{n+1} starts after T_n has committed; that is, the two transactions are not concurrent, then the result is immediate; there cannot be any constraint violation with serial transactions which operate correctly in isolation. So, assume that $\{T_n, T_{n+1}\}$ forms a concurrent pair. Let u_n and u_{n+1} be the updates which T_n and T_{n+1} perform on $\Gamma_n^{(w)}$ and $\Gamma_{n+1}^{(w)}$, respectively, and let $\mathbf{S}_{n-1} = \mathbf{T} \setminus \{T_n, T_{n+1}\}$, $\mathbf{S}_n = \mathbf{T} \setminus \{T_n\}$, and $\mathbf{S}_{n+1} = \mathbf{T} \setminus \{T_{n+1}\}$, with $<_{\mathbf{S}_{n-1}}$, $<_{\mathbf{S}_n}$, and $<_{\mathbf{S}_{n+1}}$ be the schedules obtained by restricting $<_{\mathbf{T}}$ to the transactions in $\mathbf{S}_{n-1}$, $\mathbf{S}_n$, and $\mathbf{S}_{n+1}$, respectively. Then by the inductive hypothesis, each of $<_{\mathbf{S}_{n-1}}$, $<_{\mathbf{S}_n}$, and $<_{\mathbf{S}_{n+1}}$ is constraint preserving. If at least one of $\mathsf{Lift}^{+}_{\mathbf{D}}\langle\langle \Gamma_{T_n}^{(w)}, \{u_n\}\rangle\rangle$ and $\mathsf{Lift}^{+}_{\mathbf{D}}\langle\langle \Gamma_{T_{n+1}}^{(w)}, \{u_{n+1}\}\rangle\rangle$ is the identity update (for example, if one of the liftings was not defined or did not result in a legal state), then the corresponding transaction may be removed from $<_{\mathbf{T}}$ to obtain one of $<_{\mathbf{S}_n}$ or $<_{\mathbf{S}_{n+1}}$, without any change in the semantics, since an identity transaction has no effect. In that case, the result follows from the inductive hypothesis. So, assume that both $\mathsf{Lift}_{\mathbf{D}}\langle\langle \Gamma_{T_n}^{(w)}, \{u_n\}\rangle\rangle(N)\downarrow$ and $\mathsf{Lift}_{\mathbf{D}}\langle\langle \Gamma_{T_{n+1}}^{(w)}, \{u_{n+1}\}\rangle\rangle(N)\downarrow$, with $N = \mathsf{AfterCmt}_{\langle <_{\mathbf{T}}:M\rangle}\langle T_{n-1}\rangle$ if $n \geq 2$ and $N = M$ if $n = 1$. Then $(\mathsf{Lift}_{\mathbf{D}}\langle\langle \Gamma_{T_n}^{(w)}, \{u_n\}\rangle\rangle \circ \mathsf{Lift}_{\mathbf{D}}\langle\langle \Gamma_{T_{n+1}}^{(w)}, \{u_{n+1}\}\rangle\rangle)(N)\downarrow \in \mathsf{LDB}(\mathbf{D})$, since by assumption $\{T_1, T_2\}$ forms a write commuting pair. However, it is easy to see that in that case the semantics of re-inserting T_n and T_{n+1} into $<_{\mathbf{S}}$ is just to perform that composed update, which establishes that $\mathsf{AfterCmt}_{\langle <_{\mathbf{T}}:M\rangle}\langle T_{n+1}\rangle \in \mathsf{LDB}(\mathbf{D})$, as required. □

Definition 4.12 (Guard views and guarded black-box transactions). The property of write commutativity is an abstract one. A useful, concrete class of transactions with that property may be obtained via the notion of a guard for an updateable object $\langle \Gamma, \mathbf{u}\rangle$. Such a guard is a view Γ' with the property that if, for a given $M \in \mathsf{LDB}(\mathbf{D})$ and any $u \in \mathbf{u}$, the projection $\lambda\langle \Gamma, \Gamma^{(w)}\rangle(u)$ of u onto $\Gamma^{(w)}$ restricted to M may be lifted to $\mathbf{D}$ iff it may be lifted to $\Gamma^{(w)} \vee \Gamma'$. Thus, a guard view reduces the global test for lifting to all of $\mathbf{D}$ to the much more local test of lifting to just the write view and its guard. Formally, given $\langle \Gamma, \mathbf{u}\rangle \in \mathsf{FUpdObj}(\mathcal{V})$, $\Gamma' \in \mathcal{V}$ is a *guard view for* $\langle \Gamma, \mathbf{u}\rangle$ if $\Gamma^{(w)} \wedge \Gamma' = \bot$ and for every $M \in \mathsf{LDB}(\mathbf{D})$, $M \in \mathsf{Compat}_{\mathbf{D}}\langle\langle \Gamma, \mathbf{u}\rangle\rangle$ iff $(\gamma^{(w)} \vee \gamma')(M) \in \mathsf{Compat}_{\Gamma^{(w)} \vee \Gamma'}\langle \mathsf{Proj}_{\langle \Gamma^{(w)} \vee \Gamma' | M\rangle}\langle\langle \Gamma, \mathbf{u}\rangle\rangle\rangle$. The set of all guard views for $\langle \Gamma, \mathbf{u}\rangle$ is denoted $\mathsf{Guards}_{\mathcal{V}}\langle \Gamma, \mathbf{u}\rangle$.

A *guarded black-box transaction* is represented by a pair $\langle\langle \Gamma, \mathbf{u}\rangle, \Gamma'\rangle$ in which $\langle \Gamma, \mathbf{u}\rangle \in \mathsf{FUpdObj}(\mathcal{V})$ and $\Gamma' \in \mathcal{V}$ is a guard for $\langle \Gamma, \mathbf{u}\rangle$. It is convenient to have a notation for guarded black-box transactions which extends that of Definition 4.6. To that end, if T is such a transaction, then its guard will be denoted $\Gamma_T^{(g)}$. Thus, T is represented by $\langle\langle \Gamma_T, \mathbf{u}_T\rangle, \Gamma_T^{(g)}\rangle$. The set of all guarded black-box transactions over $\mathcal{V}$ is denoted $\mathsf{GBBTrans}_{\mathcal{V}}$.

Examples 4.13 (Guards). Returning to the context $\mathbf{E}_2$ of Examples 4.3, a guard for $\langle \Omega_{x_i}, \mathbf{v}_{x_i} \rangle$ is Ω_{y_i}. Indeed, to verify that an update to x_i is legal, only the value of y_i need be checked; the state of the rest of the database is irrelevant. Similarly, a guard for $\langle \Omega_{y_i}, \mathbf{v}_{y_i} \rangle$ is Ω_{x_i}. It is only the write view of an update, and not its read-only view, which affects the definition of a guard. Thus, a guard of $\langle \Omega_{x_i} \vee \Omega_{z_i}, \mathbf{v}_{x_i z_i} \rangle$ is Ω_{y_i}, and a guard of $\langle \Omega_{y_i} \vee \Omega_{z_i}, \mathbf{v}_{y_i z_i} \rangle$ is Ω_{x_i}.

The guard view need not be disjoint from the read-only view. For example, given the update set $\mathbf{v}_{x_i y_i}$ on $\Omega_{x_i} \vee \Omega_{y_i}$ defined by the update rule $x_i \leftarrow x_i - y_i$, the view Ω_{y_i} is a guard for $\langle \Omega_{x_i} \vee \Omega_{y_i}, \mathbf{v}_{x_i y_i} \rangle$. However, $\Omega^{(r)}_{x_i y_i} = \Omega_{y_i}$ as well, so the guard and the read-only view are the same in this case.

Observation 4.14 (Guards always exist). *Given $\langle \Gamma, \mathbf{u} \rangle \in \mathsf{UpdObj}(\mathcal{V})$, the complement $\overline{\Gamma^{(w)}}$ of the associated write view is always a guard view for Γ. Thus, every updateable object has a guard.* □

Definition 4.15 (Minimal and least guards). Let $\langle \Gamma, \mathbf{u} \rangle \in \mathsf{FUpdObj}(\mathcal{V})$. Then $\Gamma' \in \mathsf{Guards}_{\mathcal{V}}\langle \Gamma, \mathbf{u} \rangle$ is is a *minimal guard view* for $\langle \Gamma, \mathbf{u} \rangle$ if for any guard Γ'' for $\langle \Gamma, \mathbf{u} \rangle$, if $\Gamma'' \preceq_{\mathrm{D}} \Gamma'$, then $\Gamma'' = \Gamma'$. A unique minimal guard view is *least*.

It is always desirable to choose a minimal guard, because it will be the independence of the guard view of one transaction from the write view of another which will prove to be the critical property in characterizing schedules which are constraint preserving.

Example 4.16 (Least guards need not exist). While a minimal guard may always be chosen, it is not the case that least guards always exist. For example, if the constraint $y_i = z_i$ is added to the schema $\mathbf{E}_2$ of Examples 4.3, then both Ω_{y_i} and Ω_{z_i} are guards for $\langle \Omega_{x_i}, \mathbf{v}_{x_i} \rangle$. Of course, these are effectively the same. In general, it can be shown that the choice of a minimal guard does not matter. Roughly, the reason is that if a transaction T_1 writes a guard of another transaction T_2 in way which affects which updates T_2 may perform legally, then it must write all guards of T_2. Otherwise, some guards would allow updates which others would not, which is not consistent with the definition of guard.

Definition 4.17 (Independent and conflicting pairs of guarded transactions). Two transactions are guard independent if at least one does not write the guard of the other. In other words, they do not each have a read-write dependency on the other, with respect to reading the guard view only. More formally, the two element set $\{T_1, T_2\} \subseteq \mathsf{GBBTrans}_{\mathcal{V}}$ forms a *guard-independent* pair if $\{T_1, T_2\}$ is nonoverlapping and at least one of $\Gamma^{(w)}_{T_2} \wedge \Gamma^{(g)}_{T_1} = \bot$ and $\Gamma^{(w)}_{T_1} \wedge \Gamma^{(g)}_{T_2} = \bot$ holds. A pair $\{T_1, T_2\}$ which is not guard independent is *guard-conflicting*, and T_1 and T_2 are then said to be *in guard conflict* with each other.

Examples 4.18 (Independent and conflicting pairs). Continuing with Examples 4.13, $\{\tau_{x_i}, \tau_{y_i}\}$ form a guard-conflicting pair, since each reads the guard of the other. On the other hand, for distinct i and j, τ_{x_i} and τ_{x_j} are guard independent, since neither writes the guard of the other. To illustrate the interesting middle ground, define $\tau_{x'_i}$ to be the transaction on Ω_{x_i} which implements the update rule $x_i \leftarrow x_i + 50$. Then $\{\tau_{x'_i}, \tau_{y_i}\}$ forms a guard independent pair, since the least guard of $\tau_{x'_i}$ is $\bot$; i.e., its update is always legal. As will be shown next, guard independence implies write commutativity, and this example illustrates the intuition behind this. Even though $\tau_{x'_i}$ writes the guard of τ_{y_i}, which is Ω_{x_i} itself, it does so in a "harmless" way. Because $\tau_{x'_i}$ does not read y_i, it cannot make any updates whose legality depends upon the value of y_i.

Proposition 4.19 (Guard independence $\Rightarrow$ write commutativity). *Every $\{T_1, T_2\} \in \mathsf{GBBTrans}_{\mathcal{V}}$ which is guard independent forms a write-commuting pair.*

Proof. Let $\{T_1, T_2\} \subseteq \mathsf{GBBTrans}_{\mathcal{V}}$. Without loss of generality, assume that $\Gamma_{T_2}^{(w)} \wedge \Gamma_{T_1}^{(g)} = \bot$. Let $M \in \mathsf{LDB}(\mathbf{D})$ and let any $u_1 \in \mathsf{Proj}_{\langle \Gamma_1^{(w)} | M \rangle}\langle\langle \Gamma_1, \mathbf{u}_1 \rangle\rangle$, $u_2 \in \mathsf{Proj}_{\langle \Gamma_2^{(w)} | M \rangle}\langle\langle \Gamma_2, \mathbf{u}_2 \rangle\rangle$, with both $\mathsf{Lift}_{\mathbf{D}}\langle\langle \Gamma_{T_1}^{(w)}, \{u_1\}\rangle\rangle(M)\downarrow \in \mathsf{LDB}(\mathbf{D})$ and $\mathsf{Lift}_{\mathbf{D}}\langle\langle \Gamma_{T_2}^{(w)}, \{u_2\}\rangle\rangle(M)\downarrow \in \mathsf{LDB}(\mathbf{D})$. It is immediate that $\mathsf{Lift}_{\mathbf{D}}\langle\langle \Gamma_{T_2}^{(w)}, \{u_2\}\rangle\rangle \circ \mathsf{Lift}_{\mathbf{D}}\langle\langle \Gamma_{T_1}^{(w)}, \{u_1\}\rangle\rangle(M)\downarrow \in \mathsf{LDB}(\mathbf{D})$, since $\{T_1, T_2\}$ is nonoverlapping and T_2 does not write the guard of T_1, so the update which T_2 performs does not affect the legality of the update which T_1 performs.

For the opposite direction, first note that it is the case that $\mathsf{Lift}_{\mathbf{D}}\langle\langle \Gamma_{T_1}^{(w)}, \{u_1\}\rangle\rangle \circ \mathsf{Lift}_{\mathbf{D}}\langle\langle \Gamma_{T_2}^{(w)}, \{u_2\}\rangle\rangle(M)\downarrow \in \mathsf{DB}(\mathbf{D})$ since the write views are nonoverlapping; i.e., $\Gamma_1^{(w)} \wedge \Gamma_2^{(w)} = \bot$. The only question is whether the result is in $\mathsf{LDB}(\mathbf{D})$. However, again since $\Gamma_1^{(w)} \wedge \Gamma_2^{(w)} = \bot$, the two compositions must be identical; i.e., $\mathsf{Lift}_{\mathbf{D}}\langle\langle \Gamma_{T_1}^{(w)}, \{u_1\}\rangle\rangle \circ \mathsf{Lift}_{\mathbf{D}}\langle\langle \Gamma_{T_2}^{(w)}, \{u_2\}\rangle\rangle(M) = \mathsf{Lift}_{\mathbf{D}}\langle\langle \Gamma_{T_2}^{(w)}, \{u_2\}\rangle\rangle \circ \mathsf{Lift}_{\mathbf{D}}\langle\langle \Gamma_{T_1}^{(w)}, \{u_1\}\rangle\rangle(M)$. This shows that the composition $\mathsf{Lift}_{\mathbf{D}}\langle\langle \Gamma_{T_1}^{(w)}, \{u_1\}\rangle\rangle \circ \mathsf{Lift}_{\mathbf{D}}\langle\langle \Gamma_{T_2}^{(w)}, \{u_2\}\rangle\rangle(M) \in \mathsf{LDB}(\mathbf{D})$, as required. Hence $\{T_1, T_2\}$ forms a write commuting pair. □

The main theorem of this paper may now be established.

Theorem 4.20 (Guard independence guarantees constraint preservation). *Let $\mathbf{T}$ be a finite subset of $\mathsf{GBBTrans}_{\mathcal{V}}$, and let $<_{\mathbf{T}}$ be an SI-schedule for $\mathbf{T}$. If every concurrent pair of $<_{\mathbf{T}}$ is guard independent, then $<_{\mathbf{T}}$ is constraint preserving.*

Proof. The proof follows immediately from Theorem 4.11 and Proposition 4.19. □

Discussion 4.21 (Constraint-Preserving Snapshot Isolation (CPSI)). In an SI-schedule $<_{\mathbf{T}}$, there is an *rw-dependency* from T_1 to T_2, written $T_1 \xrightarrow{\mathsf{rw}} T_2$, if T_2 writes the read set of T_1. Within the context of the formalism of this paper, this translates to $\Gamma_{T_1}^{(w)} \wedge (\Gamma_{T_2}^{(r)} \vee \Gamma_{T_2}^{(g)}) \neq \bot$, since to operate correctly, a transaction T must read both its update view Γ_T (in order to know which update to execute) and its guard view $\Gamma_T^{(g)}$ (in order to determine whether that update is legal). In the implementation of SSI, as described in [5], the critical notion is the *dangerous structure*, which consists of two consecutive read-write dependencies of concurrent pairs in the conflict graph; that is, two dependencies of the form $T_1 \xrightarrow{\mathsf{rw}} T_2$ and $T_2 \xrightarrow{\mathsf{rw}} T_3$ with $\{T_1, T_2\}$ and $\{T_2, T_3\}$ concurrent pairs. The absence of such a pair of dependencies is sufficient, but not necessary, for an SI-schedule to be serializable. Necessity requires that it be part of a cycle in the conflict graph. Working with this same model, the results of this paper show that for such a dangerous structure to lead to a constraint violation, it must be the case that $T_1 = T_3$. Thus, a much simpler test suffices if only constraint violation is to be flagged.

An approach with fewer false positives may be obtained by working only with guard reads. More precisely, say that there is a *gw-dependency* from T_1 to T_2 if T_2 writes the guard of T_1; i.e., $\Gamma_2^{(w)} \wedge \Gamma_{T_1}^{(g)} \neq \bot$, and write $T_1 \xrightarrow{\text{gw}} T_2$ to denote this. Call $\{T_1, T_2\}$ a *dangerous gw-pair* if it forms a concurrent pair for which both $T_1 \xrightarrow{\text{gw}} T_2$ and $T_2 \xrightarrow{\text{gw}} T_1$ hold. Theorem 4.20 guarantees that an SI-schedule will be constraint preserving in the absence of such pairs. This strategy, called *constraint-preserving snapshot isolation (CPSI)*, has false positives only to the extent that two transactions could each write the guard of the other without causing a constraint violation. This is of course possible, but the general assumption in transaction management is that the manager only knows which objects are read and written, not how they are written or how the writes are used. With that understanding, there would be no false positives, since there is always some update to the guard which would cause a constraint violation.

A correct implementation would of course require that the system be able to identify which reads of a transaction are to the guard. This could be done, for example, in a context of fixed transactions for business processes by having such guards known to the transaction manager.

Example 4.22 (Dangerous structures which do not result in constraint violations). To illustrate the ways that the multiversion conflict graph may contain dangerous structures yet be free of dangerous gw-pairs, return to the context of $\mathbf{E}_2$, as described in Examples 4.3, and consider three concurrent transactions: τ_1 operates on $\Omega_{x_1} \vee \Omega_{x_2}$ via the rule $x_1 \leftarrow x_1 - x_2$, τ_2 operates on $\Omega_{x_1} \vee \Omega_{x_2}$ via the rule $x_2 \leftarrow x_2 - x_1$, and τ_3 operates on $\Omega_{x_2} \vee \Omega_{y_1}$ via the rule $y_1 \leftarrow y_1 + |x_2|$. It is assumed that each transaction performs the given update if it would not result in a constraint violation (when run in isolation), and performs the identity update otherwise. The multiversion conflict graph for these three transactions is shown in Fig. 3. Note in particular that although τ_3 reads x_2, it uses only its absolute value in the computation of the new value for y_1, and so Ω_{x_2} is not in its guard.

Observe that this graph contains two cycles. The first, between τ_1 and τ_2, involves only rw-edges. The second, between τ_2 and τ_3, involves one gw-edge and one rw-edge. Although both of these cycles define dangerous structures, as do the sequences $\tau_1 \xrightarrow{\text{rw}} \tau_2 \xrightarrow{\text{gw}} \tau_3$ and $\tau_3 \xrightarrow{\text{rw}} \tau_2 \xrightarrow{\text{rw}} \tau_1$, none represents a dangerous gw-pair. Since a constraint violation can occur only if there is a (two-vertex) cycle consisting of gw-edges, no constraint violation is possible when running under SI, provided that each transaction individually respects all constraints. even though the result need not be serializable. Thus, while SSI and even PSSI would force at least one of these transactions to terminate, with CPSI all may run to completion.

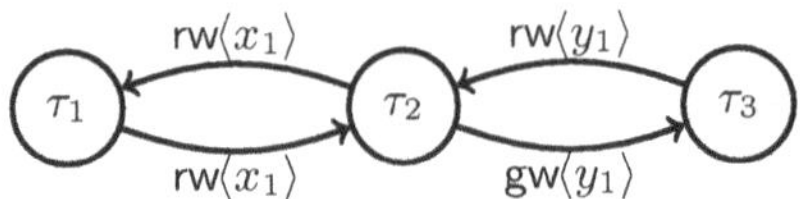

Fig. 3. An SI conflict graph with dangerous structures but no dangerous gw-pairs

5 Conclusions and Further Directions

A method for identifying conflicts leading to violations of integrity constraints in transactions whose concurrency is governed by snapshot isolation has been presented. In contrast to methods for ensuring full serializability, the method of identification involves only pairs of transactions, and may be tested fully. without concern for false positives. It promises to have application in settings in which aborting and or delaying the execution of transactions is not a viable option.

There are several key areas for further work on this subject.

STRATEGIES FOR REVISING TRANSACTIONS: The motivation for this work arose from earlier studies on cooperative updates [13,11]. The focus there is particularly upon interactive, long-running business processes in which abort and restart for transactions is not an option. Rather, the best strategy in such settings would seem to be to identify methods for cooperative revision of updates in the case of conflict. The current work is constitutes a substantial step in that direction, in that the conflicts which are considered are between pairs of transactions, rather than large sets. The goal of exploiting the current work in that context is a subject for further study.

INTEGRATION WITH WORK ON INDEPENDENCE AND OVERLAP: In [10], the foundations for a theory of structured data objects for transactions is developed. These structured objects have both writeable parts and read-only parts, with the read-only parts allowed to overlap, even for writeable objects. As that work was also motivated by work on cooperative updates, an integration of those results with the ideas of this paper would likely prove a fruitful area for study.

References

1. Adya, A., Liskov, B., O'Neil, P.E.: Generalized isolation level definitions. In: Lomet, D.B., Weikum, G. (eds.) Proceedings of the 16th International Conference on Data Engineering, San Diego, California, USA, February 28-March 3, pp. 67–78 (2000)
2. Berenson, H., Bernstein, P.A., Gray, J., Melton, J., O'Neil, E.J., O'Neil, P.E.: A critique of ANSI SQL isolation levels. In: Proceedings of the 1995 ACM SIGMOD International Conference on Management of Data, San Jose, California, May 22-25, pp. 1–10 (1995)
3. Bernstein, P., Newcomer, E.: Principles of Transaction Processing, 2nd edn. Morgan Kaufmann (2009)

4. Breitbart, Y., Georgakopoulos, D., Rusinkiewicz, M., Silberschatz, A.: On rigorous transaction scheduling. IEEE Trans. Software Eng. 17(9), 954–960 (1991)
5. Cahill, M.J., Röhm, U., Fekete, A.D.: Serializable isolation for snapshot databases. ACM Trans. Database Syst. 34(4) (2009)
6. Date, C.J.: A Guide to the SQL Standard. Addison-Wesley (1997); (with Hugh Darwen)
7. Davey, B.A., Priestly, H.A.: Introduction to Lattices and Order, 2nd edn. Cambridge University Press (2002)
8. Fekete, A., Liarokapis, D., O'Neil, E.J., O'Neil, P.E., Shasha, D.: Making snapshot isolation serializable. ACM Trans. Database Syst. 30(2), 492–528 (2005)
9. Hegner, S.J.: An order-based theory of updates for closed database views. Ann. Math. Art. Intell. 40, 63–125 (2004)
10. Hegner, S.J.: A model of independence and overlap for transactions on database schemata. In: Catania, B., Ivanović, M., Thalheim, B. (eds.) ADBIS 2010. LNCS, vol. 6295, pp. 204–218. Springer, Heidelberg (2010)
11. Hegner, S.J.: A simple model of negotiation for cooperative updates on database schema components. In: Kiyoki, Y., Tokuda, T., Heimbrger, A., Jaakkola, H., Yoshida, N. (eds.) Frontiers in Artificial Intelligence and Applications XX 2011, pp. 154–173. IOS Press (2011)
12. Hegner, S.J.: Invariance properties of the constant-complement view-update strategy. In: Schewe, K.-D., Thalheim, B. (eds.) SDKB 2011. LNCS, vol. 7693, pp. 118–148. Springer, Heidelberg (2013)
13. Hegner, S.J., Schmidt, P.: Update support for database views via cooperation. In: Ioannidis, Y., Novikov, B., Rachev, B. (eds.) ADBIS 2007. LNCS, vol. 4690, pp. 98–113. Springer, Heidelberg (2007)
14. Papadimitriou, C.: The Theory of Database Concurrency Control. Computer Science Press (1986)
15. Ports, D.R.K., Grittner, K.: Serializable snapshot isolation in PostgreSQL. Proc. VLDB Endowment 5(12), 1850–1861 (2012)
16. Revilak, S., O'Neil, P.E., O'Neil, E.J.: Precisely serializable snapshot isolation (PSSI). In: Proceedings of the 27th International Conference on Data Engineering, ICDE 2011, Hannover, Germany, April 11-16, pp. 482–493 (2011)
17. Silberschatz, A., Korth, H.F., Sudarshan, S.: Database System Concepts, 6th edn. McGraw Hill (2011)
18. Weikum, G., Vossen, G.: Transactional Information Systems. Morgan Kaufmann (2002)

Implication and Axiomatization of Functional Constraints on Patterns with an Application to the RDF Data Model

Jelle Hellings[1], Marc Gyssens[1], Jan Paredaens[2], and Yuqing Wu[3,⋆]

[1] Hasselt University and Transnational University of Limburg
{jelle.hellings,marc.gyssens}@uhasselt.be
[2] University of Antwerp
jan.paredaens@uantwerpen.be
[3] Indiana University
yuqwu@cs.indiana.edu

Abstract. Akhtar et al. introduced equality-generating constraints and functional constraints as an initial step towards dependency-like integrity constraints for RDF data [1]. Here, we focus on functional constraints. The usefulness of functional constraints is not limited to the RDF data model. Therefore, we study the functional constraints in the more general setting of relations with arbitrary arity. We show that a chase algorithm for functional constraints can be normalized to a more specialized *symmetry-preserving chase* algorithm. This symmetry-preserving chase algorithm is subsequently used to construct a sound and complete axiomatization for the functional constraints. This axiomatization is in particular applicable in the RDF data model, solving a major open problem of Akhtar et al.

Keywords: functional constraints, chase algorithm, axiomatization.

1 Introduction

Usually, data is subject to integrity constraints implied by the semantics of the data. Formalizing these constraints can help reasoning over the data and help identifying inconsistencies in the data. As such, formal constraints play a major role in database management systems that automatically maintain integrity of the data and optimize query evaluation.

For the relational data model, many types of constraints have been investigated. Among the simplest constraints are the functional dependencies [2]. Functional dependencies play an important role in the well-known Boyce-Codd normal form [3] and in relational schema normalization in general. Besides the functional dependencies, many other dependencies have been investigated (see,

⋆ Yuqing Wu carried out part of her work during a sabbatical visit to Hasselt University with a Senior Visiting Postdoctoral Fellowship of the Research Foundation Flanders (FWO).

C. Beierle and C. Meghini (Eds.): FoIKS 2014, LNCS 8367, pp. 250–269, 2014.

e.g., [4,5]). One of these, the equality-generating dependencies [4], is a natural generalization of the functional dependencies.

For the RDF and XML graph data models, a large body of work on the integrity of data focuses on the schema of the data. Examples are RDF Schema and, for the XML data model, DTDs and XSDs. The usage of dependency-like constraints is less common for these data-models although initial steps have been made (e.g. [1,6,7,8,9,10,11,12,13]).

An example of dependency-like constraints for the RDF data model are the *equality-generating constraints* and the *functional constraints* of Akhtar et al. [1,14]. Equality-generating constraints specify patterns that can occur in RDF data, together with equalities that should hold on these patterns. As such, the equality-generating constraints are similar to the equality-generating dependencies of Beeri and Vardi [4], to the equality-generating fragment of the implication dependencies of Fagin [15], and to the full equality-generating dependencies of Wijsen [16]. In these dependencies, the generality of patterns, which allow constants and are untyped, is only matched by the full equality-generating dependencies of Wijsen.

Functional constraints are a generalization of functional dependencies on ternary RDF relations and have the form

$$(P, L \rightarrow R),$$

where P specifies a pattern in the RDF data and L and R are sets of variables occurring in this pattern. Their semantics is comparable to that of the functional dependencies: if two parts of the RDF data match the pattern and are equal on L, then they must also be equal on R.

Example 1. Consider the family tree visualized in Figure 1.

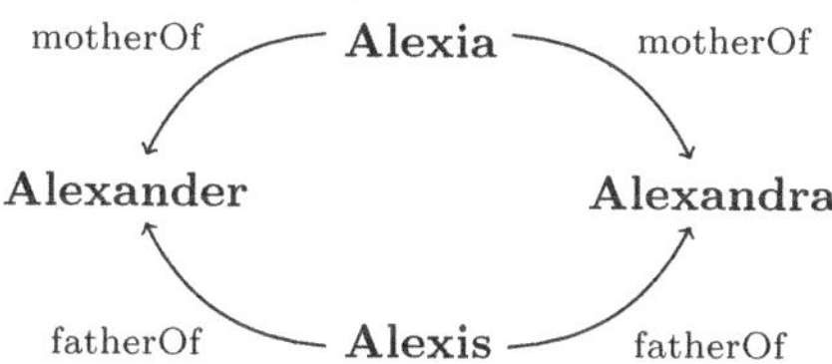

Fig. 1. Simplified visualization of an RDF representation of a small family tree

On this data, the constraint "a child only has one biological father and mother" holds. This constraint can be expressed by the functional constraints $(\{(\$p, \mathit{fatherOf}, \$c)\}, \$c \rightarrow \$p)$ and $(\{(\$p, \mathit{motherOf}, \$c)\}, \$c \rightarrow \$p)$. The stronger constraint "children have only one biological parent", which can be expressed by $(\{(\$p, \$t, \$c)\}, \$c \rightarrow \$p)$, does not hold on this data.

The functional constraints are subsumed by the equality-generating constraints of Akhtar et al. [1]. Although we shall sometimes refer to equality-generating constraints to describe the general context of this research, the focus here is on functional constraints. We shall consider functional constraints on relations of arbitrary arity, as the restriction to ternary patterns, as used in the RDF data model, is non-essential.

Functional constraints allow the expression of several types of integrity constraints; these include traditional functional dependencies [2], context-dependent functional dependencies, and constraints on the structure of graphs (described by an edge relation), as illustrated by the following examples.

Example 2. Consider the following relation schema for storing personal information: *PI*(*name*, *ssn*, *address*, *number*, *city*, *postal-code*, *country*), where *ssn* is the social security number. It is natural to add the functional dependency *ssn* → *name* to this scheme. We can express this functional dependency by the functional constraint $(\{(\$na, \$s, \$a, \$nu, \$ci, \$p, \$co)\}, \$s \rightarrow \$na)$.

Many integrity constraints are context-dependent. The functional constraints can use patterns and constants in patterns to restrict the context of a standard functional dependency to a subset of the relation.

Example 3. The information represented by postal codes is context-dependent. In the Netherlands, the postal code and house number uniquely identify an address, but this is not the case in Belgium. We thus use a constant for the country to make the functional dependency *postal-code*, *number* → *address*, *city* context-dependent: $(\{(\$na, \$s, \$a, \$nu, \$ci, \$p, \mathtt{NL})\}, \$p\$nu \rightarrow \$a\$ci)$.

Observe that functional constraints are not the only generalization of the functional dependencies which allow the expression of context-dependent functional dependencies. Other examples include conditional functional dependencies [17] and qualified functional dependencies [18]. The conditional functional dependencies define functional dependencies over a tableau with constants and blanks. The qualified functional dependencies allow the specification of views in which functional dependencies should hold. Patterns are conceptually related to tableaux and to views as tableau queries. Even though functional constraints, conditional functional dependencies, and qualified functional dependencies are related in this way, functional constraints on the one hand and conditional and qualified functional dependencies on the other hand are incomparable, as is argued next.

Example 4. The constraint "Ireland does not have postal codes" can obviously not be expressed as a functional constraint. By using constants in the right-hand side, however, we can express it as the conditional functional dependency $(\{(\$na, \$s, \$a, \$nu, \$ci, \$p, \mathtt{IE})\}, \emptyset \rightarrow [\$p = \mathtt{null}])$.[1]

The use of free variables and constants in patterns cannot be simulated by the tableaux or views used in conditional and qualified functional dependencies, however.

[1] We adapted the original notation of conditional functional dependencies to better match our notation of functional constraints.

Because of the use of free variables and constants in patterns, patterns may also match specific structures in the relation. This is particularly useful if the underlying relation represents a graph. In this setting, functional constraints may impose structural constraints.

Example 5. Let *Edge*(*from*, *to*) be a binary relation schema representing the edge relations of a graph. The functional constraint $(\{(\$n, \$n)\}, \emptyset \to \$n)$ expresses that there is at most one node with a self-loop. The pattern $\{(\$n, \$m), (\$m, \$n)\}$ in the functional constraint $(\{(\$n, \$m), (\$m, \$n)\}, \$n \to \$m)$ matches cycles (closed paths) of length 2 (including self-loops). Consider two pairs of such cycles starting in node v. By the constraint, the second node in both cycles must be equal, and thus the latter constraint expresses that every node v is part of at most one cycle of length 2.

For the functional dependencies in the relational data model, a sound and complete axiomatization is long known [19]. Akhtar et al. presented a sound and complete axiomatization for the equality-generating constraints in the RDF data model [1]. As functional constraints are subsumed by equality-generating constraints, this axiomatization can also be used for the inference of functional constraints only. In this case, intermediate inference steps can generate equality-generating constraints that are not necessarily equivalent to functional constraints, unfortunately. Akhtar et al. identified the existence of a sound and complete axiomatization of functional constraints (not including other types of constraints) as a major open problem. On the one hand, the Armstrong axiomatization for the functional dependencies [19] can be generalized to the setting of functional constraints. This generalization, however, lacks the reasoning power over patterns necessary for a complete axiomatization. On the other hand, there is no straightforward way to specialize the axiomatization of the equality-generating constraints to functional constraints only.

In this paper, we present a sound and complete axiomatization for the functional constraints over relations of arbitrary arity. In particular, the case of ternary relations yields a sound and complete axiomatization for the functional constraints in the RDF data model, thereby positively solving the open problem of Akhtar et al. [1].

The key insight that led to the breakthrough is that the chase algorithm for equality-generating constraints [1]—which is a variation of the standard chase algorithm [20,21]—can be normalized to a more specialized, symmetry-preserving, chase algorithm when applied to functional constraints only. The main idea behind the symmetry-preserving chase algorithm is that, due to their semantics, chases for functional constraints always start with tableaux that are symmetric. We prove that during such chases one can always maintain this symmetry in the tableau. Such a symmetry-preserving chase can be described as a sequence of inferences of functional constraints, which in turn leads to the sound and complete axiomatization.

Organization. In Section 2, we present the necessary definitions used throughout this paper. In Section 3, we introduce generalized functional constraints and equality-generating constraints. In Section 4, the chase algorithm for equality-generating constraints is specialized to functional constraints and subsequently normalized to the symmetry-preserving chase algorithm. In Section 5, we present a sound axiomatization for the functional constraints which suffices to simulate every symmetry-preserving chase, and which is therefore also complete. In Section 6, we conclude on our findings and discuss directions for future work.

2 Preliminaries

Functional and equality-generating constraints [1] have originally been introduced in the context of the RDF data model. In this model, RDF data are usually represented by a single ternary relation. In the Introduction, we have already argued that functional and equality-generating constraints are useful in a wider range of data models. We therefore generalize functional and equality-generating constraints to relations of arbitrary arity. The following notations and definitions will be used throughout the paper.

We consider disjoint infinitely enumerable sets $\mathcal{U}$ and $\mathcal{V}$ of *constants* and *variables*, respectively. For distinction, we usually prefix variables by "\$". A *term* is either a constant or a variable. Hence, the set $\mathcal{T}$ of all terms equals $\mathcal{U} \cup \mathcal{V}$. A *tuple* of arity n is a sequence $(t_1, \ldots, t_n)$ of terms. A *pattern* of arity n is a finite set of tuples of arity n. If P is a pattern, then $\mathcal{V}_P$ denotes the set of all variables in P. A *relation* $\mathcal{R}$ of arity n is a pattern of arity n with $\mathcal{V}_{\mathcal{R}} = \emptyset$.

We define the *domain*, *range*, and *inverse* of a function f in the usual way and denote these by domain(f), range(f), and f^{-1}, respectively. Two functions f and g *agree* on a set S, denoted by $f =_S g$, if $f(x) = g(x)$ for all $x \in S$. The *restriction* of a function f to a set S is defined as $f|_S = \{(x, y) \mid x \in S, y = f(x)\}$. The *identity* on a set S is defined as $\mathrm{id}_S = \{(s, s) \mid s \in S\}$. The *extension with identity* of a function f to a set S, $S \cap \mathrm{domain}(f) = \emptyset$, is $f \cup \mathrm{id}_S$.

The *term-based renaming function* $\phi_{a_1 \hookleftarrow b_1, \ldots, a_i \hookleftarrow b_i}$, $a_1, b_1, \ldots, a_i, b_i \in \mathcal{T}$, is the function on $\mathcal{T}$ for which $\phi_{a_1 \hookleftarrow b_1, \ldots, a_i \hookleftarrow b_i}(b_j) = a_j$, $j = 1, \ldots, i$, and which is the identity elsewhere. Likewise, the *function-based renaming function* $\Phi_{f \hookleftarrow g}$, with f a function and g an injective function on the same set of variables, is the function on $\mathcal{T}$ for which $\Phi_{f \hookleftarrow g}(g(\$v)) = f(\$v), \$v \in \mathrm{domain}(g) = \mathrm{domain}(f)$, and which is the identity elsewhere. Notice that this function is well defined due to the injectivity of g.

A function f on terms is extended to tuples, patterns, and sets in the following natural way: for a tuple $(t_1, \ldots, t_n)$, $f((t_1, \ldots, t_n)) = (f(t_1), \ldots, f(t_n))$, and, for a set S, $f(S) = \{f(s) \mid s \in S\}$.

For two patterns P and Q, a function $e : \mathcal{V}_P \cup \mathcal{U} \to \mathcal{T}$ is an embedding of P into Q if $e|_{\mathcal{U}} = \mathrm{id}_{\mathcal{U}}$ and $e(P) \subseteq Q$.

We finally review some notation and terminology that can be applied to any type of constraint. "Relation $\mathcal{R}$ *satisfies* constraint C" is denoted by $\mathcal{R} \models C$. A relation $\mathcal{R}$ *satisfies* a set of constraints $\mathcal{C}$, denoted by $\mathcal{R} \models \mathcal{C}$, if, for every $C \in \mathcal{C}$, $\mathcal{R} \models C$. If $\mathcal{C}_1$ and $\mathcal{C}_2$ are sets of constraints then $\mathcal{C}_1$ *implies* $\mathcal{C}_2$, denoted by $\mathcal{C}_1 \models \mathcal{C}_2$, if, for every relation $\mathcal{R}$ with $\mathcal{R} \models \mathcal{C}_1$, we have $\mathcal{R} \models \mathcal{C}_2$. For a set of constraints $\mathcal{C}$ and a single constraint C, we write $\mathcal{C} \models C$ for $\mathcal{C} \models \{C\}$. The sets of constraints $\mathcal{C}_1$ and $\mathcal{C}_2$ are *equivalent*, denoted by $\mathcal{C}_1 \equiv \mathcal{C}_2$, if $\mathcal{C}_1 \models \mathcal{C}_2$ and $\mathcal{C}_2 \models \mathcal{C}_1$. If, in this notation, $\mathcal{C}_i$, $i = 1$ and/or $i = 2$, is a singleton set $\{C_i\}$, we write C_i for $\mathcal{C}_i$, as before. "Constraint C can be derived from set of constraints $\mathcal{C}$ using the set of *axioms* $\mathfrak{R}$" is denoted by $\mathcal{C} \vdash_{\mathfrak{R}} C$. We usually omit $\mathfrak{R}$ if $\mathfrak{R}$ is clear from the context. The set $\mathfrak{R}$ is *sound* if, for all sets of constraints $\mathcal{C}$ and for all single constraints C, $\mathcal{C} \vdash_{\mathfrak{R}} C$ implies $\mathcal{C} \models C$; it is *complete* if, for all sets of constraints $\mathcal{C}$ and for all single constraints C, $\mathcal{C} \models C$ implies $\mathcal{C} \vdash_{\mathfrak{R}} C$. A set of axioms is an *axiomatization* if it is sound, complete, and recursive.

3 Functional Constraints

We formally define functional constraints on n-ary relations.

Definition 1. *A* functional constraint *is a pair* $(P, L \to R)$, *where* P *is a nonempty pattern and* $L, R \subseteq \mathcal{V}_P$.

If $C = (P, L \to R)$ is a functional constraint, then P is the pattern of C, L is the left-hand side of C, and R is the right-hand side of C.

Definition 2. *Let* $\mathcal{R}$ *be a relation and let* $C = (P, L \to R)$ *be a functional constraint. Then* $\mathcal{R}$ satisfies C *if, for every pair of embeddings* e_1 *and* e_2 *of* P *into* $\mathcal{R}$ *with* $e_1 =_L e_2$, *we have* $e_1 =_R e_2$.

As already mentioned, the functional constraints are a strict subclass of the equality-generating constraints, and the functional constraints are a generalization of the functional dependencies. Below, we formalize these relationships in our setting. This allows us to apply results for equality-generating constraints to functional constraints, and to generalize results for functional dependencies to functional constraints.

3.1 Equality-Generating Constraints

We formally define equality-generating constraints on n-ary relations.

Definition 3. *An* equality-generating constraint *is a pair* (P, E), *where* P *is a nonempty pattern and* E *is a set of equalities of the form* $t_1 = t_2$ *with* $t_1, t_2 \in \mathcal{V}_P \cup \mathcal{U}$.

Definition 4. *Let* $\mathcal{R}$ *be a relation and let* $C = (P, E)$ *be an equality-generating constraint. Then* $\mathcal{R}$ satisfies C *if, for every embedding* e *of* P *into* $\mathcal{R}$ *and every equality* $(t_1 = t_2) \in E$, *we have* $e(t_1) = e(t_2)$.

Akhtar et al. [1] already showed that every functional constraint can be written as an equality-generating constraint. Adopted to our setting, their result is as follows:

Proposition 1. *Let* $C_{\text{FC}} = (P, L \to R)$ *be a functional constraint. Let* $f_1, f_2 : \mathcal{V}_P \to \mathcal{V}$ *be injections with* $f_1 =_L f_2$ *and* $\text{range}(f_1|_{\mathcal{V}_P \setminus L}) \cap \text{range}(f_2|_{\mathcal{V}_P \setminus L}) = \emptyset$. *Let* $C_{\text{EGC}} = ((f_1 \cup \text{id}_{\mathcal{U}})(P) \cup (f_2 \cup \text{id}_{\mathcal{U}})(P), \{f_1(\$r) = f_2(\$r) \mid \$r \in R\})$. *Then* $C_{\text{FC}} \equiv C_{\text{EGC}}$.

3.2 Functional Dependencies

We assume familiarity with the functional dependencies of Codd [2,5].

Proposition 2. *Let $C = L \to R$ be a functional dependency over the relation schema $\mathcal{R} = (A_1, \ldots, A_n)$ with $L, R \subseteq \{A_1, \ldots, A_n\}$. Consider the functional constraint $C_{\text{FC}} = (\{(A_1, \ldots, A_n)\}, L \to R)$, in which the attribute names are assumed to be variables. Then $C \equiv C_{\text{FC}}$.*

The functional dependencies have a well-known axiomatization in the form of *Armstrong's axioms*, consisting of the three axioms *reflexivity*, *augmentation*, and *transitivity* [19]. We generalize Armstrong's axioms to our setting of the functional constraints.

Proposition 3 (Reflexivity). *Let P be a pattern. If $R \subseteq L \subseteq \mathcal{V}_P$, then $(P, L \to R)$.*

Proof (soundness). Let e_1 and e_2 be embeddings of P into a relation $\mathcal{R}$ with $e_1 =_L e_2$. We have $R \subseteq L$ and hence also $e_1 =_R e_2$. □

Proposition 4 (Augmentation). *If $(P, L \to R)$ and $V \subseteq \mathcal{V}_P$, then $(P, L \cup V \to R \cup V)$.*

Proof (soundness). Let e_1 and e_2 be embeddings of P into a relation $\mathcal{R}$ satisfying $(P, L \to R)$. If we have $e_1 =_{L \cup V} e_2$, then we have $e_1 =_L e_2$ and $e_1 =_V e_2$. By $e_1 =_L e_2$ and $(P, L \to R)$, we also have $e_1 =_R e_2$ and hence $e_1 =_{R \cup V} e_2$. □

Proposition 5 (Transitivity). *If $(P, V_1 \to V_2)$ and $(P, V_2 \to V_3)$, then $(P, V_1 \to V_3)$.*

Proof (soundness). Let e_1 and e_2 be embeddings of P into a relation $\mathcal{R}$ satisfying $(P, V_1 \to V_2)$ and $(P, V_2 \to V_3)$. If $e_1 =_{V_1} e_2$, then, by $(P, V_1 \to V_2)$, we have $e_1 =_{V_2} e_2$, and, by $(P, V_2 \to V_3)$, we have $e_1 =_{V_3} e_2$. □

Since Armstrong's axioms also hold for functional constraints, it follows that the well-known decomposition and union rules also hold for functional constraints.

Lemma 1. *Let $C_{\text{FC}} = (P, L \to R)$ be a functional constraint. Then*

$$C_{\text{FC}} \equiv \{(P, L \to \$r) \mid \$r \in R\}.$$

From now on, we assume that every functional constraint has at most one variable in its right-hand side. By Lemma 1, all our results generalize to arbitrary functional constraints.

4 Chasing Functional Constraints

For equality-generating constraints, a chase-based algorithm is known to decide implication [1]. We use the relationship between functional and equality-generating constraints described in Proposition 1 to construct a chase-based algorithm that decides implication of functional constraints, shown as Algorithm 1.

The entries in the tableau constructed in Algorithm 1 can be either constants of $\mathcal{U}$ or dedicated tableau variables, which we shall denote by capitals. These tableau variables intuitively correspond to the variables in the pattern of the target constraint. To this purpose, we assume the existence of an infinitely enumerable set $\mathfrak{V}$ of tableau variables. Further, we assume that $\mathfrak{V}$ is disjoint from both $\mathcal{U}$ and $\mathcal{V}$.[2] We generalize embeddings in a straightforward way to also allow embeddings from and to tableaux.

Algorithm 1. Chase for functional constraints

Input: A set of functional constraints $\mathcal{C} = \{(P_i, L_i \rightarrow \$r_i) \mid 1 \leq i \leq n\}$
A functional constraint $C_{\text{FC}} = (P, L \rightarrow \$r)$
Output: $\mathcal{C} \models C_{\text{FC}}$
1: let $f_1, f_2 : \mathcal{V}_P \rightarrow \mathfrak{V}$ be injections with $f_1 =_L f_2$ and
range$(f_1|_{\mathcal{V}_P \setminus L}) \cap$ range$(f_2|_{\mathcal{V}_P \setminus L}) = \emptyset$
2: $\mathfrak{T} \leftarrow (f_1 \cup \text{id}_{\mathcal{U}})(P) \cup (f_2 \cup \text{id}_{\mathcal{U}})(P)$
3: **while** there exist functional constraint $(P_i, L_i \rightarrow \$r_i) \in \mathcal{C}$ and
embeddings e_1, e_2 of P_i into $\mathfrak{T}$ with $e_1 =_{L_i} e_2$ and $e_1(\$r_i) \neq e_2(\$r_i)$ **do**
4: /* *equalize* $e_1(\$r_i)$ *and* $e_2(\$r_i)$ *in* $\mathfrak{T}$ */
5: **if** $e_2(\$r_i) \in \mathfrak{V}$ **then**
6: replace all occurrences of $e_2(\$r_i)$ in $\mathfrak{T}$ by $e_1(\$r_i)$
7: **else if** $e_1(\$r_i) \in \mathfrak{V}$ **then**
8: replace all occurrences of $e_1(\$r_i)$ in $\mathfrak{T}$ by $e_2(\$r_i)$
9: **else** /* $e_1(\$r_i), e_2(\$r_i) \in \mathcal{U}$ *and* $e_1(\$r_i) \neq e_2(\$r_i)$ */
10: **return** TRUE
11: **end if**
12: **end while**
13: **return** $\mathfrak{T} \models C_{\text{FC}}$

In Algorithm 1, we refer to lines 5–8 as *equalization steps*, to lines 9–10 as *inconsistency termination*, and to line 13 as *regular termination*. Inconsistency termination indicates that the pattern P is inconsistent with the functional constraints in $\mathcal{C}$, and, hence, that the implication under consideration is voidly true. The following example illustrates the case of inconsistency termination.

Example 6. Consider the set of functional constraints $\mathcal{C} = \{(\{(\$a, \$b)\}, \$a \rightarrow \$b)\}$. If the functional constraints in this set $\mathcal{C}$ hold on a relation $\mathcal{R}$, then no embedding of the pattern $P = \{(\$a, \textit{Constant}_1), (\$a, \textit{Constant}_2)\}$ with $\textit{Constant}_1 \neq$

[2] The distinction between $\mathfrak{V}$ and $\mathcal{V}$ is not necessary, as these sets of variables are not used in the same context. For clarity, however, we use different sets of variables.

$Constant_2$ into relation $\mathcal{R}$ is possible, and, hence, every functional constraint on the pattern P holds. This is reflected by Algorithm 1: if a functional constraint on P is chased by $\mathcal{C}$, then inconsistency termination results and TRUE is returned.

Theorem 1. *Algorithm 1 is correct: it returns* TRUE *if and only if* $\mathcal{C} \models C_{\text{FC}}$ *holds.*

Proof (sketch). Algorithm 1 implicitly translates the target functional constraint C_{FC} to an equality-generating constraint. Indeed, at line 2, a tableau for the pattern $P_{\text{EGC}} = (f_1 \cup \text{id}_{\mathcal{U}})(P) \cup (f_2 \cup \text{id}_{\mathcal{U}})(P)$ is constructed. By Proposition 1, P_{EGC} is the pattern used by the equality-generating constraint equivalent to C_{FC}.

At line 3, considering two embeddings e_1 and e_2 of P_i into $\mathfrak{T}$ with $e_1 =_{L_i} e_2$ is equivalent to considering one embedding of the pattern of the equality-generating constraint equivalent to $(P_i, L_i \rightarrow \$r_i)$. Hence, Algorithm 1 can be interpreted as a chase for an equality-generating constraint with equality-generating constraints. Therefore the correctness of Algorithm 1 follows directly from the correctness of the chase algorithm for equality-generating constraints [1]. □

So, Algorithm 1 is essentially a chase algorithm for equality-generating constraints. As a consequence, it is to be expected that intermediate tableaux produced by this algorithm do not always correspond to non-trivial functional constraints. Hence, the corresponding functional constraints are not always relevant to answering $\mathcal{C} \models C_{\text{FC}}$. Example 7, below, shows that this is indeed not always the case.

Example 7. We apply Algorithm 1 to the set of functional constraints

$$\mathcal{C} = \{(\{(\$a, \$b, \$c)\}, \$a \rightarrow \$c), (\{(\$a, \$b, \$c), (\$a, \$d, e)\}, \$b \rightarrow \$a)\}$$

and the target functional constraint $C_{\text{FC}} = (\{(\$a, \$b, \$c), (\$a, \$b, e)\}, \$b \rightarrow \$a)$. We initially have the tableau

$$\{(A_1, B, C_1), (A_1, B, e), (A_2, B, C_2), (A_2, B, e)\}.$$

We can apply $(\{(\$a, \$b, \$c)\}, \$a \rightarrow \$c)$ to the first two tuples in this tableau, yielding the tableau

$$\{(A_1, B, e), (A_2, B, C_2), (A_2, B, e)\}.$$

We can use Proposition 1 to search for a functional constraint with such a pattern when translated to an equality-generating constraint. Let $C = (P, L \rightarrow R)$ be such a functional constraint. It is easily verified that the only way to achieve this is by relating A_1, A_2, B, C_2 to distinct variables $\$a_1, \$a_2, \$b, \$c_2 \in \mathcal{V}_P$ for which $L = \{\$a_1, \$a_2, \$b, \$c_2\}$. Since L contains all variables present in the pattern, it follows that C must be trivial. Hence, the tableau we obtained does not correspond to a functional constraint relevant to answering $\mathcal{C} \models C_{\text{FC}}$.

Example 7 also illustrates the main problem of Algorithm 1. While the initial chase tableau exhibits a certain symmetry, this symmetry is lost after performing the equalization. As a consequence, only trivial functional constraints can be associated with the resulting tableau. Luckily, Algorithm 1 is non-deterministic in the equalization steps it performs. We shall take advantage of this to show the existence of a *symmetry-preserving chase*, which we define formally in Definition 7. The steps performed by symmetry-preserving chases are closely related to sound derivation steps for functional constraint in a way that shall be made precise in Section 5. Before we can introduce symmetry-preserving chases, we need some additional terminology.

Definition 5. *Let T be a tableau. A* tableau state *of T is a 4-tuple consisting of a pattern P', a set of variables $L' \subseteq \mathcal{V}_{P'}$, and injections $g_1, g_2 : \mathcal{V}_{P'} \to \mathfrak{V}$ with $g_1 =_{L'} g_2$, $\text{range}(g_1|_{\mathcal{V}_{P'} \setminus L'}) \cap \text{range}(g_2|_{\mathcal{V}_{P'} \setminus L'}) = \emptyset$, and $T = (g_1 \cup \text{id}_{\mathcal{U}})(P') \cup (g_2 \cup \text{id}_{\mathcal{U}})(P')$.*

Given a tableau T, we denote a tableau state of T such as in Definition 5 by $\mathbf{S}_T(P', L', g_1, g_2)$, this to emphasize the relationship between the tableau and the corresponding tableau state.

We can easily construct tableau states $\mathbf{S}_T(P', L', g_1, g_2)$ for every tableau T. We simply map every tableau variable from $\mathfrak{V}$ used in T to a unique variable, yielding the pattern P', and pick $L' = \mathcal{V}_{P'}$. Finally, $g_1 = g_2$ maps each variable in $\mathcal{V}_{P'}$ to the tableau variable in $\mathfrak{V}$ it represents.

Example 8. A tableau state for the tableau $\{(A_1, B, e), (A_2, B, C_2), (A_2, B, e)\}$ of Example 7 is $\mathbf{S}_{\mathfrak{T}}(P', L', g_1, g_2)$ with $P' = \{(\$a_1, \$b, e), (\$a_2, \$b, \$c_2), (\$a_2, \$b, e)\}$, $L' = \{\$a_1, \$a_2, \$b, \$c_2\}$, and $g_1 = g_2$ the injective functions mapping $\$a_1$ to A_1, $\$a_2$ to A_2, $\$b$ to B, and $\$c_2$ to C_2.

Tableau states enjoy the following useful properties.

Lemma 2. *Let $\mathbf{S}_T(P', L', g_1, g_2)$ be a state of tableau T. Then*

1. *The pattern $(g_1 \cup \text{id}_{\mathcal{U}})(P')$ is isomorphic to the pattern $(g_2 \cup \text{id}_{\mathcal{U}})(P')$.*
2. *For any tuple $t \in T$, also $\Phi_{g_1 \leftrightarrow g_2}(t) \in T$ and $\Phi_{g_2 \leftrightarrow g_1}(t) \in T$.*
3. *$\mathbf{S}_T(P', L', g_2, g_1)$ is also a tableau state of T.*

Proof. We have Lemma 2(1) as g_1 and g_2 are injections. Lemma 2(2) follows from Lemma 2(1), $g_1 =_{L'} g_2$, and $\text{range}(g_1|_{\mathcal{V}_{P'} \setminus L'}) \cap \text{range}(g_2|_{\mathcal{V}_{P'} \setminus L'}) = \emptyset$. Lemma 2(3), finally, follows immediately from Definition 5. □

Observe that the initial tableau in Algorithm 1 has state $\mathbf{S}_{\mathfrak{T}}(P, L, f_1, f_2)$. We already noted that this initial tableau exhibits some symmetry due to the semantics of functional constraints. We would like that, after a sequence of equalization steps, the resulting tableau exhibits a similar symmetry. What we mean by this is made precise in Definition 6, minding that a sequence of equalization steps can be viewed as a mapping on tableau entries that maps a tableau into the tableau resulting from the equalization steps.

Definition 6. *Let m be a mapping on tableau entries, mapping a tableau T into a tableau $m(T)$. The mapping m is* symmetry-preserving *on T if there exists a tableau state $\mathbf{S}_T(P', L', g_1, g_2)$ of T and $\mathbf{S}_{m(T)}(P'', L'', g_1', g_2')$ of $m(T)$ such that $m((g_1 \cup \text{id}_{\mathcal{U}})(P')) = (g_1' \cup \text{id}_{\mathcal{U}})(P'')$ and $m((g_2 \cup \text{id}_{\mathcal{U}})(P')) = (g_2' \cup \text{id}_{\mathcal{U}})(P'')$.*

Definition 6 is visualized in Figure 2. By Lemma 2(1), $(g_1 \cup \mathrm{id}_{\mathcal{U}})(P')$ and $(g_2 \cup \mathrm{id}_{\mathcal{U}})(P')$ are isomorphic, and so are $(g'_1 \cup \mathrm{id}_{\mathcal{U}})(P'') = m((g_1 \cup \mathrm{id}_{\mathcal{U}})(P'))$ and $(g'_2 \cup \mathrm{id}_{\mathcal{U}})(P'') = m((g_2 \cup \mathrm{id}_{\mathcal{U}})(P'))$. Hence, we can say that m preserves the isomorphism between $(g_1 \cup \mathrm{id}_{\mathcal{U}})(P')$ and $(g_2 \cup \mathrm{id}_{\mathcal{U}})(P')$, explaining why we call m "symmetry-preserving".

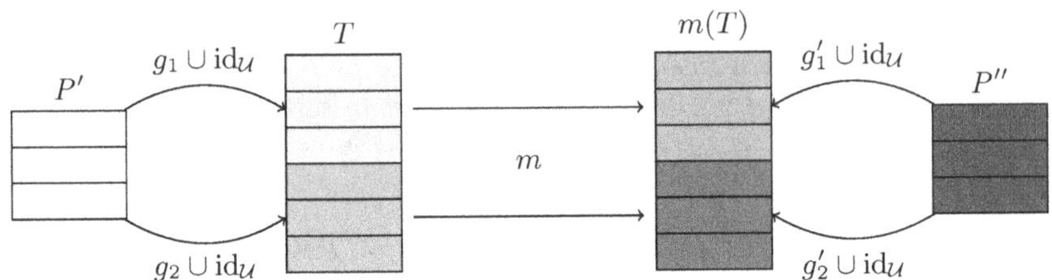

Fig. 2. Visualization of Definition 6

Example 7 shows that not all sequences of equalization steps preserve symmetry. However, if an equalization step is possible in Algorithm 1, then also a sequence of at most two equalization steps is possible which does preserve symmetry. Moreover, all equalization steps concerned use the same constraint. This is shown next.

Theorem 2. *Let $\mathfrak{T} := T$ be the tableau of Algorithm 1 at line 3. If it is possible to perform an equalization step using the functional constraint $C_i \in \mathcal{C}$, then it is also possible to perform a sequence of at most two equalization steps, both using C_i, such that the composition of these equalization steps yields a symmetry-preserving mapping on T.*

Proof (sketch). Let $\mathbf{S}_T(P', L', g_1, g_2)$ be a tableau state of T.

If it is possible to perform an equalization with $C_i = (P_i, L_i \rightarrow \$r_i)$, then, by Lemma 2(3), we may assume, without loss of generality, that there exist terms $t_1, t_2 \in \mathcal{V}_{P'} \cup \mathcal{U}$ such that $e_1(\$r_i) = (g_1 \cup \mathrm{id}_{\mathcal{U}})(t_1)$, and either $e_2(\$r_i) = (g_1 \cup \mathrm{id}_{\mathcal{U}})(t_2)$ or $e_2(\$r_i) = (g_2 \cup \mathrm{id}_{\mathcal{U}})(t_2)$. Observe that t_1 and t_2 cannot both be constants. We now distinguish a number of cases. In each case, we suffice with providing the required sequence of at most two equalization steps and the resulting tableau $\mathfrak{T} := T'$, together with a tableau state $\mathbf{S}_{T'}(P'', L'', g'_1, g'_2)$.[3] Using the provided tableau states for T and T', it is straightforward to verify that the composition of the equalization steps is a symmetry-preserving mapping.

First, we consider all the cases where one of t_1 and t_2 is a variable, and the other a constant. Since the roles of e_1 and e_2 are interchangeable, we may assume, without loss of generality, that $t_1 = \$v_1$ is a variable and $t_2 = u_2$ is a constant. From the above, it follows that, in all these cases, $e_1(\$r_i) = g_1(\$v_1)$ and $e_2(\$r_i) = u_2$.

[3] From the provided tableau state $\mathbf{S}_{T'}(P'', L'', g'_1, g'_2)$, it follows implicitly what P'', L'', g'_1, and g'_2 are.

1. $\$v_1 \in L'$. Performing the equalization step using C_i, e_1, and e_2 results in the tableau $T' = \phi_{u_2 \leftrightarrow g_1(\$v_1)}(T)$ with state

$$\mathbf{S}_{T'}\left(\phi_{u_2 \leftrightarrow \$v_1}(P'), L' \setminus \{\$v_1\}, g_1|_{\mathcal{V}_{P'} \setminus \{\$v_1\}}, g_2|_{\mathcal{V}_{P'} \setminus \{\$v_1\}}\right).$$

2. $\$v_1 \notin L'$. By Lemma 2(2), the functions $\varepsilon_1 = \Phi_{g_2 \leftrightarrow g_1} \circ e_1$ and $\varepsilon_2 = \Phi_{g_2 \leftrightarrow g_1} \circ e_2$ are embeddings of P_i into T. Since $(g_1 \cup \mathrm{id}_{\mathcal{U}})(\$v_1) = g_1(\$v_1) = e_1(\$r_i) \neq e_2(\$r_i) = u_2 = (g_1 \cup \mathrm{id}_{\mathcal{U}})(u_2)$, we have, by Lemma 2(1), that $\varepsilon_1(\$r_i) = g_2(\$v_1) = (g_2 \cup \mathrm{id}_{\mathcal{U}})(\$v_1) \neq (g_2 \cup \mathrm{id}_{\mathcal{U}})(u_2) = u_2 = \varepsilon_2(\$r_i)$. The equalization step using C_i, e_1, and e_2 on T only affects tuples in $(g_1 \cup \mathrm{id}_{\mathcal{U}})(P')$ as $e_1(\$r_i) \notin \mathrm{range}(g_2)$. Hence, after the equalization step, ε_1 and ε_2 are embeddings of P_i into the resulting tableau with $\varepsilon_1(\$r_i) \neq \varepsilon_2(\$r_i)$, and, by construction, we have $\varepsilon_1 =_{L'} \varepsilon_2$. Therefore, we can perform a second equalization step using C_i, ε_1, and ε_2. Performing this second equalization step results in the tableau $T' = \phi_{u_2 \leftrightarrow g_1(\$v_1), u_2 \leftrightarrow g_2(\$v_1)}(T)$ with state

$$\mathbf{S}_{T'}\left(\phi_{u_2 \leftrightarrow \$v_1}(P'), L', g_1|_{\mathcal{V}_{P'} \setminus \{\$v_1\}}, g_2|_{\mathcal{V}_{P'} \setminus \{\$v_1\}}\right).$$

Next, we consider all the cases where $t_1 = \$v_1$ and $t_2 = \$v_2$ are both variables, and where $e_2(\$r_i) = g_1(\$v_2)$. Observe that $\$v_1 \neq \$v_2$ since $e_1(\$r_i) \neq e_2(\$r_i)$.

3. Both $\$v_1$ and $\$v_2$ are in L'. Performing the equalization step with C_i, e_1, and e_2 results in the tableau $T' = \phi_{g_1(\$v_1) \leftrightarrow g_2(\$v_2)}(T)$ with state

$$\mathbf{S}_{T'}\left(\phi_{\$v_1 \leftrightarrow \$v_2}(P'), L' \setminus \{\$v_2\}, g_1|_{\mathcal{V}_{P'} \setminus \{\$v_2\}}, g_2|_{\mathcal{V}_{P'} \setminus \{\$v_2\}}\right).$$

4. At least one of $\$v_1$ and $\$v_2$ is not in L'. Since the roles of e_1 and e_2 are interchangeable, we may assume, without loss of generality, that $\$v_2 \notin L'$. As in Case 2, we can perform a second equalization step following the equalization step with C_i, e_1, and e_2. Performing this second equalization step results in the tableau $T' = \phi_{g_1(\$v_1) \leftrightarrow g_1(\$v_2), g_2(\$v_1) \leftrightarrow g_2(\$v_2)}(T)$ with state

$$\mathbf{S}_{T'}\left(\phi_{\$v_1 \leftrightarrow \$v_2}(P'), L', g_1|_{\mathcal{V}_{P'} \setminus \{\$v_2\}}, g_2|_{\mathcal{V}_{P'} \setminus \{\$v_2\}}\right).$$

Finally, we consider all the cases where $t_1 = \$v_1$ and $t_2 = \$v_2$ are both variables, and where $e_2(\$r_i) = g_2(\$v_2)$.

5. $\$v_1 = \$v_2 = \$v$. As $g_1 =_{L'} g_2$ and $e_1(\$r_i) \neq e_2(\$r_i)$, we must have $\$v \notin L'$. The equalization step using C_i, e_1, and e_2 results in the tableau $T' = \phi_{g_1(\$v) \leftrightarrow g_2(\$v)}(T)$ with state

$$\mathbf{S}_{T'}\left(P', L' \cup \{\$v\}, \phi_{g_1(\$v) \leftrightarrow g_2(\$v)} \circ g_1, \phi_{g_1(\$v) \leftrightarrow g_2(\$v)} \circ g_2\right).$$

6. $\$v_1 \neq \$v_2$. By Lemma 2(2), $\varepsilon_1 = \Phi_{g_1 \leftrightarrow g_2} \circ e_1$ and $\varepsilon_2 = \Phi_{g_1 \leftrightarrow g_2} \circ e_2$ are embeddings of P_i into T. By construction and the injectivity of g_1, we have $\varepsilon_1(\$r_i) = (g_1 \cup \mathrm{id}_{\mathcal{U}})(\$v_1) \neq (g_1 \cup \mathrm{id}_{\mathcal{U}})(\$v_2) = \varepsilon_2(\$r_i)$ and $\varepsilon_1 =_{L'} \varepsilon_2$. Instead of performing the equalization using C_i, e_1, and e_2, we perform the equalization using C_i, ε_1, and ε_2. Hence, Case 6 has been reduced to Cases 3 and 4. □

We refer to each sequence of at most two equalization steps from tableau T to tableau T', considered in the proof of Theorem 2, as a *symmetry-preserving step*. We refer to the symmetry-preserving step in Case i, $1 \leq i \leq 5$, in the proof of Theorem 2 as the symmetry-preserving step of type i.[4]

Definition 7. *Executions of Algorithm 1 consisting of a sequence of symmetry-preserving steps and in which inconsistency termination occurs if no equalization steps can be performed, are called* symmetry-preserving chases.

Based on Definition 7 and on Theorem 2 we specialize Algorithm 1 to a symmetry-preserving chase algorithm, shown as Algorithm 2. Notice that we use the non-deterministic nature of Algorithm 1 to delay inconsistency termination to the latest-possible moment. By delaying inconsistency termination, we are able to perform equalization steps until no such step is possible anymore, and only then, when necessary, perform inconsistency termination.

Algorithm 2. Symmetry-preserving chase for functional constraints

Input: A set of functional constraints $\mathcal{C} = \{(P_i, L_i \to \$r_i) \mid 1 \leq i \leq n\}$
A functional constraint $C_{\text{FC}} = (P, L \to \$r)$
Output: $\mathcal{C} \models C_{\text{FC}}$

let $f_1, f_2 : \mathcal{V}_P \to \mathfrak{V}$ be injections with $f_1 =_L f_2$ and
  $\text{range}(f_1|_{\mathcal{V}_P \setminus L}) \cap \text{range}(f_2|_{\mathcal{V}_P \setminus L}) = \emptyset$
$\mathfrak{T} \leftarrow (f_1 \cup \text{id}_{\mathcal{U}})(P) \cup (f_2 \cup \text{id}_{\mathcal{U}})(P)$
/* $\mathbf{S}_{\mathfrak{T}}(P, L, f_1, f_2)$ *is a tableau state of* $\mathfrak{T}$ */
while an equalization step can be performed using functional constraint
  $(P_i, L_i \to \$r_i) \in \mathcal{C}$ and embeddings e_1, e_2 of P_i into $\mathfrak{T}$ with
  $e_1 =_{L_i} e_2$ and $e_1(\$r_i) \neq e_2(\$r_i)$ **do**
  perform the corresponding symmetry-preserving step
   (cf. the proof of Theorem 2)
end while
if *inconsistency termination* **then**
  return TRUE
else
  return $\mathfrak{T} \models C_{\text{FC}}$
end if

Theorem 2 now immediately yields the following.

Corollary 1. *Algorithm 2 is correct: it returns* TRUE *if and only if* $\mathcal{C} \models C_{\text{FC}}$ *holds.*

5 Axiomatization for the Functional Constraints

Let $\mathcal{C}$ be a set of functional constraints and let $C_{\text{FC}} = (P, L \to \$r)$ be a single functional constraint for which $\mathcal{C} \models C_{\text{FC}}$. By simulating a symmetry-preserving

[4] We have no symmetry-preserving step of type 6, as Case 6 in the proof of Theorem 2 has been reduced to Cases 3 and 4.

chase for $\mathcal{C} \models C_{\text{FC}}$ (Algorithm 2), by a derivation of functional constraints using sound derivation rules, we construct an axiomatization for the functional constraints which must be complete by Corollary 1.

First, we consider the (base) cases where the chase terminates immediately without performing symmetry-preserving steps. By the *restricted reflexivity axiom*, below, we mean the specialization of the reflexivity axiom in which only functional constraints are derived with at most one variable in the right-hand side.

Lemma 3. *If only regular termination is possible in a symmetry-preserving chase for $\mathcal{C} \models C_{\text{FC}}$, then C_{FC} can be derived using the restricted reflexivity axiom.*

Proof. Consider a symmetry-preserving chase for $\mathcal{C} \models C_{\text{FC}}$. If initially only regular termination is possible, then this chase is also a successful symmetry-preserving chase for $\emptyset \models C_{\text{FC}}$. It follows that $\mathfrak{T} \models C_{\text{FC}}$, with $\mathfrak{T}$ the initial tableau constructed in lines 1–2 of Algorithm 2. This implies $f_1(\$r) = f_2(\$r)$, which in turn implies $\$r \in L$. Hence, C_{FC} can be derived using the restricted reflexivity axiom. □

For the case where initially only inconsistency termination is possible, we introduce the inconsistency axiom, of which we prove the soundness next.

Proposition 6 (Inconsistency). *If $(P', L' \to \$r')$, if there exist two embeddings of P' into a pattern P which agree on $L' \in \mathcal{V}_{P'}$ and map $\$r'$ to different constants of $\mathcal{U}$, and if $\$r \in \mathcal{V}_P$, then $(P, L \to \$r)$.*

Proof (soundness). Let e be an embedding of P into a relation $\mathcal{R}$ satisfying $(P', L' \to \$r')$. Let h_1 and h_2 be two embeddings of P' into P satisfying the conditions of Proposition 6. Then, clearly, $\varepsilon_1 = e \circ h_1$ and $\varepsilon_2 = e \circ h_2$ are embeddings of P' into $\mathcal{R}$ with $\varepsilon_1 =_{L'} \varepsilon_2$ and $\varepsilon_1(\$r_i) \neq \varepsilon_2(\$r_i)$. Hence, if there is an embedding of P into $\mathcal{R}$, then there exist two embeddings e_1 and e_2 of P' into $\mathcal{R}$ that agree on L', but not on $\$r'$. Hence, embeddings e_1 and e_2 show that $\mathcal{R}$ violates the functional constraint $(P', L' \to \$r')$, a contradiction. We conclude that there is no embedding of P into $\mathcal{R}$, as a consequence of which $\mathcal{R}$ voidly satisfies $(P, L \to \$r)$. □

We observe that the inconsistency axiom can be used in Example 6 to derive $(P, L \to \$r)$ from $\mathcal{C}$. We now generalize this observation.

Lemma 4. *If initially only inconsistency termination is possible in a symmetry-preserving chase for $\mathcal{C} \models C_{\text{FC}}$, then C_{FC} can be derived from $\mathcal{C}$ using the inconsistency axiom.*

Proof. Consider a symmetry-preserving chase for $\mathcal{C} \models C_{\text{FC}}$. If inconsistency termination is possible, then there exists a functional constraint $C_i = (P_i, L_i \to \$r_i) \in \mathcal{C}$ and embeddings e_1, e_2 of P_i into $\mathfrak{T}$ with $e_1 =_{L_i} e_2$, $e_1(\$r_i) \neq e_2(\$r_i)$, and $e_1(\$r_i), e_2(\$r_i) \in \mathcal{U}$. The embeddings e_1 and e_2 map P_i into $\mathfrak{T}$ and the function $({f_1}^{-1} \cup {f_2}^{-1} \cup \text{id}_{\mathcal{U}})$, which is well defined, maps $\mathfrak{T}$ into P. Hence,

$h_1 = ({f_1}^{-1} \cup {f_2}^{-1} \cup \mathrm{id}_{\mathcal{U}}) \circ e_1$ and $h_2 = ({f_1}^{-1} \cup {f_2}^{-1} \cup \mathrm{id}_{\mathcal{U}}) \circ e_2$ are embeddings of P_i into P with $h_1 =_{L_i} h_2$, $h_1(\$r_i) \neq h_2(\$r_i)$, and $h_1(\$r_i), h_2(\$r_i) \in \mathcal{U}$. Hence, C_{FC} can be derived from C_i using the inconsistency axiom. □

Next, consider the case where the chase for $\mathcal{C} \models C_{\mathrm{FC}}$ initially performs a symmetry-preserving step. We introduce the axioms *pattern-modification* and *left-modification* to deal with this case.

Proposition 7 (Pattern-modification). *Let P be a pattern, $L \subseteq \mathcal{V}_P$, $t \in \mathcal{V}_P \cup \mathcal{U}$, and $\$r, \$v \in \mathcal{V}_P$. If $(P', L' \to \$r')$, and $(\phi_{t \leftarrow \$v}(P), \phi_{t \leftarrow \$v}(L) \cap \mathcal{V}_P \to \{\phi_{t \leftarrow \$v}(\$r)\} \cap \mathcal{V}_P)$, and if there exists two embeddings of P' into P which agree on L' and map $\$r'$ to t and $\$v$, respectively, then $(P, L \to \$r)$.*

Proof (soundness). Let e be an embedding of P into a relation $\mathcal{R}$ satisfying $(P', L' \to \$r')$ and $(\phi_{t \leftarrow \$v}(P), \phi_{t \leftarrow \$v}(L) \cap \mathcal{V}_P \to \{\phi_{t \leftarrow \$v}(\$r)\} \cap \mathcal{V}_P)$. Let h_1 and h_2 be two embeddings of P' into P satisfying the conditions of Proposition 7. Then, $\varepsilon_1 = e \circ h_1$ and $\varepsilon_2 = e \circ h_2$ are embeddings of P' into $\mathcal{R}$ with $\varepsilon_1 =_{L'} \varepsilon_2$. By $(P', L' \to \$r')$, we have $\varepsilon_1(\$r') = \varepsilon_2(\$r')$, and hence $e(t) = e(\$v)$. Hence, $e|_{\mathrm{domain}(e) \setminus \{\$v\}}$ is an embedding of $\phi_{t \leftarrow \$v}(P)$ into $\mathcal{R}$.

Now, let e_1 and e_2 be two embeddings of P into $\mathcal{R}$ with $e_1 =_L e_2$. From the above, $\varepsilon_1 = e_1|_{\mathrm{domain}(e_1) \setminus \{\$v\}}$ and $\varepsilon_2 = e_2|_{\mathrm{domain}(e_2) \setminus \{\$v\}}$ are embeddings of $\phi_{t \leftarrow \$v}(P)$ into $\mathcal{R}$ satisfying $\varepsilon_1 =_{\phi_{t \leftarrow \$v}(L)} \varepsilon_2$, and hence, also $\varepsilon_1 =_{\phi_{t \leftarrow \$v}(L) \cap \mathcal{V}_P} \varepsilon_2$. By $(\phi_{t \leftarrow \$v}(P), \phi_{t \leftarrow \$v}(L) \cap \mathcal{V}_P \to \{\phi_{t \leftarrow \$v}(\$r)\} \cap \mathcal{V}_P)$, we have $\varepsilon_1 =_{\{\phi_{t \leftarrow \$v}(\$r)\} \cap \mathcal{V}_P} \varepsilon_2$, and, hence, we have $\varepsilon_1 =_{\{\phi_{t \leftarrow \$v}(\$r)\}} \varepsilon_2$. As $e_1(t) = e_1(\$v)$ and $e_2(t) = e_2(\$v)$, we also have $e_1(\$r) = e_2(\$r)$, even if $\$r = \$v$. □

Generally speaking, the pattern-modification axiom modifies the pattern of a functional constraint. More specifically, the axiom generalizes the pattern of a constraint due to constraints imposed by other functional constraints.

Example 9. Consider the set of functional constraints

$$\mathcal{C} = \{(\{(\$a, \$b, \$c)\}, \$a \to \$c), (\{(\$a, \$b, \$c), (\$a, \$b, e)\}, \$b \to \$a)\}$$

and the target functional constraint $C_{\mathrm{FC}} = (\{(\$a, \$b, e)\}, \$b \to \$a)$. We can derive C_{FC} from $\mathcal{C}$ by using the the embeddings h_1 and h_2 mapping $\{(\$a, \$b, \$c)\}$ to $\{(\$a, \$b, \$c)\}$ and $\{(\$a, \$b, e)\}$, respectively, and by picking $t = e$ and $\$v = \$c$. Indeed, due to the constraint imposed by $(\{(\$a, \$b, \$c)\}, \$a \to \$c)$, we are able to generalize $(\{(\$a, \$b, \$c), (\$a, \$b, e)\}, \$b \to \$a)$ to C_{FC}.

Proposition 8 (Left-modification). *Let P be a pattern, $L \subseteq \mathcal{V}_P$, $\$v \in \mathcal{V}_P$, and let $i_1, i_2 : \mathcal{V}_P \to \mathcal{V}$ be injective functions with $i_1(\$v) \neq i_2(\$v)$, $i_1 =_L i_2$, and $\mathrm{range}(i_1|_{\mathcal{V}_P \setminus L}) \cap \mathrm{range}(i_2|_{\mathcal{V}_P \setminus L}) = \emptyset$. If $(P', L' \to \$r')$, and $(P, L \cup \{\$v\} \to \$r)$, and if there exist two embeddings from P' into $(i_1 \cup \mathrm{id}_{\mathcal{U}})(P) \cup (i_2 \cup \mathrm{id}_{\mathcal{U}})(P)$ which agree on L' and map $\$r'$ to $i_1(\$v)$ and $i_2(\$v)$, respectively, then $(P, L \to \$r)$.*

Proof (soundness). Let e_1 and e_2 be two embeddings of P into a relation $\mathcal{R}$ satisfying $(P', L' \to \$r')$, $(P, L \cup \{\$v\} \to \$r)$, and $e_1 =_L e_2$. Let h_1 and h_2 be two embeddings of P' into $(i_1 \cup \mathrm{id}_{\mathcal{U}})(P) \cup (i_2 \cup \mathrm{id}_{\mathcal{U}})$ satisfying the conditions of Proposition 8. Since $i_1 =_L i_2$, $e_1 =_L e_2$, and $i_1 \cup \mathrm{id}_{\mathcal{U}}$ and $i_2 \cup \mathrm{id}_{\mathcal{U}}$ are injections whose range only overlap on $L \cup \mathcal{U}$, the function $f = \Phi_{e_1 \hookleftarrow i_1 \cup \mathrm{id}_{\mathcal{U}}} \circ \Phi_{e_2 \hookleftarrow i_2 \cup \mathrm{id}_{\mathcal{U}}}$ is well-defined. Hence, the functions $\varepsilon_1 = f \circ h_1$ and $\varepsilon_2 = f \circ h_2$ are embeddings of P' into $\mathcal{R}$ with $\varepsilon_1 =_{L'} \varepsilon_2$. By construction, we have $\varepsilon_1(\$r') = e_1(\$v)$ and $\varepsilon_2(\$r') = e_2(\$v)$. Hence, by $(P', L' \rightarrow \$r')$, we have $e_1(\$v) = e_2(\$v)$, and thus $e_1 =_{L \cup \{\$v\}} e_2$. By $(P, L \cup \{\$v\} \rightarrow \$r)$ and $e_1 =_{L \cup \{\$v\}} e_2$, we conclude $e_1(\$r) = e_2(\$r)$. □

The left-modification axiom generalizes a functional constraint by removing a variable from its left-hand side. This as a consequence of constraints imposed by other functional constraints.

Example 10. Consider the set of functional constraints

$$\mathcal{C} = \{(\{(\$a, \$b, \$c), (d, \$e, \$f)\}, \$c \rightarrow \$f), \\ (\{(\$a, \$b, \$c), (d, \$e, \$f)\}, \{\$a, \$f\} \rightarrow \$b)\}$$

and the target functional constraint $C_{\mathrm{FC}} = (\{(\$a, \$b, \$c), (d, \$e, \$f)\}, \$a \rightarrow \$b)$. We pick i_1 and i_2 such that:

$$(i_1 \cup \mathrm{id}_{\mathcal{U}})(\{(\$a, \$b, \$c), (d, \$e, \$f)\}) = \{(\$a, \$b_1, \$c_1), (d, \$e_1, \$f_1)\}$$
$$(i_2 \cup \mathrm{id}_{\mathcal{U}})(\{(\$a, \$b, \$c), (d, \$e, \$f)\}) = \{(\$a, \$b_2, \$c_2), (d, \$e_2, \$f_2)\}.$$

We can derive C_{FC} from $\mathcal{C}$ by picking the embeddings h_1 and h_2 such that:

$$h_1(\{(\$a, \$b, \$c), (d, \$e, \$f)\}) = \{(\$a, \$b_1, \$c_1), (d, \$e_1, \$f_1)\}$$
$$h_2(\{(\$a, \$b, \$c), (d, \$e, \$f)\}) = \{(\$a, \$b_1, \$c_1), (d, \$e_2, \$f_2)\}.$$

Indeed, due to the constraint imposed by $(\{(\$a, \$b, \$c), (d, \$e, \$f)\}, \$c \rightarrow \$f)$, we are able to generalize $(\{(\$a, \$b, \$c), (d, \$e, \$f)\}, \{\$a, \$f\} \rightarrow \$b)$ to C_{FC}. We notice that there is a relation between the left-modification axiom and the well-known multivalued dependencies [22]. In this example, the possible embeddings of the pattern $\{(\$a, \$b, \$c), (d, \$e, \$f)\}$ can be represented by a relational table T with schema $R(A, B, C, E, F)$. Due to $\mathcal{C}$, the functional dependencies $C \rightarrow F$ and $AF \rightarrow B$ hold on T. Due to the construction of T, also the multivalued dependency $A \twoheadrightarrow EF$ holds. Indeed, by using well-known derivation rules for functional dependencies and multivalued dependencies, we conclude $A \rightarrow B$.

We claim that the pattern-modification axiom simulates the symmetry-preserving steps of type 1–4, and the left-modification axiom the symmetry-preserving steps of type 5. Before proving that this is indeed the case, we introduce an auxiliary derivation rule. We emphasize that this rule is not part of our axiomatization. We shall only use its soundness to simplify the proof of Lemma 6.

Lemma 5 (Embedding). *If $(P', L' \rightarrow \$r')$ and h is an embedding from P' into P, then $(P, h(L') \cap \mathcal{V}_P \rightarrow \{h(\$r')\} \cap \mathcal{V}_P)$.*

Proof (soundness). Let e_1 and e_2 be embeddings of P into a relation $\mathcal{R}$ satisfying $(P', L' \to \$r')$. Then, $\varepsilon_1 = e_1 \circ h$ and $\varepsilon_2 = e_2 \circ h$ are embeddings of P' into $\mathcal{R}$. If $e_1 =_{h(L') \cap \mathcal{V}_P} e_2$, then $e_1 =_{h(L')} e_2$ and $\varepsilon_1 =_{L'} \varepsilon_2$, as embeddings always agree on constants. By $(P', L' \to \$r')$, we have $\varepsilon_1(\$r') = \varepsilon(\$r')$. As a consequence, we have $e_1 =_{h(\{\$r'\})} e_2$ and, hence, also $e_1 =_{h(\{\$r'\}) \cap \mathcal{V}_P} e_2$. □

The embedding rule explicitly maps functional constraints to different patterns, whereas the chase algorithm implicitly uses embeddings to deal with different patterns.

Example 11. If $(\{(\$a, \$b)\}, \$a \to \$b)$ holds, then trivially also $(\{(\$c, \$d)\}, \$c \to \$d)$ holds. We can derive $(\{(\$c, \$d)\}, \$c \to \$d)$ from $(\{(\$a, \$b)\}, \$a \to \$b)$ by using the embedding rule with the embedding that maps $\$a$ to $\$c$ and $\$b$ to $\$d$.

We now prove that symmetry-preserving steps can indeed be simulated by the pattern-modification and left-modification axioms.

Lemma 6. *Consider a successful symmetry-preserving chase for $\mathcal{C} \models C_{\text{FC}}$. If the chase starts with a symmetry-preserving step, using the functional constraint $C_i \in \mathcal{C}$ and resulting in tableau T' with tableau state $\mathbf{S}_{T'}(P', L', g_1, g_2)$, then there exists a functional constraint $C = (P', L' \to \$r')$ such that*

1. *the remainder of the chase starting from tableau T' is a successful symmetry-preserving chase for $\mathcal{C} \models C$.*
2. *we can derive C_{FC} from C_i and C using the pattern-modification and left-modification axioms.*

Proof. Let $\mathfrak{T} := T$ be the initial tableau in Algorithm 2. We assume that the initial symmetry-preserving step using $C_i = (P_i, L_i \to \$r_i)$ equalizes with the embeddings e_1 and e_2 satisfying $e_1 =_{L_i} e_2$ and $e_1(\$r_i) \neq e_2(\$r_i)$. The embeddings e_1 and e_2 map P_i into T and the function $({f_1}^{-1} \cup {f_2}^{-1} \cup \text{id}_{\mathcal{U}})$, which is well defined, maps T into P. Hence, $h_1 = ({f_1}^{-1} \cup {f_2}^{-1} \cup \text{id}_{\mathcal{U}}) \circ e_1$ and $h_2 = ({f_1}^{-1} \cup {f_2}^{-1} \cup \text{id}_{\mathcal{U}}) \circ e_2$ are embeddings of P_i into P with $h_1 =_{L_i} h_2$. Since the roles of f_1 and f_2 are interchangeable, we may assume, without loss of generality, that $e_1(\$r_1) = (f_1 \cup \text{id}_{\mathcal{U}})(t_1)$ and either $e_2(\$r_1) = (f_1 \cup \text{id}_{\mathcal{U}})(t_2)$ or $e_2(\$r_1) = (f_2 \cup \text{id}_{\mathcal{U}})(t_2)$. Here, t_1 and t_2 are terms of $\mathcal{V}_P \cup \mathcal{U}$ which are not both constants. We now distinguish two cases.

1. *The symmetry-preserving step is of type 1–4.* Without loss of generality, we may assume that $t_2 = \$v_2 \in \mathcal{V}_P$. The symmetry-preserving step results in a tableau $\mathfrak{T} := T' = \phi_{(f_1 \cup \text{id}_{\mathcal{U}})(t_1) \leftrightarrow f_1(\$v_2), (f_2 \cup \text{id}_{\mathcal{U}})(t_1) \leftrightarrow f_2(\$v_2)}(T)$ with state
$$\mathbf{S}_{T'}\left(\phi_{t_1 \leftrightarrow \$v_2}(P), L \setminus \{\$v_2\}, f_1|_{\mathcal{V}_P \setminus \{\$v_2\}}, f_2|_{\mathcal{V}_P \setminus \{\$v_2\}}\right).$$
Let $C = (\phi_{t_1 \leftrightarrow \$v_2}(P), \phi_{t_1 \leftrightarrow \$v_2}(L) \cap \mathcal{V}_P \to \{\phi_{t_1 \leftrightarrow \$v_2}(\$r)\} \cap \mathcal{V}_P)$. Clearly, T' is an initial tableau for the symmetry-preserving chase for $\mathcal{C} \models C$. It follows that the remainder of the chase for $\mathcal{C} \models C_{\text{FC}}$ is a successful chase for $\mathcal{C} \models C$, as C can be derived from C_{FC} using the embedding rule with embedding $\phi_{t \leftrightarrow \$w}$. By construction of h_1 and h_2, we have $h_1(\$r_i) = t_1$ and $h_2(\$r_i) = \$v_2$. Hence, C_{FC} can be derived from C_i and C using the pattern-modification axiom.

2. *The symmetry-preserving step is of type 5.* Then $t_1 = t_2 = \$v \in \mathcal{V}_P$. The symmetry-preserving step results in a tableau $\mathfrak{T} = T' = \phi_{f_1(\$v) \hookleftarrow f_2(\$v)}(T)$ with state

$$\mathbf{S}_{T'}\left(P, L \cup \{\$v\}, \phi_{f_1(\$v) \hookleftarrow f_2(\$v)} \circ f_1, \phi_{f_1(\$v) \hookleftarrow f_2(\$v)} \circ f_2\right).$$

Let $C = (P, L \cup \{\$v\} \rightarrow \$r)$. Clearly, T' is an initial tableau for the symmetry-preserving chase for $\mathcal{C} \models C$. It follows that the remainder of the chase for $\mathcal{C} \models C_{\mathrm{FC}}$ is a successful chase for $\mathcal{C} \models C$ as C can be derived from C_{FC} using a straightforward application of the reflexivity and transitivity axioms. Observe that $t_1 = f_1(\$v)$ and $t_2 = f_2(\$v)$ together with the embeddings e_1 and e_2 satisfy the conditions of Proposition 8, which allows the derivation of C_{FC} from C_i and C using the left-modification axiom. □

As a consequence of Corollary 1, Lemmas 3–6 yield a sound and complete axiomatization of the functional constraints.

Theorem 3. *The restricted reflexivity, inconsistency, pattern-modification, and left-modification axioms constitute an axiomatization for the functional constraints with at most one variable in their right-hand side.*

Proof. We have already proven soundness of the axioms and it is straightforward that the axioms are recursive, hence we only need to verify that the axioms are complete. Let $\mathcal{C}$ be a set of functional constraints and C_{FC} be a functional constraint with $\mathcal{C} \models C_{\mathrm{FC}}$. By Corollary 1, there exists a successful symmetry-preserving chase for $\mathcal{C} \models C_{\mathrm{FC}}$. We must prove, which we shall do by induction on the number of symmetry-preserving steps performed in this chase, that $\mathcal{C} \vdash C_{\mathrm{FC}}$. The base case is that no symmetry-preserving steps are performed, i.e., that the chase terminates immediately. Then $\mathcal{C} \vdash C_{\mathrm{FC}}$ follows from Lemma 3 and 4.

As inductive hypothesis, we assume that the existence of a successful symmetry-preserving chase for $\mathcal{C}' \models C'_{\mathrm{FC}}$ with $i \geq 0$ symmetry-preserving steps ($\mathcal{C}'$ a set of functional constraints and C'_{FC} a single functional constraint) yields $\mathcal{C}' \vdash C'_{\mathrm{FC}}$. For the inductive step, assume that the successful symmetry-preserving chase for $\mathcal{C} \models C_{\mathrm{FC}}$ has $i+1$ symmetry-preserving steps. Assume that the first symmetry-preserving step uses $C_i \in \mathcal{C}$. By Lemma 6, there exists a functional constraint $C = (P', L' \rightarrow \$r')$ such that $\{C_i, C\} \vdash C_{\mathrm{FC}}$ and such that the remainder of the chase is a successful symmetry-preserving chase for $\mathcal{C} \models C$. As this chase has only i symmetry-preserving steps, the inductive hypothesis yields $\mathcal{C} \vdash C$. We thus conclude that $\mathcal{C} \vdash C_{\mathrm{FC}}$, which completes the proof. □

Using Lemma 1 we generalize Theorem 3 to functional constraints with arbitrary sets of variables in their right-hand side.

Corollary 2. *The inconsistency, pattern-modification, and left-modification axioms together with the reflexivity, augmentation, and transitivity axioms constitute an axiomatization for the functional constraints.*

Moreover, we have the following (proof omitted).

Theorem 4. *The axiomatization of the functional constraints is no longer complete if one of the axioms reflexivity, augmentation, transitivity, inconsistency, pattern-modification, or left-modification is removed.*

6 Conclusions and Directions for Future Work

Starting from functional and equality-generating constraints for the RDF data model, we studied functional constraints on arbitrary relations. As our first result, we proved the existence of a symmetry-preserving chase for the functional constraints. Using the symmetry-preserving chase, we derived a sound and complete axiomatization for the functional constraints. This solves a major open problem in the work on functional constraints for the RDF data model.

We believe that our work provides a promising formal basis for reasoning about functional constraints. As for future work, one remaining open problem is the existence of Armstrong relations [15,19] for the functional constraints. Another avenue of research concerns generalizations of functional constraints. In particular, adding constants to the right-hand side of functional constraints would result in a very powerful class of constraints that generalizes both the functional constraints and the conditional functional dependencies [17]. Finally, it is unknown what the complexity of working with functional constraints is, as compared with the functional dependencies and equality-generating constraints.

References

1. Akhtar, W., Cortés-Calabuig, Á., Paredaens, J.: Constraints in RDF. In: Schewe, K.-D., Thalheim, B. (eds.) SDKB 2010. LNCS, vol. 6834, pp. 23–39. Springer, Heidelberg (2011)
2. Codd, E.F.: Relational completeness of data base sublanguages. Technical Report RJ 987, IBM Research Laboratory, San Jose, California (1972)
3. Codd, E.F.: Recent investigations in relational data base systems. Information Processing 74, 1017–1021 (1974)
4. Beeri, C., Vardi, M.: The implication problem for data dependencies. In: Even, S., Kariv, O. (eds.) Automata, Languages and Programming. LNCS, vol. 115, pp. 73–85. Springer, Heidelberg (1981)
5. Abiteboul, S., Hull, R., Vianu, V.: Foundations of Databases. Addison-Wesley (1995)
6. Lausen, G., Meier, M., Schmidt, M.: SPARQLing constraints for RDF. In: Proceedings of the 11th International Conference on Extending Database Technology: Advances in Database Technology, EDBT 2008, pp. 499–509 (2008)
7. Hartmann, S., Link, S.: More functional dependencies for XML. In: Advances in Databases and Information Systems. In: Kalinichenko, L., Manthey, R., Thalheim, B., Wloka, U. (eds.) ADBIS 2003. LNCS, vol. 2798, pp. 355–369. Springer, Heidelberg (2003)
8. Buneman, P., Davidson, S., Fan, W., Hara, C., Tan, W.C.: Keys for XML. Computer Networks 39(5), 473–487 (2002)
9. Hartmann, S., Link, S.: Efficient reasoning about a robust XML key fragment. ACM Transactions on Database Systems 34(2), 10:1–10:33 (2009)

10. Vincent, M.W., Liu, J., Mohania, M.: The implication problem for 'closest node' functional dependencies in complete XML documents. Journal of Computer and System Sciences 78(4), 1045–1098 (2012)
11. Arenas, M., Libkin, L.: A normal form for XML documents. ACM Transactions on Database Systems 29(1), 195–232 (2004)
12. Calbimonte, J.P., Porto, F., Keet, C.M.: Functional dependencies in OWL ABOX. In: XXIV Simpósio Brasileiro de Banco de Dados, pp. 16–30 (2009)
13. Yu, Y., Heflin, J.: Extending functional dependency to detect abnormal data in RDF graphs. In: Aroyo, L., Welty, C., Alani, H., Taylor, J., Bernstein, A., Kagal, L., Noy, N., Blomqvist, E. (eds.) ISWC 2011, Part I. LNCS, vol. 7031, pp. 794–809. Springer, Heidelberg (2011)
14. Cortés-Calabuig, A., Paredaens, J.: Semantics of constraints in RDFS. In: Proceedings of the 6th Alberto Mendelzon International Workshop on Foundations of Data Management, pp. 75–90 (2012)
15. Fagin, R.: Horn clauses and database dependencies. Journal of the ACM 29(4), 952–985 (1982)
16. Wijsen, J.: Database repairing using updates. ACM Transactions on Database Systems 30(3), 722–768 (2005)
17. Fan, W., Geerts, F., Jia, X., Kementsietsidis, A.: Conditional functional dependencies for capturing data inconsistencies. ACM Transactions on Database Systems 33(2), 6:1–6:48 (2008)
18. He, Q., Ling, T.W.: Extending and inferring functional dependencies in schema transformation. In: Proceedings of the Thirteenth ACM International Conference on Information and Knowledge Management, CIKM 2004, pp. 12–21. ACM (2004)
19. Armstrong, W.W.: Dependency structures of data base relationships. Information Processing 74, 580–583 (1974)
20. Aho, A.V., Beeri, C., Ullman, J.D.: The theory of joins in relational databases. ACM Transactions on Database Systems 4(3), 297–314 (1979)
21. Beeri, C., Vardi, M.Y.: A proof procedure for data dependencies. Journal of the ACM 31(4), 718–741 (1984)
22. Beeri, C., Fagin, R., Howard, J.H.: A complete axiomatization for functional and multivalued dependencies in database relations. In: Proceedings of the 1977 ACM SIGMOD International Conference on Management of Data, SIGMOD 1977, pp. 47–61 (1977)

View-Based Tree-Language Rewritings for XML

Laks V.S. Lakshmanan[1] and Alex Thomo[2]

[1] University of British Columbia, Vancouver, BC, Canada
laks@cs.ubc.ca
[2] University of Victoria, Victoria, BC, Canada
thomo@cs.uvic.ca

Abstract. We study query rewriting using views (QRV) for XML. Our queries and views are regular tree languages (RTLs) represented by tree automata over marked alphabets, where the markers serve as "node selectors". We formally define query rewriting using views for RTLs and give an automata-based algorithm to compute the maximally contained rewriting. The formalism we use is equal in power with Monadic Second Order (MSO) logic, and our algorithm for computing QRV is the first to target this expressive class. Furthermore we prove a tight lower bound, thus showing that our algorithm is optimal. Another strength of our automata-based approach is that we are able to cast computing QRV into executing a sequence of intuitive operations on automata, thus rendering our approach practical as it can be easily implemented utilizing off-the-shelf automata toolboxes. Finally, we generalize our framework to account for more complex queries in the spirit of the `FOR` clause in XQuery. For this generalization as well, we give an optimal algorithm for computing the maximally contained rewriting of queries using views.

Keywords: XML, View-Based Rewriting, Tree Automata.

1 Introduction

Query rewriting using views (QRV) is a fundamental problem that finds wide applications in query optimization, data integration, data warehousing, security, and other critical database services. In this paper we study QRV for node-selecting queries and views, over XML trees. As there are several variants of views for answering queries we begin by illustrating the classical QRV problem we focus on in this paper.

Example 1. Suppose we have a very large collection of movies such as imdb.com, organized in a super-tree, with each movie being a sub-tree containing title, year, and characters. Suppose the character nodes branch out into actors playing the character. Sometimes, a movie contains characters played by more than one actor. We call these characters "multi-actor characters (MAC)". Consider a query which returns all the movie sub-trees having a MAC. Such movies are a small minority; their number is about 50. Assume that the result of this query is materialized into a view. Obviously this view is of tremendous help in answering

C. Beierle and C. Meghini (Eds.): FoIKS 2014, LNCS 8367, pp. 270–289, 2014.

some new queries, such as "find all the actors playing a MAC". We can rewrite this new query into "find all the actors playing a MAC in a movie having a MAC" and answer it on the view-extension instead of accessing the original database. The difference in performance is huge; using the view materialization takes 50 accesses, whereas using the original database takes hundreds of thousands of accesses (there are about 700,000 movies as per imdb.com).

Due to its importance, QRV has been explored for fragments of XPath (cf. [5,31,16,8,2,30]) and fragments of XQuery (cf. [32,24,4]). In this paper, we study the problem for queries and views represented by tree automata, which provide for a more general approach that can elegantly capture fine grained structure along vertical and horizontal axes. These queries can be easily specified using a DTD-like syntax, rendering them user-friendly devices for querying XML. While the techniques and tools introduced in the aforementioned works are interesting and well-founded for computing QRV for XPath and XQuery, they do not seem to extend to dealing with the more general setting of queries and views represented by tree automata, which have desirable properties with respect to expressivity in querying XML. In regard to expressivity, Neven and Schwentick [20,21] and Schwentick [26] argue for formalisms as expressive as monadic second order logic (MSO) for specifying node-selecting queries. This is sometimes called a "golden standard" against which query formalisms should be measured up.

In this paper, we study the QRV problem for queries and views represented by automata equivalent in power to MSO. It is worth pointing out that being able to handle this target expressivity is one of the strengths of our approach to QRV. Another strength is the fact that computing QRV is cast into executing a sequence of intuitive operations on automata. This makes our solution quite practical, as it can be easily implemented on top of the readily available automata toolboxes (e.g. [12,9]).

Our queries and views are first described as sets of tree-position pairs in order to facilitate the definitions and understanding of rewritings. In practice the queries and views are specified by finite tree automata over alphabets with marked symbols that serve the purpose of selecting nodes in XML trees. More specifically, for our constructions we use colors as markers, which makes the development easier. This formalism is equivalent to the formalism of querying trees using automata with boolean markings (ABM) [cf. [28], further developed in [23]].

Automata over colored alphabets provide us with critical advantages in computing QRV. The first advantage is the ability to express and construct a series of intermediate languages for obtaining query rewritings. Being able to use multiple colors for marking regions of interest in the trees of these intermediate languages is crucial to our approach. Our language constructions are also facilitated by the one-way nature of the automata we use. The second advantage is the ability to determinize the automata. This is key to our main construction for query rewriting.

Summarizing, we provide an algorithm for QRV for the general case where queries and views are implemented in an automata framework equal in power to MSO, to our knowledge, for the first time. Our algorithm runs in *singly-exponential time*. We show that QRV in our case is EXPTIME-hard, thus showing the optimality of our algorithm. Additionally, we show how to extend our results to reasoning about QRV for the more general case of queries returning a forest of trees as an answer in the spirit of the `FOR` clause in XQuery. For this generalization as well, we provide an optimal algorithm to compute query rewritings using views.

We make the following contributions.

1. We define tree-pattern operators to cleanly express view-based rewriting as the solution to a view-query equation (Section 4).
2. Next, we present an algorithm for computing query rewritings using views. We define languages *over colored alphabets* and present a series of intermediate automata constructions, which we believe are of independent interest. We prove an EXPTIME lower bound for QRV, and show that our algorithm is optimal (Section 5).
3. We generalize our results to the case where queries select more than one subtree and produce a set of forests as output. We show that we are still able to compute view-based rewritings in singly-exponential time, thus being again optimal with respect to the above lower bound (Sections 6, 7, and 8).

2 Related Works

Automata theory has long been recognized as a useful tool for providing elegant solutions to challenging problems on XML (cf. [19,21,14,22,26,17,7]).

Two prominent automata-based approaches for querying XML data that have been proposed are (bottom-up) finite tree automata with selecting states (FTAS) introduced by Neven ([19], p. 128) and Frick et al. ([14]), and query automata (QA) introduced by Neven and Schwentick ([21]). Both formalisms are shown to capture precisely the queries definable in MSO and thus are the natural candidates for us to study the QRV problem in. However, two important issues prevent us from using either of these formalisms directly for our study of QRV: (1) Deterministic FTAS are too weak to express all queries expressible by non-deterministic FTAS; determinism is a key property that our techniques and algorithms rely on. (2) On the other hand, QA are deterministic, but two-way, and thus can go up and down a target tree multiple times, a feature that makes reasoning about QRV difficult.

Another nice standard for querying XML is Propositional Dynamic Logic (PDL) for trees ([1,6]), which corresponds to Regular XPath. The latter was shown by ten Cate and Segoufin ([27]) to be not expressively complete for MSO. Regular XPath was extended later to μXPath by Calvanese et al. ([7]) to obtain the full power of MSO. It may be possible to use automata formalisms, such the one in [7] capturing μXPath, for QRV–this is an avenue we leave open for future exploration.

We close this section by briefly discussing two other works on query rewriting. Fan et al. ([11]) study rewritings of regular XPath queries defined over *virtual* views. This is a different problem from QRV that we consider here. The views in our case are materialized and the queries are over the original database. It would be interesting to see how our techniques could be used for the problem studied by Fan et al., considering an alternate query formalism such as the one we propose that is complete for MSO.

Thomo and Venkatesh ([29]) use Visibly Pushdown Automata to model and rewrite (XML) schemas by using other schemas. Again, this problem is different from the problem we study in this paper.

3 Automata

We consider finite ordered trees – simply called *trees*. Also, we consider the trees to be *unranked*, which is to say that the nodes of the tree have an arbitrary (but finite) arity (cf. [10], p. 200). The nodes of the trees are labeled by symbols drawn from a fixed alphabet Σ. We denote by Υ the set of all trees over Σ. We use $a, b, \ldots, e$ to denote labels in Σ and $x, y, \ldots$, possibly with subscripts, to denote tree nodes. We denote by r_t the root of tree t, by σ_x the label of node x, and by N_t the set of nodes of a tree $t \in \Upsilon$.

Definition 1. *A non-deterministic, bottom-up,* finite tree automaton *(FTA) over Σ is a quadruple $\mathcal{A} = (S, \Sigma, F, \Delta)$, where S is a finite set of states, $F \subseteq S$ is a set of final states, and Δ is a finite set of transition rules of the form $H \xrightarrow{a} s$, where $H \subseteq S^*$ is a regular language over S, $a \in \Sigma$, and $s \in S$.*

Here, H is called a *horizontal* language. For simplicity, we blur the distinction between regular languages over S and the regular expressions used to specify them.

Definition 2. *A tree t is accepted by an FTA $\mathcal{A}$ if there exists a mapping $\mu : N_t \to S$ such that:*

1. *If $\mu(x) = s$, then there is a transition rule $H \xrightarrow{\sigma_x} s$ in Δ with $\mu(x_1) \ldots \mu(x_n) \in H$, where $x_1, \ldots, x_n$ are* all *the children of x in t in order.*
2. *If x is a leaf and $\mu(x) = s$, then there is a transition rule $H \xrightarrow{\sigma_x} s$ in Δ with $\epsilon \in H$.*
3. *$\mu(r_t) \in F$.*

A mapping μ as above specifies an *accepting run* of $\mathcal{A}$ on t. An accepting run can be considered to be a tree of the same shape as t whose nodes are labeled by states given by the mapping μ.

We denote by $L(\mathcal{A})$ the set (language) of trees accepted by $\mathcal{A}$. A tree language L is said to be accepted (or recognized) by an FTA $\mathcal{A}$ if $L = L(\mathcal{A})$. Tree languages recognized by FTAs are called *regular tree languages* (RTLs).

Example 2. Consider a collection of trees representing movies having, among other elements, one or more characters, each played by one or more actors. Let us use m, t, y, c, *and* a *to abbreviate* movie, title, year, character, *and* actor, *respectively.*

An FTA accepting movies having at least one character played by two or more actors is

$$\mathcal{A} = (\{s, s_{\mathtt{m}}, s_{\mathtt{c}}, s_{\mathtt{a}}\}, \Sigma, \{s_{\mathtt{m}}\}, \Delta)$$

where $\Sigma \supset \{\mathtt{m}, \mathtt{t}, \mathtt{y}, \mathtt{c}, \mathtt{a}\}$ *and* Δ *has the following transition rules:*

$$s^* s_{\mathtt{c}} s^* \xrightarrow{\mathtt{m}} s_{\mathtt{m}}$$
$$s^* s_{\mathtt{a}} s_{\mathtt{a}} s^* \xrightarrow{\mathtt{c}} s_{\mathtt{c}}$$
$$s^* \xrightarrow{\mathtt{a}} s_{\mathtt{a}}$$
$$s^* \xrightarrow{\Sigma} s.$$

The last transition rule is a shorthand for saying that "we can go from s^* *to* s *on any symbol of* Σ*."*

Let now t *be the movie tree given in Figure 1, left. Clearly,* t *is accepted by* $\mathcal{A}$*. We assume that all the textual (unstructured data) have been identified (labeled) with a special symbol, say* d $\in \Sigma$*. An accepting run of* $\mathcal{A}$ *on* t *is given in the same figure, right.*

We describe in Section 4 how to modify $\mathcal{A}$ *to* query *for those movies having a character played by two or more actors.* □

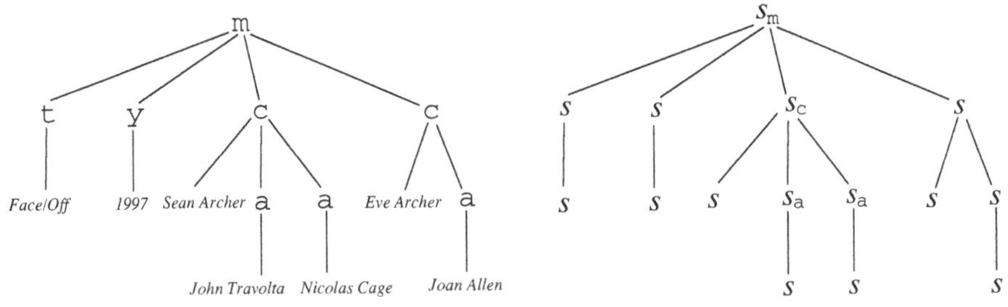

Fig. 1. A tree t and an accepting run of $\mathcal{A}$ on t

It is worth noting that an FTA can be specified using extended DTDs (EDTDs), a syntax that may be more familiar to a user (cf. [10], p. 233).

Consider again Definition 1. The FTA $\mathcal{A}$ is called *deterministic* if $H_1 \cap H_2 = \emptyset$ for all transitions $H_1 \xrightarrow{a} s_1$ and $H_2 \xrightarrow{a} s_2$, where $s_1 \neq s_2$. In such a case, for any tree t, there can be at most one accepting run of $\mathcal{A}$ on t.

FTAs that never get "stuck" are called *complete*. Given an FTA, it can be verified that the determinization procedure of [10] (p. 204) produces a deterministic FTA that is complete. Given a tree t, a complete and deterministic FTA (CDFTA) $\mathcal{A}$ can read t in only one way, i.e., there is exactly one run of $\mathcal{A}$ on t.

An FTA is *normalized* if for each symbol-state pair (a, s), there is at most one transition $H \xrightarrow{a} s$. Each FTA can be transformed into a normalized FTA by unioning the left-hand sides of the transition rules labeled by the same symbol and having the same state on the right-hand side. If the FTA is deterministic, then it will remain so after this transformation.

Intersection. It is a well known fact that if L_1 and L_2 are RTLs, so is $L_1 \cap L_2$. However, the proof is typically done by encoding unranked trees into binary ones (cf. [10], p. 209). We can instead use a direct construction which preserves completeness and determinism. Given FTAs $\mathcal{A}_1$ and $\mathcal{A}_2$ for L_1 and L_2, respectively, we can construct an FTA $\mathcal{A}$ which recognizes $L_1 \cap L_2$. Furthermore, if $\mathcal{A}_1$ and $\mathcal{A}_2$ are complete and deterministic, so is $\mathcal{A}$ (see [15]).

Targetedness. Here we also introduce our notion of "targeted" FTAs. Targeted FTAs turn out to be important in computing rewritings of queries using views.

Definition 3. *A normalized FTA is* targeted *if each state is the target (right side) of at most one transition rule.*

We present the following result, whose proof can be found in [15].

Theorem 1. *For each FTA $\mathcal{A}$ we can obtain an equivalent targeted FTA $\mathcal{A}\uparrow$ in PTIME.*

Furthermore, the procedure preserves determinism, i.e. we have that if $\mathcal{A}$ is deterministic, then so is $\mathcal{A}\uparrow$.

4 Queries, Views, and Rewritings

We consider here (tree node) positions given by Dewey-style strings in $\mathbb{N}^*$. If t is a tree, we denote by $pos(t)$ the set of all t's positions. $pos(t)$ is prefix-closed and contains ϵ as the root position. Given a position $x \in pos(t)$, we define t_x to be the subtree of t rooted at x.

Definition 4. *A* pattern *is a tree-position pair (p, x), where $x \in pos(p)$.*

Let $\Upsilon^{\mathsf{x}} = \{(p, x) \mid p \in \Upsilon, x \in pos(p)\}$.

Definition 5. *A* tree query (TQ) *Q is a subset of Υ^{x}.*

When a query has only one pattern (p, x), we will blur the distinction between $\{(p, x)\}$ and (p, x).

Recall that Υ is the set of all trees over Σ. We call them *target trees*. Let $t \in \Upsilon$ be a target tree.

Definition 6. *The answer to Q on $t \in \Upsilon$ is $ans(Q, t) = \{t_x : (t, x) \in Q\}$.*

For two queries Q_1, Q_2, we define containment and equivalence as follows.

Definition 7.

1. $Q_1 \sqsubseteq Q_2$ *if* $ans(Q_1, t) \subseteq ans(Q_2, t)$ *for each* $t \in \Upsilon$.
2. $Q_1 \equiv Q_2$ *if* $Q_1 \sqsubseteq Q_2$ *and* $Q_2 \sqsubseteq Q_1$.

We show that

Theorem 2. *If* $Q_1 \subseteq Q_2$ *then* $Q_1 \sqsubseteq Q_2$.

Corollary 1. *If* $Q_1 = Q_2$ *then* $Q_1 \equiv Q_2$.

Regular TQs. Let $\hat{\Sigma} = \{\hat{a} : a \in \Sigma\}$ be an alphabet of marked symbols. Given (p, x), consider $\hat{p}$ on $\Sigma \cup \hat{\Sigma}$ that is the same as p, but with the node at position x being marked by the corresponding symbol in $\hat{\Sigma}$. Now a query Q is a *regular tree query* (RTQ) if set $\{\hat{p} \mid (p, x) \in Q\}$ is regular, i.e. given by a tree automaton.

Example 3. Let us revisit Example 2. Assume that the collection of movie trees is organized into a super-tree t with a root labeled by r, and the movie trees as sub-trees of the root. Now let Q be a query that asks for all the movie sub-trees having at least one character played by two or more actors. It can be verified that this query can be given by the following automaton.

$$\mathcal{A} = (\{s, s_{\mathtt{t}}, s_{\mathtt{m}}, s_{\mathtt{c}}, s_{\mathtt{a}}\}, \Sigma \cup \hat{\Sigma}, \{s_{\mathtt{t}}\}, \Delta)$$

where Δ has the following transition rules

$$\begin{aligned} s^* s_{\mathtt{m}} s^* &\xrightarrow{\mathtt{r}} s_{\mathtt{t}} \\ s^* s_{\mathtt{c}} s^* &\xrightarrow{\hat{\mathtt{m}}} s_{\mathtt{m}} \\ s^* s_{\mathtt{a}} s_{\mathtt{a}} s^* &\xrightarrow{\mathtt{c}} s_{\mathtt{c}} \\ s^* &\xrightarrow{\mathtt{a}} s_{\mathtt{a}} \\ s^* &\xrightarrow{\Sigma} s. \square \end{aligned}$$

Views and Rewritings. First we give a simple example to build up the intuition.

Example 4. Consider a view definition containing only the pattern tree given in Figure 2 [left] and a query containing only the pattern tree given in the same figure [middle]. Clearly, the view is useful in answering the query. Intuitively, the pattern tree we need to use to extract the answer to the query using the view is shown in the same figure [right]. This is nothing else but the "rewriting" of the query using the view in this example.

We now make precise a rewriting of a query Q using a view V.

Recall sets $\Upsilon, \Upsilon^{\mathsf{x}}$. Furthermore, let $\Upsilon^{\mathsf{x},\mathsf{y}}$ be the set of all tree-position-position triples (p, x, y), such that $x, y \in pos(p)$, and x is a proper prefix of y, i.e. x is a proper ancestor of y. We restrict x to be a proper prefix of y because, as we

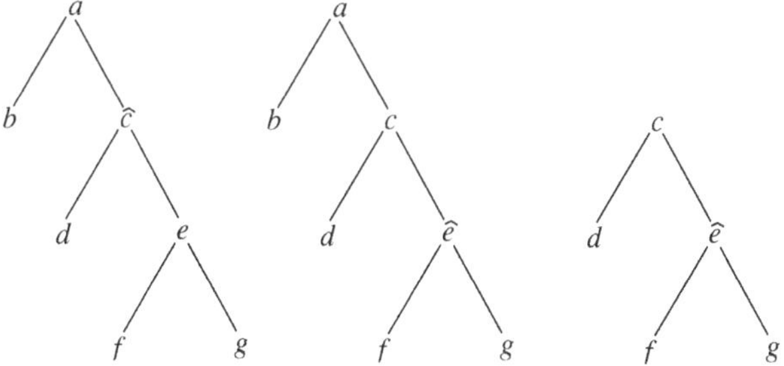

Fig. 2. A simple view, a simple query, and the rewriting. Pattern positions are given by ˆ.

show in the full version [15], the queries and views can be transformed so that their specified positions (markings) never coincide in matching patterns.

Let $(p,x),(q,y) \in \Upsilon^{\mathsf{x}}$. We define

$$(p,x) \star (q,y) = \begin{cases} (p, xy) & \text{if } p_x = q \text{ and } y \neq \epsilon \\ \text{undefined} & \text{otherwise} \end{cases}$$

and

$$(p,x) \bigstar (q,y) = \begin{cases} (p, x, xy) & \text{if } p_x = q \text{ and } y \neq \epsilon \\ \text{undefined} & \text{otherwise.} \end{cases}$$

where xy is the concatenation of x and y.

In simple words, for $(p,x) \star (q,y)$ and $(p,x) \bigstar (q,y)$ to be defined, the subtree of p rooted at x must *be identical to* q structure-wise. Also, structure-wise, $(p,x) \star (q,y)$ and $(p,x) \bigstar (q,y)$ are the same as (p,x).

We note that, when defined, $(p,x) \star (q,y) \in \Upsilon^{\mathsf{x}}$ and $(p,x) \bigstar (q,y) \in \Upsilon^{\mathsf{x},\mathsf{y}}$. Referring to Figure 2 from the previous example, let the view be the tree on the left and the query be the tree in the middle. Then the tree on the right is a rewriting. If we regard the view on the left as (p,x) and the query in the middle as $(p,x) \star (q,y)$, then the rewriting needed to answer the query is (q,y), which is indeed the tree on the right. Therefore, finding the rewriting is solving a $\star$ equation with the rewriting as the unknown.

For two sets $L_1, L_2 \subseteq \Upsilon^{\mathsf{x}}$, we define $L_1 \star L_2$ and $L_1 \bigstar L_2$ in the natural way. Also, we blur the distinction between a set of one element and the element itself. We note that $L_1 \star L_2 \subseteq \Upsilon^{\mathsf{x}}$ and $L_1 \bigstar L_2 \subseteq \Upsilon^{\mathsf{x},\mathsf{y}}$.

We will use $\star$ to define the rewriting and $\bigstar$ to reason about some steps in the construction for the rewriting.

In the following we use Greek letters υ and ξ to denote tree-position pairs.

Definition 8. *The* maximally contained rewriting (MCR) *of* $Q \subseteq \Upsilon^{\mathsf{x}}$ *using* $V \subseteq \Upsilon^{\mathsf{x}}$ *is*

$$R = \{\xi \in \Upsilon^{\mathsf{x}} : V \star \xi \subseteq Q\}.$$

We also define set X of "bad" patterns as

$$X = \{\xi \in \Upsilon^{\mathsf{x}} : \text{ there exists } v \in V \text{ such that } v \star \xi \in Q^c\}$$

where $Q^c = \Upsilon^{\mathsf{x}} \setminus Q$. Set X will be crucial in our construction. Observe that X is not equal to $\Upsilon^{\mathsf{x}} \setminus R$. The latter also contains patterns that cannot be $\star$-ed with V's patterns. Now let us define

$$Y = \{\xi \in \Upsilon^{\mathsf{x}} : \text{ there exists } v \in V \text{ such that } v \star \xi \in Q\}.$$

We have that

Proposition 1. $R = Y \setminus X$.

Example 5. Consider the following queries and views on chain trees, where the positions have been represented as ˆ over the corresponding nodes. The chains should be read backwards to simulate a bottom-up processing.

1. *Let* $Q = \{aac\hat{d}e, bbc\hat{d}e, ccc\hat{d}e\}$ *and* $V = \{aa\hat{c}de, bb\hat{c}de\}$. *We have* $c\hat{d}e \in R$ *because* $V \star c\hat{d}e = \{aac\hat{d}e, bbc\hat{d}e\} \subseteq Q$. *In fact* $c\hat{d}e$ *is the only chain in* R.
2. *Let* $Q = \{aac\hat{d}e, ccc\hat{d}e\}$ *and* $V = \{aa\hat{c}de, bb\hat{c}de\}$. *We have* $c\hat{d}e \notin R$ *because* $bbc\hat{d}e \in V \star c\hat{d}e$, *but* $bbc\hat{d}e \notin Q$. *Notice,* $c\hat{d}e$ *is in* X *and* Y. □

Let $T_{V,t}$ be the *materialized answer* (MA) to V on $t \in \Upsilon$, i.e., $T_{V,t} = ans(V,t)$. We answer Q using V by computing

$$ans(R, T_{V,t}) = \bigcup_{v \in T_{V,t}} ans(R, v)$$

where R is the MCR of Q using V. We have that

Theorem 3. $ans(R, T_{V,t}) \subseteq ans(Q,t)$ *for each* $t \in \Upsilon$.

Proof. From Definition 8 and Theorem 2, we have $ans(V \star R, t) \subseteq ans(Q,t)$. Now the claim follows from this equality which we prove next

$$ans(V \star R, t) = ans(R, T_{V,t}).$$

"$\supseteq$": Let $z \in ans(R, T_{V,t})$. Then there exists $(q,y) \in R$ and $q \in T_{V,t}$ such that $q_y = z$. Since $q \in T_{V,t}$, there exists $(t,x) \in V$ such that $t_x = q$. Clearly, $q_y = z \in ans((t,x) \star (q,y), t)$. By $(t,x) \in V$ and $(q,y) \in R$ we obtain $z \in ans(V \star R, t)$ as required.

"$\subseteq$": Let $z \in ans(V \star R, t)$. There exists $(t,xy) = (t,x) \star (q,y) \in V \star R$, where $(t,x) \in V$ and $(q,y) \in R$, such that $ans((t,xy),t) = t_{xy} = z$. Consequently, $t_x \in T_{V,t}$ and $z \in ans(R, T_{V,t})$. □

5 Computing the MCR

The previous section refers to general queries and views. In this section we show that when they are RTQs, we can effectively compute QRV and show that it is an RTQ as well.

First we define the inverse of the $\star$ operation. Let $\upsilon, \upsilon' \in \Upsilon^{\mathbf{x}}$. We define

$$\upsilon \circ \upsilon' = \begin{cases} \xi & \text{if } \upsilon' = \upsilon \star \xi \\ \text{undefined} & \text{otherwise} \end{cases}$$

This definition is lifted to subsets of $\Upsilon^{\mathbf{x}}$ in the natural way. Now, it can be verified that

Proposition 2. *$X = V \circ Q^c$ and $Y = V \circ Q$.*

Once we obtain X and Y, we can obtain R as $Y \setminus X$ (see Proposition 1). In the rest of the section, we present an automata-based solution for computing

$$K = J \circ J'$$

when J and J' are RTQs.

Tree-position pair-sets facilitate the definitions of queries and rewritings. However, when we work with automata, we need to talk in terms of languages that the automata recognize. Here, we introduce colors for marking the positions in pattern trees. Colors are similar to Boolean markings of [28] and [23] (or to our previous ^ marking), however, they make the presentation easier.

Thereto, we consider the "red" and "blue" alphabets $\Sigma^{\mathbf{r}} = \{a^{\mathbf{r}} : a \in \Sigma\}$ and $\Sigma^{\mathbf{b}} = \{a^{\mathbf{b}} : a \in \Sigma\}$ in addition to the alphabet Σ. We refer to the elements of Σ as being "black" and to Σ as the "black" alphabet. We refer to nodes as black, red, or blue nodes if their symbol is black, red, or blue.

We define $\mathbf{b}$ as a (unary) operator that given $(p, x) \in \Upsilon^{\mathbf{x}}$ returns a tree over $\Sigma \cup \Sigma^{\mathbf{b}}$ that is isomorphic to p with the node at position x colored blue.

Likewise, we define $\mathbf{r}$ as a (unary) operator that given $(p, y) \in \Upsilon^{\mathbf{y}}$ returns a tree over $\Sigma \cup \Sigma^{\mathbf{r}}$ that is isomorphic to p with the node at position y colored red. [$\Upsilon^{\mathbf{y}}$ is the same as $\Upsilon^{\mathbf{x}}$ and is used for notation parallelism.]

Furthermore, we define $\mathbf{br}$ as a (unary) operator that given $(p, x, y) \in \Upsilon^{\mathbf{x},\mathbf{y}}$ returns a tree over $\Sigma \cup \Sigma^{\mathbf{b}} \cup \Sigma^{\mathbf{r}}$ that is isomorphic to p with the node at position x colored blue and the node at the position y colored red.

Based on these operators, we define

$$\begin{aligned} \Upsilon^{\mathbf{b}} &= \{\mathbf{b}(p, x) : (p, x) \in \Upsilon^{\mathbf{x}}\} \\ \Upsilon^{\mathbf{r}} &= \{\mathbf{r}(p, y) : (p, y) \in \Upsilon^{\mathbf{y}}\} \\ \Upsilon^{\mathbf{b},\mathbf{r}} &= \{\mathbf{br}(p, x, y) : (p, x, y) \in \Upsilon^{\mathbf{x},\mathbf{y}}\}. \end{aligned}$$

$\Upsilon^{\mathbf{b}}$, $\Upsilon^{\mathbf{r}}$, $\Upsilon^{\mathbf{b},\mathbf{r}}$ are *languages of trees* over $\Sigma \cup \Sigma^{\mathbf{b}}$, $\Sigma \cup \Sigma^{\mathbf{r}}$, $\Sigma \cup \Sigma^{\mathbf{b}} \cup \Sigma^{\mathbf{r}}$, respectively.

Clearly, there is a one-to-one correspondence between the elements of $\Upsilon^{\mathbf{x}}$, $\Upsilon^{\mathbf{y}}$, $\Upsilon^{\mathbf{x},\mathbf{y}}$, and the elements of $\Upsilon^{\mathbf{b}}$, $\Upsilon^{\mathbf{r}}$, $\Upsilon^{\mathbf{b},\mathbf{r}}$, respectively. Therefore, we will blur the distinction between elements and subsets of $\Upsilon^{\mathbf{x}}$, $\Upsilon^{\mathbf{y}}$, $\Upsilon^{\mathbf{x},\mathbf{y}}$, and elements and subsets of $\Upsilon^{\mathbf{b}}$, $\Upsilon^{\mathbf{r}}$, $\Upsilon^{\mathbf{b},\mathbf{r}}$, respectively.

We will use (sans-serif) p, q to refer to colored patterns of $\Upsilon^{\mathbf{b}}$, $\Upsilon^{\mathbf{r}}$, $\Upsilon^{\mathbf{b},\mathbf{r}}$.

Let $\mathsf{p} \in \Upsilon^{\mathbf{b}}$ and $\mathsf{q} \in \Upsilon^{\mathbf{r}}$ with their corresponding $(p, x) \in \Upsilon^{\mathbf{x}}$ and $(q, y) \in \Upsilon^{\mathbf{y}}$. We define

$$\mathsf{p} \star \mathsf{q} = \mathbf{r}((p, x) \star (q, y))$$
$$\mathsf{p} \bigstar \mathsf{q} = \mathbf{br}((p, x) \bigstar (q, y)).$$

We have $\mathsf{p} \star \mathsf{q} \in \Upsilon^{\mathbf{r}}$ and $\mathsf{p} \bigstar \mathsf{q} \in \Upsilon^{\mathbf{b},\mathbf{r}}$ (when they are defined). We extend $\star$ and $\bigstar$ to languages in the natural way.

In particular, from now on, we will consider $J \subseteq \Upsilon^{\mathbf{b}}$, $J' \subseteq \Upsilon^{\mathbf{r}}$, $K \subseteq \Upsilon^{\mathbf{r}}$, $J \star K \subseteq \Upsilon^{\mathbf{r}}$, and $J \bigstar K \subseteq \Upsilon^{\mathbf{b},\mathbf{r}}$.

In order to aid us in the development, we also define

$\Phi^{\mathbf{r}}$ to be the set of all trees having all nodes black, except for the root which is red, and

$\Phi^{\mathbf{b},\mathbf{r}}$ to be the set of all trees having all nodes black, except for the root which is blue and another node which is red.

It can be verified that the languages we defined so far, $\Upsilon^{\mathbf{r}}$, $\Upsilon^{\mathbf{b}}$, $\Upsilon^{\mathbf{b},\mathbf{r}}$, $\Phi^{\mathbf{r}}$, and $\Phi^{\mathbf{b},\mathbf{r}}$ are all RTLs. Also observe that from the definition of the $\circledast$ operation, $K \subseteq \Upsilon^{\mathbf{r}} \setminus \Phi^{\mathbf{r}}$.

If p is a tree over $\Sigma \cup \Sigma^{\mathbf{r}} \cup \Sigma^{\mathbf{b}}$, we denote by $\mathsf{p}^{\neg\mathbf{b}}$ the tree over $\Sigma \cup \Sigma^{\mathbf{r}}$ that is the same as p, but with the blue nodes turned black. For a language L over $\Sigma \cup \Sigma^{\mathbf{r}} \cup \Sigma^{\mathbf{b}}$, we define

$$L^{\neg\mathbf{b}} = \{\mathsf{p}^{\neg\mathbf{b}} : \mathsf{p} \in L\}.$$

If L is an RTL, we can construct an FTA for L and then an FTA for $L^{\neg\mathbf{b}}$ by changing all the blue symbols in the transitions of the FTA for L to black. We define similarly $\mathsf{p}^{\neg\mathbf{r}}$ and $L^{\neg\mathbf{r}}$.

5.1 Auxiliaries

In our construction we will need the following items

$$\begin{aligned} B_L &= \{\mathsf{p} \in \Upsilon^{\mathbf{b},\mathbf{r}} : \mathsf{p}^{\neg\mathbf{b}} \in L\} \text{ for } L \subseteq \Upsilon^{\mathbf{r}} \\ B'_L &= \{\mathsf{p} \in \Upsilon^{\mathbf{b},\mathbf{r}} : \mathsf{p}^{\neg\mathbf{r}} \in L\} \text{ for } L \subseteq \Upsilon^{\mathbf{b}} \\ C_L &= \{\mathsf{p} \in \Phi^{\mathbf{b},\mathbf{r}} : \mathsf{p}^{\neg\mathbf{b}} \in L\} \text{ for } L \subseteq \Upsilon^{\mathbf{r}} \setminus \Phi^{\mathbf{r}}. \end{aligned}$$

These languages are easy to construct when L is RTL. We show how to do that for B_L. The other ones are similar. Let $\mathcal{A} = (S, \Sigma \cup \Sigma^{\mathbf{r}}, F, \Delta)$ be an FTA for L. We construct FTA $\mathcal{B} = (S, \Sigma \cup \Sigma^{\mathbf{b}} \cup \Sigma^{\mathbf{r}}, F, \Delta_{\mathcal{B}})$, where

$$\Delta_{\mathcal{B}} = \Delta \cup \{H \xrightarrow{a^{\mathbf{b}}} s : H \xrightarrow{a} s \text{ in } \Delta \text{ and } a \in \Sigma\}.$$

Proposition 3. *Given an FTA for L, an FTA for B_L can be constructed in polynomial time. Furthermore, if the FTA for L is a complete and deterministic FTA (CDFTA), then a CDFTA for B_L can be constructed in polynomial time as well.*

Proof. The statements follow from the observation that $L(\mathcal{B}) \cap \Upsilon^{\mathbf{b},\mathbf{r}} = B_L$, and the facts that (1) $\mathcal{B}$ is constructed in polynomial time from $\mathcal{A}$ and is a CDFTA when $\mathcal{A}$ is such, (2) an intersection CDFTA is computable in polynomial time when supplied with CDFTAs as input. □

We also get exactly the same facts for B'_L and C_L as those stated for B_L in the above proposition.

5.2 The Algorithm

Here we construct an FTA for C_K [K, we are interested in, is $C_K^{\neg\mathbf{b}}$, easily computed once we have an automaton for C_K].

Consider B_J and $B'_{J'}$. From FTAs for B_J and $B'_{J'}$, we construct an FTA for $B_J \cap B'_{J'}$. Let $\mathcal{D} = (S, \Sigma \cup \Sigma^{\mathbf{b}} \cup \Sigma^{\mathbf{r}}, F, \Delta)$ be this FTA. We also transform it to be targeted. Observe that $L(\mathcal{D}) \subseteq \Upsilon^{\mathbf{b},\mathbf{r}}$.

Now, we construct FTA $\mathcal{E} = (S, \Sigma \cup \Sigma^{\mathbf{b}} \cup \Sigma^{\mathbf{r}}, F_{\mathcal{E}}, \Delta)$, where

$$F_{\mathcal{E}} = \{s \in S : \text{ there exists } H \xrightarrow{a^{\mathbf{b}}} s \text{ in } \Delta\}.$$

Clearly, $L(\mathcal{E}) \subseteq \Phi^{\mathbf{b},\mathbf{r}}$. This is true because $\mathcal{D}$ is targeted. We have

Theorem 4. $L(\mathcal{E}) = C_K$.

Proof. "$\subseteq$": Let $\mathsf{q} \in L(\mathcal{E}) \subseteq \Phi^{\mathbf{b},\mathbf{r}}$. By the construction of $\mathcal{E}$, there exists $\mathsf{p} \in B_J \cap B'_{J'} \subseteq \Upsilon^{\mathbf{b},\mathbf{r}}$, with the blue node at a position x, such that $\mathsf{p}_x = q$.

Since $\mathsf{p} \in B_J \cap B'_{J'}$, $\mathsf{p}^{\neg\mathbf{r}} \in J$ and $\mathsf{p}^{\neg\mathbf{b}} \in J'$. Also, since $\mathsf{p} \in \Upsilon^{\mathbf{b},\mathbf{r}}$, $(\mathsf{p}^{\neg\mathbf{b}})_x \in \Upsilon^{\mathbf{r}} \setminus \Phi^{\mathbf{r}}$. We have $\mathsf{p}^{\neg\mathbf{r}} \star (\mathsf{p}^{\neg\mathbf{b}})_x = \mathsf{p}^{\neg\mathbf{b}}$, i.e. $(\mathsf{p}^{\neg\mathbf{b}})_x \in J \circ J' = K$, which means $\mathsf{p}_x = q \in C_K$.

"$\supseteq$": Let $\mathsf{q} \in C_K \subseteq \Phi^{\mathbf{b},\mathbf{r}}$. We have $\mathsf{q}^{\neg\mathbf{b}} \in K \subseteq \Upsilon^{\mathbf{r}} \setminus \Phi^{\mathbf{r}}$. By the definition of K, there exists $\mathsf{p} \in J$ such that $\mathsf{p} \star \mathsf{q}^{\neg\mathbf{b}}$ is defined and $\mathsf{p} \star \mathsf{q}^{\neg\mathbf{b}} \in J'$. This, coupled with the fact that $\mathsf{q}^{\neg\mathbf{b}} \in \Upsilon^{\mathbf{r}} \setminus \Phi^{\mathbf{r}}$, implies $\mathsf{p} \bigstar \mathsf{q}^{\neg\mathbf{b}} \in B_J$. By the definition of "$\bigstar$" and $B'_{J'}$, $\mathsf{p} \bigstar \mathsf{q}^{\neg\mathbf{b}} \in B'_{J'}$, too, therefore, $\mathsf{p} \bigstar \mathsf{q}^{\neg\mathbf{b}} \in B_J \cap B'_{J'}$, i.e., $\mathsf{p} \bigstar \mathsf{q}^{\neg\mathbf{b}}$ is accepted by $\mathcal{D}$. By the construction of $\mathcal{E}$, we have that q is accepted by $\mathcal{E}$. □

5.3 Complexity

Let J and J' be given by non-deterministic FTAs. It can be verified that

Proposition 4. *$K = J \circ J'$ can be computed in polynomial time.*

Therefore, $Y = V \circ Q$ can be computed in polynomial time. However, we also need to compute $X = V \circ Q^c$, and then X^c, in order to compute R as $Y \setminus X$. If we are not careful, we can incur a double-exponential penalty. Here we show that it can be done instead in single-exponential time.

For this let us refer to the steps of computing C_K. Suppose J and J' are given by CDFTAs. By Proposition 3 we can obtain CDFTAs for B_J and $B'_{J'}$ in polynomial time. From these CDFTAs we obtain a CDFTA $\mathcal{D}$ for $B_J \cap B'_{J'}$ in polynomial time as well. Automaton $\mathcal{E}$ (representing C_K) is the same as $\mathcal{D}$, but with a different set of final states, so $\mathcal{E}$ is a CDFTA as well.

Since $\mathcal{E}$ is a CDFTA we can compute $(C_K)^c$ from it in polynomial time.

Now, we have that for X and Y

1. $C_X, C_Y \subseteq \Phi^{\mathbf{b},\mathbf{r}}$
2. $Y \setminus X = (C_Y \setminus C_X)^{\neg\mathbf{b}} = (C_Y \cap (C_X)^c)^{\neg\mathbf{b}}$.

Clearly, $\cap$ and $\neg\mathbf{b}$ are polynomial.

Let a query Q and a view V be specified as non-deterministic FTAs. We have the following theorems.

Theorem 5. *The MCR R of Q using V can be computed in exponential time.*

Proof. The proof follows from the above reasoning and the fact that to construct CDFTAs for Q (and then Q^c), and V takes exponential time. □

Theorem 6. *Computing the MCR of Q using V is EXPTIME-hard.*

Proof. We present a reduction from the problem of inclusion for RTLs (represented by non-deterministic FTAs with horizontal languages represented by NFAs) which is EXPTIME-complete (cf. [10], Th. 8.5.9). Let L_1 and L_2 be such RTLs over some alphabet Σ. Let $\# \notin \Sigma$. We construct V from L_1 by adding to the rightmost leaf of trees in L_1 a subtree $\#^{\mathbf{b}}(\#)$. Similarly, we construct Q from L_2 by adding to the rightmost leaf of trees in L_2 a subtree $\#(\#^{\mathbf{r}})$. Now, we have that $L_1 \subseteq L_2$ if and only if the MCR of Q using V contains the single node pattern $\#^{\mathbf{r}}$. The latter can be checked in time polynomial in the size of an FTA for the rewriting.

In conclusion, if computing the MCR of Q using V were not EXPTIME-hard, then we would be able to decide the language inclusion for RTLs in less than EXPTIME, which is not possible in the worst case. □

6 k-ary Queries

So far, our attention has been focused on unary queries. In this section, we consider k-ary queries, for $k \geq 1$. Miklau and Suciu [18] motivate the study of containment for k-ary tree pattern queries by observing that the results can be applied in the context of optimizing the `FOR` clause of XQuery expressions, where multiple variables can be defined using XPath variables. The idea is that they can be captured in a single tree pattern[1] whose distinguished variables correspond to the variables bound in the `FOR` clause. Inspired by this, we consider tree-position vector queries whose vector contains k positions, for some arbitrary, but *fixed* $k \geq 1$. Such queries return a tuple (forest) of sub-trees from a target tree.

[1] Patterns considered in [18] have child and descendant edges, wildcards, and no order.

On the theoretical side, it has been shown that k-ary queries can be encoded by unary queries (see [25] and [3]). However, this is done by going through MSO formulas. Therefore, going from a k-ary query represented by an automaton to an encoding by unary queries and then back to (combination of) automata would incur non elementary complexity. Even if the complexity were not so, we still would be left with a combination of unary queries for which we needed to compute QRV in terms of another combination of unary queries which corresponds to the view. This is something that does not follow from the rewriting mechanism we developed in the previous part of the paper. The latter comment also applies to other k-ary query formalisms, defined in terms of compositions of unary queries such as the formalism introduced in [13].

Definition 9. *A k-*pattern *is a tree-position vector pair $(p, [x_1, \ldots, x_k])$, where*

1. $x_1, \ldots, x_k \in pos(p)$
2. $x_1 < \ldots < x_k$
3. $\nexists i, j \in [1, k]$, *such that x_i is a prefix of x_j.*

We also write a pattern as $(p, \mathbf{x})$, where $\mathbf{x} = [x_1, \ldots, x_k]$. A position vector possessing the three properties in above definition is called *proper*. Whenever we refer to a position vector, we assume it to be proper. We let $\Upsilon^{k\mathbf{x}}$ denote the set of all k-patterns.

Given a tree t and $\mathbf{x} = [x_1, \ldots, x_k]$, we define $t_{\mathbf{x}} = (t_{x_1}, \ldots, t_{x_k})$ as the tuple (forest) of sub-trees of t rooted at the positions in $\mathbf{x}$. We call a forest of k trees a k-forest, and denote by Ξ^k the set of all k-forests.

Definition 10. *A k-tree query (k-TQ) is a subset of $\Upsilon^{k\mathbf{x}}$. If $\{\hat{p} : (p, \mathbf{x}) \in Q\}$*[2] *is in addition* regular, *we say that Q is a* regular tree query (RTQ).

When a query has only one pattern $(p, \mathbf{x})$, we blur the distinction between $\{(p, \mathbf{x})\}$ and $(p, \mathbf{x})$.

Definition 11. *The answer to a k-query Q on t is $ans(Q, t) = \{t_{\mathbf{x}} : (t, \mathbf{x}) \in Q\}$.*

As for unary queries, it can be shown that for k-tree queries, set containment implies query containment.

Example 6. Let us recall the movie example in Section 4. Now let Q be a query given by $\mathcal{A} = (\{s, s_{\mathtt{t}}, s_{\mathtt{m}}, s_{\mathtt{c}}, s_{\mathtt{a}}\}, \Sigma \cup \hat{\Sigma}, \{s_{\mathtt{t}}\}, \Delta)$, where Δ has the following transition rules

$$s^* s_{\mathtt{m}} s^* \xrightarrow{\mathtt{r}} s_{\mathtt{t}}$$
$$s^* s_{\mathtt{c}} s^* \xrightarrow{\mathtt{m}} s_{\mathtt{m}}$$
$$s^* s_{\mathtt{a}} s^* s_{\mathtt{a}} s^* \xrightarrow{\mathtt{c}} s_{\mathtt{c}}$$
$$s^* \xrightarrow{\hat{\mathtt{a}}} s_{\mathtt{a}}$$
$$s^* \xrightarrow{\Sigma} s.$$

[2] Similarly as for unary queries, $\hat{p}$ denotes the pattern on $\Sigma \cup \hat{\Sigma}$ that is the same as p, but with the nodes at positions of $\mathbf{x}$ being marked by the corresponding symbols in $\hat{\Sigma}$.

Clearly, $Q \subseteq \Upsilon^{2\mathbf{x}}$, and $ans(Q,t)$ consists of all 2-forests of (sub)tree pairs for actors who have played the same character together in some movie. □

7 k-ary Views and Rewritings

We define a $k, m\mathbf{y}$-forest to be a tuple $((q_1, \mathbf{y}_1), \ldots, (q_k, \mathbf{y}_k))$ of patterns, such that $|\mathbf{y}_1| + \ldots + |\mathbf{y}_k| = m$, where $|\mathbf{y}_i|$, for $i \in [1,k]$, is the dimension of vector $\mathbf{y}_i$. We let $\Xi^{k,m\mathbf{y}}$ be the set of all $k, m\mathbf{y}$-forests.

Furthermore, we define a $k\mathbf{x}, m\mathbf{y}$-pattern to be a tuple of the form

$$(p, [(x_1, \mathbf{y}_1), \ldots, (x_k, \mathbf{y}_k)])$$

where $|\mathbf{y}_1| + \ldots + |\mathbf{y}_k| = m$, and x_i is a proper[3] prefix of positions in $\mathbf{y}_i$, for $i \in [1,k]$. We let $\Upsilon^{k\mathbf{x},m\mathbf{y}}$ be the set of all $k\mathbf{x}, m\mathbf{y}$-patterns.

Now, let $(p, [x_1, \ldots, x_k]) \in \Upsilon^{k\mathbf{x}}$ and $((q_1, \mathbf{y}_1), \ldots, (q_k, \mathbf{y}_k)) \in \Xi^{k,m\mathbf{y}}$. We define

$$(p, [x_1, \ldots, x_k]) \star ((q_1, \mathbf{y}_1), \ldots, (q_k, \mathbf{y}_k)) =$$

$$\begin{cases} (p, [x_1\mathbf{y}_1, \ldots, x_k\mathbf{y}_k]) \text{ if } p_{x_i} = q_i \text{ and } \mathbf{y}_i \neq \epsilon, \forall i \in [1,k] \\ \text{undefined} \quad\quad\quad\quad \text{otherwise} \end{cases}$$

and

$$(p, [x_1, \ldots, x_k]) \bigstar ((q_1, \mathbf{y}_1), \ldots, (q_k, \mathbf{y}_k)) =$$

$$\begin{cases} (p, [(x_1, x_1\mathbf{y}_1), \ldots, (x_k, x_k\mathbf{y}_k)]) \\ \quad\quad\quad\quad \text{if } p_{x_i} = q_i \text{ and } \mathbf{y}_i \neq \epsilon, \forall i \in [1,k] \\ \text{undefined} \quad\quad\quad \text{otherwise} \end{cases}$$

where $x_i\mathbf{y}_i$, for $i \in [1,k]$, is the tuple obtained from $\mathbf{y}_i$ by prepending x_i to each of its positions.

We note that, when defined

$$(p, [x_1, \ldots, x_k]) \star ((q_1, \mathbf{y}_1), \ldots, (q_k, \mathbf{y}_k)) \in \Upsilon^{m\mathbf{y}}$$
$$(p, [x_1, \ldots, x_k]) \bigstar ((q_1, \mathbf{y}_1), \ldots, (q_k, \mathbf{y}_k)) \in \Upsilon^{k\mathbf{x},m\mathbf{y}}.$$

[For ease of exposition we are using $\Upsilon^{m\mathbf{y}}$ instead of $\Upsilon^{m\mathbf{x}}$.]

For $L_1 \subseteq \Upsilon^{k\mathbf{x}}$, $L_2 \subseteq \Xi^{k,m\mathbf{y}}$ we define $L_1 \star L_2$ and $L_1 \bigstar L_2$ in the natural way. Again we blur the distinction between a set of one element and the element itself.

Definition 12. *The* maximally contained rewriting (MCR) *of* $Q \subseteq \Upsilon^{m\mathbf{y}}$ *using* $V \subseteq \Upsilon^{k\mathbf{x}}$ *is*

$$R = \{\xi \in \Xi^{k,m\mathbf{y}} : V \star \xi \subseteq Q\}.$$

[3] By assuming properly transformed query and view (as we show in the full version [15]), we do not need to consider the case when some x_i coincides with some position in $\mathbf{y}_i$.

(ξ being in $\Xi^{k,m\mathbf{y}}$ is of the form $((q_1, \mathbf{y}_1), \ldots, (q_k, \mathbf{y}_k))$.)
We also define set X of "bad" patterns as

$$X = \{\xi \in \Xi^{k,m\mathbf{y}} : \text{ there exists } \upsilon \in V \text{ s.t. } \upsilon \star \xi \in Q^c\}$$

where $Q^c = \Upsilon^{m\mathbf{y}} \setminus Q$. ($\upsilon$ being in $\Upsilon^{k\mathbf{x}}$ is of the form $(p, [x_1, \ldots, x_k])$.) Now we define

$$Y = \{\xi \in \Xi^{k,m\mathbf{y}} : \text{ there exists } \upsilon \in V \text{ s.t. } \upsilon \star \xi \in Q\}.$$

Notice that R, X, and Y are subsets of $\Xi^{k,m\mathbf{y}}$. We have that

Proposition 5. $R = Y \setminus X$.

Dummy roots. We turn forests into trees by introducing a "dummy" root labeled by a special symbol $\Diamond \notin \Sigma$.

Given $(t_1, \ldots, t_k) \in \Xi^k$, we define

$$\Diamond(t_1, \ldots, t_k)$$

to be the tree obtained by making the roots of $t_1, \ldots, t_k$ to be children of a dummy new root labeled by $\Diamond$.

Likewise, given $((q_1, \mathbf{y}_1), \ldots, (q_k, \mathbf{y}_k)) \in \Xi^{k,m\mathbf{y}}$, we define

$$\Diamond((q_1, \mathbf{y}_1), \ldots, (q_k, \mathbf{y}_k))$$

to be the tree obtained by making the roots of $q_1, \ldots, q_k$ to be children of a dummy new root labeled by $\Diamond$.

If L is a subset of Ξ^k or $\Xi^{k,m\mathbf{y}}$, we define $\Diamond L$ in the natural way by turning each forest in L into a tree as described.

Let $F \subseteq \Xi^k$ and $L \subseteq \Xi^{k,m\mathbf{y}}$. We define

$$ans(L, F) = \bigcup_{\upsilon \in F} ans(\Diamond L, \Diamond \upsilon).$$

Let $F_{V,t} = ans(V, t) \subseteq \Xi^k$ be the materialization of V on $t \in \Upsilon$, We answer Q using V by $ans(R, F_{V,t})$. Now, by generalizing the reasoning in the proof of Theorem 3 it can be verified that

Theorem 7. *$ans(R, F_{V,t}) \subseteq ans(Q, t)$ for each $t \in \Upsilon$.*

8 Computing the MCR of an m-ary Query Using a k-ary View

Analogus to the case of unary queries, we first define the inverse of the $\star$ operation.

Let $\upsilon \in \Upsilon^{k\mathbf{x}}$ and $\upsilon' \in \Upsilon^{m\mathbf{x}}$. We define

$$\upsilon \circledast \upsilon' = \begin{cases} \xi & \text{if } \upsilon' = \upsilon \star \xi \\ \text{undefined otherwise} \end{cases}$$

This definition is lifted to subsets of $\Upsilon^{\mathbf{x}}$ in the natural way. Now, it can be verified that as in the case of unary queries

Proposition 6. *$X = V \circ Q^c$ and $Y = V \circ Q$.*

Once we obtain X and Y, we can obtain R as $Y \setminus X$. In the rest of the section, we present an automata-based solution for computing

$$K = J \circ J'$$

when J and J' are RTQs.

Here we again use colors in order to work with languages and automata. We define $\mathbf{b}$ as a (unary) operator that given $(p, \mathbf{x}) \in \Upsilon^{k\mathbf{x}}$ returns a tree over $\Sigma \cup \Sigma^{\mathbf{b}}$ that is isomorphic to p with the nodes at the positions of $\mathbf{x}$ colored blue.

Likewise, we define $\mathbf{r}$ as a (unary) operator that given $(p, \mathbf{y}) \in \Upsilon^{m\mathbf{y}}$ returns a tree over $\Sigma \cup \Sigma^{\mathbf{r}}$ that is isomorphic to p with the nodes at the positions of $\mathbf{y}$ colored red.

Furthermore, we define $\mathbf{br}$ as a (unary) operator that given $(p, (x_1, \mathbf{y}_1), \ldots, (x_k, \mathbf{y}_k)) \in \Upsilon^{k\mathbf{x},m\mathbf{y}}$ returns a tree over $\Sigma \cup \Sigma^{\mathbf{b}} \cup \Sigma^{\mathbf{r}}$ that is isomorphic to p with the nodes at positions $x_1, \ldots, x_k$ colored blue and the nodes at the positions of $\mathbf{y}_1, \ldots, \mathbf{y}_k$ colored red.

Based on these operators, we define

$$\begin{aligned}
\Upsilon^{k\mathbf{b}} &= \{\mathbf{b}(p, \mathbf{x}) : (p, \mathbf{x}) \in \Upsilon^{k\mathbf{x}}\} \\
\Upsilon^{m\mathbf{r}} &= \{\mathbf{r}(p, \mathbf{y}) : (p, \mathbf{x}) \in \Upsilon^{m\mathbf{y}}\} \\
\Upsilon^{k\mathbf{b},m\mathbf{r}} &= \{\mathbf{br}(p, (x_1, \mathbf{y}_1), \ldots, (x_k, \mathbf{y}_k)) : \\
&\qquad (p, (x_1, \mathbf{y}_1), \ldots, (x_k, \mathbf{y}_k)) \in \Upsilon^{k\mathbf{x},m\mathbf{y}}\} \\
\Xi^{k,m\mathbf{r}} &= \{(\mathbf{r}(q_1, \mathbf{y}_1), \ldots, \mathbf{r}(q_k, \mathbf{y}_k)) : \\
&\qquad ((q_1, \mathbf{y}_1), \ldots, (q_k, \mathbf{y}_k)) \in \Xi^{k,m\mathbf{y}}\}.
\end{aligned}$$

Clearly, there is a one-to-one correspondence between the elements of $\Upsilon^{k\mathbf{x}}$, $\Upsilon^{m\mathbf{y}}$, $\Upsilon^{k\mathbf{x},m\mathbf{y}}$, $\Xi^{k,m\mathbf{y}}$ and the elements of $\Upsilon^{k\mathbf{b}}$, $\Upsilon^{m\mathbf{r}}$, $\Upsilon^{k\mathbf{b},m\mathbf{r}}$, $\Xi^{k,m\mathbf{r}}$, respectively. Therefore, we will blur the distinction between elements and subsets of $\Upsilon^{k\mathbf{x}}$, $\Upsilon^{m\mathbf{y}}$, $\Upsilon^{k\mathbf{x},m\mathbf{y}}$, $\Xi^{k,m\mathbf{y}}$ and elements and subsets of $\Upsilon^{k\mathbf{b}}$, $\Upsilon^{m\mathbf{r}}$, $\Upsilon^{k\mathbf{b},m\mathbf{r}}$, $\Xi^{k,m\mathbf{r}}$, respectively.

We will use (sans-serif) p, q to refer to colored patterns of $\Upsilon^{k\mathbf{b}}$, $\Upsilon^{m\mathbf{r}}$, $\Upsilon^{k\mathbf{b},m\mathbf{r}}$, and $\mathbf{q} = (\mathsf{q}_1, \ldots, \mathsf{q}_k)$ to refer to colored forests of $\Xi^{k,m\mathbf{r}}$.

Let $\mathsf{p} \in \Upsilon^{k\mathbf{b}}$ and $\mathbf{q} = (\mathsf{q}_1, \ldots, \mathsf{q}_k) \in \Xi^{k,m\mathbf{r}}$ with their corresponding $(p, \mathbf{x}) \in \Upsilon^{k\mathbf{x}}$ and $((q_1, \mathbf{y}_1), \ldots, (q_k, \mathbf{y}_k)) \in \Xi^{k,m\mathbf{y}}$. We define

$$\begin{aligned}
\mathsf{p} \star \mathbf{q} &= \mathbf{r}((p, \mathbf{x}) \star ((q_1, \mathbf{y}_1), \ldots, (q_k, \mathbf{y}_k))) \\
\mathsf{p} \bigstar \mathbf{q} &= \mathbf{br}((p, \mathbf{x}) \bigstar ((q_1, \mathbf{y}_1), \ldots, (q_k, \mathbf{y}_k))).
\end{aligned}$$

We extend $\star$ and $\bigstar$ to languages and forests in the natural way.

In particular, from now on, we will consider $J \subseteq \Upsilon^{k\mathbf{b}}$, $J' \subseteq \Upsilon^{m\mathbf{r}}$, $K \subseteq \Xi^{k,m\mathbf{r}}$, $J \star K \subseteq \Upsilon^{m\mathbf{r}}$, and $J \bigstar K \subseteq \Upsilon^{k\mathbf{b},m\mathbf{r}}$.

We further define a few more notions. Let

$\Phi^{k,m\mathbf{r}}$ be the subset of $\Xi^{k,m\mathbf{r}}$ forests having *at least one* tree that has its root red.

$\Phi^{k\mathbf{b},m\mathbf{r}}$ be the set of all k-forests over $\Sigma \cup \Sigma^{\mathbf{b}} \cup \Sigma^{\mathbf{r}}$, such that: (1) the roots of all patterns in the forest are blue, (2) *exactly* m nodes in total are red, and (3) all the other nodes are black.

Observe that from the definition of the $\circ$ operation, $K \subseteq \Xi^{k,m\mathbf{r}} \setminus \Phi^{k,m\mathbf{r}}$.

We define the operators $(.)^{\neg\mathbf{b}}$ and $(.)^{\neg\mathbf{r}}$, for patterns, sets of patterns, forests, and sets of forests, over $\Sigma \cup \Sigma^{\mathbf{b}} \cup \Sigma^{\mathbf{r}}$, in a manner similar to Section 5.

8.1 Auxiliaries

In our construction we will need the following items

$$
\begin{aligned}
B_L &= \{\mathsf{p} \in \Upsilon^{k\mathbf{b},m\mathbf{r}} : \mathsf{p}^{\neg\mathbf{b}} \in L\} \text{ for } L \subseteq \Upsilon^{m\mathbf{r}} \\
B'_L &= \{\mathsf{p} \in \Upsilon^{k\mathbf{b},m\mathbf{r}} : \mathsf{p}^{\neg\mathbf{r}} \in L\} \text{ for } L \subseteq \Upsilon^{k\mathbf{b}} \\
C_L &= \{\mathbf{q} \in \Phi^{k\mathbf{b},m\mathbf{r}} : \mathbf{q}^{\neg\mathbf{b}} \in L\} \text{ for } L \subseteq \Xi^{k,m\mathbf{r}} \setminus \Phi^{k,m\mathbf{r}}.
\end{aligned}
$$

In plain language C_L is the set of all forests in $\Phi^{k\mathbf{b},m\mathbf{r}}$ obtained from the forests of L after making the root of their trees blue.

The above languages are easy to construct when L ($\Diamond L$ in the third definition) is RTL. For instance, for B_L, we first construct an automaton $\mathcal{B}$ exactly as in Subsection 5.1, and then compute B_L as $L(\mathcal{B}) \cap \Upsilon^{k\mathbf{b},m\mathbf{r}}$. Similar statements as those made in Proposition 3 can be made here as well.

8.2 The Algorithm

In order to be compatible with the framework of FTAs (which work on trees, not forests), here we construct a CDFTA for $\Diamond C_K$. Set $\Diamond K$ (we are interested in) is easily computed once we have an automaton for $\Diamond C_K$.

Consider B_J and $B'_{J'}$. From CDFTAs for B_J and $B'_{J'}$, we construct a CDFTA for $B_J \cap B'_{J'}$. Let $\mathcal{D} = (S_{\mathcal{D}}, \Sigma \cup \Sigma^{\mathbf{b}} \cup \Sigma^{\mathbf{r}}, F_{\mathcal{D}}, \Delta_{\mathcal{D}})$ be this CDFTA. We also transform it to be targeted. Observe that $L(\mathcal{D}) \subseteq \Upsilon^{k\mathbf{b},m\mathbf{r}}$. Let

$$S^{\mathbf{b}} = \{s \in S_{\mathcal{D}} : \text{ there exists } H \xrightarrow{a^{\mathbf{b}}} s \text{ in } \Delta_{\mathcal{D}}\}.$$

We denote $(S^{\mathbf{b}})^k$ by $S^{k\mathbf{b}}$. Next, we construct FTA $\mathcal{E} = (S_{\mathcal{E}}, \Sigma \cup \Sigma^{\mathbf{b}} \cup \Sigma^{\mathbf{b}} \cup \{\Diamond\}, F_{\mathcal{E}}, \Delta_{\mathcal{E}})$ where

$$
\begin{aligned}
S_{\mathcal{E}} &= S_{\mathcal{D}} \cup \{s_{\text{final}}, s_{\text{garbage}}\} \\
F_{\mathcal{E}} &= \{s_{\text{final}}\} \\
\Delta_{\mathcal{E}} &= \Delta_{\mathcal{D}} \cup \{S^{k\mathbf{b}} \xrightarrow{\Diamond} s_{\text{final}}\} \cup \{S^*_{\mathcal{E}} \setminus S^{k\mathbf{b}} \xrightarrow{\Diamond} s_{\text{garbage}}\} \cup \\
&\quad \{(s_{\text{garbage}})^* \xrightarrow{a} s_{\text{garbage}} : a \in \Sigma \cup \Sigma^{\mathbf{b}} \cup \Sigma^{\mathbf{r}}\}.
\end{aligned}
$$

Since k is fixed, an automaton for $S^*_{\mathcal{E}} \setminus S^{k\mathbf{b}}$ can be constructed in polynomial time in the size of $S^{k\mathbf{b}}$ and $S_{\mathcal{E}}$. We can show that the FTA $\mathcal{E}$ constructed above is a CDFTA.

Clearly, $L(\mathcal{E}) \subseteq \Diamond\Phi^{k\mathbf{r},m\mathbf{r}}$. This is true because $\mathcal{D}$ is targeted. Now, by generalizing the reasoning in the proof of Theorem 4, we can show that

Theorem 8. *$L(\mathcal{E}) = \Diamond C_{\mathcal{K}}$. Furthermore, $\mathcal{E}$ is complete and deterministic.*

Complexity. It can be verified that, similarly as in the case of unary queries (Subsection 5.3), computing R can be done in singly-exponential time when starting with non-deterministic FTAs for for Q and V. By Theorem 6, which applies to the case of $k = m = 1$, it follows that our constructions are optimal.

9 Conclusions

In this paper, we studied the problem of rewriting queries using views for XML data, focusing on tree-selecting queries equal in power to MSO. Colored tree languages and automata provided us a critical ground for implementing rewriting queries using views. We developed a singly-exponential algorithm for deriving the maximally contained rewriting of a query using a view, and then extended it to an algorithm for the general case where the query can be an m-ary query and the view can be a k-ary view. The latter class of queries is useful in optimizing the `FOR` clause of XQuery expressions. We showed the problem is EXPTIME-hard, thus showing our algorithms are optimal. Allowing our query formalism to also involve selection, join and restructuring is an avenue worth exploring. Extending the reasoning about QRV for this wider class is an interesting open problem.

References

1. Afanasiev, L., Blackburn, P., Dimitriou, I., Gaiffe, B., Goris, E., Marx, M., de Rijke, M.: PDL for ordered trees. Journal of Applied Non-Classical Logics 15(2), 115–135 (2005)
2. Afrati, F.N., Chirkova, R., Gergatsoulis, M., Kimelfeld, B., Pavlaki, V., Sagiv, Y.: On rewriting XPath queries using views. In: EDBT (2009)
3. Arenas, M., Barceló, P., Libkin, L.: Combining temporal logics for querying XML documents. In: Schwentick, T., Suciu, D. (eds.) ICDT 2007. LNCS, vol. 4353, pp. 359–373. Springer, Heidelberg (2006)
4. Arion, A., Benzaken, V., Manolescu, I., Papakonstantinou, Y.: Structured materialized views for XML queries. In: VLDB (2007)
5. Balmin, A., Özcan, F., Beyer, K.S., Cochrane, R., Pirahesh, H.: A framework for using materialized XPath views in XML query processing. In: VLDB (2004)
6. Calvanese, D., De Giacomo, G., Lenzerini, M., Vardi, M.Y.: An automata-theoretic approach to regular XPath. In: Gardner, P., Geerts, F. (eds.) DBPL 2009. LNCS, vol. 5708, pp. 18–35. Springer, Heidelberg (2009)
7. Calvanese, D., Giacomo, G.D., Lenzerini, M., Vardi, M.Y.: Node selection query languages for trees. In: AAAI (2010)
8. Cautis, B., Deutsch, A., Onose, N.: XPath rewriting using multiple views: Achieving completeness and efficiency. In: WebDB (2008)
9. Claves, P., Jansen, D., Holtrup, S.J., Mohr, M., Reis, A., Schatz, M., Thesing, I.: LETHAL: Library for working with finite tree and hedge automata (2009), `http://lethal.sf.net`

10. Comon, H., Dauchet, M., Gilleron, R., Löding, C., Jacquemard, F., Lugiez, D., Tison, S., Tommasi, M.: Tree automata techniques and applications (2007)
11. Fan, W., Geerts, F., Jia, X., Kementsietsidis, A.: Rewriting regular XPath queries on XML views. In: ICDE (2007)
12. Filiot, E.: Ranked and unranked tree automata libraries (grappa), http://www.grappa.univ-lille3.fr/$\sim$filiot/tata
13. Filiot, E., Niehren, J., Talbot, J.-M., Tison, S.: Composing monadic queries in trees. In: PLAN-X (2006)
14. Frick, M., Grohe, M., Koch, C.: Query evaluation on compressed trees (extended abstract). In: LICS (2003)
15. Lakshmanan, L.V.S., Thomo, A.: View-based tree-language rewritings for XML (2013), http://webhome.cs.uvic.ca/~thomo/tarewfull.pdf
16. Lakshmanan, L.V.S., Wang, H., Zhao, Z.J.: Answering tree pattern queries using views. In: VLDB (2006)
17. Libkin, L., Sirangelo, C.: Reasoning about XML with temporal logics and automata. J. Applied Logic 8(2), 210–232 (2010)
18. Miklau, G., Suciu, D.: Containment and equivalence for a fragment of XPath. JACM 51(1), 2–45 (2004)
19. Neven, F.: Design and Analysis of Query Languages for Structured Documents–A Formal and Logical Approach. PhD thesis. Limburgs Universitair Centrum (1999)
20. Neven, F., Schwentick, T.: Expressive and efficient pattern languages for tree-structured data. In: PODS (2000)
21. Neven, F., Schwentick, T.: Query automata over finite trees. TCS 275(1-2), 633–674 (2002)
22. Neven, F., Schwentick, T.: On the complexity of XPath containment in the presence of disjunction, DTDs, and variables. Logical Methods in Computer Science 2(3) (2006)
23. Niehren, J., Planque, L., Talbot, J.-M., Tison, S.: N-ary queries by tree automata. In: DBPL (2005)
24. Onose, N., Deutsch, A., Papakonstantinou, Y., Curtmola, E.: Rewriting nested XML queries using nested views. In: SIGMOD Conf. (2006)
25. Schwentick, T.: On diving in trees. In: Nielsen, M., Rovan, B. (eds.) MFCS 2000. LNCS, vol. 1893, pp. 660–669. Springer, Heidelberg (2000)
26. Schwentick, T.: Automata for XML - a survey. J. Comput. Syst. Sci. 73(3), 289–315 (2007)
27. ten Cate, B., Segoufin, L.: XPath, transitive closure logic, and nested tree walking automata. In: PODS (2008)
28. Thatcher, J.W., Wright, J.B.: Generalized finite automata theory with an application to a decision problem of second-order logic. Mathematical Systems Theory 2(1), 57–81 (1968)
29. Thomo, A., Venkatesh, S.: Rewriting of VPLs for XML data integration. In: CIKM (2008)
30. Wang, J., Li, J., Yu, J.X.: Answering tree pattern queries using views: A revisit. In: EDBT (2011)
31. Xu, W., Özsoyoglu, Z.M.: Rewriting XPath queries using materialized views. In: VLDB (2005)
32. Yu, C., Popa, L.: Constraint-based XML query rewriting for data integration. In: SIGMOD (2004)

EHC: Non-parametric Editing by Finding Homogeneous Clusters

Stefanos Ougiaroglou* and Georgios Evangelidis

Department of Applied Informatics, School of Information Sciences,
University of Macedonia, 156 Egnatia str, GR-54006 Thessaloniki, Greece
{stoug,gevan}@uom.gr

Abstract. Editing is a crucial data mining task in the context of k-Nearest Neighbor classification. Its purpose is to improve classification accuracy by improving the quality of training datasets. To obtain such datasets, editing algorithms try to remove noisy and mislabeled data as well as smooth the decision boundaries between the discrete classes. In this paper, a new fast and non-parametric editing algorithm is proposed. It is called Editing through Homogeneous Clusters (EHC) and is based on an iterative execution of a clustering procedure that forms clusters containing items of a specific class only. Contrary to other editing approaches, EHC is independent of input (tuning) parameters. The performance of EHC is experimentally compared to three state-of-the-art editing algorithms on ten datasets. The results show that EHC is faster than its competitors and achieves high classification accuracy.

Keywords: k-NN classification, clustering, editing, noisy items.

1 Introduction

Classification is a traditional data mining problem that has attracted the interest of many researchers in the past decades [11]. Classification algorithms (or classifiers) attempt to assign unclassified items to a class from a set of predefined classes. Classifiers can be divided into eager and instance-based (or lazy) classifiers. Contrary to eager classifiers, lazy classifiers do not build any classification model that is then used to classify new items. Instead they use the training set (TS) as the classification model.

A popular lazy classification method is k-Nearest Neighbor (k-NN) classifier [5]. It is simple, very easy to implement and has many applications. The k-NN classifier works as follows: for each new item x, it searches TS and retrieves the k nearest items to x according to a distance metric. The class of x is determined by a majority vote, i.e., the most common class among the classes of the k nearest neighbors. Possible ties during voting can be resolved either randomly or by assigning x to the class of the nearest neighbor.

* S. Ougiaroglou is supported by a scholarship from State Scholarships Foundation of Greece (I.K.Y.)

C. Beierle and C. Meghini (Eds.): FoIKS 2014, LNCS 8367, pp. 290–304, 2014.

k-NN classifier is considered to be an effective classifier. However, it has some weaknesses. The first one is high computational cost since it must compute all distances between each unclassified item and all items in TS. In cases of large datasets, this drawback renders its use a time-consuming procedure and in some cases even prohibitive. Another weakness is large storage requirements for storing TS. A third weakness is that k-NN classifier is a noise-sensitive method. Classification accuracy depends on the level of noise in TS. Usage of high k values extend the examined neighbourhood and thus can partially remedy this drawback. However, this implies a high number of trial-and-error executions to determine the appropriate k value and that noise is uniformly distributed in TS (otherwise, dynamic determination of k should be adopted [19]).

Data Reduction Techniques (DRTs) [26,8,15,25,28,13,10,4,17] can effectively deal with the aforementioned weaknesses. They can be divided into Prototype Selection (PS) [8] and Prototype Abstraction (PA) [26] algorithms. PS algorithms select items from TS whereas PA algorithms generate items by summarizing similar items from TS and use them as prototypes.

PS algorithms are divided into condensing and editing algorithms. PA and PS-condensing algorithms aim to built a small representative set (condensing set) of the initial TS. Usage of a Condensing Set (CS) has the benefits of low computational cost and storage requirements while accuracy remains at high levels. PS-editing algorithms aim to improve accuracy rather than achieve high reduction rates. To achieve this, they remove noisy and mislabelled items and smooth the decision boundaries (see Figure 1). Ideally, a PS-editing algorithm builds an Edited training Set (ES) without overlaps between the classes.

The reduction rates of many PA and PS-condensing algorithms depend on the level of noise in TS. The higher the level of noise, the lower the reduction rates achieved. Therefore, effective application of such algorithms implies removal of noise from the data, i.e., application of an editing algorithm beforehand [6,15]. Hence, editing has a double goal: accuracy improvement and effective application of PA and PS-condensing algorithms. We should mention that some condensing algorithms, such as IB3 [1], integrate the idea of editing into their reduction procedures. These algorithms are called hybrid (see [8,26] for details).

Although PS-editing algorithms contribute in obtaining high quality training data, they constitute a costly preprocessing step. Moreover, most PS-editing algorithms are parametric, i.e., the user defines the values of certain input (tuning) parameters. This implies time-consuming trial-and-error procedures to tune the parameters. These observations are behind the motivation of this paper. The contribution is the development and evaluation of a fast, non-parametric PS-editing algorithm that is based on a k-means clustering [16] procedure that forms homogeneous clusters. The proposed algorithm is called Editing through Homogeneous Clusters (EHC), leads to accurate k-NN classifiers and has low preprocessing cost.

The rest of the paper is organized as follows: Section 2 reviews the most well-known editing algorithms. Section 3 presents the proposed EHC algorithm.

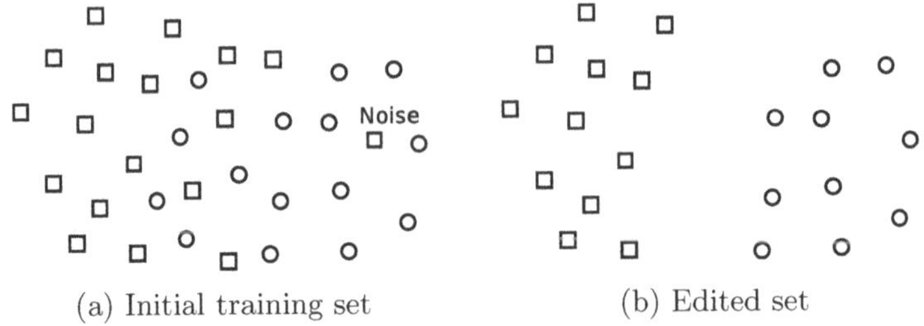

(a) Initial training set (b) Edited set

Fig. 1. Smoothing decision boundaries and removing noisy items

Performance evaluation experiments are presented in Section 4 and, finally, Section 5 concludes the paper.

2 Editing Algorithms

2.1 The Edited Nearest Neighbor (ENN) Rule

The reference editing algorithm is Wilson's Edited Nearest Neighbor (ENN) rule [29]. It constitutes the base of all other editing algorithms. ENN-rule is very simple. Algorithm 1 contains the pseudocode of the algorithm. Initially, the edited set (ES) is set to be equal to the TS (line 1). For each item x of TS, the algorithm scans TS and retrieves its k nearest neighbors (line 3). If x is misclassified by the majority vote of the retrieved nearest neighbors, it is removed from ES (lines 4–7). ENN-rule considers wrongly classified items to be noisy or close-border items and, thus, they must be removed. Note that, in each algorithm iteration, ENN-rule searches for nearest neighbors in the original TS and not in the "under construction" ES.

Algorithm 1. ENN-rule

Input: TS, k
Output: ES

1: $ES \leftarrow TS$
2: **for** each $x \in TS$ **do**
3: $\quad NNs \leftarrow$ find the k nearest to x neighbors in $TS - \{x\}$
4: $\quad majorClass \leftarrow$ find the most common class of NNs
5: $\quad$ **if** $x_{class} \neq majorClass$ **then**
6: $\quad\quad ES \leftarrow ES - \{x\}$
7: $\quad$ **end if**
8: **end for**
9: **return** ES

Obviously, the cost of editing depends on the size of TS. In cases of large datasets, ENN-rule is a time-consuming algorithm. ENN-rule must compute all distances between the items of TS. Therefore, $\frac{N*(N-1)}{2}$ distances must be estimated, where N is the number of items in TS.

A crucial issue that should be addressed is the determination of the value of k that defines the size of the examined neighborhood. [28,9,17] consider $k = 3$ to be a typical value. This is adopted in many papers (e.g. [20]), whereas, other papers use $k = 3$ and additional k values (e.g., [23,12]). In some cases, researchers determine the value of k that achieves the best performance through trial-and-error procedures (e.g., [27]). In [29], the impact of k is discussed in detail. Furthermore, in [12], a large number of k values are experimentally evaluated. It turns out that the best value of k depends on the dataset at hand and should be determined by considering the distribution of items in the multidimensional space. Even the best value of k may not be optimal and it may remove non-noisy items (see [9]) or keep noisy items. This happens because ENN-rule uses a unique k value for the entire TS. Different k values may be optimal for different regions in space.

2.2 All k-NN

All-kNN [24] is a popular variation of ENN-rule. It iteratively executes ENN-rule with different k values (see Algorithm 2). All-kNN adopts $kmax$ as an upper limit for the value of k. Initially, ES is set to be the whole TS (line 1). For each item x in TS (line 2), All-kNN applies k-NN classifier on the items of TS (lines 6–7), initially, with $k = 1$ and tries to remove x from ES in a way similar to ENN-rule. If x is misclassified, it is removed and the procedure continues with the next item (lines 8–11). Otherwise, k is incremented by one (line 12) and the algorithm retries to remove x. If the item is not removed after $kmax$ iterations (line 5), x remains in the final ES and All-kNN continues with the next item.

Since All-kNN uses more than one values for k, it removes more items than ENN-rule. Although All-kNN is an iterative version of ENN-rule, an efficient implementation of it does not re-compute the same distances again and again. Therefore, All-kNN computes as many distances as ENN-rule and is parametric, too. The value of $kmax$ must be defined by the user. This usually implies tuning through a trial-and-error procedure. M. Garcia-Borroto et al. consider $kmax = 7$ or $kmax = 9$ to be appropriate values [9].

2.3 Multiedit

Multiedit [7] is another well-known editing approach (see Algorithm 3). Initially, ES is set to be equal to TS (line 1). Then, TS is divided into n random subsets, $s_1, s_2, \ldots, s_n$ (line 5). The algorithm continues by applying ENN-rule over each item $x \in s_i$ (line 7) of each subset s_i (line 6), but searching for the single nearest neighbor (1-NN) in the module n following subset, i.e., $s_{(i+1)modn}$ (line 8). The misclassified items are removed from ES (line 10). Then, TS is set to be ES (line 20) and the whole process is repeated. Multiedit continues until the last R iterations produce no editing (lines 15–19, line 21).

Algorithm 2. All-kNN

Input: $TS, kmax$
Output: ES

```
ES ← TS
for each x ∈ TS do
    k ← 1
    flag ← FALSE
    while (k ≤ kmax) and (flag == FALSE) do
        NNs ← find the k nearest to x neighbors in TS − {x}
        majorClass ← find the most common class in NNs
        if x_class ≠ majorClass then
            ES ← ES − {x}
            flag ← TRUE
        end if
        k ← k + 1
    end while
end for
return ES
```

Here, parameter k is not used since multiedit utilizes 1-NN classifier during editing. However, parameters n and R influence the resulting ES. Parameter $n \geq 3$ defines the number of subsets. In many papers (e.g., [9,23]), $n = 3$ is either adopted or proposed. Parameter R defines the number of non-editing iterations. In [9], $R = 2$ is suggested as an appropriate value. Nevertheless, the best values for these parameters can not be determined without tuning through a trial-end-error procedure.

Multiedit usually achieves higher reduction rates than ENN-rule. It can successfully remove noisy, outlier and close-border items. However, it may also remove non-noisy items. If items of two or more classes are close to each other, multiedit may eliminate entire classes [9]. Another drawback of multiedit is that it is based on a random formation of subsets, i.e., repeated applications may build a completely different ES from the same TS.

Multedit is usually more time-consuming than ENN-rule. However, it may compute even fewer than $\frac{N*(N-1)}{2}$ distances. An efficient implementation of multiedit does not compute a distance more than once. However, since the distances that have been already computed should be available until the end of the execution, such an implementation requires more memory. In cases where each distance is computed more than once, the computational cost of the algorithm highly depends on the value of R.

2.4 Other Editing Algorithms

Subsections 2.1, 2.2 and 2.3 presented in detail three state-of-the-art editing algorithms that we use for comparison purposes in our experimental study in Section 4. Many more editing approaches have been proposed in the literature.

Algorithm 3. Multiedit

```
Input: TS, n, R
Output: ES
ES ← TS
r ← 0
repeat
   flag ← FALSE
   S ← set of n random subsets, s_1, s_2, ..., s_n of TS
   for each s_i ∈ S do
      for each x ∈ s_i do
         nn ← find the nearest neighbor in s_(i+1)modn
         if x_class ≠ nn_class then
            ES ← ES − {x}
            flag ← TRUE
         end if
      end for
   end for
   if flag = FALSE then
      r ← r + 1
   else
      r ← 0
   end if
   TS ← ES
until r == R {until the last R iterations do not edit data}
return ES
```

EENProb and ENNth [27] are extensions of ENN-rule. Both retrieve the k nearest neighbors, and then perform editing based on probability estimations. Repeated ENN (RENN) rule [24] is also a variation of ENN-rule. Actually, it is quite similar to All-kNN. RENN-rule applies ENN-rule in an iterative way until each item's majority of k nearest items have the same class. In [12], another simple variation of ENN is proposed. It places an item in ES, if all its k nearest neighbors have the same class label with it (distance ties increase the value of k).

Sanchez et al. proposed two editing algorithms that are based on geometric information provided by proximity graphs [23]. They are also based on the concept of removal of misclassified items. To the best of our knowledge, they are the only non-parametric editing algorithms. Nevertheless, the type of proximity graphs used influence the resulting ES. In [23], two types of proximity graphs were used. Consequently, four editing approaches were obtained and evaluated. From this point of view, even these algorithms can be characterized to be parametric methods.

k-NCN editing and its iterative version [20] are also based on ENN-rule. Particularly, they use k nearest centroid neighborhood classifier [22] instead of k-NN classifier. Both are based on the following simple idea: the appropriate neighborhood that should be examined for each item is defined by taking into consideration not only its nearest neighbors but also the symmetrical distribution of neighbors around it.

In [3,20] a depuration algorithm is proposed for editing training data. In addition to removing some items from TS, the algorithm also changes the class labels of some items. To achieve this, it uses two input parameters (see [3] or [20] for details). [14] considers and evaluates editing approaches based on the depuration algorithm and proposes the Neural Network Ensemble Editing (NNEE). This method is also parametric. NNEE trains a neural network ensemble that is then used to relabel some items. Last by not least, a recent paper [21] proposes the use of local support vector machines for noise reduction. Like the other methods, its performance depends on parameter tuning.

3 Editing through Homogeneous Clusters (EHC) Algorithm

As we already mentioned in Section 2, PS-editing algorithms either extend ENN-rule or are based on the same idea. The proposed EHC algorithm follows a completely different, non-parametric strategy in order to remove noisy, mislabeled and close-border data items. Actually, it is based on RHC [18], a PA algorithm we have recently proposed. EHC iteratively applies k-means clustering on TS until all constructed clusters contain items of a specific class only, i.e., they are homogeneous. In the process, EHC removes all the clusters that contain only one item. We call these clusters one-item clusters.

Initially, EHC considers TS to be a non-homogeneous cluster. The algorithm computes a mean item for each class (class-mean) by averaging the corresponding items in the non-homogeneous cluster. If the cluster contains items from c classes, EHC computes c means. Then, it applies k-means clustering on the cluster, using the class-means as initial means, and builds c clusters. k-means clustering is recursively applied on the items of each non-homogeneous cluster built. One-item clusters are removed.

Two examples that demonstrate the operation of EHC are depicted in Figures 2 and 3. More specifically, Figure 2 demonstrates how EHC identifies and removes a close-border item, while Figure 3 demonstrates how the algorithm removes a noisy item. Note that non-homogeneous clusters are depicted with dashed borders. EHC identifies and removes outliers in a similar way.

Of course, EHC may assign a typical data item (non-noisy, non-close-border) to an one-item cluster and remove it. For instance, suppose that a non-homogeneous cluster with two items is built. EHC will remove both items even when one of them belongs to the major class of the region.

Algorithm 4 describes a possible implementation of EHC. It utilizes a queue data structure Q in order to hold the unprocessed clusters. Initially, ES is set to be TS (line 1) and Q includes the whole TS as one unprocessed cluster (lines 2–3). In each algorithm iteration, cluster C is taken from the head of Q and is examined (line 5). If C is homogeneous (line 6), the algorithm counts the items in C and if C is a one-item cluster, its item is removed from ES (lines 7–9). If C

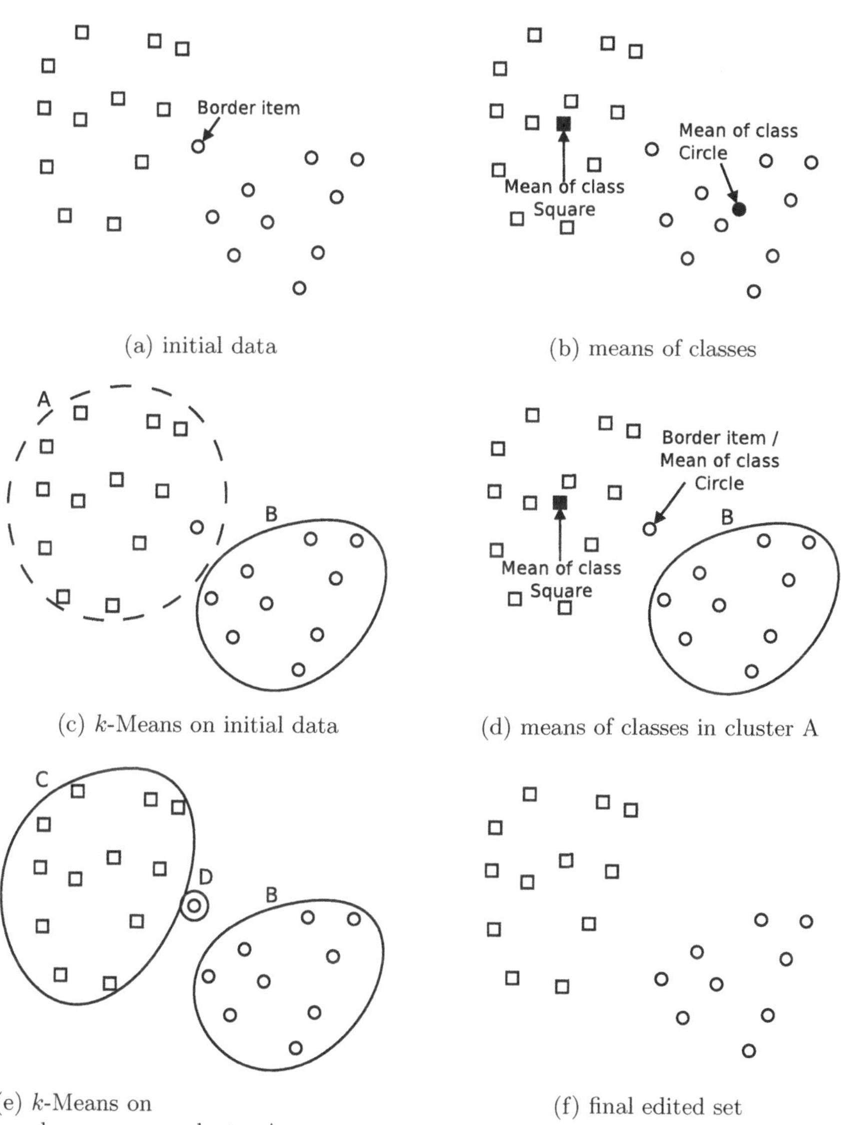

(a) initial data

(b) means of classes

(c) k-Means on initial data

(d) means of classes in cluster A

(e) k-Means on non-homogeneous cluster A

(f) final edited set

Fig. 2. EHC: Smoothing decision boundaries

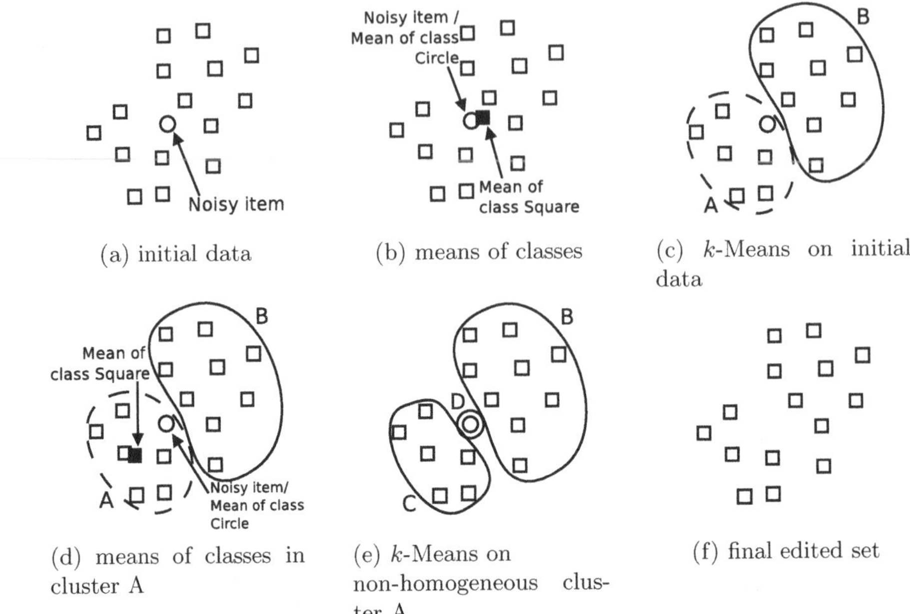

(a) initial data

(b) means of classes

(c) k-Means on initial data

(d) means of classes in cluster A

(e) k-Means on non-homogeneous cluster A

(f) final edited set

Fig. 3. EHC: Removing noisy items

is a non-homogeneous cluster, the class means for all the classes present in it are computed and added to set R (lines 11–14). Set R and cluster C are the input parameters to k-means clustering (line 15). The returned clusters are enqueued in Q (lines 16–18). The loop continues as long as there are non-homogeneous clusters (line 20).

Concerning the computational cost, we can easily conclude that EHC is a fast algorithm. It uses the fast k-means clustering algorithm that is also sped-up by considering as initial means the means of the classes that are present in each cluster. One expects that the resulting clusters are quickly consolidated and the cost is lower than when opting for random means initialization. It is worth mentioning that contrary to all other editing methods, EHC does not compute distances between "real" items. It computes distances between items and mean items. Moreover, contrary to ENN-rule and some of its variations that compute a fixed number of distances regardless the item distribution in the multidimensional space, the number of distances computed by EHC is difficult to predict in advance. It exclusively depends on the item distribution in the data space. Finally, the main advantage of the proposed method is that it is non-parametric. Therefore, there is no need for time-consuming trial-end-error procedures. Finally, note that EHC builds the same ES regardless of data ordering.

Algorithm 4. EHC

```
Input: TS
Output: ES
ES ← TS
Q ← ∅
Enqueue(Q, TS)
repeat
   C ← Dequeue(Q)
   if C is homogeneous then
      if |C| = 1 then
         ES ← ES − C
      end if
   else
      R ← ∅ {R is the set of class means}
      for each class M in C do
         R ← R ∪ mean_of(M)
      end for
      Clusters ← K-MEANS(C, R)
      for each cluster Cl ∈ Clusters do
         Enqueue(Q, Cl)
      end for
   end if
until IsEmpty(Q) {until all constructed clusters are homogeneous}
return ES
```

4 Performance Evaluation

4.1 Experimental Setup

The proposed EHC algorithm was coded in C and evaluated on ten datasets. We downloaded eight datasets from KEEL dataset repository[1] [2]. Their main characteristics are shown in Table 1. Initially, we did not know the level of noise in each dataset. After our experimentation, we realized that LIR is an almost noise-free dataset and LS and PH have low levels of noise. Since, we wanted to test how editing behaves on noise-free datasets, we decided to include these datasets in our experimentation. Moreover, we built two additional datasets by adding 10% random noise in LS and PH. We refer to these datasets as LS-n and PH-n respectively. Practically, we changed the class label of each item with a probability of 0.1. No other data transformation was performed. No dataset included missing values. Finally, euclidean distance was adopted as the distance metric.

For comparison purposes, we coded the three state-of-the-art algorithms presented in detail in Section 2 (ENN-rule [29], All-kNN [24], Multiedit [7]). We coded and used a non-optimized implementation of multiedit that may recompute same distances more than once.

[1] `http://sci2s.ugr.es/keel/datasets.php`

Table 1. Dataset details

Dataset	Size (items)	Attributes	Classes
Magic Gamma Telescope (MGT)	19020	10	2
Landsat Satellite (LS)	6435	36	6
Phoneme (PH)	5404	5	2
Letter Image Recognition (LIR)	20000	16	26
Banana (BN)	5300	2	2
Ecoli (ECL)	336	7	8
Pima (PM)	768	8	2
Yeast (YS)	1484	8	10

An important issue that we had to address was the tuning of the parameters of the aforementioned methods. For all of them, we adopted the settings proposed in [9]. In particular, we used $k = 3$ for ENN-rule, $k = 7$ and $k = 9$ for All-kNN and $n = 3$ and $R = 2$ for multiedit. These settings are very common in many experimental studies in the literature. In addition, we used $k = 5$ for ENN-rule and $n = 5$ for multiedit. Of course, we also measured and present the performance of the conventional 1-NN classifier (classification without editing).

The four editing algorithms were compared to each other in terms of two main criteria: classification accuracy and preprocessing (editing) cost. The latter was estimated by counting the distances computed by each algorithm. Accuracy measurements were estimated by executing 1-NN classifier on the edited sets. For each algorithm and dataset, we report the average accuracy and cost measurements obtained via a five-fold cross validation. We used the pairs of training/testing sets distributed by KEEL repository. Although the reduction rates achieved by each method do not indicate the best performing algorithm, they reveal the percentage of data that is considered as noise by each algorithm. Therefore reduction rates were estimated and are reported.

4.2 Comparisons

The performance measurements of our experimental study are presented in Table 2. Each table cell contains three measurements that correspond to the execution of an editing approach on a particular dataset. The three measurements are: accuracy (Acc), reduction rate (RR) and preprocessing cost (PC). The best measurements are in bold.

As we expected, EHC is the fastest approach. It achieves very low average PC measurements compared to its competitors (see the last row of the table). EHC computes the fewest distances in nine out of ten datasets. Furthermore, we observe that the cost gains are very high for large datasets. Finally, as we mentioned in Section 3, EHC computes a completely different number of distances for LS, LS-n and PH, PH-n. Here, we should mention that multiedit would have computed as many distances as ENN-rule and All-kNN had we used a more efficient implementation.

Table 2. Experimental measurements Accuracy (Acc(%)), Reduction Rate (RR(%)) and Preprocessing Cost (PC (millions of distance computations))

Dataset		1-NN	ENN (k=3)	ENN k=5)	Multiedit (n=3, R=2)	Multiedit (n=5, R=2)	All$k$NN (k=7)	All$k$NN (k=9)	EHC
MGT	Acc	78.144	80.44	80.57	76.75	75.26	80.76	**80.86**	79.52
	RR	-	20.08	19.20	39.98	42.36	29.67	30.38	10.70
	PC	-	115.76	115.76	2,839.55	1,447.93	115.76	115.76	**4.08**
LS	Acc	**90.60**	90.30	90.43	86.79	86.03	90.12	90.16	**90.55**
	RR	-	9.07	9.27	24.13	26.17	13.92	14.51	3.11
	PC	-	13.25	13.25	266.22	139.53	13.25	13.25	**1.69**
PH	Acc	**90.10**	88.14	87.53	80.77	79.72	86.55	86.23	**89.06**
	RR	-	11.25	11.93	34.14	36.91	17.92	19.30	7.36
	PC	-	9.35	9.35	166.22	53.71	9.35	9.35	**0.66**
LIR	Acc	**95.83**	94.98	94.87	70.94	58.35	94.28	94.00	**95.23**
	RR	-	4.33	4.44	43.43	56.59	7.31	7.97	3.95
	PC	-	127.99	127.99	7,214.38	2,900.53	127.99	127.99	**41.85**
BN	Acc	86.906	89.36	89.55	89.83	**90.38**	89.509	89.79	88.60
	RR	-	11.53	10.98	20.12	21.64	17.10	17.51	10.65
	PC	-	8.99	8.99	106.69	60.26	8.99	8.99	**0.56**
ECL	Acc	79.781	81.57	81.86	63.10	46.11	81.26	80.66	**82.16**
	RR	-	20.45	20.45	47.29	60.15	28.63	30.48	17.01
	PC	-	0.036	0.036	0.100	0.055	0.036	0.036	**0.035**
PM	Acc	68.358	71.87	71.75	71.36	68.89	72.65	**73.30**	70.32
	RR	-	30.16	29.43	53.07	58.96	45.56	46.24	16.59
	PC	-	0.19	0.19	0.51	0.26	0.19	0.19	**0.06**
YS	Acc	52.156	56.47	57.07	52.90	50.54	58.29	**58.42**	54.45
	RR	-	45.73	43.89	74.34	80.93	59.90	61.25	29.58
	PC	-	0.70	0.70	1.19	**0.58**	0.70	0.70	0.84
LS-n	Acc	82.067	89.96	**90.13**	86.70	85.86	89.74	89.79	89.67
	RR	-	20.21	18.08	37.90	40.25	30.01	30.57	13.80
	PC	-	13.25	13.25	131.95	94.98	13.25	13.25	**8.06**
PH-n	Acc	81.884	87.71	87.25	80.63	79.66	86.20	85.83	**88.40**
	RR	-	21.78	20.21	34.14	36.91	32.36	33.46	22.29
	PC	-	9.35	9.35	166.22	53.71	9.35	9.35	**4.35**
AVG	ACC	80.23	83.08	**83.10**	75.98	72.08	82.93	82.90	82.80
	RR	-	19.46	18.79	40.85	46.09	28.24	29.17	13.50
	PC	-	29.89	29.89	1,089.30	475.15	29.89	29.89	**6.22**

Concerning accuracy measurements, we observe that the proposed algorithm is comparable to ENN-rule and All-kNN. Multiedit has the worst accuracy, especially for LIR and ECL, where its accuracy is unacceptable. This happens because multiedit removes data that should not be removed. Although the differences in accuracy between EHC, ENN and All-kNN are not statistically significant, we observe that EHC has the highest Acc measurements in half the datasets. However, ENN-rule has the highest average Acc measurement.

For LIR, LS and PH that contain low levels of noise, all editing approaches seem to negatively affect accuracy since conventional 1-NN classifier achieves the highest Acc measurements. However, in all these cases, EHC is the most accurate editing algorithm. In contrast, in the rest seven datasets, most of the editing approaches achieve higher Acc measurements than conventional-1NN classifier. Therefore, it appears that editing constitutes a necessary preprocessing step.

The proposed algorithm has the lowest reduction rate. EHC removes items by using the strict criterion of one-item clusters. For datasets with extremely high levels of noise (e.g. 30% or more), it is not certain that EHC will improve classification accuracy like ENN-rule with an appropriate k value does. On the other hand, EHC is not expected to negatively affect classification accuracy as much as the other methods do.

5 Conclusions

Classification accuracy achieved by k-NN classifier strongly depends on the quality of the available training data. Noisy and mislabeled data as well as outliers and overlaps between regions of different classes are the reasons of bad classification performance for the particular classifier. Editing algorithms can improve classification accuracy by removing such data. In this paper, we presented a short review of editing algorithms. Then, we proposed a non-parametric algorithm, called Editing through Homogeneous Clusters (EHC), which follows a completely different strategy than the other editing approaches. EHC is based on a clustering procedure that forms homogeneous clusters in the training data. The clusters that contain only one item are considered redundant (they contain noisy, outlier or close-border items) and are removed. An experimental study with ten datasets showed that the proposed algorithm is very fast and achieves comparable classification accuracy to the state-of-the-art methods.

References

1. Aha, D.W., Kibler, D., Albert, M.K.: Instance-based learning algorithms. Mach. Learn. 6(1), 37–66 (1991), http://dx.doi.org/10.1023/A:1022689900470
2. Alcalá-Fdez, J., Fernández, A., Luengo, J., Derrac, J., García, S.: Keel data-mining software tool: Data set repository, integration of algorithms and experimental analysis framework. Multiple-Valued Logic and Soft Computing 17(2-3), 255–287 (2011)
3. Barandela, R., Gasca, E.: Decontamination of training samples for supervised pattern recognition methods. In: Ferri, F.J., Iñesta, J.M., Amin, A., Pudil, P. (eds.) SSPR&SPR 2000. LNCS, vol. 1876, pp. 621–630. Springer, Heidelberg (2000)
4. Brighton, H., Mellish, C.: Advances in instance selection for instance-based learning algorithms. Data Min. Knowl. Discov. 6(2), 153–172 (2002), http://dx.doi.org/10.1023/A:1014043630878
5. Dasarathy, B.V.: Nearest neighbor (NN) norms: NN pattern classification techniques. IEEE Computer Society Press (1991)
6. Dasarathy, B.V., Snchez, J.S., Townsend, S.: Nearest neighbour editing and condensing tools synergy exploitation. Pattern Analysis & Applications 3(1), 19–30 (2000), http://dx.doi.org/10.1007/s100440050003

7. Devijver, P.A., Kittler, J.: On the edited nearest neighbor rule. In: Proceedings of the Fifth International Conference on Pattern Recognition. The Institute of Electrical and Electronics Engineers (1980)
8. Garcia, S., Derrac, J., Cano, J., Herrera, F.: Prototype selection for nearest neighbor classification: Taxonomy and empirical study. IEEE Trans. Pattern Anal. Mach. Intell. 34(3), 417–435 (2012), http://dx.doi.org/10.1109/TPAMI.2011.142
9. García-Borroto, M., Villuendas-Rey, Y., Carrasco-Ochoa, J.A., Martínez-Trinidad, J.F.: Using maximum similarity graphs to edit nearest neighbor classifiers. In: Bayro-Corrochano, E., Eklundh, J.-O. (eds.) CIARP 2009. LNCS, vol. 5856, pp. 489–496. Springer, Heidelberg (2009)
10. Grochowski, M., Jankowski, N.: Comparison of instance selection algorithms ii. results and comments. In: Rutkowski, L., Siekmann, J.H., Tadeusiewicz, R., Zadeh, L.A. (eds.) ICAISC 2004. LNCS (LNAI), vol. 3070, pp. 580–585. Springer, Heidelberg (2004)
11. Han, J., Kamber, M., Pei, J.: Data Mining: Concepts and Techniques. The Morgan Kaufmann Series in Data Management Systems. Elsevier Science (2011)
12. Hattori, K., Takahashi, M.: A new edited k-nearest neighbor rule in the pattern classification problem. Pattern Recognition 33(3), 521–528 (2000), http://www.sciencedirect.com/science/article/pii/S0031320399000680
13. Grochowski, M., Jankowski, N.: Comparison of instances seletion algorithms i. algorithms survey. In: Rutkowski, L., Siekmann, J.H., Tadeusiewicz, R., Zadeh, L.A. (eds.) ICAISC 2004. LNCS (LNAI), vol. 3070, pp. 598–603. Springer, Heidelberg (2004)
14. Jiang, Y., Zhou, Z.-H.: Editing training data for knn classifiers with neural network ensemble. In: Yin, F.-L., Wang, J., Guo, C. (eds.) ISNN 2004. LNCS, vol. 3173, pp. 356–361. Springer, Heidelberg (2004)
15. Lozano, M.: Data Reduction Techniques in Classification processes (Phd Thesis). Universitat Jaume I (2007)
16. McQueen, J.: Some methods for classification and analysis of multivariate observations. In: Proc. of 5th Berkeley Symp. on Math. Statistics and Probability, pp. 281–298. University of California Press, Berkeley (1967)
17. Olvera-López, J.A., Carrasco-Ochoa, J.A., Martínez-Trinidad, J.F., Kittler, J.: A review of instance selection methods. Artif. Intell. Rev. 34(2), 133–143 (2010), http://dx.doi.org/10.1007/s10462-010-9165-y
18. Ougiaroglou, S., Evangelidis, G.: Efficient dataset size reduction by finding homogeneous clusters. In: Proceedings of the Fifth Balkan Conference in Informatics, BCI 2012, pp. 168–173. ACM, New York (2012), http://doi.acm.org/10.1145/2371316.2371349
19. Ougiaroglou, S., Nanopoulos, A., Papadopoulos, A.N., Manolopoulos, Y., Welzer-Druzovec, T.: Adaptive k-nearest-neighbor classification using a dynamic number of nearest neighbors. In: Ioannidis, Y., Novikov, B., Rachev, B. (eds.) ADBIS 2007. LNCS, vol. 4690, pp. 66–82. Springer, Heidelberg (2007)
20. Sánchez, J.S., Barandela, R., Marqués, A.I., Alejo, R., Badenas, J.: Analysis of new techniques to obtain quality training sets. Pattern Recogn. Lett. 24(7), 1015–1022 (2003), http://dx.doi.org/10.1016/S0167-8655(02)00225-8
21. Segata, N., Blanzieri, E., Delany, S.J., Cunningham, P.: Noise reduction for instance-based learning with a local maximal margin approach. J. Intell. Inf. Syst. 35(2), 301–331 (2010), http://dx.doi.org/10.1007/s10844-009-0101-z

22. Snchez, J., Pla, F., Ferri, F.: On the use of neighbourhood-based non-parametric classifiers. Pattern Recognition Letters 18(11–13), 1179–1186 (1997), http://www.sciencedirect.com/science/article/pii/S0167865597001128
23. Snchez, J., Pla, F., Ferri, F.: Prototype selection for the nearest neighbour rule through proximity graphs. Pattern Recognition Letters 18(6), 507–513 (1997), http://www.sciencedirect.com/science/article/pii/S0167865597000354
24. Tomek, I.: An experiment with the edited nearest-neighbor rule. IEEE Transactions on Systems, Man, and Cybernetics 6, 448–452 (1976)
25. Toussaint, G.: Proximity graphs for nearest neighbor decision rules: Recent progress. In: 34th Symposium on the INTERFACE, pp. 17–20 (2002)
26. Triguero, I., Derrac, J., Garcia, S., Herrera, F.: A taxonomy and experimental study on prototype generation for nearest neighbor classification. Trans. Sys. Man Cyber Part C 42(1), 86–100 (2012), http://dx.doi.org/10.1109/TSMCC.2010.2103939
27. Vázquez, F., Sánchez, J.S., Pla, F.: A stochastic approach to wilson's editing algorithm. In: Marques, J.S., de la Pérez Blanca, N., Pina, P. (eds.) IbPRIA 2005. LNCS, vol. 3523, pp. 35–42. Springer, Heidelberg (2005)
28. Wilson, D.R., Martinez, T.R.: Reduction techniques for instance-basedlearning algorithms. Mach. Learn. 38(3), 257–286 (2000), http://dx.doi.org/10.1023/A:1007626913721
29. Wilson, D.L.: Asymptotic properties of nearest neighbor rules using edited data. IEEE Trans. on Systems, Man, and Cybernetics 2(3), 408–421 (1972)

A Logic for Specifying Stochastic Actions and Observations

Gavin Rens[1], Thomas Meyer[1], and Gerhard Lakemeyer[2]

[1] Centre for Artificial Intelligence Research, University of KwaZulu-Natal, and CSIR Meraka, South Africa
{grens,tmeyer}@csir.co.za
[2] RWTH Aachen University, Germany
gerhard@cs.rwth-aachen.de

Abstract. We present a logic inspired by partially observable Markov decision process (POMDP) theory for specifying agent domains where the agent's actuators and sensors are noisy (causing uncertainty). The language features modalities for actions and predicates for observations. It includes a notion of probability to represent the uncertainties, and the expression of rewards and costs are also catered for. One of the main contributions of the paper is the formulation of a sound and complete decision procedure for checking validity of sentences: a tableau method which appeals to solving systems of equations. The tableau rules eliminate propositional connectives, then, for all open branches of the tableau tree, systems of equations are generated and checked for feasibility. This paper presents progress made on previously published work.

1 Introduction

Imagine a robot that is in need of an oil refill. There is an open can of oil on the floor within reach of its gripper. If there is nothing else in the robot's gripper, it can grab the can (or miss it, or knock it over) and it can drink the oil by lifting the can to its mouth and pouring the contents in (or miss its mouth and spill). The robot may also want to confirm whether there is anything left in the oil-can by weighing its contents with its 'weight' sensor. And once holding the can, the robot may wish to replace it on the floor. In situations where the oil-can is full, the robot gets five units of reward for gabbing the can, and it gets ten units of reward for a drink action.

In order for robots and intelligent agents in stochastic domains to reason about actions and observations, they must first have a *representation* or *model* of the domain over which to reason. For example, a robot may need to represent available knowledge about its `grab` action in its current situation. It may need to represent that when 'grabbing' the oil-can, there is a 5% chance that it will knock over the oil-can. As another example, if the robot has access to information about the weight of an oil-can, it may want to represent the fact that the can weighs heavy 90% of the time in 'situation A', but that it is heavy 98% of the time in 'situation B'.

C. Beierle and C. Meghini (Eds.): FoIKS 2014, LNCS 8367, pp. 305–323, 2014.

The oil-drinking domain is (partially) formalized as follows. The robot has the set of (intended) actions $\mathcal{A} = \{\texttt{grab}, \texttt{drink}, \texttt{weigh}, \texttt{replace}\}$ with expected meanings. The robot can make observations only from the set $\Omega = \{\texttt{obsNil}, \texttt{obsLight}, \texttt{obsMedium}, \texttt{obsHeavy}\}$. Intuitively, when the robot performs a `weigh` action (i.e., it activates its 'weight' sensor) it will perceive either `obsLight`, `obsMedium` or `obsHeavy`; for other actions, it will perceive `obsNil`. The robot experiences its world (domain) through three Boolean features: $\mathcal{F} = \{\texttt{full}, \texttt{drank}, \texttt{holding}\}$ meaning respectively that the oil-can is full, that the robot has drunk the oil and that it is currently holding something in its gripper. Given a formalization BK of our scenario, the robot may have the following queries:

- If the oil-can is empty and I'm not holding it, is there a 0.9 probability that I'll be holding it after grabbing it, and a 0.1 probability that I'll have missed it? That is, does $(\neg\texttt{full} \wedge \neg\texttt{holding}) \rightarrow ([\texttt{grab}]_{0.9}(\neg\texttt{full} \wedge \texttt{holding}) \wedge [\texttt{grab}]_{0.1}(\neg\texttt{full} \wedge \neg\texttt{holding}))$ follow from BK?
- If the oil-can is not full, I've drunk the oil and I'm holding the can, is there a 0.7 probability of perceiving the can is light, given I weighed it? That is, does $(\neg\texttt{full} \wedge \texttt{drank} \wedge \texttt{holding}) \rightarrow (\texttt{obsLight} \mid \texttt{weigh} : 0.7)$ follow from BK?

Modal logic is considered to be well suited to reasoning about beliefs and changing situations [6,14]. Partially observable Markov decision process (POMDP) theory [30,17] has proven to be a good general framework for formalizing dynamic stochastic systems. Our goal is to integrate logic with stochastic actions and observations, taking the semantics of POMDPs in particular. To our knowledge, there exists no such logic; see the discussion of related work below. This paper though, concerns work that is a step towards that goal. Here we present the Specification Logic of Actions and Observations with Probability (SLAOP). With SLAOP, POMDP models can be represented compactly.

The present version of SLAOP is an extension of the Specification Logic of Actions with Probability (SLAP) [24,27], but with an improved completeness proof due to a new decision procedure. SLAP is extended with (i) notions of rewards and action costs, (ii) a notion of equality between actions and observations and (iii) observations for dealing with perception/sensing. To establish a correspondence between POMDPs and SLAOP, SLAOP must view observations as objects at the same semantic level as actions. We make use of the results of [26] to add observations as first-class objects.

A *preliminary* version of SLAOP has been presented at a doctoral consortium [23]. Since then, significant progress has been made. We mention only some of the major changes. Firstly, the present version of SLAOP inherits the $\Box$ operator from SLAP, which is important for marking sentences as globally applicable axioms. The preliminary version of SLAOP had no $\Box$ operator. Another change is, instead of the predicate $(\varsigma \mid \alpha : q)$ used in the present version, a modal operator $[\varsigma \mid \alpha]_q\varphi$ with a slightly different definition was used in the 'old' SLAOP. $[\varsigma \mid \alpha]_q\varphi$ can be read 'The probability of perceiving ς in a world in which φ holds is equal to q, given α was performed.' It turned out that specifying φ creates unwanted interactions with the modal operator $[\alpha]_q\varphi$ for specifying transition probabilities. Moreover, we have determined that $(\varsigma \mid \alpha : q)$ (with the given meaning; cf. § 3.2) is sufficient for specifying perception probabilities (cf. § 5). Last and most importantly, the decision procedure of the preliminary version relied on many intricate tableau rules; relying on the solvability of systems of inequalities (as in the present version) is much cleaner and the decidability of such systems carries over to help prove the decidability of SLAOP. The decision procedure for the previous version of SLAOP was not proven complete. The current version is proved complete and terminating.

A formal approach how to specify probabilistic transition models with SLAP has been published [24], and there, a solution to the frame problem for SLAP is also presented. That frame solution can easily be employed for SLAOP. A decision procedure for validity checking in SLAP is presented in a journal article [27]. The procedure for SLAP is simpler than for SLAOP because it does not use the 'label assignment' approach (cf. § 4.3). We opted for a decision procedure with label assignments for SLAOP because the proof of completeness is then easier to understand (see the accompanying technical report [25]), and sentences of a certain form which are not allowed in SLAP are allowed in SLAOP (see § 4.3).

Related work is discussed next. Then Section 3 presents the syntax and semantics of SLAOP. Section 4 presents the two-phase decision procedure. Section 5 provides examples of application of the decision procedure. Some concluding remarks are made in Section 6.

2 Related Work

Several frameworks exist for reasoning about probabilistic inference in static domains [1,9,13,16,29,33]. Here, a "static domain" is a domain in which the physical state of the system does not change, although the state of *information* of various agents in the system may change. In SLAOP, the focus is more on how stochastic actions change the physical state of a system. Some of these logics are concerned with how knowledge changes as new information is gained, however, the information received is not seen as an observation *object*. Moreover, they do not express the probability with which the received information was expected in the current situation. That is, they take the new information as certain. SLAOP can express the fact that information (in the form of observation objects) may be incorrect to some degree. This ability of SLAOP is carried over from the SLAP logic [24,27].

Poole's Independent Choice Logic using the situation calculus (ICL_{SC}) [20] is a relatively rich framework, with acyclic logic programs which may contain variables, quantification and function symbols. For certain applications, SLAOP may be preferred due to its comparative simplicity. And because SLAOP's semantics is very close to that of standard POMDP theory, it may be easier to understand by people familiar with POMDPs. Finally, decidability of inferences made in the ICL_{SC} are, in general, not guaranteed.

Bonet and Geffner [3] present a framework with heuristic search algorithms for modeling and solving MDPs and POMDPs. It seems similar to the ICL_{SC}

in its application area. Their framework uses high-level logical representations, but it is not presented as a logic, nor does it empoy logical entailment.

DTGolog [5] is a programming language, rather than a logic, and it does not deal with stochastic observations.

PODTGolog [22] is another logic programming framework which does deal with stochastic observations, but it does not have a well defined semantics.

Many popular frameworks for reasoning about action, employ or are based on the situation calculus [22]. Reified situations make the meaning of formulae perspicuous. However, the situation calculus seems too rich and expressive for our purposes, and it would be desirable to remain decidable, hence the restriction to a propositional modal framework. The validity problem for SLAOP is decidable, which sets it apart from first-order logics for reasoning about action (including the situation calculus) or reasoning with probabilities (including BHL's approach [2] and $\mathcal{ESP}$ [11]). In other words, having a decidable formalism to reason about POMDP's is considered an asset and would set us apart from other more expressive logical formalisms addressing action and sensing under uncertainty.

Iocchi *et al.* [15] present a logic called $\mathcal{E}+$ for reasoning about agents with sensing, qualitative nondeterminism and probabilistic uncertainty in action outcomes. Planning with sensing and uncertain actions is also dealt with. Noisy sensing is not dealt with, that is, sensing actions are deterministic. They mention that although they would like to be able to represent action rewards and costs as in POMDPs, $\mathcal{E}+$ does not yet provide the facilities.

There are some logics that come closer to what we desire [8,34,11,33], that is, they incorporate notions of probability, but they were not created with POMDPs in mind and typically do not take observations as first-class objects. On the other hand, there are formalisms for specifying POMDPs that employ logic-based representation [4,35,28], but they are not defined entirely as logics. Our work is to bring the representation of and reasoning about POMDPs *totally* into the logical arena. One is then in very familiar territory and new opportunities for the advancement in reasoning about POMDPs may be opened up.

Systems of linear inequalities are at the heart of Nilsson's probabilistic logic [19], which has been extended with stochastic actions by Thiébaux *et al.* [32]. Fagin, Halpern and Megiddo [10] use a similar idea to prove that the axiomatization of their logic for reasoning about probabilities is complete. None of these deals with observations.

3 Specification Logic of Actions and Observations with Probability

First we present the syntax of SLAOP, then we state its semantics.

3.1 Syntax

The vocabulary of our language contains six sorts of objects of interest:

1. a finite set of *fluents* (alias, *propositional atoms*) $\mathcal{F} = \{f_1, \ldots, f_n\}$,
2. a finite set of names of atomic *actions* $\mathcal{A} = \{\alpha_1, \ldots, \alpha_n\}$,
3. a finite set of names of atomic *observations* $\Omega = \{\varsigma_1, \ldots, \varsigma_n\}$,
4. all *real numbers* $\mathbb{R}$,[1]
5. a countable set of *action variables* $V_{\mathcal{A}} = \{v_1^{\alpha}, v_2^{\alpha}, \ldots\}$,
6. a countable set of *observation variables* $V_{\Omega} = \{v_1^{\varsigma}, v_2^{\varsigma}, \ldots\}$.

From now on, we denote $\mathbb{R} \cap [0,1]$ as $\mathbb{R}_{[0,1]}$. We shall refer to elements of $\mathcal{A} \cup \Omega$ as *constants*. We are going to work in a multi-modal setting, in which we have modal operators $[\alpha]_q$, one for each $\alpha \in \mathcal{A}$ and $q \in \mathbb{R}_{[0,1]}$, and predicates $(\varsigma \mid \alpha : q)$, one for each pair in $\Omega \times \mathcal{A}$ and $q \in \mathbb{R}_{[0,1]}$.

Definition 1. *Let $f \in \mathcal{F}$, $\alpha \in (\mathcal{A} \cup V_{\mathcal{A}})$, $\varsigma \in (\Omega \cup V_{\Omega})$, $v \in (V_{\mathcal{A}} \cup V_{\Omega})$, $q \in \mathbb{R}_{[0,1]}$ and $r \in \mathbb{R}$. The language of SLAOP, denoted $\mathcal{L}_{SLAOP}$, is the least set of Ψ defined by the grammar:*

$$\varphi ::= f \mid \top \mid \neg\varphi \mid \varphi \wedge \varphi.$$

$$\Phi ::= \varphi \mid \alpha = \alpha \mid \varsigma = \varsigma \mid Reward(r) \mid Cost(\alpha, r) \mid [\alpha]_q\varphi \mid (\varsigma \mid \alpha : q) \mid (\forall v)\Phi \mid \neg\Phi \mid \Phi \wedge \Phi.$$

$$\Psi ::= \Phi \mid \Box\Phi \mid \neg\Psi \mid \Psi \wedge \Psi.$$

The scope of quantifier $(\forall v)$ is determined in the same way as is done in first-order logic. A variable v' appearing in a formula Ψ is said to be bound by quantifier $(\forall v)$ if and only if v' is the same variable as v and is in the scope of $(\forall v)$. If a variable is not bound by any quantifier, it is free. In $\mathcal{L}_{SLAOP}$, variables are not allowed to be free; they are always bound.

(For SLAP, $\Phi ::= \varphi \mid [\alpha]_q\varphi \mid \neg\Phi \mid \Phi \wedge \Phi$.) Note that formulae with nested modal operators of the form $\Box\Box\Phi$, $\Box\Box\Box\Phi$, $[\alpha]_q[\alpha]_q\varphi$ and $[\alpha]_q[\alpha]_q[\alpha]_q\varphi$ et cetera are not in $\mathcal{L}_{SLAOP}$. 'Single-step' or 'flat' formulae are sufficient to *specify* action transitions probabilities, that is, for specifying a transition model. To reason about the effects of sequences of actions, nesting may be appropriate, but SLAOP is not for reasoning at that level. As usual, we treat $\bot, \vee, \rightarrow$ and $\leftrightarrow$ as abbreviations. $\rightarrow$ and $\leftrightarrow$ have the weakest bindings and $\neg$ the strongest; parentheses enforce or clarify the scope of operators conventionally.

The definition of a POMDP reward function $R(a, s)$ may include not only the reward value of state s, but it may deduct the cost of performing a in s. It will be convenient for the person specifying a POMDP using SLAOP to be able to specify action costs independently from the rewards of states, because these two notions are not necessarily connected. To specify rewards and execution costs in SLAOP, we require *Reward* and *Cost* as special predicates. $Reward(r)$ can be read 'The reward for being in the current situation is r units' and we read $Cost(\alpha, c)$ as 'The cost for executing α is c units'.

[1] In SLAP [27] and the previous version of SLAOP [23], rational numbers were used. Due to our completeness proof relying on Tarski's quantifier elimination method [31] which involves real numbers, we use real numbers here.

$[\alpha]_q\varphi$ is read 'The probability of reaching a φ-world after executing α, is equal to q'. $[\alpha]$ abbreviates $[\alpha]_1$. $(\varsigma \mid \alpha : q)$ is read 'The probability of perceiving ς, given α was performed, is q'.

$\langle\alpha\rangle\varphi$ abbreviates $\neg[\alpha]_0\varphi$ and is read 'It is possible to reach a world in which φ holds after executing α'. Note that $\langle\alpha\rangle\varphi$ does not mean $\neg[\alpha]\neg\varphi$. $[\alpha]_q\varphi$ and $\neg[\alpha]_q\varphi$ are referred to as *dynamic literals*. $(\varsigma \mid \alpha : q)$ and $\neg(\varsigma \mid \alpha : q)$ are referred to as *perception literals*.

One reads $\Box\Phi$ as 'Φ holds in every possible world'. We require the $\Box$ operator to mark certain information (sentences) as holding in *all* possible worlds—essentially, the axioms which model the domain of interest. $(\forall v^{\alpha})$ is to be read 'For all actions' and $(\forall v^{\varsigma})$ is to be read 'For all observations'. $(\forall v)\Phi$ (where $v \in (V_{\mathcal{A}} \cup V_{\Omega})$) can be thought of as a syntactic shorthand for the finite conjunction of Φ with the variables replaced by the constants of the right sort (cf. Def. 3 for the formal definition). $(\exists v)\Phi$ abbreviates $\neg(\forall v)\neg\Phi$.

3.2 Semantics

SLAOP extends SLAP. SLAP structures are non-standard: They have the form $\langle W, R\rangle$, where W is a *finite* set of worlds such that each world assigns a truth value to each atomic proposition, and R is a binary relation on W. Moreover, SLAP is multi-modal in that there are multiple accessibility relations. Intuitively, when talking about some world w, we mean a set of features (*propositions*) that the agent understands and that describes a state of affairs in the world or that describes a possible, alternative world. Let $w : \mathcal{F} \mapsto \{0,1\}$ be a total function that assigns a truth value to each fluent. Let C be the set of all possible functions w. We call C the *conceivable worlds*.

SLAP structures are comparable to Markov decision processes (MDPs) [21] without reward functions, whereas SLAOP structures are comparable to POMDPs (with reward functions). A POMDP model is a tuple $\langle S, A, T, R, \Omega, O, b^0\rangle$; S is a finite set of states the agent can be in; A is a finite set of actions the agent can choose to execute; T is the function defining the probability of reaching one state from another for each action; R is a function giving the expected immediate reward gained by the agent for any state and agent action; Ω is a finite set of observations the agent can experience of its world; O is a function giving a probability distribution over observations for any state and action performed to reach that state; b^0 is the initial probability distribution over all states in S.

A SLAOP structure is a 'translation' of a POMDP model, except for the initial belief-state b^0.[2]

Definition 2. *A SLAOP structure is a tuple $\mathcal{S} = \langle W, R, O, N, Q, U\rangle$ such that*

1. $W \subseteq C$ a non-empty set of possible worlds.

[2] Specification of the initial belief-state is required at a higher level of reasoning. It is left for future work.

2. *$R : \mathcal{A} \mapsto R_\alpha$, where $R_\alpha : (W \times W) \mapsto \mathbb{R}_{[0,1]}$ is a total function from pairs of worlds into the reals; That is, R is a mapping that provides an accessibility relation R_α for each action $\alpha \in \mathcal{A}$; For every $w^- \in W$, it is required that either $\sum_{w^+ \in W} R_\alpha(w^-, w^+) = 1$ or $\sum_{w^+ \in W} R_\alpha(w^-, w^+) = 0$.*
3. *O is a nonempty finite set of observations;*
4. *$N : \Omega \mapsto O$ is a bijection that associates to each name in Ω, a unique observation in O;*
5. *$Q : \mathcal{A} \mapsto Q_\alpha$, where $Q_\alpha : (W \times O) \mapsto \mathbb{R}_{[0,1]}$ is a total function from pairs in $W \times O$ into the reals; That is, Q is a mapping that provides a perceivability relation Q_α for each action $\alpha \in \mathcal{A}$; For all $w^-, w^+ \in W$: if $R_\alpha(w^-, w^+) > 0$, then $\sum_{o \in O} Q_\alpha(w^+, o) = 1$, that is, there is a probability distribution over observations in a reachable world; Else if $R_\alpha(w^-, w^+) = 0$, then $\sum_{o \in O} Q_\alpha(w^+, o) = 0$;*
6. *U is a pair $\langle Re, Co \rangle$, where $Re : W \mapsto \mathbb{R}$ is a reward function and Co is a mapping that provides a cost function $Co_\alpha : C \mapsto \mathbb{R}$ for each $\alpha \in \mathcal{A}$.*

Note that the set of possible worlds may be the whole set of conceivable worlds.

R_α defines the transition probability $pr \in \mathbb{R}_{[0,1]}$ between worlds w^+ and world w^- via action α. If $R_\alpha(w^-, w^+) = 0$, then w^+ is said to be *inaccessible* or *not reachable* via α performed in w^-, else if $R_\alpha(w^-, w^+) > 0$, then w^+ is said to be *accessible* or *reachable* via action α performed in w^-. If for some w^-, $\sum_{w^+ \in W} R_\alpha(w^-, w^+) = 0$, we say that α is *inexecutable* in w^-.

Q_α defines the observation probability $pr \in \mathbb{R}_{[0,1]}$ of observation o perceived in world w^+ after the execution of action α. Assuming w^+ is accessible, if $Q_\alpha(w^+, o) > 0$, then o is said to be *perceivable* in w^+, given α, else if $Q_\alpha(w^+, o) = 0$, then o is said to be *unperceivable* in w^+, given α. The definition of perceivability relations implies that there is always at least one possible observation in any world reached due to an action.

Because N is a bijection, it follows that $|O| = |\Omega|$. (We take $|X|$ to be the cardinality of set X.) The value of the reward function $Re(w)$ is a real number representing the reward an agent gets for being in or getting to the world w. It must be defined for each $w \in C$. The value of the cost function $Co(\alpha, w)$ is a real number representing the cost of executing α in the world w. It must be defined for each action $\alpha \in \mathcal{A}$ and each $w \in C$.

Definition 3 (Truth Conditions). *Let $\mathcal{S}$ be a SLAOP structure, with $\alpha, \alpha' \in \mathcal{A}$, $q, pr \in \mathbb{R}_{[0,1]}$ and $r \in \mathbb{R}$. Let $f \in \mathcal{F}$ and let Φ be any sentence in $\mathcal{L}_{SLAOP}$. We say Φ is* satisfied *at world w in structure $\mathcal{S}$ (written $\mathcal{S}, w \models \Phi$) if and only if the following holds:*

$\mathcal{S}, w \models \top$ for all $w \in W$;
$\mathcal{S}, w \models f \iff w(f) = 1$ for $w \in W$;
$\mathcal{S}, w \models \neg\Psi \iff \mathcal{S}, w \not\models \Psi$;
$\mathcal{S}, w \models \Psi \wedge \Psi' \iff \mathcal{S}, w \models \Psi$ and $\mathcal{S}, w \models \Psi'$;
$\mathcal{S}, w \models (\alpha = \alpha') \iff \alpha, \alpha' \in \mathcal{A}$ are the same element;
$\mathcal{S}, w \models (\varsigma = \varsigma') \iff \varsigma, \varsigma' \in \Omega$ are the same element;

$$\mathcal{S}, w \models Reward(r) \iff Re(w) = r;$$
$$\mathcal{S}, w \models Cost(\alpha, r) \iff Co_\alpha(w) = r;$$
$$\mathcal{S}, w \models [\alpha]_q \varphi \iff \sum_{w' \in W, \mathcal{S}, w' \models \varphi} R_\alpha(w, w') = q;$$
$$\mathcal{S}, w \models (\varsigma \mid \alpha : q) \iff Q_\alpha(w, N(\varsigma)) = q;$$
$$\mathcal{S}, w \models \Box \Phi \iff \text{for all } w' \in W, \mathcal{S}, w' \models \Phi;$$
$$\mathcal{S}, w \models (\forall v^\alpha)\Phi \iff \mathcal{S}, w \models \Phi|^{v^\alpha}_{\alpha_1} \wedge \ldots \wedge \Phi|^{v^\alpha}_{\alpha_n};$$
$$\mathcal{S}, w \models (\forall v^\varsigma)\Phi \iff \mathcal{S}, w \models \Phi|^{v^\varsigma}_{\varsigma_1} \wedge \ldots \wedge \Phi|^{v^\varsigma}_{\varsigma_n},$$

where we write $\Phi|^v_c$ to mean the formula Φ with all variables $v \in (V_\mathcal{A} \cup V_\Omega)$ appearing in it replaced by constant $c \in \mathcal{A} \cup \Omega$ of the right sort.

A formula φ is *valid* in a SLAOP structure (denoted $\mathcal{S} \models \varphi$) if $\mathcal{S}, w \models \varphi$ for every $w \in W$. φ is *SLAOP-valid* (denoted $\models \varphi$) if φ is true in every structure $\mathcal{S}$. If $\models \theta \leftrightarrow \psi$, we say θ and ψ are *semantically equivalent* (abbreviated $\theta \equiv \psi$). φ is *satisfiable* if $\mathcal{S}, w \models \varphi$ for some $\mathcal{S}$ and $w \in W$. A formula that is not satisfiable is *unsatisfiable* or a *contradiction*. The truth of a propositional formula depends only on the world in which it is evaluated. We may thus write $w \models \varphi$ instead of $\mathcal{S}, w \models \varphi$ when φ is a propositional formula.

Let $\mathcal{K}$ be a finite subset of $\mathcal{L}_{SLAOP}$. We say that ψ is a *local semantic consequence* of $\mathcal{K}$ (denoted $\mathcal{K} \models \psi$) if for all structures $\mathcal{S}$, and all $w \in W$ of $\mathcal{S}$, if $\mathcal{S}, w \models \bigwedge_{\theta \in \mathcal{K}} \theta$ then $\mathcal{S}, w \models \psi$. We shall also say that $\mathcal{K}$ *entails* ψ whenever $\mathcal{K} \models \psi$. If $\{\theta\} \models \psi$ then we simply write $\theta \models \psi$. In fact, $\mathcal{K} \models \Psi$ if and only if $\models \bigwedge_{\theta \in \mathcal{K}} \theta \rightarrow \Psi$ (i.e., $\mathcal{K}$ entails Ψ iff $\bigwedge_{\theta \in \mathcal{K}} \theta \rightarrow \Psi$ is SLAOP-valid).

If there exists a world $w \in C$ such that $w \models \delta$, where δ is a propositional formula, and for all $w' \in C$, if $w' \neq w$ then $w' \not\models \delta$, we say that δ is *definitive* (then, δ defines a world; δ is a *complete propositional theory*). Let $Def(\varphi)$ be all the definitive formulae which entail φ, that is, $Def(\varphi) = \{\delta \in \mathcal{L}_{SLAOP} \mid \delta \text{ is definitive and } \delta \models \varphi\}$.

4 Decision Procedure for SLAOP Entailment

In this section we describe a decision procedure which has two phases: creation of a tableau tree (the *tableau* phase) which essentially eliminates propositional connectives, then a phase which checks for inconsistencies given possible mappings from 'labels' (of the tableau calculus) to worlds (the *label assignment* phase). Particularly, in the label assignment phase, solutions for systems of inequalities (equations and disequalities) are sought.

4.1 The Tableau Phase

The necessary definitions and terminology are given next.

A *labeled formula* is a pair (x, Ψ), where $\Psi \in \mathcal{L}_{SLAOP}$ is a formula and x is an integer called the *label* of Ψ. A *node* Γ^j_k with superscript j (the *branch* index) and subscript k (the *node* index), is a set of labeled formulae. The initial node, that is, Γ^0_0, to which the tableau rules must be applied, is called the *trunk*.

Definition 4. *A* tree *T is a set of nodes. A tree must include Γ_0^0 and only nodes resulting from the application of* tableau rules *to the trunk and subsequent nodes. If one has a tree with trunk $\Gamma_0^0 = \{(0, \Psi)\}$, we'll say one has* a tree for Ψ.

When we say '...where x is a fresh integer', we mean that x is the smallest positive integer of the right sort (formula label or branch index) not yet used in the node to which the incumbent tableau rule will be applied.

A tableau rule applied to node Γ_k^j creates one or more new nodes; its child(ren). If it creates one child, then it is identified as Γ_{k+1}^j. If Γ_k^j creates a second child, it is identified as $\Gamma_0^{j'}$, where j' is a fresh integer. That is, for every child created beyond the first, a new branch is started.

A node Γ is a *leaf* node of tree T if no tableau rule has been applied to Γ in T. A *branch* is the set of nodes on a path from the trunk to a leaf node. Note that nodes with different branch indexes may be on the some path.

Definition 5. *Γ is* higher *on a branch than Γ' if and only if Γ is an ancestor of Γ'.*

A node Γ is *closed* if $(x, \bot) \in \Gamma$ for any $x \geq 0$. It is *open* if it is not closed. A branch is closed if and only if its leaf node is closed. A tree is closed if all of its branches are closed, else it is open.

A preprocessing step occurs, where all (sub)formulae of the form $(\forall v^\alpha)\Phi$ and $(\forall v^\varsigma)\Phi$ are replaced by, respectively, $(\Phi|_{\alpha_1}^{v^\alpha} \wedge \ldots \wedge \Phi|_{\alpha_n}^{v^\alpha})$ and $(\Phi|_{\varsigma_1}^{v^\varsigma} \wedge \ldots \wedge \Phi|_{\varsigma_n}^{v^\varsigma})$. The occurrence of $(\exists v^\varsigma)\neg(v^\varsigma \mid \alpha : 0)$ in rule obs (below) is only an abbreviation for the semantically equivalent formula without a quantifier and variables.

The tableau rules for SLAOP follow. A rule may only be applied to an open leaf node. To constrain rule application to prevent trivial re-applications of rules, a rule may not be applied to a formula if it has been applied to that formula higher in the tree, as in Definition 5. For example, if rule $\Box$ were applied to $\{(0, \Box p_1), (1, \neg[go]_0 p_2)\} \subset \Gamma_3^2$, then it may not be applied to $\{(0, \Box p_1), (1, \neg[go]_0 p_2)\} \subset \Gamma_4^2$.

Let Γ_k^j be a leaf node.

- rule $\bot$: If Γ_k^j contains (n, Φ) and $(n, \neg\Phi)$, then create node $\Gamma_{k+1}^j = \Gamma_k^j \cup \{(n, \bot)\}$.
- rule $\neg$: If Γ_k^j contains $(n, \neg\neg\Phi)$, then create node $\Gamma_{k+1}^j = \Gamma_k^j \cup \{(n, \Phi)\}$.
- rule $\wedge$: If Γ_k^j contains $(n, \Phi \wedge \Phi')$, then create node $\Gamma_{k+1}^j = \Gamma_k^j \cup \{(n, \Phi), (n, \Phi')\}$.
- rule $\vee$: If Γ_k^j contains $(n, \neg(\Phi \wedge \Phi'))$, then create node $\Gamma_{k+1}^j = \Gamma_k^j \cup \{(n, \neg\Phi)\}$ and node $\Gamma_0^{j'} = \Gamma_k^j \cup \{(n, \neg\Phi')\}$, where j' is a fresh integer.
- rule $=$: If Γ_k^j contains $(n, c = c')$ and c and c' are distinct constants, or if Γ_k^j contains $(n, \neg(c = c'))$ and c and c' are identical constants, then create node $\Gamma_{k+1}^j = \Gamma_k^j \cup \{(n, \bot)\}$.
- rule $\Diamond\varphi$: If Γ_k^j contains $(0, \neg[\alpha]_0\varphi)$ or $(0, [\alpha]_q\varphi)$ for $q > 0$, then create node $\Gamma_{k+1}^j = \Gamma_k^j \cup \{(n, \varphi)\}$, where n is a fresh integer.
- rule obs: If Γ_k^j contains $(x, \neg[\alpha]_0\varphi)$ or $(x, [\alpha]_q\varphi)$ for $q > 0$ and some x, then create node $\Gamma_{k+1}^j = \Gamma_k^j \cup \{(0, \Box(\delta_1 \rightarrow (\exists v^\varsigma)\neg(v^\varsigma \mid \alpha : 0)) \vee \Box(\delta_2 \rightarrow (\exists v^\varsigma)\neg(v^\varsigma \mid \alpha : 0)) \vee \cdots \vee \Box(\delta_n \rightarrow (\exists v^\varsigma)\neg(v^\varsigma \mid \alpha : 0)))\}$, where $\delta_i \in Def(\varphi)$.
- rule $\Box$: If Γ_k^j contains $(0, \Box\Phi)$ and (n, Φ') for any $n \geq 0$, and if it does not yet contain (n, Φ), then create node $\Gamma_{k+1}^j = \Gamma_k^j \cup \{(n, \Phi)\}$.

- rule $\Diamond$: If Γ^j_k contains $(0, \neg\Box\Phi)$, then create node $\Gamma^j_{k+1} = \Gamma^j_k \cup \{(n, \neg\Phi)\}$, where n is a fresh integer.

One might wonder why there is not a rule to deal with the case when Γ^j_k contains $(x, [\alpha]_q\varphi)$ and $(x, \neg[\alpha]_q\varphi)$, or no rule for when Γ^j_k contains $(x, (\varsigma \mid \alpha : r))$ and $(x, (\varsigma \mid \alpha : r'))$ where $r \neq r'$. As will be seen in Section 4.2, these and similar cases are dealt with.

Definition 6. *A branch is* saturated *if and only if any rule that can be applied to its leaf node has been applied. A tree is* saturated *if and only if all its branches are saturated.*

Once the tableau phase is completed, inconsistencies are sought for each open branch of the saturated tree. Depending on the results, certain branches may become closed. Depending on the final structure and contents of the tree, the sentence for which the tree was created can be determined as valid or not. Before the second phase can be explained, we need to explain how a system of inequalities (SI) can be generated from a set of dynamic and perception literals.

4.2 Systems of Inequalities

Definition 7. $W(\Gamma, n) \stackrel{def}{=} \{w \in C \mid w \models \ell$ *for all* $(n, \ell) \in \Gamma$ *where* ℓ *is a propositional literal*$\}$. $W(\Gamma) \stackrel{def}{=} \bigcup_{x\in\{0,1,\ldots,n'\}} W(\Gamma, x)$, *where* n' *is the largest label mentioned in* Γ.

Let $n = |W(\Gamma)|$. Let $W(\Gamma)^\# = (w_1, w_2, \ldots, w_n)$ be an ordering of the worlds in $W(\Gamma)$. With each world $w_k \in W(\Gamma)^\#$, we associate a real variable $pr^\alpha_k \in \mathbb{R}_{[0,1]}$. One can generate

$$c_{i,1}pr^\alpha_1 + c_{i,2}pr^\alpha_2 + \cdots + c_{i,n}pr^\alpha_n = q_i \text{ and } c_{i,1}pr^\alpha_1 + c_{i,2}pr^\alpha_2 + \cdots + c_{i,n}pr^\alpha_n \neq q_i,$$

for a formulae $(x, [\alpha]_{q_i}\varphi_i) \in \Gamma$, respectively, $(x, \neg[\alpha]_{q_i}\varphi_i) \in \Gamma$ such that $c_{i,k} = 1$ if $w_k \models \varphi_i$, else $c_{i,k} = 0$, where x represents a label.

Adding an equation

$$pr^\alpha_1 + pr^\alpha_2 + \cdots + pr^\alpha_n = \lceil pr^\alpha_1 + pr^\alpha_2 + \cdots + pr^\alpha_n \rceil$$

will ensure that either $\sum_{w^+\in W(\Gamma)} R_\alpha(w^-, w^+) = 1$ or $\sum_{w^+\in W(\Gamma)} R_\alpha(w^-, w^+) = 0$, as stated in Definition 2 on page 311.

Let $m = |\Omega|$. Let $\Omega^\# = (\varsigma_1, \varsigma_2, \ldots, \varsigma_m)$ be an ordering of the observations in Ω. With each observation in $\varsigma_j \in \Omega^\#$, we associate a real variable pr^ς_j.

One can generate

$$pr^\sigma_j = q_j \text{ and } pr^\sigma_j \neq q_j$$

for a formula $(x, (\sigma_j \mid \alpha : q_j)) \in \Gamma$, respectively, $(x, \neg(\sigma_j \mid \alpha : q_j) \in \Gamma$, where $\sigma_j \in \Omega^\#$ and $pr^\sigma_j \in \{pr^\varsigma_1, \ldots, pr^\varsigma_2, \ldots, pr^\varsigma_m\}$.

Adding an equation

$$pr^\varsigma_1 + pr^\varsigma_2 + \cdots + pr^\varsigma_2 + \cdots + pr^\varsigma_m = \lceil pr^\varsigma_1 + pr^\varsigma_2 + \cdots + pr^\varsigma_2 + \cdots + pr^\varsigma_m \rceil.$$

ensures that either $\sum_{o \in O} Q_\alpha(w^+, o) = 1$ or $\sum_{o \in O} Q_\alpha(w^+, o) = 0$, as stated in Definition 2 on page 311.

Let $\Delta(\alpha)$ be a set of dynamic literals mentioning α and let $\Omega(\alpha)$ be a set of perception literals involving α. Let $S(\Delta(\alpha))$ and $S(\Omega(\alpha))$ be the systems formed from $\Delta(\alpha)$, respectively, $\Omega(\alpha)$. Let $\mathbf{v}$ be the vector of all variables mentioned in $S(\Delta(\alpha))$ or $S(\Omega(\alpha))$. $Z(\Delta(\alpha))$ and $Z(\Omega(\alpha))$ denote the solution set for $S(\Delta(\alpha))$, respectively, $S(\Omega(\alpha))$. It is the set of all solutions of the form $(s_1^\alpha, s_2^\alpha, \dots, s_n^\alpha)$, respectively, $(s_1^\varsigma, s_2^\varsigma, \dots, s_m^\varsigma)$, where assigning s_i^α to $pr_i^\alpha \in \mathbf{v}$ for $i = 1, 2, \dots, n$, respectively, assigning s_j^ς to $pr_j^\varsigma \in \mathbf{v}$ for $j = 1, 2, \dots, m$ solves all the (in)equalities in $S(\Delta(\alpha))$, respectively, $S(\Omega(\alpha))$ simultaneously. An SI is *feasible* if and only if its solution set is not empty.

Suppose $\Delta(\mathtt{replace})$ contains $[\mathtt{replace}]_{0.43}(\mathtt{full} \wedge \neg\mathtt{holding})$ and $\neg[\mathtt{replace}]_{0.43}(\mathtt{full} \wedge \neg\mathtt{holding})$. Then $S(\Delta(\mathtt{replace}))$ will contain

$$
\begin{aligned}
0 + pr_2^\alpha + 0 + pr_4^\alpha + 0 + 0 + 0 + 0 &= 0.43 \\
0 + pr_2^\alpha + 0 + pr_4^\alpha + 0 + 0 + 0 + 0 &\neq 0.43.
\end{aligned}
$$

This system is clearly infeasible, and the whole system $S(\Delta(\mathtt{replace}))$ of which this one is a subsystem is, by extension, also infeasible. As will be seen in the next subsection, a node for which an infeasible system can be generated will be recognized as closed.

Suppose $\Omega(\mathtt{weigh})$ contains $(\mathtt{obsHeavy}|\mathtt{weigh} : 0.56)$ and $(\mathtt{obsHeavy}|\mathtt{weigh} : 0.55)$. Then $S(\Omega(\mathtt{weigh}))$ will contain

$$
\begin{aligned}
pr_4^\varsigma &= 0.56 \\
pr_4^\varsigma &= 0.55,
\end{aligned}
$$

where $\Omega^\# = \{\mathtt{obsNil}, \mathtt{obsLight}, \mathtt{obsMedium}, \mathtt{obsHeavy}\}$. This system is clearly infeasible, and thus also $S(\Delta(\mathtt{replace}))$.

The interested reader can refer to the technical report [25] for a more thorough explication of the generation of SIs.

4.3 The Label Assignment Phase

Given two formulae $(x, \Phi), (x', \Phi') \in \Gamma$ such that Φ contradicts Φ', if x and x' represent the same world, then Γ should close. But if $x \neq x'$, one must determine whether x and x' can be made to represent different worlds. In other words, one must check whether there is a 'proper' assignment of worlds to labels such that no contradictions occur.

In SLAP, sentences of the form $\neg\Box\Phi$ are not in the language. The reason is that the decision procedure for SLAP [27] would not notice certain contradictions which may occur due to such sentences being allowed. In SLAOP, sentences of the form $\neg\Box\Phi$ are in the language, because the label assignment procedure described below picks up the contradictions which may occur.

Informally, x mentioned in Γ could represent any one of the worlds in $W(\Gamma, x)$. Now suppose $(x, \Phi), (x', \Phi') \in \Gamma$ such that Φ contradicts Φ' and $W(\Gamma, x) =$

$\{w_1, w_2\}$ and $W(\Gamma, x') = \{w_2, w_3\}$. Assuming that Φ and Φ' do not involve the $\Box$ operator, it is conceivable that there exists a structure $\mathcal{S}$ such that (i) $\mathcal{S}, w_1 \models \Phi$ and $\mathcal{S}, w_2 \models \Phi'$, (ii) $\mathcal{S}, w_1 \models \Phi$ and $\mathcal{S}, w_3 \models \Phi'$ or (iii) $\mathcal{S}, w_2 \models \Phi$ and $\mathcal{S}, w_3 \models \Phi'$. But to have $\mathcal{S}, w_2 \models \Phi$ and $\mathcal{S}, w_2 \models \Phi'$ is inconceivable. Hence, if it were the case that, for example, $W(\Gamma, x) = \{w_2\}$ and $W(\Gamma, x') = \{w_2\}$, then we would have found a contradiction and Γ should be made closed.

To formalize the process, some more definitions are required:

- $SoLA(\Gamma) \stackrel{def}{=} \{(0{:}w^0, 1{:}w^1, \ldots, x'{:}w^{x'}) \mid w^x \in W(\Gamma, x)\}$, where 0, 1, ..., x' are all the labels mentioned in Γ. We shall call an element of $SoLA(\Gamma)$ a *label assignment*. $LA(\Gamma)$ denotes an element of $SoLA(\Gamma)$.
- $E(\Gamma, x) \stackrel{def}{=} \{(x, \Phi) \in \Gamma \mid \Phi$ is $Reward(r)$ or $\neg Reward(r)$ or $Cost(\alpha, c)$ or $\neg Cost(\alpha, c)$ for some/any constants r and c and some/any action $\alpha\}$.
- $E(\Gamma, LA, w) \stackrel{def}{=} \bigcup_{x:w \in LA(\Gamma)} E(\Gamma, x)$.
- $F(\Gamma, \alpha, x) \stackrel{def}{=} \{[\alpha]_q\varphi \mid (x, [\alpha]_q\varphi) \in \Gamma\} \cup \{\neg[\alpha]_q\varphi \mid (x, \neg[\alpha]_q\varphi) \in \Gamma\}$.
- $F(\Gamma, \alpha, LA, w) \stackrel{def}{=} \bigcup_{x:w \in LA(\Gamma)} F(\Gamma, \alpha, x)$.
- $G(\Gamma, \alpha, x) \stackrel{def}{=} \{(\varsigma \mid \alpha : q) \mid (x, (\varsigma \mid \alpha : q)\varphi) \in \Gamma\} \cup \{\neg(\varsigma \mid \alpha : q) \mid (x, \neg(\varsigma \mid \alpha : q)) \in \Gamma\}$.
- $G(\Gamma, \alpha, LA, w) \stackrel{def}{=} \bigcup_{x:w \in LA(\Gamma)} G(\Gamma, \alpha, x)$.

After the tableau phase has completed, the label assignment phase begins. For each leaf node Γ^j_k of an open branch, do the following.

Do the following for every $LA \in SoLA(\Gamma^j_k)$. If one of the following two cases holds, then mark LA as "unsat".

- For some $w \in W(\Gamma^j_k)$, $E(\Gamma^j_k, LA, w)$ contains
  - $Reward(r)$ and $Reward(r')$ such that $r \neq r'$, or
  - $Reward(r)$ and $\neg Reward(r)$, or
  - $Cost(\alpha, c)$ and $Cost(\alpha, c')$ (same action α) such that $c \neq c'$, or
  - $Cost(\alpha, c)$ and $\neg Cost(\alpha, c)$ (same action α).
- For some action $\alpha \in \mathcal{A}$ and some $w \in W(\Gamma^j_k)$, $Z(F(\Gamma^j_k, \alpha, LA, w)) = \emptyset$ or $Z(G(\Gamma^j_k, \alpha, LA, w)) = \emptyset$.

If every $LA \in SoLA(\Gamma^j_k)$ is marked as "unsat", then create new leaf node $\Gamma^j_{k+1} = \Gamma^j_k \cup \{(0, \bot)\}$.

That is, if for all logically correct ways of assigning possible worlds to labels (i.e., for all the label assignments in $SoLA(\Gamma^j_k)$), no assignment (LA) satisfies all formulae in Γ^j_k, then Γ^j_k is unsatisfiable.

Definition 8. *A tree is called* finished *after the label assignment phase is completed.*

Definition 9. *If a tree for $\neg\Psi$ is closed, we write $\vdash \Psi$. If there is a finished tree for $\neg\Psi$ with an open branch, we write $\not\vdash \Psi$.*

Theorem 1 (Decidability). *Determining whether a sentence is SLAOP-valid is decidable.*

Proof. The proof is sketched; an accompanying technical report [25] presents the full proof. The proof shows that the decision procedure is sound, complete and terminating, thus decidable.

Soundness. If $\vdash \Psi$ then $\models \Psi$. (Contrapositively, if $\not\models \Psi$ then $\not\vdash \Psi$.) Let $\psi = \neg\Psi$. Then $\not\vdash \Psi$ if and only if the tree for ψ is open. And

$$\begin{aligned} \not\models \Psi &\iff \text{not } (\forall \mathcal{S})\ \mathcal{S} \models \Psi \\ &\iff \text{not } (\forall \mathcal{S}, w)\ \mathcal{S}, w \models \Psi \\ &\iff (\exists \mathcal{S}, w)\ \mathcal{S}, w \models \psi. \end{aligned}$$

For the soundness proof, it thus suffices to show that if there exists a structure $\mathcal{S}$ and w in it such that $\mathcal{S}, w \models \psi$, then the tree rooted at $\Gamma_0^0 = \{(0, \psi)\}$ is open. This is shown using induction on the height of a node in a tableau tree, and looking at each tableau rule and the label assignment phase.

Completeness. If $\models \Psi$ then $\vdash \Psi$. (Contrapositively, if $\not\vdash \Psi$ then $\not\models \Psi$.) Let $\psi = \neg\Psi$. Then $\not\vdash \Psi$ means that there is an open branch of a finished tree for ψ. And

$$\begin{aligned} \not\models \Psi &\iff (\exists \mathcal{S})\ \mathcal{S} \not\models \Psi \\ &\iff (\exists \mathcal{S}, w)\ \mathcal{S}, w \not\models \Psi \\ &\iff (\exists \mathcal{S}, w)\ \mathcal{S}, w \models \psi. \end{aligned}$$

For the completeness proof, it thus suffices to construct for some open branch of a finished tree for $\psi \in \mathcal{L}_{SLAOP}$, a SLAOP structure $\mathcal{S} = \langle W, R, O, N, Q, U \rangle$ in which there is a world $w \in W$ in $\mathcal{S}$ such that ψ is satisfied in $\mathcal{S}$ at w. That is, we show (i) how to construct a structure $\mathcal{S}$ from the information contained in the leaf node Γ of any open branch of a finished tree and (ii) that for all $(x, \Phi) \in \Gamma$, $\mathcal{S}, w \models \Phi$ for $x{:}w \in LA$, for the label assignment LA which is known to exist. Point (ii) relies on induction on the structure of the formulae in Γ.

Termination. Finally, by showing that all trees will become saturated and that the label assignment phase always terminates, it follows that the whole procedure terminates. In particular, rule obs cannot cause cycles because $\Box(\delta_1 \rightarrow (\exists v^{\varsigma})\neg(v^{\varsigma} \mid \alpha : 0)) \vee \Box(\delta_2 \rightarrow (\exists v^{\varsigma})\neg(v^{\varsigma} \mid \alpha : 0)) \vee \cdots \vee \Box(\delta_n \rightarrow (\exists v^{\varsigma})\neg(v^{\varsigma} \mid \alpha : 0))$ is not dynamic; it can thus not make rule obs applicable again. That is, rule obs can only cause other rules to become applicable; rules which add $\bot$ to the new node, and rules with the subformula property.

$$\Gamma_0^0 = \{(0, \bigwedge_{\Phi \in BK} \Box\Phi \wedge \neg(\neg f \wedge d \wedge \neg h \rightarrow [g]_{0.9}(\neg f \wedge h))\}$$

$$\downarrow \text{ nf}$$

$$(0, \neg\neg(\neg f \wedge d \wedge \neg h) \wedge \neg[g]_{0.9}(\neg f \wedge h)) \in \Gamma_1^0$$

$$\downarrow \wedge$$

$$(0, \neg\neg(\neg f \wedge d \wedge \neg h)), (0, \neg[g]_{0.9}(\neg f \wedge h)) \in \Gamma_2^0$$

$$\downarrow \neg$$

$$(0, \neg f \wedge d \wedge \neg h) \in \Gamma_3^0$$

$$\downarrow \wedge$$

$$(0, \neg f), (0, d), (0, \neg h) \in \Gamma_4^0$$

$$\downarrow \Box$$

$$(0, \neg(\neg f \wedge \neg h) \vee ([g]_{0.9} h \wedge [g]_{0.1} \neg h \wedge [g]\neg f)) \in \Gamma_5^0$$

$$\downarrow \vee$$

$$(0, ([g]_{0.9} h \wedge [g]_{0.1} \neg h \wedge [g]\neg f) \in \Gamma_0^3$$

$$\downarrow \wedge$$

$$(0, [g]_{0.9} h), (0, [g]_{0.1} \neg h), (0, [g]\neg f) \in \Gamma_1^3$$

$$\downarrow \Diamond\varphi$$

$$(1, h), (2, \neg h), (3, \neg f) \in \Gamma_2^3$$

$$\downarrow \Box$$

$$(1, f \vee h \vee ([g]_{0.9} h \wedge [g]_{0.1} \neg h)), (1, f \vee h \vee [g]\neg f), (2, f \vee h \vee ([g]_{0.9} h \wedge [g]_{0.1} \neg h)),$$
$$(2, f \vee h \vee [g]\neg f), (3, f \vee h \vee ([g]_{0.9} h \wedge [g]_{0.1} \neg h)), (3, f \vee h \vee [g]\neg f) \in \Gamma_3^3$$

$$\downarrow \vee$$

$$(1, [g]_{0.9} h), (1, [g]_{0.1} \neg h), (1, [g]\neg f), (2, [g]_{0.9} h), (2, [g]_{0.1} \neg h), (2, [g]\neg f),$$
$$(3, [g]_{0.9} h), (3, [g]_{0.1} \neg h), (3, [g]\neg f) \in \Gamma_0^8$$

Fig. 1. One branch of a tree for proving that $\{\Box\Phi \mid \Phi \in BK\}$ entails $\neg f \wedge d \wedge \neg h \rightarrow [g]_{0.9}(\neg f \wedge h)$

5 Examples

The following abbreviations for constants will be used: `grab` $:= g$, `weigh` $:= w$, `full` $:= f$, `drank` $:= d$, `holding` $:= h$, `obsHeavy` $:= oH$, `obsMedium` $:= oM$ and `obsLight` $:= oL$.

In Figures 1 and 2, the vertices represent nodes and the arcs represent the application of tableau rules. Arcs are labeled with the rule they represent, except when branching occurs, in which case, the $\vee$ rule was applied. The figures show how the vertices relate to the corresponding nodes. The reader should keep in mind that the node corresponding to a vertex v contains all the labeled formulae in vertices above v on the same branch—the vertices show only the elements of nodes which are 'added' to a node due to the application of some rule. An exception is the top vertex of a tree, which is the trunk and not the result of any rule application.

In order to show the development of the tree, some liberties were taken with respect to rule application: In some cases, rule application is not shown, that is, from parent node to child node, a formula may be 'processed' more than is possible by the application of the rule represented by the arc from parent to

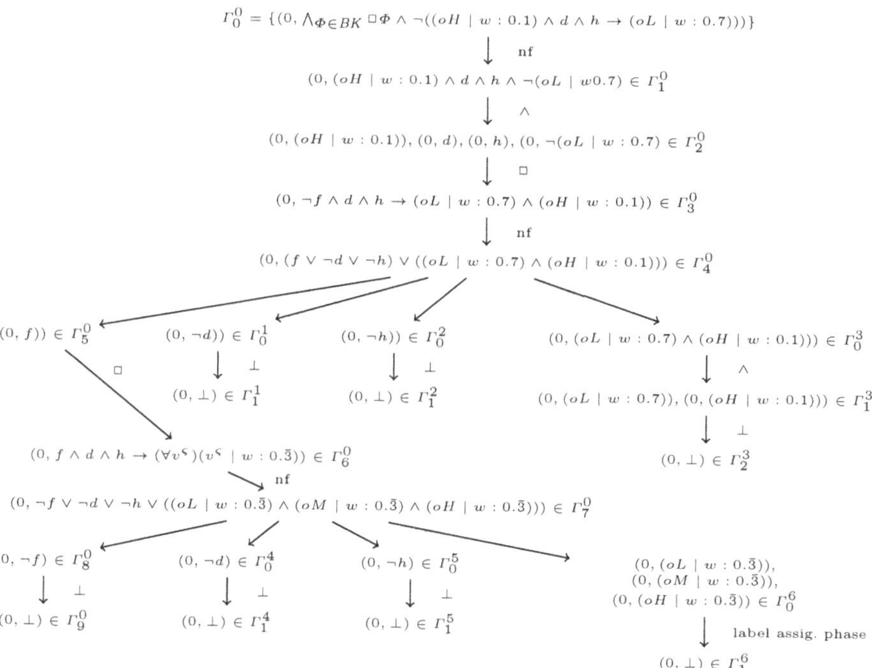

Fig. 2. A tree for proving that $\{\Box\Phi \mid \Phi \in BK\}$ entails $(oH \mid w : 0.1) \wedge d \wedge h \rightarrow (oL \mid w : 0.7)$

child in the figure. The arc labeled "nf" denotes *normal forming*: translating abbreviations into symbols in the language.

Suppose the following *domain axioms*[3] are part of the robot's background knowledge BK for the oil-drinking scenario.

$$f \wedge d \wedge h \rightarrow (\forall v^{\varsigma})(v^{\varsigma} \mid w : 0.\bar{3})$$
$$f \wedge \neg d \wedge h \rightarrow (oL \mid w : 0.1) \wedge (oH \mid w : 0.7)$$
$$((f \wedge \neg d) \vee (\neg f \wedge d)) \wedge h \rightarrow (oM \mid w : 0.2)$$
$$\neg f \wedge d \wedge h \rightarrow (oL \mid w : 0.7) \wedge (oH \mid w : 0.1)$$
$$\neg f \wedge \neg d \wedge h \rightarrow (oL \mid w : 0.5) \wedge (oM \mid w : 0.3) \wedge (oH \mid w : 0.2)$$
$$\neg f \wedge \neg h \rightarrow [g]_{0.9}h \wedge [g]_{0.1}\neg h \wedge [g]\neg f.$$

For the first example, we claim that $\{\Box\Phi \mid \Phi \in BK\} \models \neg f \wedge d \wedge \neg h \rightarrow [g]_{0.9}(\neg f \wedge h)$. Figure 1 shows only one branch of a tree for

$$\bigwedge_{\Phi \in BK} \Box\Phi \wedge \neg(\neg f \wedge d \wedge \neg h \rightarrow [g]_{0.9}(\neg f \wedge h)). \tag{1}$$

[3] Only the last of these sentences can be expressed in SLAP. Notice the compact representation of the perception probabilities in the first sentence, due to quantification.

For the claim to hold, the tree for (1) must close. We'll only show that the branch in Figure 1 closes. The leaf node of the branch is open and must thus be considered in the label assignment phase.

For clarity, denote w_1 as 111 where $w_1 \models f \wedge d \wedge h$, w_2 as 110 where $w_2 \models f \wedge d \wedge \neg h$, ..., w_8 as 000 where $w_8 \models \neg f \wedge \neg d \wedge \neg h$. We shall refer to the leaf node as Γ. Observe that $W(\Gamma, 0) = \{010\}$, $W(\Gamma, 1) = \{111, 101, 011, 001\}$, $W(\Gamma, 2) = \{110, 100, 010, 000\}$ and $W(\Gamma, 3) = \{011, 010, 001, 000\}$, and that $W(\Gamma) = \{111, 101, 011, 001, 110, 100, 010, 000\} = C$. Observe that 0:010 is in every label assignment in $SoLA(\Gamma)$. Note that $F(\Gamma, grab, 0) \subseteq F(\Gamma, \texttt{grab}, LA, 010)$ for all $LA \in SoLA(\Gamma)$. And note that $F(\Gamma, grab, 0)$ equals

$$\{[\texttt{grab}]_{0.9}\texttt{holding}, [\texttt{grab}]_{0.1}\neg\texttt{holding}, [\texttt{grab}]\neg\texttt{full}, \neg[\texttt{grab}]_{0.9}(\neg\texttt{full} \wedge \texttt{holding})\}.$$

The system generated from $F(\Gamma, \texttt{grab}, 0)$ is

$$\begin{array}{l} 0 + 0 + 0 + 0 + pr_5^\alpha + 0 + pr_7^\alpha + 0 = 0.9 \\ 0 + pr_2^\alpha + 0 + pr_4^\alpha + 0 + pr_6^\alpha + 0 + pr_8^\alpha = 0.1 \\ 0 + 0 + 0 + 0 + pr_5^\alpha + pr_6^\alpha + pr_7^\alpha + pr_8^\alpha = 1 \\ pr_1^\alpha + 0 + pr_3^\alpha + 0 + pr_5^\alpha + 0 + pr_7^\alpha + 0 \neq 0.9 \\ pr_1^\alpha + pr_2^\alpha + pr_3^\alpha + pr_4^\alpha + pr_5^\alpha + pr_6^\alpha + pr_7^\alpha + pr_8^\alpha = 1. \end{array}$$

Due to $pr_5^\alpha + pr_6^\alpha + pr_7^\alpha + pr_8^\alpha = 1$ (3rd equation), it must be the case that $pr_5^\alpha + pr_7^\alpha \neq 0.9$ (4th inequation). But it is required by the first equation that $pr_5^\alpha + pr_7^\alpha = 0.9$, which forms a contradiction. Thus, for every label assignment, there exists an action and a world w—that is, 010—for which $Z(F(\Gamma, \texttt{grab}, LA, w) = \emptyset$ and the branch closes.

For the second example, we claim that $\{\Box\Phi \mid \Phi \in BK\} \models (oH \mid w : 0.1) \wedge d \wedge h \rightarrow (oL \mid w : 0.7)$. Figure 2 shows the closed tree for

$$\bigwedge_{\Phi \in BK} \Box\Phi \wedge \neg((oH \mid w : 0.1) \wedge d \wedge h \rightarrow (oL \mid w : 0.7)).$$

The arc labelled "label assig. phase" means that for all label assignments, the SI generated for a set of formulae will include $(oH \mid w : 0.1)$ and $(oH \mid w : 0.\bar{3})$, which will cause all SIs to be infeasible. Hence, the label assignment phase will create a new node containing $(0, \bot)$ at the end of the branch.

6 Conclusion

A decidable logic with a semantics closely related to partially observable Markov decision processes (POMDPs) was presented. The logic a step towards the definition of a logic for reasoning about an agent's belief-states and expected future rewards, where the agent's actions and observations are stochastic.

Two examples were provided in this paper, which give an indication of how SLAOP-validity is computed. In a sequent paper, we would like to explain the formal approach of how SLAOP is used to give a complete specification of a domain.

Predicate $(\varsigma \mid \alpha : q)$ is useful for specifying the probability of perceiving an observation in the 'current' world. However, it would be useful to query the probability q of ending in a φ-world after executing action α in the 'current' world and then perceiving ς in the φ-world. To make such queries possible, one could add a modal operator with the following definition. $\mathcal{S}, w \models [\alpha{+}\varsigma : q]\varphi \iff \sum_{w' \in W, \mathcal{S}, w' \models \varphi} R_\alpha(w, w') \times Q_\alpha(w', N(\varsigma)) \geq q$.

Informally, sentences of the form $[\alpha]_q\varphi$ and $(\varsigma \mid \alpha : q)$ have a meaning 'probability is exactly q.' In future, to make the language more expressive, the syntax and semantics of these kinds of sentences can be replaced with sentences which have a meaning 'probability is less that, less than or equal to, etc. q.'

An important next step for SLAOP would be to add the ability to express sequences of actions, and then evaluate the part of the sentence occurring after the sequence.

For specifying a domain in SLAOP, the question of what world an agent is in does not arise. But due to partial observability, after the agent has executed a few actions, the agent will only have an (uncertain) *belief* about which world it is in, as opposed to (certain) *knowledge* of where it is. For an agent to reason with beliefs, the notion of an epistemic or belief state needs to be added to SLAOP.

We would also like to add a notion of the expected value of a sequence of actions, and then be able to determine whether the expected value is less than, less than or equal to, etc. some given value. Generating POMDP policies is also on the cards for the future of SLAOP.

The complexity of the decision procedure has not been analysed. Our focus for SLAOP is mainly decidability. Evaluation of the systems of equations in the SI phase has the potential for being very expensive. These are linear systems of equations; one could thus investigate Linear Programming methods [7,18,12] to optimize the evaluation of the systems.

We feel that presenting a decidability result for a new class of logics is not trivial. Even though the entailment problem in SLAOP—as presented in this paper—may be intractable, it is important to have a decision procedure as a launchpad for tackling the computational complexity. We would like to implement some extended version of SLAOP. Determining the complexity of an optimized entailment decision procedure may be attempted before an implementation, though.

References

1. Bacchus, F.: Representing and Reasoning with Uncertain Knowledge. MIT Press, Cambridge (1990)
2. Bacchus, F., Halpern, J.Y., Levesque, H.J.: Reasoning about noisy sensors and effectors in the situation calculus. Artificial Intelligence 111(1-2), 171–208 (1999)
3. Bonet, B., Geffner, H.: Planning and control in artificial intelligence: A unifying perspective. Applied Intelligence 14(3), 237–252 (2001)
4. Boutilier, C., Poole, D.: Computing optimal policies for partially observable decision processes using compact representations. In: Proc. of 13th Natl. Conf. on Artificial Intelligence, pp. 1168–1175 (1996)

5. Boutilier, C., Reiter, R., Soutchanski, M., Thrun, S.: Decision-theoretic, high-level agent programming in the situation calculus. In: Proceedings of the 17th National Conference on Artificial Intelligence (AAAI 2000) and of the 12th Conference on Innovative Applications of Artificial Intelligence (IAAI 2000), pp. 355–362. AAAI Press, Menlo Park (2000)
6. Chellas, B.: Modal Logic: An introduction. Cambridge University Press, Cambridge (1980)
7. Dantzig, G.B.: Linear Programming and Extensions. Princeton University Press (1963&1998)
8. De Weerdt, M., De Boer, F., Van der Hoek, W., Meyer, J.J.: Imprecise observations of mobile robots specified by a modal logic. In: Proc. of Fifth Annual Conference of the Advanced School for Computing and Imaging (ASCI 1999), pp. 184–190 (1999)
9. Fagin, R., Halpern, J.Y.: Reasoning about knowledge and probability. Journal of the ACM 41(2), 340–367 (1994)
10. Fagin, R., Halpern, J.Y., Megiddo, N.: A logic for reasoning about probabilities. Information and Computation 87, 78–128 (1990)
11. Gabaldon, A., Lakemeyer, G.: $\mathcal{ESP}$: A logic of only-knowing, noisy sensing and acting. In: Proc. of 22nd Natl. Conf. on Artificial Intelligence (AAAI 2007), pp. 974–979. AAAI Press (2007)
12. Gass, S.I.: Linear programming: Methods and applications, 5th edn. Dover Publications (2010)
13. Halpern, J.Y.: Reasoning about Uncertainty. The MIT Press, Cambridge (2003)
14. Hughes, G., Cresswell, M.: A New Introduction to Modal Logic, Routledge, New York (1996)
15. Iocchi, L., Lukasiewicz, T., Nardi, D., Rosati, R.: Reasoning about actions with sensing under qualitative and probabilistic uncertainty. ACM Transactions on Computational Logic 10(1), 5:1–5:41 (2009)
16. Kooi, B.: Probabilistic dynamic epistemic logic. Journal of Logic, Language and Information 12(4), 381–408 (2003)
17. Monahan, G.E.: A survey of partially observable Markov decision processes: Theory, models, and algorithms. Management Science 28(1), 1–16 (1982)
18. Murty, K.G.: Linear programming, revised edn. John Wiley and sons (1983)
19. Nilsson, N.: Probabilistic logic. Artificial Intelligence 28, 71–87 (1986)
20. Poole, D.: Decision theory, the situation calculus and conditional plans. Linköping Electronic Articles in Computer and Information Science 8(3) (1998)
21. Puterman, M.: Markov Decision Processes: Discrete Dynamic Programming. Wiley, New York (1994)
22. Reiter, R.: Knowledge in action: logical foundations for specifying and implementing dynamical systems. MIT Press, Massachusetts (2001)
23. Rens, G., Lakemeyer, G., Meyer, T.: A logic for specifying agent actions and observations with probability. In: Kersting, K., Toussaint, M. (eds.) Proc. of 6th Starting AI Researchers' Symposium (STAIRS 2012). Frontiers in Artificial Intelligence and Applications, vol. 241, pp. 252–263. IOS Press (2012), `http://www.booksonline.iospress.nl/Content/View.aspx?piid=31509`
24. Rens, G., Meyer, T., Lakemeyer, G.: On the logical specification of probabilistic transition models. In: Proc. of 11th Intl. Symposium on Logical Formalizations of Commonsense Reasoning, COMMONSENSE 2013 (May 2013)

25. Rens, G., Meyer, T., Lakemeyer, G.: A sound and complete decision procedure for SLAOP. Tech. rep., Centre for Artificial Intelligence Research (University of KwaZulu-Natal, and CSIR Meraka), South Africa (July 2013), http://www.cair.za.net/sites/default/files/outputs/The-SLAOP-decision-procedure.pdf
26. Rens, G., Varzinczak, I., Meyer, T., Ferrein, A.: A logic for reasoning about actions and explicit observations. In: Li, J. (ed.) AI 2010. LNCS (LNAI), vol. 6464, pp. 395–404. Springer, Heidelberg (2010)
27. Rens, G., Meyer, T., Lakemeyer, G.: SLAP: Specification logic of actions with probability. Journal of Applied Logic (2013), www.sciencedirect.com/science/article/pii/S157086831300075X, http://dx.doi.org/10.1016/j.jal.2013.09.001
28. Sanner, S., Kersting, K.: Symbolic dynamic programming for first-order POMDPs. In: Proc. of 24th Natl. Conf. on Artificial Intelligence (AAAI 2010), pp. 1140–1146. AAAI Press (2010)
29. Shirazi, A., Amir, E.: Probabilistic modal logic. In: Proc. of 22nd Natl. Conf. on Artificial Intelligence (AAAI 2007), pp. 489–494. AAAI Press (2007)
30. Smallwood, R., Sondik, E.: The optimal control of partially observable Markov processes over a finite horizon. Operations Research 21, 1071–1088 (1973)
31. Tarski, A.: A decision method for elementary algebra and geometry. Tech. rep., The RAND Corporation, Santa Monica, Calif. (1957)
32. Thiébaux, S., Hertzberg, J., Schoaff, W., Schneider, M.: A stochastic model of actions and plans for anytime planning under uncertainty. International Journal of Intelligent Systems 10(2), 155–183 (1995)
33. Van Benthem, J., Gerbrandy, J., Kooi, B.: Dynamic update with probabilities. Studia Logica 93(1), 67–96 (2009)
34. Van Diggelen, J.: Using Modal Logic in Mobile Robots. Master's thesis, Cognitive Artificial Intelligence, Utrecht University (2002)
35. Wang, C., Schmolze, J.: Planning with POMDPs using a compact, logic-based representation. In: Proc. of 17th IEEE Intl. Conf. on Tools with Artif. Intell (ICTAI 2005), pp. 523–530. IEEE Computer Society, Los Alamitos (2005)

Belief Revision in Structured Probabilistic Argumentation

Paulo Shakarian[1], Gerardo I. Simari[2], and Marcelo A. Falappa[3]

[1] Department of Electrical Engineering and Computer Science
U.S. Military Academy, West Point, NY, USA
paulo@shakarian.net

[2] Department of Computer Science, University of Oxford, United Kingdom
gerardo.simari@cs.ox.ac.uk

[3] Departamento de Ciencias e Ingeniería de la Computación
Universidad Nacional del Sur, Bahía Blanca, Argentina
mfalappa@cs.uns.edu.ar

Abstract. In real-world applications, knowledge bases consisting of all the information at hand for a specific domain, along with the current state of affairs, are bound to contain contradictory data coming from different sources, as well as data with varying degrees of uncertainty attached. Likewise, an important aspect of the effort associated with maintaining knowledge bases is deciding what information is no longer useful; pieces of information (such as intelligence reports) may be outdated, may come from sources that have recently been discovered to be of low quality, or abundant evidence may be available that contradicts them. In this paper, we propose a probabilistic structured argumentation framework that arises from the extension of Presumptive Defeasible Logic Programming (PreDeLP) with probabilistic models, and argue that this formalism is capable of addressing the basic issues of handling contradictory and uncertain data. Then, to address the last issue, we focus on the study of non-prioritized belief revision operations over probabilistic PreDeLP programs. We propose a set of rationality postulates – based on well-known ones developed for classical knowledge bases – that characterize how such operations should behave, and study a class of operators along with theoretical relationships with the proposed postulates, including a representation theorem stating the equivalence between this class and the class of operators characterized by the postulates.

1 Introduction and Related Work

Decision-support systems that are part of virtually any kind of real-world application must be part of a framework that is rich enough to deal with several basic problems: (i) handling contradictory information; (ii) answering abductive queries; (iii) managing uncertainty; and (iv) updating beliefs. *Presumptions* come into play as key components of answers to abductive queries, and must be maintained as elements of the knowledge base; therefore, whenever candidate answers to these queries are evaluated, the (in)consistency of the knowledge base

C. Beierle and C. Meghini (Eds.): FoIKS 2014, LNCS 8367, pp. 324–343, 2014.

together with the presumptions being made needs to be addressed via belief revision operations.

In this paper, we begin by proposing a framework that addresses items (i)–(iii) by extending Presumptive DeLP [1] (PreDeLP, for short) with probabilistic models in order to model uncertainty in the application domain; the resulting framework is a general-purpose probabilistic argumentation language that we will refer to as Probabilistic PreDeLP(P-PreDeLP, for short).

In the second part of this paper, we address the problem of updating beliefs – item (iv) above – in P-PreDeLP knowledge bases, focusing on the study of non-prioritized belief revision operations. We propose a set of rationality postulates characterizing how such operations should behave – these postulates are based on the well-known postulates proposed in [2] for non-prioritized belief revision in classical knowledge bases. We then study a class of operators and their theoretical relationships with the proposed postulates, concluding with a representation theorem.

Related Work. Belief revision studies changes to knowledge bases as a response to *epistemic inputs*. Traditionally, such knowledge bases can be either belief sets (sets of formulas closed under consequence) [3,4] or belief bases [5,2] (which are not closed); since our end goal is to apply the results we obtain to real-world domains, here we focus on belief bases. In particular, as motivated by requirements (i)–(iv) above, our knowledge bases consist of logical formulas over which we apply argumentation-based reasoning and to which we couple a probabilistic model. The connection between belief revision and argumentation was first studied in [6]; since then, the work that is most closely related to our approach is the development of the explanation-based operators of [7].

The study of argumentation systems together with probabilistic reasoning has recently received a lot attention, though a significant part has been in the combination between the two has been in the form of probabilistic abstract argumentation [8,9,10,11]. There have, however, been several approaches that combine structured argumentation with models for reasoning under uncertainty; the first of such approaches to be proposed was [12], and several others followed, such as the possibilistic approach of [13], and the probabilistic logic-based approach of [14]. The main difference between these works and our own is that here we adopt a bipartite knowledge base, where one part models the knowledge that is not inherently probabilistic – uncertain knowledge is modeled separately, thus allowing a clear separation of interests between the two kinds of models. This approach is based on a similar one developed for ontological languages in the Semantic Web (see [15], and references within).

Finally, to the best of our knowledge, this is the first paper in which the combination of structured argumentation, probabilistic models, and belief revision has been addressed in conjunction.

Table 1. Examples of the kind of information that could be represented in the two different models in a cyber-security application domain

Probabilistic Model (EM)	Analytical Model (AM)
"Malware X was compiled on a system using the English language."	"Malware X was compiled on a system in English-speaking country Y."
"County Y and country Z are currently at war."	"Country Y has a motive to launch a cyber-attack against country Z
"Malware W and malware X were created in a similar coding style."	"Malware W and malware X are related.

2 Preliminaries

The Probabilistic PreDeLP (P-PreDeLP, for short) framework is composed of two separate models of the world. The first is called the *environmental model* (referred to as "EM"), and is used to describe the probabilistic knowledge that we have about the domain. The second one is called the *analytical model* (referred to as "AM"), and is used to analyze competing hypotheses that can account for a given phenomenon – what we will generally call queries. The AM is composed of a classical (that is, non-probabilistic) PreDeLP program in order to allow for contradictory information, giving the system the capability to model competing explanations for a given query.

Two Kinds of Uncertainty. In general, the EM contains knowledge such as evidence, uncertain facts, or knowledge about agents and systems. The AM, on the other hand, contains ideas that a user may conclude based on the information in the EM. Table 1 gives some examples of the types of information that could appear in each of the two models in a cyber-security application. Note that a knowledge engineer (or automated system) could assign a probability to statements in the EM column, whereas statements in the AM column can be either true or false depending on a certain combination (or several possible combinations) of statements from the EM. There are thus two kinds of uncertainty that need to be modeled: probabilistic uncertainty and uncertainty arising from defeasible knowledge. As we will see, our model allows both kinds of uncertainty to coexist, and also allows for the combination of the two since defeasible rules and presumptions (that is, defeasible facts) can also be annotated with probabilistic events.

In the rest of this section, we formally describe these two models, as well as how knowledge in the AM can be annotated with information from the EM – these annotations specify the conditions under which the various statements in the AM can potentially be true.

Basic Language. We assume sets of variable and constant symbols, denoted with $\mathbf{V}$ and $\mathbf{C}$, respectively. In the rest of this paper, we will use capital letters to represent variables (e.g., X, Y, Z), while lowercase letters represent constants. The next component of the language is a set of n-ary predicate symbols; the EM and AM use separate sets of predicate symbols, denoted with $\mathbf{P}_{EM}, \mathbf{P}_{AM}$, respectively – the two models can, however, share variables and constants. As usual, a *term* is composed of either a variable or constant. Given terms $t_1, ..., t_n$ and n-ary predicate symbol p, $p(t_1, ..., t_n)$ is called an *atom*; if $t_1, ..., t_n$ are constants, then the atom is said to be *ground*. The sets of all ground atoms for EM and AM are denoted with $\mathbf{G}_{EM}$ and $\mathbf{G}_{AM}$, respectively.

Given set of ground atoms, a *world* is any subset of atoms – those that belong to the set are said to be *true* in the world, while those that do not are *false*. Therefore, there are $2^{|\mathbf{G}_{EM}|}$ possible worlds in the EM and $2^{|\mathbf{G}_{AM}|}$ worlds in the AM. These sets are denoted with $\mathcal{W}_{EM}$ and $\mathcal{W}_{AM}$, respectively. In order to avoid worlds that do not model possible situations given a particular domain, we include *integrity constraints* of the form $\mathsf{oneOf}(\mathcal{A}')$, where $\mathcal{A}'$ is a subset of ground atoms. Intuitively, such a constraint states that any world where more than one of the atoms from set $\mathcal{A}'$ appears is invalid. We use $\mathbf{IC}_{EM}$ and $\mathbf{IC}_{AM}$ to denote the sets of integrity constraints for the EM and AM, respectively, and the sets of worlds that conform to these constraints is denoted with $\mathcal{W}_{EM}(\mathbf{IC}_{EM}), \mathcal{W}_{AM}(\mathbf{IC}_{AM})$, respectively.

Finally, logical formulas arise from the combination of atoms using the traditional connectives ($\wedge$, $\vee$, and $\neg$). As usual, we say a world w *satisfies* formula (f), written $w \models f$, iff: (i) If f is an atom, then $w \models f$ iff $f \in w$; (ii) if $f = \neg f'$ then $w \models f$ iff $w \not\models f'$; (iii) if $f = f' \wedge f''$ then $w \models f$ iff $w \models f'$ and $w \models f''$; and (iv) if $f = f' \vee f''$ then $w \models f$ iff $w \models f'$ or $w \models f''$. We use the notation $form_{EM}, form_{AM}$ to denote the set of all possible (ground) formulas in the EM and AM, respectively.

2.1 Probabilistic Model

The EM or environmental model is largely based on the probabilistic logic of [16], which we now briefly review.

Definition 1. *Let f be a formula over $\mathbf{P}_{EM}$, $\mathbf{V}$, and $\mathbf{C}$, $p \in [0,1]$, and $\epsilon \in [0, \min(p, 1-p)]$. A* probabilistic formula *is of the form $f : p \pm \epsilon$. A set $\mathcal{K}_{EM}$ of probabilistic formulas is called a* probabilistic knowledge base.

In the above definition, the number ϵ is referred to as an *error tolerance*. Intuitively, probabilistic formulas are interpreted as "formula f is true with probability between $p - \epsilon$ and $p + \epsilon$" – note that there are no further constraints over this interval apart from those imposed by other probabilistic formulas in the knowledge base. The uncertainty regarding the probability values stems from the fact that certain assumptions (such as probabilistic independence) may not be suitable in the environment being modeled.

Example 1. Consider the following set $\mathcal{K}_{EM}$:

$$
\begin{array}{lll}
f_1 = a : 0.8 \pm 0.1 & f_4 = d \wedge e \quad\;\; : 0.7 \pm 0.2 & f_7 = k : 1 \pm 0 \\
f_2 = b : 0.2 \pm 0.1 & f_5 = f \wedge g \wedge h : 0.6 \pm 0.1 & \\
f_3 = c : 0.8 \pm 0.1 & f_6 = i \vee \neg j \quad\; : 0.9 \pm 0.1 &
\end{array}
$$

Throughout the paper, we also use $\mathcal{K}'_{EM} = \{f_1, f_2, f_3\}$ ■

A set of probabilistic formulas describes a set of possible probability distributions Pr over the set $\mathcal{W}_{EM}(\mathsf{IC}_{EM})$. We say that probability distribution Pr *satisfies* probabilistic formula $f : p \pm \epsilon$ iff: $p - \epsilon \leq \sum_{w \in \mathcal{W}_{EM}(\mathsf{IC}_{EM})} \mathsf{Pr}(w) \leq p + \epsilon$. We say that a probability distribution over $\mathcal{W}_{EM}(\mathsf{IC}_{EM})$ *satisfies* $\mathcal{K}_{EM}$ iff it satisfies all probabilistic formulas in $\mathcal{K}_{EM}$.

Given a probabilistic knowledge base and a (non-probabilistic) formula q, the *maximum entailment* problem seeks to identify real numbers p, ϵ such that all valid probability distributions Pr that satisfy $\mathcal{K}_{EM}$ also satisfy $q : p \pm \epsilon$, and there does not exist p', ϵ' s.t. $[p - \epsilon, p + \epsilon] \supset [p' - \epsilon', p' + \epsilon']$, where all probability distributions Pr that satisfy $\mathcal{K}_{EM}$ also satisfy $q : p' \pm \epsilon'$. In order to solve this problem we must solve the linear program defined below.

Definition 2. *Given a knowledge base $\mathcal{K}_{EM}$ and a formula q, we have a variable x_i for each $w_i \in \mathcal{W}_{EM}(\mathsf{IC}_{EM})$.*

- *For each $f_j : p_j \pm \epsilon_j \in \mathcal{K}_{EM}$, there is a constraint of the form:*

$$p_j - \epsilon_j \leq \sum_{w_i \in \mathcal{W}_{EM}(\mathsf{IC}_{EM}) \text{ s.t. } w_i \models f_j} x_i \leq p_j + \epsilon_j.$$

- *We also have the constraint:* $\sum_{w_i \in \mathcal{W}_{EM}(\mathsf{IC}_{EM})} x_i = 1$.
- *The objective is to minimize the function:* $\sum_{w_i \in \mathcal{W}_{EM}(\mathsf{IC}_{EM}) \text{ s.t. } w_i \models q} x_i$.

We use the notation $\mathsf{EP\text{-}LP\text{-}MIN}(\mathcal{K}_{EM}, q)$ *to refer to the value of the objective function in the solution to the* $\mathsf{EM\text{-}LP\text{-}MIN}$ *constraints.*

The next step is to solve the linear program a second time, but instead maximizing the objective function (we shall refer to this as EM-LP-MAX) – let ℓ and u be the results of these operations, respectively. In [16], it is shown that $\epsilon = \frac{u-\ell}{2}$ and $p = \ell + \epsilon$ is the solution to the maximum entailment problem. We note that although the above linear program has an exponential number of variables in the worst case (i.e., no integrity constraints), the presence of constraints has the potential to greatly reduce this space. Further, there are also good heuristics (cf. [17,18]) that have been shown to provide highly accurate approximations with a reduced-size linear program.

Example 2. Consider KB $\mathcal{K}'_{EM}$ from Example 1 and a set of ground atoms restricted to those that appear in that program; we have the following worlds:

$$\begin{array}{llll} w_1 = \{a, b, c\} & w_2 = \{a, b\} & w_3 = \{a, c\} & w_4 = \{b, c\} \\ w_5 = \{b\} & w_6 = \{a\} & w_7 = \{c\} & w_8 = \emptyset \end{array}$$

and suppose we wish to compute the probability for formula $q = a \vee c$. For each formula in $\mathcal{K}_{EM}$ we have a constraint, and for each world above we have a variable. An objective function is created based on the worlds that satisfy the query formula (in this case, worlds $w_1, w_2, w_3, w_4, w_6, w_7$). Solving $\mathsf{EP\text{-}LP\text{-}MAX}(\mathcal{K}'_{EM}, q)$ and $\mathsf{EP\text{-}LP\text{-}MIN}(\mathcal{K}'_{EM}, q)$, we obtain the solution 0.9 ± 0.1. ∎

3 Argumentation Model

For the analytical model (AM), we choose a structured argumentation framework [19] due to several characteristics that make such frameworks highly applicable to many domains. Unlike the EM, which describes probabilistic information about the state of the real world, the AM must allow for competing ideas. Therefore, it must be able to represent contradictory information. The algorithmic approach we shall later describe allows for the creation of *arguments* based on the AM that may "compete" with each other to answer a given query. In this competition – known as a *dialectical process* – one argument may defeat another based on a *comparison criterion* that determines the prevailing argument. Resulting from this process, certain arguments are *warranted* (those that are not *defeated* by other arguments) thereby providing a suitable explanation for the answer to a given query.

The transparency provided by the system can allow knowledge engineers to identify potentially incorrect input information and fine-tune the models or, alternatively, collect more information. In short, argumentation-based reasoning has been studied as a natural way to manage a set of inconsistent information – it is the way humans settle disputes. As we will see, another desirable characteristic of (structured) argumentation frameworks is that, once a conclusion is reached, we are left with an explanation of how we arrived at it and information about why a given argument is warranted; this is very important information for users to have. In the following, we first recall the basics of the underlying argumentation framework used, and then go on to introduce the analytical model (AM).

3.1 Defeasible Logic Programming with Presumptions (PreDeLP)

Defeasible Logic Programming with Presumptions (PreDeLP) [1] is a formalism combining logic programming with defeasible argumentation; it arises as an extension of classical DeLP [20] with the possibility of having presumptions, as described below – since this capability is useful in many applications, we adopt this extended version in this paper. In this section, we briefly recall the basics of PreDeLP; we refer the reader to [20,1] for the complete presentation.

The formalism contains several different constructs: facts, presumptions, strict rules, and defeasible rules. Facts are statements about the analysis that can always be considered to be true, while presumptions are statements that may or may not be true. Strict rules specify logical consequences of a set of facts or presumptions (similar to an implication, though not the same) that must always occur, while defeasible rules specify logical consequences that may be assumed to be true when no contradicting information is present. These building blocks are used in the construction of *arguments*, and are part of a PreDeLP program, which is a set of facts, strict rules, presumptions, and defeasible rules. Formally, we use the notation $\Pi_{AM} = (\Theta, \Omega, \Phi, \Delta)$ to denote a PreDeLP program, where Ω is the set of strict rules, Θ is the set of facts, Δ is the set of defeasible rules, and Φ is the set of presumptions. In Figure 1, we provide an example Π_{AM}. We now define these constructs formally.

$\Theta : \theta_{1a} = p$	$\theta_{1b} = q$	$\theta_2 = r$	
$\Omega : \omega_{1a} = \neg s \leftarrow t$	$\omega_{1b} = \neg t \leftarrow s$	$\omega_{2a} = s \leftarrow p, u, r, v$	$\omega_{2b} = t \leftarrow q, w, x, v$
$\Phi : \phi_1 = y \prec$	$\phi_2 = v \prec$	$\phi_3 = \neg z \prec$	
$\Delta : \delta_{1a} = s \prec p$	$\delta_{1b} = t \prec q$	$\delta_2 = s \prec u$	$\delta_3 = s \prec r, v$
$\delta_4 = u \prec y$	$\delta_{5a} = \neg u \prec \neg z$	$\delta_{5b} = \neg w \prec \neg n$	

Fig. 1. An example (propositional) argumentation framework

Facts (Θ) are ground literals representing atomic information or its negation, using strong negation "$\neg$". Note that all of the literals in our framework must be formed with a predicate from the set $\mathbf{P}_{AM}$. Note that information in the form of facts cannot be contradicted. We will use the notation $[\Theta]$ to denote the set of all possible facts.

Strict Rules (Ω) represent non-defeasible cause-and-effect information that resembles an implication (though the semantics is different since the contrapositive does not hold) and are of the form $L_0 \leftarrow L_1, \ldots, L_n$, where L_0 is a ground literal and $\{L_i\}_{i>0}$ is a set of ground literals. We will use the notation $[\Omega]$ to denote the set of all possible strict rules.

Presumptions (Φ) are ground literals of the same form as facts, except that they are not taken as being true but rather defeasible, which means that they can be contradicted. Presumptions are denoted in the same manner as facts, except that the symbol $\prec$ is added.

Defeasible Rules (Δ) represent tentative knowledge that can be used if nothing can be posed against it. Just as presumptions are the defeasible counterpart of facts, defeasible rules are the defeasible counterpart of strict rules. They are of the form $L_0 \prec L_1, \ldots, L_n$, where L_0 is a ground literal and $\{L_i\}_{i>0}$ is a set of ground literals. In both strict and defeasible rules, *strong negation* is allowed in the head of rules, and hence may be used to represent contradictory knowledge.

Even though the above constructs are ground, we allow for schematic versions with variables that are used to represent sets of ground rules. We denote variables with strings starting with an uppercase letter.

Arguments. Given a query in the form of a ground atom, the goal is to derive arguments for and against it's validity – derivation follows the same mechanism of logic programming [21]. Since rule heads can contain strong negation, it is possible to defeasibly derive contradictory literals from a program. For the treatment of contradictory knowledge, PreDeLP incorporates a defeasible argumentation formalism that allows the identification of the pieces of knowledge that are in conflict and, through the previously mentioned dialectical process, decides which information prevails as warranted. This dialectical process involves

$\langle \mathcal{A}_1, s\rangle$	$\mathcal{A}_1 = \{\theta_{1a}, \delta_{1a}\}$	$\langle \mathcal{A}_2, s\rangle$	$\mathcal{A}_2 = \{\phi_1, \phi_2, \delta_4, \omega_{2a}, \theta_{1a}, \theta_2\}$
$\langle \mathcal{A}_3, s\rangle$	$\mathcal{A}_3 = \{\phi_1, \delta_2, \delta_4\}$	$\langle \mathcal{A}_4, s\rangle$	$\mathcal{A}_4 = \{\phi_2, \delta_3, \theta_2\}$
$\langle \mathcal{A}_5, u\rangle$	$\mathcal{A}_5 = \{\phi_1, \delta_4\}$	$\langle \mathcal{A}_6, \neg s\rangle$	$\mathcal{A}_6 = \{\delta_{1b}, \theta_{1b}, \omega_{1a}\}$
$\langle \mathcal{A}_7, \neg u\rangle$	$\mathcal{A}_7 = \{\phi_3, \delta_{5a}\}$		

Fig. 2. Example ground arguments from the framework of Figure 1

the construction and evaluation of arguments, building a *dialectical tree* in the process. Arguments are formally defined next.

Definition 3. *An* argument $\langle \mathcal{A}, L\rangle$ *for a literal* L *is a pair of the literal and a (possibly empty) set of the EM* ($\mathcal{A} \subseteq \Pi_{AM}$) *that provides a minimal proof for* L *meeting the following requirements: (i)* L *is defeasibly derived from* $\mathcal{A}$*; (ii)* $\Omega \cup \Theta \cup \mathcal{A}$ *is not contradictory; and (iii)* $\mathcal{A}$ *is a minimal subset of* $\Delta \cup \Phi$ *satisfying 1 and 2, denoted* $\langle \mathcal{A}, L\rangle$*.*

Literal L *is called the* conclusion *supported by the argument, and* $\mathcal{A}$ *is the* support *of the argument. An argument* $\langle \mathcal{B}, L\rangle$ *is a* subargument *of* $\langle \mathcal{A}, L'\rangle$ *iff* $\mathcal{B} \subseteq \mathcal{A}$*. An argument* $\langle \mathcal{A}, L\rangle$ *is* presumptive *iff* $\mathcal{A} \cap \Phi$ *is not empty. We will also use* $\Omega(\mathcal{A}) = \mathcal{A} \cap \Omega$, $\Theta(\mathcal{A}) = \mathcal{A} \cap \Theta$, $\Delta(\mathcal{A}) = \mathcal{A} \cap \Delta$, *and* $\Phi(\mathcal{A}) = \mathcal{A} \cap \Phi$*.*

Our definition differs slightly from that of [22], where DeLP is introduced, as we include strict rules and facts as part of arguments – the reason for this will become clear in Section 4. Arguments for our scenario are shown next.

Example 3. Figure 2 shows example arguments based on the knowledge base from Figure 1. Note that $\langle \mathcal{A}_5, u\rangle$ is a sub-argument of $\langle \mathcal{A}_2, s\rangle$ and $\langle \mathcal{A}_3, s\rangle$. ■

Given an argument $\langle \mathcal{A}_1, L_1\rangle$, counter-arguments are arguments that contradict it. Argument $\langle \mathcal{A}_2, L_2\rangle$ is said to *counterargue* or *attack* $\langle \mathcal{A}_1, L_1\rangle$ at a literal L' iff there exists a subargument $\langle \mathcal{A}, L''\rangle$ of $\langle \mathcal{A}_1, L_1\rangle$ such that the set $\Omega(\mathcal{A}_1) \cup \Omega(\mathcal{A}_2) \cup \Theta(\mathcal{A}_1) \cup \Theta(\mathcal{A}_2) \cup \{L_2, L''\}$ is contradictory.

Example 4. Consider the arguments from Example 3. The following are some of the attack relationships between them: $\mathcal{A}_1$, $\mathcal{A}_2$, $\mathcal{A}_3$, and $\mathcal{A}_4$ all attack $\mathcal{A}_6$; $\mathcal{A}_5$ attacks $\mathcal{A}_7$; and $\mathcal{A}_7$ attacks $\mathcal{A}_2$. ■

A *proper defeater* of an argument $\langle A, L\rangle$ is a counter-argument that – by some criterion – is considered to be better than $\langle A, L\rangle$; if the two are incomparable according to this criterion, the counterargument is said to be a *blocking* defeater. An important characteristic of PreDeLP is that the argument comparison criterion is modular, and thus the most appropriate criterion for the domain that is being represented can be selected; the default criterion used in classical defeasible logic programming (from which PreDeLP is derived) is *generalized specificity* [23], though an extension of this criterion is required for arguments using presumptions [1]. We briefly recall this criterion next – the first definition is for generalized specificity, which is subsequently used in the definition of presumption-enabled specificity.

Definition 4. *Let $\Pi_{AM} = (\Theta, \Omega, \Phi, \Delta)$ be a PreDeLP program and let $\mathcal{F}$ be the set of all literals that have a defeasible derivation from Π_{AM}. An argument $\langle \mathcal{A}_1, L_1 \rangle$ is* preferred to *$\langle \mathcal{A}_2, L_2 \rangle$, denoted with $\mathcal{A}_1 \succ_{PS} \mathcal{A}_2$ iff:*

(1) For all $H \subseteq \mathcal{F}$, $\Omega(\mathcal{A}_1) \cup \Omega(\mathcal{A}_2) \cup H$ is non-contradictory: if there is a derivation for L_1 from $\Omega(\mathcal{A}_2) \cup \Omega(\mathcal{A}_1) \cup \Delta(\mathcal{A}_1) \cup H$, and there is no derivation for L_1 from $\Omega(\mathcal{A}_1) \cup \Omega(\mathcal{A}_2) \cup H$, then there is a derivation for L_2 from $\Omega(\mathcal{A}_1) \cup \Omega(\mathcal{A}_2) \cup \Delta(\mathcal{A}_2) \cup H$; and

(2) there is at least one set $H' \subseteq \mathcal{F}$, $\Omega(\mathcal{A}_1) \cup \Omega(\mathcal{A}_2) \cup H'$ is non-contradictory, such that there is a derivation for L_2 from $\Omega(\mathcal{A}_1) \cup \Omega(\mathcal{A}_2) \cup H' \cup \Delta(\mathcal{A}_2)$, there is no derivation for L_2 from $\Omega(\mathcal{A}_1) \cup \Omega(\mathcal{A}_2) \cup H'$, and there is no derivation for L_1 from $\Omega(\mathcal{A}_1) \cup \Omega(\mathcal{A}_2) \cup H' \cup \Delta(\mathcal{A}_1)$.

Intuitively, the principle of specificity says that, in the presence of two conflicting lines of argument about a proposition, the one that uses more of the available information is more convincing. A classic example involves a bird, Tweety, and arguments stating that it both flies (because it is a bird) and doesn't fly (because it is a penguin). The latter argument uses more information about Tweety – it is more specific – and is thus the stronger of the two.

Definition 5 ([1]). *Let $\Pi_{AM} = (\Theta, \Omega, \Phi, \Delta)$ be a PreDeLP program. An argument $\langle \mathcal{A}_1, L_1 \rangle$ is* preferred to *$\langle \mathcal{A}_2, L_2 \rangle$, denoted with $\mathcal{A}_1 \succ \mathcal{A}_2$ iff any of the following conditions hold:*

(1) $\langle \mathcal{A}_1, L_1 \rangle$ and $\langle \mathcal{A}_2, L_2 \rangle$ are both factual arguments and $\langle \mathcal{A}_1, L_1 \rangle \succ_{PS} \langle \mathcal{A}_2, L_2 \rangle$.

(2) $\langle \mathcal{A}_1, L_1 \rangle$ is a factual argument and $\langle \mathcal{A}_2, L_2 \rangle$ is a presumptive argument.

(3) $\langle \mathcal{A}_1, L_1 \rangle$ and $\langle \mathcal{A}_2, L_2 \rangle$ are presumptive arguments, and

(a) $\Phi(\mathcal{A}_1) \subsetneq \Phi(\mathcal{A}_2)$ or,

(b) $\Phi(\mathcal{A}_1) = \Phi(\mathcal{A}_2)$ and $\langle \mathcal{A}_1, L_1 \rangle \succ_{PS} \langle \mathcal{A}_2, L_2 \rangle$.

Generally, if $\mathcal{A}, \mathcal{B}$ are arguments with rules X and Y, resp., and $X \subset Y$, then $\mathcal{A}$ is stronger than $\mathcal{B}$. This also holds when $\mathcal{A}$ and $\mathcal{B}$ use presumptions P_1 and P_2, resp., and $P_1 \subset P_2$.

Example 5. The following are some relationships between arguments from Example 3, based on Definitions 4 and 5.

$\mathcal{A}_1$ and $\mathcal{A}_6$ are incomparable (blocking defeaters);
$\mathcal{A}_6 \succ \mathcal{A}_2$, and thus $\mathcal{A}_6$ defeats $\mathcal{A}_2$;
$\mathcal{A}_5$ and $\mathcal{A}_7$ are incomparable (blocking defeaters). ■

A sequence of arguments called an *argumentation line* thus arises from this attack relation, where each argument defeats its predecessor. To avoid undesirable sequences, which may represent circular argumentation lines, in DELP an *argumentation line* is *acceptable* if it satisfies certain constraints (see [20]). A literal L is *warranted* if there exists a non-defeated argument $\mathcal{A}$ supporting L.

Clearly, there can be more than one defeater for a particular argument $\langle \mathcal{A}, L \rangle$. Therefore, many acceptable argumentation lines could arise from $\langle \mathcal{A}, L \rangle$, leading to a tree structure. The tree is built from the set of all argumentation lines rooted in the initial argument. In a dialectical tree, every node (except the root) represents a defeater of its parent, and leaves correspond to undefeated arguments. Each path from the root to a leaf corresponds to a different acceptable argumentation line. A dialectical tree provides a structure for considering all the possible acceptable argumentation lines that can be generated for deciding whether an argument is defeated. We call this tree *dialectical* because it represents an exhaustive dialectical[1] analysis for the argument in its root. For a given argument $\langle \mathcal{A}, L \rangle$, we denote the corresponding dialectical tree as $\mathcal{T}(\langle \mathcal{A}, L \rangle)$.

Given a literal L and an argument $\langle \mathcal{A}, L \rangle$, in order to decide whether or not a literal L is warranted, every node in the dialectical tree $\mathcal{T}(\langle \mathcal{A}, L \rangle)$ is recursively marked as "D" (*defeated*) or "U" (*undefeated*), obtaining a marked dialectical tree $\mathcal{T}^*(\langle \mathcal{A}, L \rangle)$ as follows:

1. All leaves in $\mathcal{T}^*(\langle \mathcal{A}, L \rangle)$ are marked as "U"s, and
2. Let $\langle \mathcal{B}, q \rangle$ be an inner node of $\mathcal{T}^*(\langle \mathcal{A}, L \rangle)$. Then $\langle \mathcal{B}, q \rangle$ will be marked as "U" iff every child of $\langle \mathcal{B}, q \rangle$ is marked as "D". The node $\langle \mathcal{B}, q \rangle$ will be marked as "D" iff it has at least a child marked as "U".

Given an argument $\langle \mathcal{A}, L \rangle$ obtained from Π_{AM}, if the root of $\mathcal{T}^*(\langle \mathcal{A}, L \rangle)$ is marked as "U", then we will say that $\mathcal{T}^*(\langle \mathcal{A}, h \rangle)$ *warrants* L and that L is *warranted* from Π_{AM}. (Warranted arguments correspond to those in the grounded extension of a Dung argumentation system [24].) There is a further requirement when the arguments in the dialectical tree contains presumptions – the conjunction of all presumptions used in even (respectively, odd) levels of the tree must be consistent. This can give rise to multiple trees for a given literal, as there can potentially be different arguments that make contradictory assumptions.

We can then extend the idea of a dialectical tree to a *dialectical forest*. For a given literal L, a dialectical forest $\mathcal{F}(L)$ consists of the set of dialectical trees for all arguments for L. We shall denote a marked dialectical forest, the set of all marked dialectical trees for arguments for L, as $\mathcal{F}^*(L)$. Hence, for a literal L, we say it is *warranted* if there is at least one argument for that literal in the dialectical forest $\mathcal{F}^*(L)$ that is labeled as "U", *not warranted* if there is at least one argument for the literal $\neg L$ in the dialectical forest $\mathcal{F}^*(\neg L)$ that is labeled as "U", and *undecided* otherwise.

4 Probabilistic PreDeLP

Probabilistic PreDeLP arises from the combination of the environmental and analytical models (Π_{EM} and Π_{AM}, respectively). Intuitively, given Π_{AM}, every element of $\Omega \cup \Theta \cup \Delta \cup \Phi$ might only hold in certain worlds in the set $\mathcal{W}_{EM}$ – that is, they are subject to probabilistic events. Therefore, we associate elements of $\Omega \cup \Theta \cup \Delta \cup \Phi$ with a formula from $form_{EM}$. For instance, we could associate formula *rainy* to fact *umbrella* to state that the latter only holds when the probabilistic event *rainy* holds; since weather is uncertain in nature, it has been modeled as part of the EM.

[1] In the sense of providing reasons for and against a position.

$af(\theta_{1a}) = af(\theta_{1b}) = k \vee (f \wedge (h \vee (e \wedge l)))$	$af(\phi_3) = b$
$af(\theta_2) = i$	$af(\delta_{1a}) = af(\delta_{1b}) = \mathsf{True}$
$af(\omega_{1a}) = af(\omega_{1b}) = \mathsf{True}$	$af(\delta_2) = \mathsf{True}$
$af(\omega_{2a}) = af(\omega_{2b}) = \mathsf{True}$	$af(\delta_3) = \mathsf{True}$
$af(\phi_1) = c \vee a$	$af(\delta_4) = \mathsf{True}$
$af(\phi_2) = f \wedge m$	$af(\delta_{5a}) = af(\delta_{5b}) = \mathsf{True}$

Fig. 3. Example annotation function

We can then compute the probabilities of subsets of $\Omega \cup \Theta \cup \Delta \cup \Phi$ using the information contained in Π_{EM}, as we describe shortly. The notion of an *annotation function* associates elements of $\Omega \cup \Theta \cup \Delta \cup \Phi$ with elements of $form_{EM}$.

Definition 6. *An* annotation function *is any function* $af : \Omega \cup \Theta \cup \Delta \cup \Phi \to form_{EM}$. *We shall use* $[af]$ *to denote the set of all annotation functions.*

We will sometimes denote annotation functions as sets of pairs $(f, af(f))$ in order to simplify the presentation. Figure 3 shows an example of an annotation function for our running example.

We now have all the components to formally define Probabilistic PreDeLP programs (P-PreDeLP for short).

Definition 7. *Given environmental model* Π_{EM}*, analytical model* Π_{AM}*, and annotation function* af*, a* probabilistic PreDeLP program *is of the form* $\mathcal{I} = (\Pi_{EM}, \Pi_{AM}, af)$. *We use notation* $[\mathcal{I}]$ *to denote the set of all possible programs.*

Given this setup, we can consider a world-based approach; that is, the defeat relationship among arguments depends on the current state of the (EM) world.

Definition 8. *Let* $\mathcal{I} = (\Pi_{EM}, \Pi_{AM}, af)$ *be a P-PreDeLP program, argument* $\langle \mathcal{A}, L \rangle$ *is* valid *w.r.t. world* $w \in \mathcal{W}_{EM}$ *iff* $\forall c \in \mathcal{A}, w \models af(c)$.

We extend the notion of validity to argumentation lines, dialectical trees, and dialectical forests in the expected way (for instance, an argumentation line is valid w.r.t. w iff all arguments that comprise that line are valid w.r.t. w). We also extend the idea of a dialectical tree w.r.t. worlds; so, for a given world $w \in \mathcal{W}_{EM}$, the dialectical (resp., marked dialectical) tree induced by w is denoted with $\mathcal{T}_w\langle \mathcal{A}, L \rangle$ (resp., $\mathcal{T}^*_w\langle \mathcal{A}, L \rangle$). We require that all arguments and defeaters in these trees to be valid with respect to w. Likewise, we extend the notion of dialectical forests in the same manner (denoted with $\mathcal{F}_w(L)$ and $\mathcal{F}^*_w(L)$, resp.). Based on these concepts we introduce the notion of *warranting scenario*.

Definition 9. *Let* $\mathcal{I} = (\Pi_{EM}, \Pi_{AM}, af)$ *be a P-PreDeLP program and* L *be a literal formed with a ground atom from* $\mathbf{G}_{AM}$*; a world* $w \in \mathcal{W}_{EM}$ *is said to be a* warranting scenario *for* L *(denoted* $w \vdash_{\mathsf{war}} L$*) iff there is a dialectical forest* $\mathcal{F}^*_w(L)$ *in which* L *is warranted and* $\mathcal{F}^*_w(L)$ *is valid w.r.t.* w.

Hence, the set of worlds in the EM where a literal L in the AM *must* be true is exactly the set of warranting scenarios – these are the "necessary" worlds: $nec(L) = \{w \in \mathcal{W}_{EM} \mid (w \vdash_{\mathsf{war}} L)\}$. Now, the set of worlds in the EM where AM literal L *can* be true is the following – these are the "possible" worlds: $poss(L) = \{w \in \mathcal{W}_{EM} \mid w \not\vdash_{\mathsf{war}} \neg L\}$. The probability distribution Pr defined over the worlds in the EM induces an upper and lower bound on the probability of literal L (denoted $\mathbf{P}_{L,\mathsf{Pr},\mathcal{I}}$) as follows:

$$\ell_{L,\mathsf{Pr},\mathcal{I}} = \sum_{w \in nec(L)} \mathsf{Pr}(w), \qquad u_{L,\mathsf{Pr},\mathcal{I}} = \sum_{w \in poss(L)} \mathsf{Pr}(w)$$

$$\ell_{L,\mathsf{Pr},\mathcal{I}} \leq \mathbf{P}_{L,\mathsf{Pr},\mathcal{I}} \leq u_{L,\mathsf{Pr},\mathcal{I}}$$

Since the EM in general does not define a single probability distribution, the above computations should be done using linear programs EP-LP-MIN and EP-LP-MAX, as described above.

4.1 Sources of Inconsistency

We use the following notion of (classical) consistency of PreDeLP programs: Π is said to be *consistent* if there does not exist ground literal a s.t. $\Pi \vdash a$ and $\Pi \vdash \neg a$. For P-PreDeLP programs, there are two main kinds of inconsistency that can be present; the first is what we refer to as EM, or Type I, (in)consistency.

Definition 10. *Environmental model Π_{EM} is* Type I consistent *iff there exists a probability distribution Pr over the set of worlds $\mathcal{W}_{EM}$ that satisfies Π_{EM}.*

We illustrate this type of consistency in the following example.

Example 6. The following formula is a simple example of an EM for which there is no satisfying probability distribution:

$$rain \vee hail : 0.3 \pm 0;$$
$$rain \wedge hail : 0.5 \pm 0.1.$$

A P-PreDeLP program using such an EM gives rise to an example of Type I inconsistency, as it arises from the fact that there is no satisfying interpretation for the EM knowledge base. ■

Assuming a consistent EM, inconsistencies can still arise through the interaction between the annotation function and facts and strict rules. We will refer to this as combined, or Type II, (in)consistency.

Definition 11. *A P-PreDeLP program $\mathcal{I} = (\Pi_{EM}, \Pi_{AM}, af)$, with $\Pi_{AM} = \langle \Theta, \Omega, \Phi, \Delta \rangle$, is* Type II consistent *iff: given any probability distribution Pr that satisfies Π_{EM}, if there exists a world $w \in \mathcal{W}_{EM}$ such that $\bigcup_{x \in \Theta \cup \Omega \mid w \models af(x)} \{x\}$ is inconsistent, then we have $\mathsf{Pr}(w) = 0$.*

Thus, any EM world in which the set of associated facts and strict rules are inconsistent (we refer to this as "classical consistency") must always be assigned a zero probability. The following is an example of this other type of inconsistency.

Example 7. Consider the EM knowledge base from Example 1, the AM presented in Figure 1 and the annotation function from Figure 3. Suppose the following fact is added to the argumentation model:

$$\theta_3 = \neg p,$$

and that the annotation function is expanded as follows:

$$af(\theta_3) = \neg k.$$

Clearly, fact θ_3 is in direct conflict with fact θ_{1a} – this does not necessarily mean that there is an inconsistency. For instance, by the annotation function, θ_{1a} holds in the world $\{k\}$ while θ_3 does not. However, if we consider the world:

$$w = \{f, h)$$

Note that $w \models af(\theta_3)$ and $w \models af(\theta_2)$, which means that, in this world, two contradictory facts can occur. Since the environmental model indicates that this world can be assigned a non-zero probability, we have a Type II inconsist program. ■

Another example (perhaps easier to visualize) in the rain/hail scenario discussed above, is as follows: suppose we have facts $f = umbrella$ and $g = \neg umbrella$, and annotation function $af(f) = rain \vee hail$ and $af(g) = wind$. Intuitively, the first fact states that an umbrella should be carried if it either rains or hails, while the second states that an umbrella should not be carried if it is windy. If the EM assigns a non-zero probability to formula $(rain \vee hail) \wedge wind$, then we have Type II inconsistency.

In the following, we say that a P-PreDeLP program is **consistent** if and only if it is both Type I and Type II consistent. However, in this paper, we focus on Type II consistency and assume that the program is Type I consistent.

4.2 Basic Operations for Restoring Consistency

Given a P-PreDeLP program that is Type II inconsistent, there are two basic strategies that can be used to restore consistency:

Revise the EM: the probabilistic model can be changed in order to force the worlds that induce contradicting strict knowledge to have probability zero.

Revise the annotation function: The annotations involved in the inconsistency can be changed so that the conflicting information in the AM does not become induced under any possible world.

It may also appear that a third option would be to adjust the AM – this is, however, equivalent to modifying the annotation function. Consider the presence of two facts in the AM: $a, \neg a$. Assuming that this causes an inconsistency (that is, there is at least one world in which they both hold), one way to resolve it would be to remove one of these two literals. Suppose $\neg a$ is removed; this would be equivalent to setting $af(\neg a) = \bot$ (where $\bot$ represents a contradiction in the language of the EM). *In this paper, we often refer to "removing elements of Π_{AM}" to refer to changes to the annotation function that cause certain elements of the Π_{AM} to not have their annotations satisfied in certain EM worlds.*

Now, suppose that Π_{EM} is consistent, but that the overall program is Type II inconsistent. Then, there must exist a set of worlds in the EM where there is a probability distribution that assigns each of them a non-zero probability. This gives rise to the following result.

Proposition 1. *If there exists a probability distribution* Pr *that satisfies* Π_{EM} *s.t. there exists a world* $w \in \mathcal{W}_{EM}$ *where* $\mathsf{Pr}(w) > 0$ *and* $\bigcup_{x \in \Theta \cup \Omega \,|\, w \models af(x)} \{x\}$ *is inconsistent (Type II inconsistency), then any change made in order to resolve this inconsistency by modifying only* Π_{EM} *yields a new EM* Π'_{EM} *such that* $\left(\bigwedge_{a \in w} a \wedge \bigwedge_{a \notin w} \neg a\right) : 0 \pm 0$ *is entailed by* Π'_{EM}.

Proposition 1 seems to imply an easy strategy of adding formulas to Π_{EM} causing certain worlds to have a zero probability. However, this may lead to Type I inconsistencies in the resulting model Π'_{EM}. If we are applying an EM-only strategy to resolve inconsistencies, this would then lead to further adjustments to Π'_{EM} in order to restore Type I consistency. However, such changes could potentially lead to Type II inconsistency in the overall P-PreDeLP program (by either removing elements of Π'_{EM} or loosening probability bounds of the sentences in Π'_{EM}), which would lead to setting more EM worlds to a probability of zero. It is easy to devise an example of a situation in which the probability mass cannot be accommodated given the constraints imposed by the AM and EM together – in such cases, it would be impossible to restore consistency by only modifying Π_{EM}. We thus arrive at the following observation:

Observation 1 *Given a Type II inconsistent P-PreDeLP program, consistency cannot always be restored via modifications to* Π_{EM} *alone.*

Therefore, due to this line of reasoning, in this paper we focus our efforts on modifications to the annotation function only. However, in the future, we intend to explore belief revision operators that consider both the annotation function (which, as we saw, captures changes to the AM) along with changes to the EM, as well as combinations of the two.

5 Revising Probabilistic PreDeLP Programs

Given a P-PreDeLP program $\mathcal{I} = (\Pi_{EM}, \Pi_{AM}, af)$, with $\Pi_{AM} = \Omega \cup \Theta \cup \Delta \cup \Phi$, we are interested in solving the problem of incorporating an epistemic input (f, af') into $\mathcal{I}$, where f is either an atom or a rule and af' is equivalent to af, except for its expansion to include f. For ease of presentation, we assume that f is to be incorporated as a fact or strict rule, since incorporating defeasible knowledge can never lead to inconsistency. As we are only conducting annotation function revisions, for $\mathcal{I} = (\Pi_{EM}, \Pi_{AM}, af)$ and input (f, af') we denote the revision as follows: $\mathcal{I} \bullet (f, af') = (\Pi_{EM}, \Pi'_{AM}, af'')$ where $\Pi'_{AM} = \Pi_{AM} \cup \{f\}$ and af'' is the revised annotation function.

Notation. We use the symbol "$\bullet$" to denote the revision operator. We also slightly abuse notation for the sake of presentation, as well as introduce notation to convert sets of worlds to/from formulas.

- $\mathcal{I} \cup (f, af')$ to denote $\mathcal{I}' = (\Pi_{EM}, \Pi_{AM} \cup \{f\}, af')$.
- $(f, af') \in \mathcal{I} = (\Pi_{AM}, \Pi_{EM}, af)$ to denote $f \in \Pi_{AM}$ and $af = af'$.
- $wld(f) = \{w \mid w \models f\}$ – the set of worlds that satisfy formula f; and
- $for(w) = \bigwedge_{a \in w} a \wedge \bigwedge_{a \notin w} \neg a$ – the formula that has w as its only model.
- $\Pi^{\mathcal{I}}_{AM}(w) = \{f \in \Theta \cup \Omega \mid w \models af(f)\}$
- $\mathcal{W}^{0}_{EM}(\mathcal{I}) = \{w \in \mathcal{W}_{EM} \mid \Pi^{\mathcal{I}}_{AM}(w) \text{ is inconsistent}\}$
- $\mathcal{W}^{I}_{EM}(\mathcal{I}) = \{w \in \mathcal{W}^{0}_{EM} \mid \exists \mathsf{Pr} \text{ s.t. } \mathsf{Pr} \models \Pi_{EM} \wedge \mathsf{Pr}(w) > 0\}$

Intuitively, $\Pi^{\mathcal{I}}_{AM}(w)$ is the subset of facts and strict rules in Π_{AM} whose annotations are true in EM world w. The set $\mathcal{W}^{0}_{EM}(\mathcal{I})$ contains all the EM worlds for a given program where the corresponding knowledge base in the AM is classically inconsistent and $\mathcal{W}^{I}_{EM}(\mathcal{I})$ is a subset of these that can be assigned a non-zero probability – the latter are the worlds where inconsistency in the AM can arise.

5.1 Postulates for Revising the Annotation Function

We now analyze the rationality postulates for non-prioritized revision of belief bases first introduced in [2] and later generalized in [25], in the context of P-PreDeLP programs. These postulates are chosen due to the fact that they are well studied in the literature for non-prioritized belief revision.

Inclusion: For $\mathcal{I} \bullet (f, af') = (\Pi_{EM}, \Pi_{AM} \cup \{f\}, af'')$, $\forall g \in \Pi_{AM}$, $wld(af''(g)) \subseteq wld(af'(g))$.

This postulate states that, for any element in the AM, the worlds that satisfy its annotation after the revision are a subset of the original set of worlds satisfying the annotation for that element.

Vacuity: If $\mathcal{I} \cup (f, af')$ is consistent, then $\mathcal{I} \bullet (f, af') = \mathcal{I} \cup (f, af')$

Consistency Preservation: If $\mathcal{I}$ is consistent, then $\mathcal{I} \bullet (f, af')$ is also consistent.

Weak Success: If $\mathcal{I} \cup (f, af')$ is consistent, then $(f, af') \in \mathcal{I} \bullet (f, af')$.

Whenever the simple addition of the input doesn't cause inconsistencies to arise, the result will contain the input.

Core Retainment: For $\mathcal{I} \bullet (f, af') = (\Pi_{EM}, \Pi_{AM} \cup \{f\}, af'')$, for each $w \in \mathcal{W}^{I}_{EM}(\mathcal{I} \cup (f, af'))$, we have $X_w = \{h \in \Theta \cup \Omega \mid w \models af''(h)\}$; for each $g \in$

$\Pi_{AM}(w) \setminus X_w$ there exists $Y_w \subseteq X_w \cup \{f\}$ s.t. Y_w is consistent and $Y_w \cup \{g\}$ is inconsistent.

For a given EM world, if a portion of the associated AM knowledge base is removed by the operator, then there exists a subset of the remaining knowledge base that is not consistent with the removed element and f.

Relevance: For $\mathcal{I} \bullet (f, af') = (\Pi_{EM}, \Pi_{AM} \cup \{f\}, af'')$, for each $w \in \mathcal{W}^I_{EM}(\mathcal{I} \cup (f, af'))$, we have $X_w = \{h \in \Theta \cup \Omega \mid w \models af''(h)\}$; for each $g \in \Pi_{AM}(w) \setminus X_w$ there exists $Y_w \supseteq X_w \cup \{f\}$ s.t. Y_w is consistent and $Y_w \cup \{g\}$ is inconsistent.

For a given EM world, if a portion of the associated AM knowledge base is removed by the operator, then there exists a superset of the remaining knowledge base that is not consistent with the removed element and f.

Uniformity 1: Let $(f, af'_1), (g, af'_2)$ be two inputs where $\mathcal{W}^I_{EM}(\mathcal{I} \cup (f, af'_1)) = \mathcal{W}^I_{EM}(\mathcal{I} \cup (g, af'_2))$; for all $w \in \mathcal{W}^I_{EM}(\mathcal{I} \cup (f, af'))$ and for all $X \subseteq \Pi_{AM}(w)$; if $\{x \mid x \in X \cup \{f\}, w \models af'_1(x)\}$ is inconsistent iff $\{x \mid x \in X \cup \{g\}, w \models af'_2(x)\}$ is inconsistent, then for each $h \in \Pi_{AM}$, we have that:

$$\{w \in \mathcal{W}^I_{EM}(\mathcal{I} \cup (f, af'_1)) \mid w \models af'_1(h) \wedge \neg af''_1(h)\} =$$

$$\{w \in \mathcal{W}^I_{EM}(\mathcal{I} \cup (g, af'_2)) \mid w \models af'_2(h) \wedge \neg af''_2(h)\}.$$

If two inputs result in the same set of EM worlds leading to inconsistencies in an AM knowledge base, and the consistency between analogous subsets (when joined with the respective input) are the same, then the models removed from the annotation of a given strict rule or fact are the same for both inputs.

Uniformity 2: Let $(f, af'_1), (g, af'_2)$ be two inputs where $\mathcal{W}^I_{EM}(\mathcal{I} \cup (f, af'_1)) = \mathcal{W}^I_{EM}(\mathcal{I} \cup (g, af'_2))$; for all $w \in \mathcal{W}^I_{EM}(\mathcal{I} \cup (f, af'))$and for all $X \subseteq \Pi_{AM}(w)$; if $\{x \mid x \in X \cup \{f\}, w \models af'_1(x)\}$ is inconsistent iff $\{x \mid x \in X \cup \{g\}, w \models af'_2(x)\}$ is inconsistent, then

$$\{w \in \mathcal{W}^I_{EM}(\mathcal{I} \cup (f, af'_1)) \mid w \models af'_1(h) \wedge af''_1(h)\} =$$

$$\{w \in \mathcal{W}^I_{EM}(\mathcal{I} \cup (g, af'_2)) \mid w \models af'_2(h) \wedge af''_2(h)\}.$$

If two inputs result in the same set of EM worlds leading to inconsistencies in an AM knowledge base, and the consistency between analogous subsets (when joined with the respective input) are the same, then the models retained in the the annotation of a given strict rule or fact are the same for both inputs.

Relationships between Postulates. There are a couple of interesting relationships among the postulates. The first is a sufficient condition for Core Retainment to be implied by Relevance.

Proposition 2. *Let $\bullet$ be an operator such that $\mathcal{I} \bullet (f, af') = (\Pi_{EM}, \Pi_{AM} \cup \{f\}, af'')$, where $\forall w \in \mathcal{W}^I_{EM}(\mathcal{I} \cup (f, af'))$, $\Pi_{AM}^{\mathcal{I}\bullet(f,af')}(w)$ is a maximal consistent subset of $\Pi_{AM}^{\mathcal{I}\cup(f,af')}(w)$. If $\bullet$ satisfies Relevance then it also satisfies Core Retainment.*

Similarly, we can show the equivalence between the two Uniformity postulates under certain conditions.

Proposition 3. *Let $\bullet$ be an operator such that $\mathcal{I} \bullet (f, af') = (\Pi_{EM}, \Pi_{AM} \cup \{f\}, af'')$ and $\forall w$, $\Pi_{AM}^{\mathcal{I}\bullet(f,af')}(w) \subseteq \Pi_{AM}^{\mathcal{I}\cup(f,af')}(w)$. Operator $\bullet$ satisfies Uniformity 1 iff it satisfies Uniformity 2.*

Given the results of Propositions 2 and 3, we will not study Core Retainment and Uniformity 2 with respect to the construction of a belief revision operator in the next section.

5.2 An Operator for P-PreDeLP Revision

In this section, we introduce an operator for revising a P-PreDeLP program. As stated earlier, any subset of Π_{AM} associated with a world in $\mathcal{W}_{EM}^{I}(\mathcal{I} \cup (f, af'))$ must be modified by the operator in order to remain consistent. So, for such a world w, we introduce a set of candidate replacement programs for $\Pi_{AM}(w)$ in order to maintain consistency and satisfy the Inclusion postulate.

$$\begin{aligned} candPgm(w, \mathcal{I}) = \{\Pi'_{AM} \mid \Pi'_{AM} \subseteq \Pi_{AM}(w) \text{ s.t. } \Pi'_{AM} \text{ is consistent and} \\ \nexists \Pi''_{AM} \subseteq \Pi_{AM}(w) \ s.t.\ \Pi''_{AM} \supset \Pi'_{AM} \text{ s.t. } \Pi''_{AM} \text{ is consistent}\} \end{aligned}$$

Intuitively, $candPgm(w, \mathcal{I})$ is the set of maximal consistent subsets of $\Pi_{AM}(w)$. Coming back to the rain/hail example presented above, we have:

Example 8. Consider the P-PreDeLP program $\mathcal{I}$ presented right after Example 7, and the following EM knowledge base:

$$\begin{aligned} rain \vee hail &: 0.5 \pm 0.1; \\ rain \wedge hail &: 0.3 \pm 0.1; \\ wind &: 0.2 \pm 0. \end{aligned}$$

Given this setup, we have, for instance:

$$candPgm(\{rain, hail, wind\}, \mathcal{I}) = \Big\{\{umbrella\}, \{\neg umbrella\}\Big\}.$$

Intuitively, this means that, since the world where *rain*, *hail*, and *wind* are all true can be assigned a non-zero probability by the EM, we must choose either *umbrella* or $\neg$*umbrella* in order to recover consistency. ■

We now show a series of intermediate results that lead up to the representation theorem (Theorem 1). First, we show how this set plays a role in showing a necessary and sufficient requirement for Inclusion and Consistency Preservation to hold together.

Lemma 1. *Given program $\mathcal{I}$ and input (f, af'), operator $\bullet$ satisfies Inclusion and Consistency Preservation iff for $\mathcal{I} \bullet (f, af') = (\Pi_{EM}, \Pi_{AM}, af'')$, for all $w \in \mathcal{W}_{EM}^{I}(\mathcal{I} \cup (f, af'))$, there exists an element $X \in candPgm(w, \mathcal{I} \cup (f, af'))$ s.t. $\{h \in \Theta \cup \Omega \cup \{f\} \mid w \models af''(h)\} \subseteq X$.*

Next, we investigate the role that the set $candPgm$ plays in showing the necessary and sufficient requirement for satisfying Inclusion, Consistency Preservation, and Relevance all at once.

Lemma 2. *Given program $\mathcal{I}$ and input (f, af'), operator $\bullet$ satisfies Inclusion, Consistency Preservation, and Relevance iff for $\mathcal{I} \bullet (f, af') = (\Pi_{EM}, \Pi_{AM}, af'')$, for all $w \in \mathcal{W}^{I}_{EM}(\mathcal{I} \cup (f, af'))$ we have $\{h \in \Theta \cup \Omega \cup \{f\} \mid w \models af''(h)\} \in candPgm(w, \mathcal{I} \cup (f, af'))$.*

The last of the intermediate results shows that if there is a consistent program where two inputs cause inconsistencies to arise in the same way, then for each world the set of candidate replacement programs (minus the added AM formula) is the same. This result will be used as a support of the satisfaction of the first Uniformity postulate.

Lemma 3. *Let $\mathcal{I} = (\Pi_{EM}, \Pi_{AM}, af)$ be a consistent program, (f_1, af'_1), (f_2, af'_2) be two inputs, and $\mathcal{I}_i = (\Pi_{EM}, \Pi_{AM} \cup \{f_i\}, af'_i)$. If $\mathcal{W}^{I}_{EM}(\mathcal{I}_1) = \mathcal{W}^{I}_{EM}(\mathcal{I}_2)$, then for all $w \in \mathcal{W}^{I}_{EM}(\mathcal{I}_1)$ and all $X \subseteq \Pi_{AM}(w)$ we have that:*

1. *If $\{x \mid x \in X \cup \{f_1\}, w \models af'_1(x)\}$ is inconsistent $\Leftrightarrow$ $\{x \mid x \in X \cup \{f_2\}, w \models af'_2(x)\}$ is inconsistent, then $\{X \setminus \{f_1\} \mid X \in candPgm(w, \mathcal{I}_1)\} = \{X \setminus \{f_2\} \mid X \in candPgm(w, \mathcal{I}_2)\}$.*
2. *If $\{X \setminus \{f_1\} \mid X \in candPgm(w, \mathcal{I}_1)\} = \{X \setminus \{f_2\} \mid X \in candPgm(w, \mathcal{I}_2)\}$ then $\{x \mid x \in X \cup \{f_1\}, w \models af'_1(x)\}$ is inconsistent $\Leftrightarrow$ $\{x \mid x \in X \cup \{f_2\}, w \models af'_2(x)\}$ is inconsistent.*

We now have the necessary tools to present the construction of our non-prioritized belief revision operator.

Construction. Before introducing the construction, we define some preliminary notation. Let $\Phi : \mathcal{W}_{EM} \rightarrow 2^{[\Theta] \cup [\Omega]}$. For each h there is a formula in $\Pi_{AM} \cup \{f\}$, where f is part of the input. Given these elements, we define:

$$newFor(h, \Phi, \mathcal{I}, (f, af')) = af'(h) \wedge \bigwedge_{w \in \mathcal{W}^{I}_{EM}(\mathcal{I} \cup (f, af')) \mid h \notin \Phi(w)} \neg for(w_i)$$

The following definition then characterizes the class of operators called **AFO** (annotation function-based operators).

Definition 12 (AF-based Operators). *A belief revision operator $\bullet$ is an "annotation function-based" (or af-based) operator ($\bullet \in$ **AFO**) iff given program $\mathcal{I} = (\Pi_{EM}, \Pi_{AM}, af)$ and input (f, af'), the revision is defined as $\mathcal{I} \bullet (f, af') = (\Pi_{EM}, \Pi_{AM} \cup \{f\}, af'')$, where:*

$$\forall h, af''(h) = newFor(h, \Phi, \mathcal{I}, (f, af'))$$

where $\forall w \in \mathcal{W}_{EM}$, $\Phi(w) \in CandPgm_{af}(w, \mathcal{I} \cup (f, af'))$.

As the main result of the paper, we now show that satisfying a key set of postulates is a necessary and sufficient condition for membership in **AFO**.

Theorem 1 (Representation Theorem). *An operator • belongs to class* **AFO** *iff it satisfies Inclusion, Vacuity, Consistency Preservation, Weak Success, Relevance, and Uniformity 1.*

Proof. (Sketch) (If) By the fact that formulas associated with worlds in the set $\mathcal{W}^{I}_{EM}(\mathcal{I} \cup (f, af'))$ are considered in the change of the annotation function, Vacuity and Weak Success follow trivially. Further, Lemma 2 shows that Inclusion, Consistency Preservation, and Relevance are satisfied while Lemma 3 shows that Uniformity 1 is satisfied.

(Only-If) Suppose BWOC that an operator • satisfies all postulates and $\bullet \notin$ **AFO**. Then, one of four conditions must hold: (i) it does not satisfy Lemma 2 or (ii) it does not satisfy Lemma 3. However, by those previous arguments, if it satisfies all postulates, these arguments must be true as well – hence a contradiction. □

6 Conclusions

We have proposed an extension of the PreDeLP language that allows sentences to be annotated with probabilistic events; such events are connected to a probabilistic model, allowing a clear separation of interests between certain and uncertain knowledge. After presenting the language, we focused on characterizing belief revision operations over P-PreDeLP KBs. We presented a set of postulates inspired in the ones presented for non-prioritized revision of classical belief bases, and then proceeded to study a construction based on these postulates and prove that the two characterizations are equivalent.

As future work, we plan to study other kinds of operators, such as more general ones that allow the modification of the EM, as well as others that operate at different levels of granularity. Finally, we are studying the application of P-PreDeLP to real-world problems in cyber security and cyber warfare domains.

Acknowledgments. The authors are partially supported by UK EPSRC grant EP/J008346/1 ("PrOQAW"), ERC grant 246858 ("DIADEM"), ARO project 2GDATXR042, DARPA project R.0004972.001, Consejo Nacional de Investigaciones Científicas y Técnicas (CONICET) and Universidad Nacional del Sur (Argentina).

The opinions in this paper are those of the authors and do not necessarily reflect the opinions of the funders, the U.S. Military Academy, or the U.S. Army.

References

1. Martinez, M.V., García, A.J., Simari, G.R.: On the use of presumptions in structured defeasible reasoning. In: Proc. of COMMA, pp. 185–196 (2012)
2. Hansson, S.: Semi-revision. J. of App. Non-Classical Logics 7(1-2), 151–175 (1997)
3. Alchourrón, C.E., Gärdenfors, P., Makinson, D.: On the logic of theory change: Partial meet contraction and revision functions. J. Sym. Log. 50(2), 510–530 (1985)

4. Gardenfors, P.: Knowledge in flux: Modeling the dynamics of epistemic states. MIT Press, Cambridge (1988)
5. Hansson, S.O.: Kernel contraction. J. Symb. Log. 59(3), 845–859 (1994)
6. Doyle, J.: A truth maintenance system. Artif. Intell. 12(3), 231–272 (1979)
7. Falappa, M.A., Kern-Isberner, G., Simari, G.R.: Explanations, belief revision and defeasible reasoning. Artif. Intell. 141(1/2), 1–28 (2002)
8. Li, H., Oren, N., Norman, T.J.: Probabilistic argumentation frameworks. In: Proc. of TAFA, pp. 1–16 (2011)
9. Thimm, M.: A probabilistic semantics for abstract argumentation. In: Proc. of ECAI 2012, pp. 750–755 (2012)
10. Hunter, A.: Some foundations for probabilistic abstract argumentation. In: Proc. of COMMA 2012, pp. 117–128 (2012)
11. Fazzinga, B., Flesca, S., Parisi, F.: On the complexity of probabilistic abstract argumentation. Proc. of IJCAI 2013 (2013)
12. Haenni, R., Kohlas, J., Lehmann, N.: Probabilistic argumentation systems. Springer (1999)
13. Chesñevar, C.I., Simari, G.R., Alsinet, T., Godo, L.: A logic programming framework for possibilistic argumentation with vague knowledge. In: Proc. of UAI 2004, pp. 76–84 (2004)
14. Hunter, A.: A probabilistic approach to modelling uncertain logical arguments. Int. J. Approx. Reasoning 54(1), 47–81 (2013)
15. Gottlob, G., Lukasiewicz, T., Martinez, M.V., Simari, G.I.: Query answering under probabilistic uncertainty in Datalog+/– ontologies. AMAI (2013)
16. Nilsson, N.J.: Probabilistic logic. Artif. Intell. 28(1), 71–87 (1986)
17. Khuller, S., Martinez, M.V., Nau, D.S., Sliva, A., Simari, G.I., Subrahmanian, V.S.: Computing most probable worlds of action probabilistic logic programs: Scalable estimation for $10^{30,000}$ worlds. AMAI 51(2-4), 295–331 (2007)
18. Simari, G.I., Martinez, M.V., Sliva, A., Subrahmanian, V.S.: Focused most probable world computations in probabilistic logic programs. AMAI 64(2-3), 113–143 (2012)
19. Rahwan, I., Simari, G.R.: Argumentation in Artificial Intelligence. Springer (2009)
20. García, A.J., Simari, G.R.: Defeasible logic programming: An argumentative approach. TPLP 4(1-2), 95–138 (2004)
21. Lloyd, J.W.: Foundations of Logic Programming, 2nd edn. Springer (1987)
22. Simari, G.R., Loui, R.P.: A mathematical treatment of defeasible reasoning and its implementation. Artif. Intell. 53(2-3), 125–157 (1992)
23. Stolzenburg, F., García, A., Chesñevar, C.I., Simari, G.R.: Computing Generalized Specificity. Journal of Non-Classical Logics 13(1), 87–113 (2003)
24. Dung, P.M.: On the acceptability of arguments and its fundamental role in nonmonotonic reasoning, logic programming and n-person games. Artif. Intell. 77, 321–357 (1995)
25. Falappa, M.A., Kern-Isberner, G., Reis, M., Simari, G.R.: Prioritized and non-prioritized multiple change on belief bases. J. Philosophical Logic 41(1), 77–113 (2012)

A Multi-granular Database Model

Loreto Bravo and M. Andrea Rodríguez

Universidad de Concepción, Chile
Edmundo Larenas 219, 4070409 Concepción, Chile
{lbravo,arodriguez}@inf.udec.cl

Abstract. Various applications require storing and handling data at different granularities due to the nature of the data, diverse data origins, and resource-constraint specifications. The work in this paper introduces *domain schemas*, which extend the concept of domain to consider different granularities and the relationships among them. Relying on this definition, we introduce a multi-granular database model and its integrity constraints and query language. In particular, we extend traditional and conditional dependency constraints to deal with data at different granularities and study the satisfiability and consistency problems associated with them. The query language corresponds to SQL extended with operators specifically design to deal with granularities. As a case study, we focus on the spatial and temporal domains, which have extensive use in the literature and highlight the notion of granularity.

1 Introduction

Granularity defines bounds to the level of detail in which data is represented. Different real applications require storing and handling data at different granularities due to the nature of the data [2], diverse data origins, and resource-constraint specifications [14,13]. The work in this paper presents a model and query language for multi-granular databases. This is done from a general perspective, which is considered relevant if we want to integrate data from different data sources and domains. Unlike other approaches, data is not necessarily stored at the finest level of granularity upon which aggregation functions derive data at other coarser levels. Also, categories are not necessarily made explicit, but they can be done on the fly depending on the domain. Even more, we do not deal with only the hierarchical relationships between granularities, but we allow relationships between granules that make possible to address cases other than the classical finer-than or coarser-than relationship between granularities. In this way, our work relates to work done in spatio-temporal granularity [3], which provides different operations and relationships between granularities target to only the spatial and temporal domains. However, unlike this previous work, we take a more general perspective of databases and complement the database schema with integrity constraints that specify valid states of a multi-granular database. The following example clarifies the problems addressed by this work.

Example 1. Consider a database of an agency that stores information about relevant disasters (natural and man-made) occurred around the world. The information stored in the system as shown in Figure 1. Events are stored at different

C. Beierle and C. Meghini (Eds.): FoIKS 2014, LNCS 8367, pp. 344–360, 2014.

levels of detail respect to time (i.e., day, month, season, holiday season, year) and location (i.e., city, country, zone). Different granularities are useful in this databases since events may occur in a large geographic area (e.g., continent) or at a particular place (i.e., building). They also may occur at a fine time granularity (i.e., hour of a day) but also they may extend during a whole season as in the case of a drought. It can also be the case that the data availability is stored at different levels of detail. For example, the first tuple in relation Natural in Figure 1 means that there was a tsunami in some part of Indonesia and not necessarily all of it at some point during 2004. The fact that Indonesia can denote part of the region and not necessarily all is called *weak semantics*. On the other hand, a *strong* interpretation would mean that there was a tsunami in all Indonesia during all 2004. In this work, we focus on the weak semantics of granules.

Natural	Where	When	Type	Death
	Indonesia	2004	tsunami	280,000
	Pakistan	Columbus 2005	earthquake	75,000
	Huascarán	1970	avalanch	20,000
	Afghanistan	winter, 2007-2008	natural	926
	Valdivia	May, 1960	earthquake	1,655
	Haiti	winter 2009-2010	earthquake	360,000
	Chile	February 27 2010	earthquake	525
	Eastern Japan	March 11 2011	tsunami	24,000

ManMade	Where	When	Type	Death
	Santiago	December 8 2010	fire	81
	Madrid	March 11 2010	terrorist attack	191
	Sao Paulo	July 17, 2007	plane crash	199
	Newtown, Conneticut	Christmas holiday	shooting	26

Fig. 1. Example of a multi-granular database about disaster events

Despite differences in data granularity, one would like to retrieve data that are related at a specific granularity. For example, one would want to retrieve data of disasters occurred during a particular year or in particular country such that the system will need to find data at the desired time and location granularity.

In addition, assume that the agency only stores one man-made event per day and country. If one wants to enforce this constraint in the database, we could add the functional dependency Where, When $\rightarrow$ Type, Death. However, this constraint does not check what is expected since, for example, it will consider consistent an instance that stores man-made disasters in different cities of the same country occurred during the same day. The constraint should consider the granularity associated with the attribute of Where, which becomes an extension to traditional functional dependency constraints. □

Our previous work in [20] addresses the integration of databases at different granularities. It describes a database model and derives a global schema for data integration. We now further develop the database model and make the following specific contributions: (1) the definition of a global schema upon which integrity constraints are formalized and (2) the definition of a language to answer general queries and check integrity constraints.

The reminder of this paper is organized as follows. We define domain schemas in Section 2 and, based on its use, the Multi-granular Database Model and dependency constraints are formalized in Section 3. The query language for this model is presented in Section 4. Sections 5 and 6 discuss implementation issues and relevant literature followed by the conclusions in Section 7.

2 Domain Schemas

A *domain schema* will be used to represent the granularities associated with a domain and the relationships between them. In order to formalize a domain schema, we will need identifiers for granularities, e.g. Month, and identifiers for the granules that belong to a granularity, e.g., January 2012. Let L and I denote the universe of possible granularities and granule identifiers, respectively. We will also need a distinguished granule $\mathsf{all} \in \mathsf{I}$ and two distinguished granularities $\top, \bot \in \mathsf{L}$. Identifiers $\top$ and $\bot$ correspond to the coarsest and finest granularity of a specific domain. In fact, $\top$ contains the single granule all that groups in it all the elements in the domain, and $\bot$ contains one granule for each element in $\mathcal{U}$.

A *domain schema* is a tuple $\Psi = (\mathcal{U}, \ell, \mathcal{I}, \mu, \tau)$, where

- $\mathcal{U}$ is the domain associated with Ψ,
- $\ell \subseteq \mathsf{L}$ is a finite set of granularity identifiers (or labels) such that $\bot, \top \in \ell$,
- $\mathcal{I} \subseteq \mathsf{I}$ is a finite set of granule identifiers (or labels) such that $\mathsf{all} \in \mathcal{I}$,
- μ is a function $\mu : (\mathcal{I} \cup \mathcal{U}) \rightarrow 2^{\mathcal{U}}$ that maps granule identifiers to subsets of the domain. In particular, $\mu(\mathsf{all}) = \mathcal{U}$ and $\mu(g) = g$ for every $g \in \mathcal{U}$.
- τ is a function $\ell \rightarrow 2^{\mathcal{I} \cup \mathcal{U}}$ such that $\tau(\top) = \{\mathsf{all}\}$, $\tau(\bot) = \mathcal{U}$, and for all $G \in \ell$ if $i, j \in \tau(G)$ then $\mu(i) \cap \mu(j) = \emptyset$.

To simplify the presentation, we will asume that for $i, j \in \mathcal{I}$, $i \neq j$ iff $\mu(i) \neq \mu(j)$, and that for $G_1, G_2 \in \ell$, $G_1 \neq G_2$ iff $\tau(G_1) \neq \tau(G_2)$.

Given a domain schema $\Psi = (\mathcal{U}, \ell, \mathcal{I}, \mu, \tau)$, a granule $c_1 \in \mathcal{I}$ is said to *map* to $c_2 \in \mathcal{I}$ if $\mu(c_i) \subseteq \mu(c_j)$. For example, a granule 'London' would map to a granule 'England' since the former is in a way generalized by the latter. A granularity $G_1 \in \ell$ is *finer or equal than* $G_2 \in \ell$, denoted by $G_1 \preceq_{\Psi} G_2$, iff for all $c_i \in \tau(G_1)$ there exists $c_j \in \tau(G_2)$ such that c_i maps to c_j. For example, if we consider granularities 'US state', 'CA province' and 'country' with their expected meaning, both state and province would be finer than country, but there would be no such relation between state and province. A granularity $G_1 \in \ell$ is *finer than* $G_2 \in \ell$, denoted by $G_1 \prec_{\Psi} G_2$ if $G_1 \preceq_{\Psi} G_2$ and $G_2 \not\preceq_{\Psi} G_1$. When clear from the context we will replace $G_1 \preceq_{\Psi} G_2$ and $G_1 \prec_{\Psi} G_2$ by $G_1 \preceq G_2$ and $G_1 \prec G_2$.

Remark 1. In what follows, we will assume that, for any pair of indices $c_i, c_j \in \mathcal{I}$, checking that c_i maps to c_j and $\mu(i) \cap \mu(j) = \emptyset$ can be done in constant time.

This can be achieved in several ways. For example, if the domain $\mathcal{U}$ has a total order and $\mu(i)$ is defined as a range checking, those relations can be determined comparing only the limits of the range. On the other hand, in several settings the granularities in the domain schema can be pre-processed to compute the relationships between indices. In this case, the cost of checking if a granule maps to another will have a tradeoff between time and space. For example, given place names as index of a spatial domain, it is possible to have a pre-processed graph connecting all pairs of granules for which an inclusion relation holds. This results in constant time checking of mappings between granules. If we want to reduce the space used, we would leave the smallest graph for which the rest of the relations can be computed by composition. In this case, the mappings can be obtained in linear time over the number of granules.

A particular instance of a domain schema is the one defined over the time domain. A *time domain* is a pair $(T; \leq)$, where T is a non-empty set of constants that represent time instants and $\leq$ is a total order in T [5,4,8]. The next example illustrates the case of multiple time granularities for representing academic activities of universities using our notation. In this setting the set of granule identifiers is an ordered set of index values.

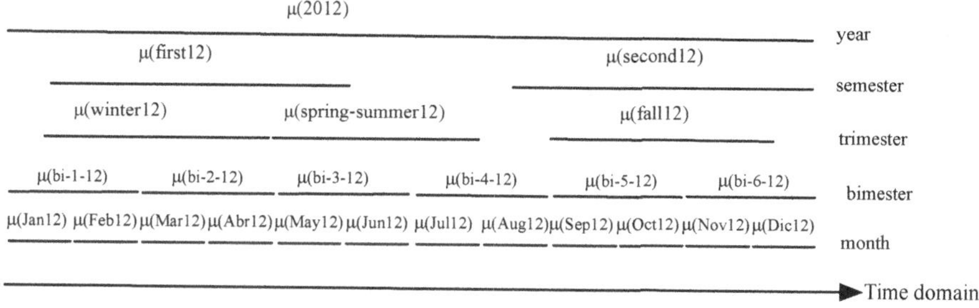

Fig. 2. Instance a of a domain schema over the time domain to represent academic activities of universities

Example 2. Consider the following domain schema $\Psi_t = (\mathcal{U}_t, \ell_t, \mathcal{I}_t, \mu_t, \tau_t)$, where $\mathcal{U}_t$ is the time domain (i.e., T of the tuple $(T; \leq)$ introduced above), $\ell_t = \{\mathsf{month}, \mathsf{bimester}, \mathsf{trimester}, \mathsf{semester}, \mathsf{year}\}$, and $\mathcal{I}_t$ is the set of indices represented graphically in Figure 2. We consider $\mathcal{U}_t$ to be a set of instants in the smallest perceptible time unit. For this example, $\tau(\bot) = \mathcal{U}_t$, $\mu(g) = g$ for every $g \in \mathcal{U}_t$, $\tau(\top) = \{\mathsf{all}\}$ and $\mu(\mathsf{all}) = \mathcal{U}_t$. The relations that hold among these granularities are $\mathsf{month} \prec \mathsf{year}$, $\mathsf{bimester} \prec \mathsf{year}$, $\mathsf{trimester} \prec \mathsf{year}$, $\mathsf{semester} \prec \mathsf{year}$, and $\mathsf{month} \prec \mathsf{bimester}$. □

The previous example could also be partially represented by a data warehousing dimension where granularities are represented as categories and granules as elements. For example, the dimension in Figure 3 represents all the relationships

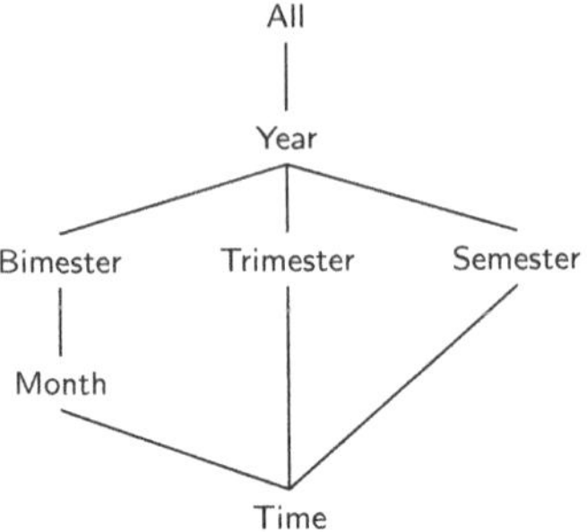

Fig. 3. Time Dimension

between granularities. However, not all the information of a dimension schema can be represented in this dimension. for example, the fact that granule bi-3-12 maps to spring-summer12 cannot be represented since, as the domains schema shows, elements in category bimester can only be connected to elements in Year.

Sometimes, specially at query time, we will want to find a suitable granularity when two attributes share the same domain schema but they are at different granularities. In order to find this granularity, we define the join operator.

Definition 1. Given a domain schema $\Psi = (\mathcal{U}, \ell, \mathcal{I}, \rho, \tau)$ and $G_1, G_2 \in \ell$, the *Join* of G_1 and G_2 in Ψ is $Join(\Psi, G_1, G_2) = G$ such that $G \in \ell$, $G_1 \preceq G, G_2 \preceq G$ and there does not exist $G' \in \ell$ such that $(G_1 \preceq G', G_2 \preceq G', G' \prec G)$. □

Thus, the join operator of G_1 and G_2 returns the finest granularity in ℓ such that it subsumes both G_1 and G_2. Note that the *Join* always exists since we always have that granularity $\top \in \ell$ that satisfies the conditions that $G_1 \preceq \top$ and $G_2 \preceq \top$.

Example 3. For the domain schema of Example 2, we have: $Join(\Psi_t, \mathsf{month}, \mathsf{semester}) = \mathsf{year}$, $Join(\Psi_t, \mathsf{month}, \mathsf{bimester}) = \mathsf{year}$ and $Join(\Psi_t, \mathsf{trimester}, \mathsf{semester}) = \mathsf{year}$. □

In the next section we formalize databases that can assign to an attribute a domain schema instead of a simple domain.

3 Multi-granular Databases

A database schema is a tuple $\Sigma = (\mathcal{M}, \mathcal{R}, Dom, Gran)$, where: (a) $\mathcal{M}$ is a set of domain schemas, (b) $\mathcal{R}$ is a set of relational schemas and (c) Given a relation $R \in \mathcal{R}$ and an attribute $A \in R$, function $Dom(R, A)$ returns a domain schema $\Psi_{RA} = (\mathcal{U}_{RA}, \ell_{RA}, \mathcal{I}_{RA}, \mu_{RA}, \tau_{RA})$ and $Gran(R, A)$ returns a granularity in ℓ_{RA}. Intuitively, *Dom* and *Gran* return, respectively, the domain schema and the granularity associated with attribute $A \in R$. Without loss of generality, we will assume that there are no schemas $\Psi_1, \Psi_2 \in \mathcal{M}$ that refer to the same domain $\mathcal{U}$.

More formally, a database instance D of a schema Σ is a finite collection of ground atoms of the form $R(c_1, \ldots, c_i, \ldots, c_l)$, where (a) $R(B_1, \ldots, B_i, \ldots, B_l) \in \mathcal{R}$, and (b) every c_i is such that for $Dom(R, B_i) = (\mathcal{U}_{RB_i}, \ell_{RB_i}, \mathcal{I}_{RB_i}, \mu_{RB_i}, \tau_{RB_i})$, $Gran(R, B_i) = G_{RB_i}$, $c_i \in \mathcal{I}_{RB_i} \cup \mathcal{U}_{RB_i}$ and one of the following holds:

- $G_{RB_i} = \top$, or
- $G_{RB_i} = \bot$ and $c_i \in \mathcal{U}_{RB_i}$
- $G_{RB_i} \in \ell$ and there exists $c_j \in \tau_{RB_i}(G_{RB_i})$ for which c_i maps to c_j.

Thus, a database instance contains in each attribute granules that can map to a granule in the granularity of the attribute. In this way, data is stored using as much detail as it is available.

In order to enforce consistency of data in a multi-granular database, traditional functional dependencies and inclusion dependencies are not enough to handle the types of inconsistencies that can arise. We now introduce extensions of functional dependencies to consider data at different granularities and also to trigger constraints conditionally to particular values of attributes.

A classic functional dependency $\mathsf{X}_1, \ldots, \mathsf{X}_\mathsf{k} \rightarrow \mathsf{X}_{\mathsf{k}+1}, \ldots, \mathsf{X}_\mathsf{n}$ is satisfied by a relation R that might contain *null* values if for every tuples t_1 and t_2 in D such that $t_1[X_i] = t_2[X_i] \neq null$ for $i \in [1, k]$, then it should hold that $t_1[X_j] = t_2[X_j]$[1] for every $j \in [k+1, n]$.

A **granular functional dependency (GFD)** extends classic functional dependency with the concept of multi-granular attributes. It enforces that tuples with the same values in a subset of attributes, at specific granularities, should have the same values in another set of attributes, at given granularities. This is formally introduced as followed.

GFD syntax. Given a database schema $\Sigma = (\mathcal{M}, \mathcal{R}, Dom, Gran)$, a Granular Functional Dependency (GFD) over a relation $R \in \mathcal{R}$ is of the form:

$$(\mathsf{X}_1, \ldots, \mathsf{X}_\mathsf{k} \rightarrow \mathsf{X}_{\mathsf{k}+1}, \ldots, \mathsf{X}_\mathsf{n}; @t_g)$$

where t_g is a pattern tuples of the form $t_g = (G_1, \ldots, G_k; G_{k+1}, \ldots, G_n)$, for every $i \in [1, n]$, X_i is an attribute in R, and G_i is either '_' or a granularity such that $Dom(R, X_i) = (\mathcal{U}_{RX_i}, \ell_{RX_i}, \mathcal{I}_{RX_i}, \mu_{RX_i}, \tau_{RX_i})$ and $G_i \in \ell_{RX_i}$.

Intuitively, a GFD will check if the functional dependency $\mathsf{X}_1, \ldots, \mathsf{X}_\mathsf{k} \rightarrow \mathsf{X}_{\mathsf{k}+1}, \ldots, \mathsf{X}_\mathsf{n}$ is satisfied at the granularity levels defined by t_g. When the pattern contains '_', the values are checked at the granularity associated with the attribute in the database schema. In order to formally provide the semantics for these constraints, we need to define the mapping of a granule identifier in a granularity.

Definition 2. Given a domain schema $\Psi = (\mathcal{U}, \ell, \mathcal{I}, \mu, \tau)$, a granule identifier $c \in \mathcal{I}$ and a granularity $G \in \ell$, the *mapping* of c in G, denoted by $\text{MAP}(c, G)$, is defined as follows:

$$\text{MAP}(c, G) ::= \begin{cases} c & \text{if } G = \bot \text{ and } c \in \mathcal{U} \\ d & \text{if } G \in \ell,\ d \in \tau(G) \text{ and } \mu(c) \text{ maps to } \mu(d) \\ null & \text{otherwise} \end{cases}$$ □

[1] The condition is satisfied even if $t_1[X_j] = t_2[X_j] = null$.

Because of the definition of granularity, the second condition can be satisfied by at most one $d \in \tau(G)$. Checking this condition requires to determine, for every $d \in \tau(G)$, if $\mu(c) \subseteq \mu(d)$. Since we have restricted to domain schemas in which checking $\mu(i) \subseteq \mu(j)$ takes constant time, the cost of this computation is $O(|\tau(G)|)$. Therefore, the following lemma holds:

Lemma 1. Given a domain schema $\Psi = (\mathcal{U}, \ell, \mathcal{I}, \rho, \tau)$, a granule $c \in \mathcal{I}$ and a granularity $G \in \ell$, the mapping $\text{MAP}(c, G)$ can be computed in linear time over $|\mathcal{I}|$. □

GFD semantics. Consider a relation R and a GFD $(\mathsf{X}_1, \ldots, \mathsf{X}_\mathsf{k} \rightarrow \mathsf{X}_{\mathsf{k}+1}, \ldots, \mathsf{X}_\mathsf{n}; @t_g)$ with $t_g = (G_1, \ldots, G_k; G_{k+1}, \ldots, G_n)$. Let $t'_g = (G'_1, \ldots, G'_k; G'_{k+1}, \ldots, G'_n)$, where $G'_i = Gran(R, X_i)$ if $G_i =$'_' and $G'_i = G_i$ otherwise. Relation R satisfies the GFD if for t'_g, every t_1, t_2 in R and every $i \in [1, k]$ such that $\text{MAP}(t_1[X_i], G'_i) = \text{MAP}(t_2[X_i], G'_i) \neq null$, it holds that $\text{MAP}(t_1[X_j], G'_j) = \text{MAP}(t_2[X_j], G'_j)$ for every $j \in [k+1, n]$.

In the same way as for classic functional dependencies, checking consistency of a database with respect to GFDs is tractable.

Proposition 1. The problem of checking if a database D is consistent with respect to GFD *ic* can be solved in polynomial time.

Proof. Let $ic = (\mathsf{X}_1, \ldots, \mathsf{X}_\mathsf{k} \rightarrow \mathsf{X}_{\mathsf{k}+1}, \ldots, \mathsf{X}_\mathsf{n}; @t_g)$. The problem can be reduced to checking satisfaction of classic functional dependencies. Indeed, we can construct a new database D' with each attribute in *ic* at the granularity defined by t_g. Database D' can be constructed efficiently since $\text{MAP}(c, G)$ is computed in polynomial time (see Lemma 1). Now, if D' satisfies the functional dependency $\mathsf{X}_1, \ldots, \mathsf{X}_\mathsf{k} \rightarrow \mathsf{X}_{\mathsf{k}+1}, \ldots, \mathsf{X}_\mathsf{n}$, then D satisfies *ic*. Note that D' can have an attribute repeated at different granularities of the corresponding domain schema. □

Given a set of constraints, the *satisfiability problem* consists in determining if there exists a non-empty instance that satisfies the constraints. As in the case for classic functional dependencies, for any set IC of GFDs defined over a schema R, there exists a non-empty instance D of R such that D satisfies the constraints in IC. Indeed, a relation with a single tuple with any value will always satisfy a set of GFDs.

A **conditional-granular functional dependency** (CGFD) is an extension to GFDs that allows us to filter the tuples over which the dependencies are checked in the same way as conditional functional dependencies [11] extend classic functional dependencies to add conditions.

CGFD syntax. Given a schema $\Sigma = (\mathcal{M}, \mathcal{R}, Dom, Gran)$ and a relation $R \in \mathcal{R}$, a Granular Conditional Functional Dependency (CGFD) over R is of the form:

$$(\mathsf{X}_1, \ldots, \mathsf{X}_\mathsf{k} \rightarrow \mathsf{X}_{\mathsf{k}+1}, \ldots, \mathsf{X}_\mathsf{n}; @t_g; t_c)$$

where, for every $i \in [1, n]$, X_i is an attribute in R and t_g, t_c are pattern tuples of the form: $t_g = (G_1, \ldots, G_k; G_{k+1}, \ldots, G_n)$ and $t_c = (c_1, \ldots, c_k; c_{k+1}, \ldots, c_n)$

where each G_i for $i \in [1,n]$ is either '_' or a granularity in the domain schema $Dom(R, X_i)$ and each $c_i \in \tau(G_i)$ or c_i='_'[2]. A value '_' in t_c represents the fact that there is no constraint about the value in that attribute. In this sense, we will say that $c_i \asymp c_j$ if either $c_i = c_j$, c_i ='_' or c_j ='_'.

Remark 2. Here we chose to force $c_i \in \tau(G_i)$. If we want to consider more general ICs, we can remove this condition and only require $c_i \in \mathcal{I}$ for $Dom(R, X_i) = (\mathcal{U}, \ell, \mathcal{I}, \mu, \tau)$. In this case, the constraint is triggered only if the value in the instance is equal to c_i. This extension would only be interesting if c_i belongs to more than one granularity.

CGFD semantics. Consider a relation R and a CGFD $(\mathsf{X}_1, \ldots, \mathsf{X}_\mathsf{k} \rightarrow \mathsf{X}_{\mathsf{k}+1}, \ldots, \mathsf{X}_\mathsf{n}; @t_g; t_c)$ such that $t_g = (G_1, \ldots, G_k; G_{k+1}, \ldots, G_n)$ and $t_c = (c_1, \ldots, c_k; c_{k+1}, \ldots, c_n)$. Let $t'_g = (G'_1, \ldots, G'_k; G'_{k+1}, \ldots, G'_n)$, where $G'_i = Gran(R, X_i)$ if G_i ='_' and $G'_i = G_i$ otherwise. Relation R satisfies the CGFD if for t'_g, every t_1, t_2 in R and $i \in [1,k]$ such that $\text{MAP}(t_1[X_i], G'_i) = \text{MAP}(t_2[X_i], G'_i) \asymp c_i$, then $\text{MAP}(t_1[X_j], G'_j) = \text{MAP}(t_2[X_j], G'_j) \asymp c_j$ for every $j \in [k+1, n]$.

By slightly modifying the proof of Proposition 1 by checking also the constants in t_c, we get that the consistency of a database with respect to CGFDs is tractable.

Proposition 2. The problem of checking if a database D is consistent with respect to CGFD *ic* can be solved in polynomial time. □

Example 4. Consider a database that tracks the number of attendees to events held at a specific location and date. Let its database schema be $\Sigma = (\mathcal{M}, \mathcal{R}, Dom, Gran)$ with (i) $\mathcal{M} = \{\Psi_\mathsf{L}, \Psi_\mathsf{D}, \Psi_\mathsf{N}\}$, where Ψ_L is the schema domain for location and contains City and Country granularities defined as expected; Ψ_D contains Date, Month and Year granularities defined as expected; and Ψ_N has as domain the natural numbers and contains no granularities except for $\bot^{\Psi_\mathsf{N}}$; (ii) $\mathcal{R}$ = {Event(Location, Date, Number)}; (iii) Dom(Event, Location) = Ψ_L, Dom(Event, Date) = Ψ_D and Dom(Event, Number) = Ψ_N; and (iv) $Gran$(Event, Location) = $City$, $Gran$(Event, Date) = $Date$ and Dom(Event, Number) = $\bot^{\Psi_\mathsf{N}}$.

The events stored in this database take place at most once a month in UK and once a year in Canada. To enforce these constraints, we need both to restrict to specific locations and to check at different levels of granularity. The instance of table Event should satisfy ic_1 : (Location, Date → Location, Number; $@t_{g_1}; t_{c_1}$) and ic_2 : (Location, Date → Location, Number; $@t_{g_2}; t_{c_2}$) where:

	Location	Date	Location	Number
t_{g_1}	Country	Year	_	_
t_{c_1}	Canada	_	_	_

	Location	Date	Location	Number
t_{g_2}	Country	Month	_	_
t_{c_2}	UK	_	_	_

Note that we need the location attribute in both sides so that the constraints are triggered considering the relevant granularity of country but they also need to enforce that the event takes place in the same city. The following instance violates both constraints:

[2] Note that this implies that there can be no *null* values in the patterns.

	Location	Date	Number
t_1	London	June 1st, 2011	12
t_2	Cambridge	March 3rd, 2012	80
t_3	Oxford	March 23rd, 2012	80
t_4	Ottawa	July 7st, 2012	250
t_5	Ottawa	March 6st, 2012	254

Tuples t_4 and t_5 violate ic_1 since they are Canadian cities for which we have events with different number of attendees. On the other hand, tuples t_2 and t_3 do not satisfy ic_2 since, for two *different* cities in the UK, there is an event in the same month. These constraints cannot be expressed using classic functional dependencies nor GFD. □

The satisfiability problem for CGFDs is NP-complete. It is NP-hard since conditional FDs [11] are a particular case of CGFDs. The problem is in NP since we can guess a database and check in polynomial time if it satisfies the constraints. It is easy to see that if a set of CGFDs is satisfiable, there always exist a database with one tuple that satisfies them.

4 Multi-granular Query Language

This section introduces a multi-granular language MSQL to express not only general queries but also queries that check integrity constraints over a database instance D of a schema $\Sigma = (\mathcal{M}, \mathcal{R}, Dom, Gran)$. Using a SQL-like syntax, a query in this language is of the form:

```
SELECT [ ALL | DISTINCT ] MSQL_expression
  [ FROM table_references ]
  [ WHERE MSQL_condition ]
  [ GROUP BY grouping_column_reference_list ]
  [ HAVING MSQL_condition ]
```

MSQL modifies the classical SQL syntax by including additional options when defining the select expression and search condition. In particular, MSQL includes operators over granules and granularities that are not found in SQL. The specification of the select expression and search condition in MSQL follows, where for simplification, we assume that elements SQL_predicate and SQL_expression include all elements of the classical SQL predicate and SQL select expression.

```
MSQL_expression ::= < SQL_expression > |
                    < granule > [ AS < column_name >][, MSQL_expression]
MSQL_condition  ::= < bool_condition > | < bool_condition > AND < bool_condition > |
                    < MSQL_condition > OR < MSQL_search_condition >
bool_condition  ::= [NOT]{< MSQL_predicate > | (< MSQL_condition >)}
                    [ IS [ NOT ]{ TRUE | FALSE | UNKNOWN }]
MSQL_predicate  ::= < SQL_predicate > | < granule_relation > (< granule >, < granule >)
granule         ::= < granule_column_name > |
                    MAP(< granule_column_name >,< granularity >)
granularity     ::= < granularity_id > | JOIN(< granularity_id >, < granularity_id >)
```

Let schema $\Psi = (\mathcal{U}, \ell, \mathcal{I}, \mu, \tau)$ be a domain schema, $x, y \in \mathcal{I}$ be granule identifiers, and $G_1, G_2 \in \ell$ be granularities. The function $\text{JOIN}(G_1, G_2)$ returns the finest granularity that can be defined in the domain that is coarser than both G_1 and G_2 (cf. Definition 1) and the function $\text{MAP}(c, G_1)$ returns the value of c at the granularity G_1 (cf. Definition 2).

In the previous specification, we have left undefined granule_relation because it may depend on the particular domain. We consider these relations to be basically set-based operators found in the context of mereology [23], which is the theory of parts and wholes. We will use the following study case to specify particular relations for the spatial and temporal domains and illustrate with them the use of the query language.

4.1 Case Study: Spatial and Temporal Domain

Mereo-topology has been an important area of research for the spatial and temporal domains. Mereo-topology is a theory, combining mereological and topological concepts, of the relations among wholes, parts, and boundaries between parts [23]. Based on this theory, Table 1 provides the semantics of the topological relations between geometries (spatial granules), which were extracted from the Open Geospatial Consortium Simple Feature Specification [17] and are currently implemented in spatial query languages. In this table, given a geometry x, $\partial(x)$ indicates its boundary, and $dim(x)$ its dimension, where $dim(x)$ is equal to 0 if x is a point, 1 if it is a curve, and 2 if it is a surface.

Table 1. Definition of topological relations by the Open Geospatial Consortium [17]

Relation	Definition
Disjoint(x, y)	True if $x \cap y = \emptyset$
Touches(x, y)	True if $x \cap y \subseteq (\partial(x) \cup \partial(y))$
Equals(x, y)	True if $x = y$
Within(x, y)|Contains(y, x)	True if $x \subseteq y$
Overlaps(x, y)	True if $x \cap y \neq \emptyset$, $x \cap y \neq x \neq y$, and $dim(x \cap y) = dim(x) = dim(y)$
Crosses(x, y)	True if $x \cap y \neq \emptyset$, $x \cap y \neq x \neq y$, and $dim(x \cap y) < max(dim(x), dim(y))$
Intersects(x, y)	True if $x \cap y \neq \emptyset$

Table 2 provides the semantics of the relations between two time intervals $x = [x_s, x_f]$ and $y = [y_s, y_f]$ (temporal granules), which are extracted from the relations in [1] and can be mapped to temporal extensions to SQL languages such as TSQL2 [22] and TQuel [21].

Granules in MSQL can be represented by geometric attributes and time intervals of spatial and temporal SQLs, respectively. In such cases, the comparison operators of MSQL over granules as binary relations are equivalent to those found in spatial and temporal languages. Although the specification of MSQL could also include the particularities for handling spatial and temporal domain, we have omitted these components of the language to keep it simpler. In addition to relations between granules, MSQL includes MAP and JOIN operators to handle granules and granularities of the schema, which cannot be mapped onto spatial and query languages.

Table 2. Definition of time interval relations [1]

Relation	Definition
Before(x,y)|After(y,x)	$x.f < y.s$
Equals(x,y)	$x.s = y.s$ and $x.f = y.f$
Meets(x,y)	$x.f = y.s$ or $y.f = x.s$
Overlaps(x,y)	$x.f > y.s$ and $y.s > x.f$ or $y.f > x.s$ and $x.s > y.f$
During(x,y)|Contains(y,x)	$x.s > y.s$ and $x.f < y.f$
Starts(x,y)|StartedBy(y,x)	$x.s = y.s$ and $x.f > y.f$
Finishes(x,y)|FinishedBy(y,x)	$x.f = y.f$ and $x.s < y.s$

Database Schema and Instance. To show the expressiveness of the language, let us consider a database of disaster events occurred around the world with schema $\Sigma = (\mathcal{M}, \mathcal{R}, Dom, Gran)$ and two relational predicates, one for natural and other for man-made disasters. This database handles multi-granular data since events may occur during a day, week, season, and so on. In addition, they occur in a place, city, region, and so on. Figure 1 shows a database instance for relations Natural(Where, When, Type, Death) $\in \mathcal{R}$ and ManMade(Where, When, Type, Death) $\in \mathcal{R}$.

In $\mathcal{M}$ there are four domain schemas of the form $\Psi_D = (\mathcal{U}_D, \ell_D, \mathcal{I}_D, \mu_D, \tau_D)$, where D is the domain of space, time, type of disaster, or natural numbers. The domain schemas for space and time are shown graphically in Figures 4 and 5, where granule identifiers are grouped into corresponding granularities of the domain and where directed lines indicate inclusion relationship between granule indices ($\subseteq$). The schema of type of disaster is essentially a taxonomy of natural and man-made disasters. In this database schema, $Gran$(Natural, Where) $= Country$, $Gran$(Natural, When) $= Year$, $Gran$(Natural, Type) $= Natural$, $Gran$(Natural, Death) $=\bot^{\mathbb{N}}$, $Gran$(ManMade, Where) $= City$, $Gran$(ManMade, When) $= Year$, $Gran$(ManMade, Type) $= ManMade$, and $Gran$(ManMade, Death) $=\bot^{\mathbb{N}}$. Although attribute Where is at granularity $Country$ in relation Natural, its values are at the level of country, zone of a country, and city. In relation ManMade, in contrast, Where is at granularity of city. Similarly, for When of relation Natural there are values in months, years, seasons[3], holidays, and days. For all these attributes, their values correspond to granule identifiers, whose mapping to the underlying domain is part of the mapping of granule indices at the attribute's granularity.

[3] Data about seasons use the north hemisphere as reference.

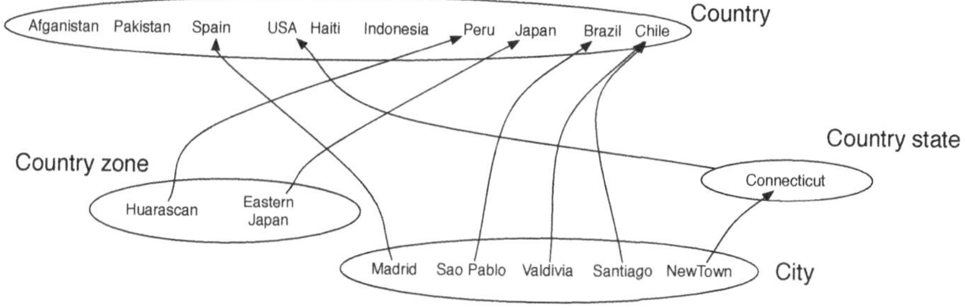

Fig. 4. Instance of the space domain schema

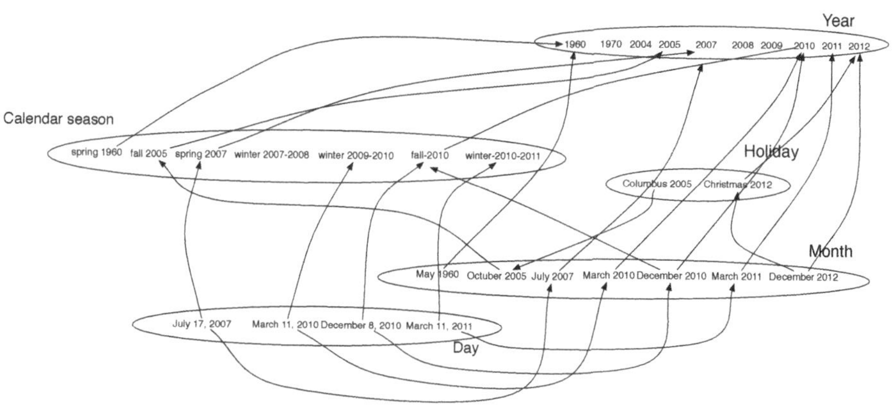

Fig. 5. Instance of the time domain schema

Even though the granularity of Where is country, it is very useful to store the finer data when available. Indeed, in the tuple ⟨Valdivia,May 1960,earthquake, 1655⟩, knowing that the earthquake was in the city of Valdivia provides more information than saying that it took place in Chile.

Queries. Some examples of queries that are of interest for this data are:

1. Find natural events occurred during year 2010:

```
SELECT Where, When, Type, Death
FROM   Natural
WHERE  During(When, 2010)
```

Answer to this query is the tuple ⟨Chile, February 27 2010, earthquake, 525⟩. Notice that the earthquake occurred during the winter 2009-2010 in Haiti is not part of the answer, since the winter 2009-2010 in not part of the year

2010. By using Overlaps instead of During in the query, the answer would also include tuple ⟨Haiti, winter 2009-2010, earthquake, 360,000⟩.

2. Find natural events of the same type that occurred during the same year in different countries:

SELECT	$d1$.Type, MAP($d1$.When, $Year$), MAP($d1$.Where, $Country$), MAP($d2$.Where, $Country$)
FROM	Natural as $d1$, Natural as $d2$
WHERE	Equals(MAP($d1$.When, $Year$), MAP($d2$.When, $Year$))AND NOT Equals(MAP ($d1$.Where, $Country$),MAP($d2$.Where, $Country$))

Answer to this query is the tuple ⟨Earthquake, 2010, Haiti, Chile⟩.

3. Find two natural events occurred in 'Haiti' in different years:

SELECT	$d1$.Where, $d1$.When, $d2$.When
FROM	Natural as $d1$, Natural as $d2$
WHERE	Within(MAP($d1$.Where, $Country$), 'Haiti') AND Within(MAP($d2$.Where, $Country$), 'Haiti') AND NOT Equals(MAP($d1$.When, $Year$),MAP($d2$.When, $Year$))

The answer to this query is empty.

4. Find man-made disasters that occur during the same year and location of an earthquake (i.e., natural disaster). Since this query retrieve data from attributes whose schema is at different granularities, we map values to the merged granularity of the attributes for comparison.

SELECT	$d2$.Where, $d2$.When, $d2$.Type,$d2$.Death
FROM	Natural as $d1$, ManMade as $d2$
WHERE	$d1$.Type = 'earthquake' AND Equals(MAP($d1$.Where,JOIN($Country, City$)), MAP($d2$.Where,JOIN($Country, City$))) AND Equals(MAP($d1$.When, $Year$),MAP($d2$.When, $Year$))

Answer to this query is the tuple ⟨Santiago, December 8 2010, fire, 81⟩.

Note that in the previous queries, and since we use names to specify where an event occurs, we can think of using string operator = instead of using Equals when comparing spatial granules that have been mapped to the same granularity. We use Equals to make clear that we are comparing granules with the semantics of a spatial domain.

5 Implementation Issues

At the implementation level of the multi-granular database model, we need to represent granules, implement operators for relations between granules, associate granules with granularities, and implement MAP and JOIN operators.

Representing granules and implementing operators for relations depend on the domain. We distinguish a representation of granules called *by-value* from a representation called *by-reference*. If the implementation is done using a representation by-value, the MAP relationships could be computed at query time by using the underlying domain. For the spatial domain, in particular, current spatial databases provide spatial data types to represent geometries in terms of boundary points of a geometry. Spatial relations are also implemented by using polynomial algorithms in terms of the number of points used to represent these geometries. These databases support spatial indexing structures for query processing, which can be used for the implementation of operators.

For a representation by-reference in the spatial domain, on the other hand, we need to have spatial relations between place names. This can be part of the data store in the database or can be handled in an external knowledge base such as the gazetteer Geonames[4], which establishes relations of inclusion. In this domain, associating granules to granularities is very common to be a type of classification. For example, given a place names, this is typically associated with a class county, city, country, and so on, which is precisely the granularity. This needs to be explicitly indicated in the database or in the external knowledge base.

A similar situation occurs for the temporal domain. A representation by-value is an interval representation, from which it is possible to calculate the level of granularity and interval relations. A representation by-reference requires additional knowledge to relate granules and to derive the level of granularity.

The implementation of the MAP and JOIN operators needs the granularity to which a granule belongs and the inclusion relations between granules. This operators require the comparison of all possible candidate granules and granularities; however, this can be efficiently calculated for some particular domains. Having the granularity given by the administration partition of regions and its respective representation by-value, spatial indices can be used to filter out candidate granules. Furthermore, this computation can be done offline and not at query time. For a representation by-reference, using a hierarchical structure such as the one used by Geonames through the inclusion relation, the implementation of MAP and JOIN operators traverses the hierarchy up to the desired level.

6 Related Work

Related work concerning multi-granular databases exists within the context of spatial and temporal databases and data warehousing. We revise here with special emphasis on the work done in the spatial and temporal domain, from where we generalize the notion of granularity.

The formalization of temporal granularity by Bettini *et. al.* [5,4] is basis for several different studies that explore temporal and spatial granularity. Granularity defines the units that quantitatively measure data with respect to the dimensions of the domain they represent. Temporal granularity is defined in terms of a mapping function from a domain of index to the time domain. Each portion of the time domain corresponding to the mapping of a granularity is

[4] `http://www.geonames.org/`

referred as the temporal granule. Temporal granule of a granularity must not overlap. A time domain is a tuple $(T, \leq)$, where T is a non-empty set of time instants and $\leq$ is a total order in T. In this context, given two time granularities Φ_1 and Φ_2, Φ_1 is said to be finer-than Φ_2 (or inversely, Φ_2 is coarser-than Φ_1), denoted by $\Phi_1 \preceq \Phi_2$, if and only if, for each granule $\Phi_1(i)$, there exists a granule $\Phi_2(j)$ such that $\Phi_1(i) \subseteq \Phi_2(j)$.

In a similar way to the definition of temporal granularity, a spatial granularity Φ can be defined by a mapping function from a domain of index to portion of the space, called spatial granules, and where the granules of a same granularity do not overlap [24]. Given two different spatial granularities Φ_1 and Φ_2 , Φ_1 is said to be finer-than Φ_2 is for all granules $\Phi_1(i)$, there exists a granule $\Phi_2(j)$ such that $\Phi_{\mathcal{I}}(i)$ is inside $\Phi_2(j)$. The work in [3] enriches the conceptualization of spatial granularity by considering that spatial granularities represent also the relations between granules. This is done with a multidigraph where vertices are granules and edges are explicit relations between granules. It also defines several relations between spatial granularities in addition to the finer-than relationship and a set of operations over spatial granularities.

The works in [24,8,3] define spatio-temporal granularity as the composition of spatial and temporal granularity. A spatio-temporal granule is therefore a tuple $STG(s,t)$, where s is a spatial index and t a temporal index of granules in the spatial SG and temporal TG granularity, respectively. In this framework, at the time instant TG(t), the spatial granule SG(s) is valid. In [24], an object fully covers, partially overlaps or not overlaps a spatial granule, which may vary at different time instants. Unlike the works in [24,8] where at each time granule there is a single spatial granule, the work in [3] assigns to each spatio-time granule a sequence of spatial granules, one for each granule in the time granularity.

ST4SQL is a spatio-temporal query language dealing with granularities [18]. This language extends SQL syntax and T4SQL [9] temporal language with different temporal and spatial semantics, and in particular, it introduces constructs to group data with respect to the temporal and spatial domain. The work in [8] proposes a multi-granular object-oriented framework that supports spatial and temporal granularity conversion. They define a system where multi-granular spatial and temporal data are defined as instances of spatial and temporal types, upon which different granularity conversion operators apply. They address conversion between granules related by inclusion and they provide a language were users can specify a particular conversion from moving one to another granularity.

Although there exists work on the formalization of spatial and spatio-temporal integrity constraint [7,16] and about modeling and querying spatial and temporal data at multiple granularities [24,8,3], to the best of our knowledge, there is no formalization of semantic integrity constraints for spatial and temporal data at multiple granularities. We investigate this new type of constraints that impose topological [10,19] and temporal interval [1] relations of multi-granular spatial and temporal data.

From a different perspective, data warehousing can be seen as a multi-granular system where conversion operators are fixed along a dimension. Although data warehousing provides efficient implementations of conversion operations along a hierarchy of granularities, it cannot fully capture the desired semantics of a multi-granular database. Homogenous data warehousing assumes data stored at the finest level of detail and where each value (granule) at this level can be mapped onto a value at a coarser level of a dimension. For heterogenous data warehousing, a particular granule of a domain cannot belong to two different granularities. Despite this disadvantages, recent work on data warehousing highlights the need of storing data at different granularities [12,15] and handling complex data objects [6].

7 Conclusions

The work in this paper proposes a database model with attributes stored at different granularities. It defines a domain schema to model the granularity of a particular domain, which is a more general approach than modeling categories in a data warehousing where data is usually stored at the finest level of detail and categories are related by an inclusion relation. It also introduces integrity constraints, analyzes the database consistency problem for these constraints, and presents a multi-granular query language.

We have left for future work the study of efficient implementations for this database model and query language, which should consider particularities of the domain.

Acknowledgements. M. Andrea Rodríguez is partially funded by FONDEF-CONICYT project D09I1185. Loreto Bravo is partially founded by FONDECYT-CONICYT N1130902.

References

1. Allen, J.: Maintaining knowledge about temporal intervals. Communications of the ACM 26(11), 832–843 (1983)
2. Belussi, A., Combi, C., Pozzani, G., Amaddeo, F., Rambaldelli, G., Salazzari, D.: Dealing with multigranular spatio-temporal databases to manage psychiatric epidemiology data. In: Computer-Based Medical Systems (CBMS), pp. 1–4 (2012)
3. Belussi, A., Combi, C., Pozzani, G.: Formal and conceptual modeling of spatio-temporal granularities. In: International Database Engineering and Applications Symposium IDEAS, pp. 275–283. ACM (2009)
4. Bettini, C., Dyreson, C.E., Evans, W.S., Snodgrass, R.T., Wang, X.S.: A glossary of time granularity concepts. In: Etzion, O., Jajodia, S., Sripada, S. (eds.) Dagstuhl Seminar 1997. LNCS, vol. 1399, pp. 406–413. Springer, Heidelberg (1998)
5. Bettini, C., Wang, X.S., Jajodia, S.: A general framework for time granularity and its application to temporal reasoning. Annals of Mathematics and Artificial Intelligence 22(1-2), 29–58 (1998)
6. Boukraâ, D., Boussaïd, O., Bentayeb, F.: Olap operators for complex object data cubes. In: Catania, B., Ivanović, M., Thalheim, B. (eds.) ADBIS 2010. LNCS, vol. 6295, pp. 103–116. Springer, Heidelberg (2010)

7. Bravo, L., Rodríguez, M.A.: Formalization and reasoning about spatial semantic integrity constraints. Data Knowl. Eng. 72, 63–82 (2012)
8. Camossi, E., Bertolotto, M., Bertino, E.: A multigranular object-oriented framework supporting spatio-temporal granularity conversions. International Journal of Geographical Information Science 20(5), 511–534 (2006)
9. Combi, C., Montanari, A., Pozzi, G.: The T4SQL temporal query language. In: ACM Conference on Information and Knowledge Management, pp. 193–202 (2007)
10. Egenhofer, M., Franzosa, R.: Point set topological relations. International Journal of Geographical Information Science 5, 161–174 (1991)
11. Fan, W., Geerts, F., Jia, X., Kementsietsidis, A.: Conditional functional dependencies for capturing data inconsistencies. ACM Trans. Database Syst. 33(2) (2008)
12. Iftikhar, N.: MMDW: A multi-dimensional and multi-granular schema for data warehousing. In: 16th Annual KES Conference. Frontiers in Artificial Intelligence and Applications, vol. 243, pp. 1211–1220. IOS Press (2012)
13. Iftikhar, N.: Ratio-based gradual aggregation of data. In: Benlamri, R. (ed.) NDT 2012, Part I. CCIS, vol. 293, pp. 316–329. Springer, Heidelberg (2012)
14. Iftikhar, N., Pedersen, T.B.: Gradual data aggregation in multi-granular fact tables on resource-constrained systems. In: Setchi, R., Jordanov, I., Howlett, R.J., Jain, L.C. (eds.) KES 2010, Part III. LNCS, vol. 6278, pp. 349–358. Springer, Heidelberg (2010)
15. Iftikhar, N., Pedersen, T.B.: Schema design alternatives for multi-granular data warehousing. In: Bringas, P.G., Hameurlain, A., Quirchmayr, G. (eds.) DEXA 2010, Part II. LNCS, vol. 6262, pp. 111–125. Springer, Heidelberg (2010)
16. del Mondo, G., Rodríguez, M.A., Claramunt, C., Bravo, L., Thibaud, R.: Modeling consistency of spatio-temporal graphs. Data Knowl. Eng. 84, 59–80 (2013)
17. OGC: OpenGIS Implementation Standard for Geographic information - Simple feature access - Part 1. Tech. rep., Open Geospatial Consortium, Inc. (2011)
18. Pozzani, G., Combi, C.: ST4SQL: A spatio-temporal query language dealing with granularities. In: Proceedings of the ACM Symposium on Applied Computing, pp. 23–25. ACM (2012)
19. Randell, D., Cui, Z., Cohn, A.: A Spatial Logic based on Regions and Connection. In: Proceedings of the 3rd International Conference on Knowledge Representation and Reasoning, pp. 165–176. Morgan Kaufmann (1992)
20. Rodríguez, M.A., Bravo, L.: Multi-granular schemas for data integration. In: 6th Alberto Mendelzon International Workshop on Foundations of Data Management. CEUR Workshop Proceedings, vol. 866, pp. 142–153. CEUR-WS.org (2012)
21. Snodgrass, R.: The temporal query language tquel. ACM Trans. Database Syst. 12(2), 247–298 (1987)
22. Snodgrass, R.T. (ed.): The TSQL2 Temporal Query Language. Kluwer (1995)
23. Varzi, A.C.: Spatial Reasoning and Ontology: Parts, Wholes, and Locations, pp. 945–1038. Springer (2007)
24. Wang, S.-S., Liu, D.-Y.: Spatio-temporal database with multi-granularities. In: Li, Q., Wang, G., Feng, L. (eds.) WAIM 2004. LNCS, vol. 3129, pp. 137–146. Springer, Heidelberg (2004)

Optimizing Computation of Repairs from Active Integrity Constraints

Luís Cruz-Filipe*

Dept. of Mathematics and Computer Science, University of Southern Denmark, Denmark
LabMag, Lisboa, Portugal

Abstract. Active integrity constraints (AICs) are a form of integrity constraints for databases that not only identify inconsistencies, but also suggest how these can be overcome. The semantics for AICs defines different types of repairs, but deciding whether an inconsistent database can be repaired is a NP- or Σ_p^2-complete problem, depending on the type of repairs one has in mind. In this paper, we introduce two different relations on AICs: an equivalence relation of *independence*, allowing the search to be parallelized among the equivalence classes, and a *precedence* relation, inducing a stratification that allows repairs to be built progressively. Although these relations have no impact on the worst-case scenario, they can make significant difference in the practical computation of repairs for inconsistent databases.

1 Introduction

Maintaining and guaranteeing database consistency is one of the major problems in knowledge management. Database dependencies have been since long a main tool in the fields of relational and deductive databases [2,3], used to express integrity constraints on databases. They formalize relationships between data in the database that need to be satisfied so that the database conforms to its intended meaning.

Whenever an integrity constraint is violated, the database must be repaired in order to regain consistency. Typically there are several sets of update actions that achieve this goal, leading to different revised consistent databases. Restricting the set of database repairs to those considered most adequate is therefore an important task. Minimality of change is commonly accepted as an essential characteristic of a repair [6,8,18], but it is not enough to narrow down the set of possible repairs sufficiently.

The most common approach to processing integrity constraints in database management systems is to use active rules (a kind of event-condition-action rules, or ECAs [17]), for which rule processing algorithms have been proposed and a procedural semantics has been defined. However, their lack of declarative

* Supported by the Danish Council for Independent Research, Natural Sciences and by Fundação para a Ciência e Tecnologia under contract PEst-OE/EEI/UI0434/2011.

C. Beierle and C. Meghini (Eds.): FoIKS 2014, LNCS 8367, pp. 361–380, 2014.

semantics makes it difficult to understand the behaviour of multiple ECAs acting together and to evaluate rule-processing algorithms in a principled way.

Active integrity constraints (AICs) [9] are special forms of production rules that encode both an integrity constraint and preferred update actions to be performed whenever the former is violated. The declarative semantics for AICs [4,5] is based on the concept of founded and justified repairs. Informally, justified repairs are the repairs that are the most strongly grounded in the given database and the given set of AICs, that is, those resulting strictly from combinations of the preferences expressed by the database designer for each of the integrity constraints and from the principle of minimal change. The operational semantics for AICs [7] allows direct computation of justified repairs by means of intuitive tree algorithms. Interaction between different AICs in a set that must be collectively satisfied makes the problem of repairing a database highly non-trivial, however, and in the worst case deciding whether a database can be repaired is NP-complete or Σ_P^2-complete on the number of AICs [5], depending on the criteria used to choose possible repairs. For this reason, it is important to be able to control the number of AICs being considered simultaneously.

In this paper we first present parallelization results that allow a set of AICs to be split in smaller, independent sets such that repairs for each smaller set can be computed independently and the results straightforwardly combined into a repair for the original set. Afterwards, we introduce a hierarchization mechanism on AICs that allows repairs to be computed progressively, starting with a small set of AICs and extending this set while simultaneously extending the computed repair. With these techniques, it is possible to speed up the problem of finding repairs significantly; and, although they do not help in the worst-case scenario, the typical structure of real-life databases indicates that parallelization and hierarchization should be widely applicable.

1.1 Related Work

When a database needs to be changed, it is necessary to find a way to make the relevant modifications while maintaining the consistency of the data. This problem, which has been the focus of intensive research for over thirty years, was extensively discussed in [1], where three main change operations were identified: insertion of new facts, deletion of existing facts, and modification of information, and the concept of "good" update was characterized.

There are two distinct scenarios where database change is required, leading to the distinction between *update* and *revision* [8,11]. An update occurs whenever the world changes and the knowledge bases needs to be changed to reflect this fact; a revision happens when new knowledge is obtained about a world that did not change. This distinction is especially relevant in deductive databases and open-world knowledge bases, where the known information is not assumed to be complete.

In spite of their differences, there are obvious similarities between updates and revisions, and in both cases one has to consider the problems that arise when the intended semantics of the database is taken into account. Typically, the changes that have to be made conflict with the integrity constraints associated with the database, and the database must be repaired in order to regain consistency. The ways in which this can be done are many, and several proposals have been around for years. One possibility is to read integrity constraints as rules that suggest possible actions to repair inconsistencies [1]; another is to express database dependencies through logic programming, namely in the setting of deductive databases [12,14,15]. A more algorithmic approach uses event-condition-action rules [16,17], where actions are triggered by specific events, and for which rule processing algorithms have been proposed and a procedural semantics has been defined.

Several algorithms for computing repairs of inconsistent databases have been proposed and studied throughout the years, focusing on the different ways integrity constraints are specified and on the different types of databases under consideration [10,12,14,15]. This multitude of approaches is not an accident: deciding whether an inconsistent database can be repaired is typically a Π_p^2- or co-Σ_p^2- complete problem, and it has been observed [8] that there is no reason to believe in the existence of general-purpose algorithms for this problem, but one should rather focus on developing more specific algorithms for particular interesting cases.

Regardless of the approach taken, when an inconsistent database can be repaired there are typically several sets of update actions that achieve this goal, leading to different revised consistent databases. Restricting the set of database repairs to those considered most adequate is therefore an important task. Among the criteria that have been proposed to obtain this restriction are minimality of change [6,8,18] – one should change as little as possible – and the common sense law of inertia [15] – one should only change something if there is a reason for it –, but these are not enough to narrow down the set of possible repairs sufficiently. Ultimately, it is usually assumed that some human interaction will be required to choose the “best” possible repair [16].

Because of the intrinsic complexity involved in the computation of repairs, techniques to split a problem in several smaller problems are of particular interest. As far as we know, this problem has received little consideration over the years. There is a reference to semantic independency in [14] that is not explored further, and syntactic precedence is used in that same paper in order to compute models – but within a scenario that is far more powerful than that of active integrity constraints. More recently, syntactic precedence between constraints was also discussed with the explicit goal of making the search for repairs more efficient [13], but the authors did not allow for cyclic dependencies. The results we prove are therefore a significant extension of previous work, and we believe they can be easily extended to different formalisms of integrity constraints.

2 Background

Active integrity constraints were originally introduced in [9] as a special type of integrity constraints, specifying not only the consistency requirements imposed upon a database, but also actions that can be taken to correct the database when such requirements are not met.

Within this framework, a database is a subset of a finite set of propositional atoms $\mathcal{A}t$. An *active integrity constraint* (AIC) is a rule of the form

$$L_1, \ldots, L_n \supset \alpha_1 \mid \ldots \mid \alpha_m$$

where $L_1, \ldots, L_n$ are literals in the language generated by $\mathcal{A}t$; $\alpha_1, \ldots, \alpha_m$ are *update actions* of the form $+a$ or $-a$, where a is an atom in the same language; and every update action must contradict some literal, i.e. if $+a$ (resp. $-a$) occurs among the α_i, then $\mathsf{not}\ a$ (resp. a) must occur among the L_i. The set $\{L_1, \ldots, L_n\}$ is the *body* of the rule and $\{\alpha_1, \ldots, \alpha_m\}$ is its *head*.

The close connection between literals and actions is made precise by means of two operators. The atom *underlying* an action α is $\mathsf{lit}(\alpha)$, defined by $\mathsf{lit}(+a) = a$ and $\mathsf{lit}(-a) = \mathsf{not}\ a$, whereas the update action *corresponding* to L is $\mathsf{ua}(L)$, defined by $\mathsf{ua}(a) = +a$ and $\mathsf{ua}(\mathsf{not}\ a) = -a$. The *dual* of a literal L, L^D, is defined as usual by $a^D = \mathsf{not}\ a$ and $(\mathsf{not}\ a)^D = a$. Using this notation, the requirement that valid AICs must satisfy can be stated as $\{\mathsf{lit}(\alpha_1), \ldots, \mathsf{lit}(\alpha_m)\}^D \subseteq \{L_1, \ldots, L_n\}$.

Being a set of propositional atoms, any database $\mathcal{I}$ induces a propositional interpretation of literals. We say that $\mathcal{I}$ *entails* literal L, $I \models L$, if L is a and $a \in \mathcal{I}$, or if L is $\mathsf{not}\ a$ and $a \notin \mathcal{I}$. Given an AIC r of the form $L_1, \ldots, L_n \supset \alpha_1, \ldots, \alpha_m$, we say that $\mathcal{I} \models r$ if $\mathcal{I} \not\models L_i$ for some i; otherwise, r is said to be *applicable* in $\mathcal{I}$. Finally, if η is a set of AICs, then $\mathcal{I} \models \eta$ iff $\mathcal{I} \models r$ for every $r \in \eta$.

The operational nature of rules is given by the notion of updating a database by a set of update actions, which captures the intuive idea conveyed above. The result of updating $\mathcal{I}$ with a set of update actions $\mathcal{U}$ is $\mathcal{I} \circ \mathcal{U}$, defined as

$$\mathcal{I} \circ \mathcal{U} = (\mathcal{I} \cup \{a \mid +a \in \mathcal{U}\}) \setminus \{a \mid -a \in \mathcal{U}\}\,.$$

In order for this definition to make sense, $\mathcal{U}$ must not contain $+a$ and $-a$ for the same atom a. A set of update actions satisfying this requirement is said to be *consistent*.

Given a set of AICs η and a database $\mathcal{I}$, a set of update actions $\mathcal{U}$ such that $\mathcal{I} \circ \mathcal{U} \models \eta$ achieves the task of making $\mathcal{I}$ consistent w.r.t. η. In general, for any given database that is inconsistent w.r.t. η there will be either none or several such $\mathcal{U}$. In order to compare different ways of repairing $\mathcal{I}$, Caroprese and Truszczyński [5] studied different semantics for AICs.

Minimality of change is commonly accepted as a desirable property [6,18]. This motivates the following notion: given a database $\mathcal{I}$ and a set of AICs η, a consistent set of update actions $\mathcal{U}$ such that (i) every action in $\mathcal{U}$ changes $\mathcal{I}$ and (ii) $\mathcal{I} \circ \mathcal{U} \models \eta$ is called a *weak repair* for $\langle \mathcal{I}, \eta \rangle$; a *repair* for $\langle \mathcal{I}, \eta \rangle$ is a weak repair for $\langle \mathcal{I}, \eta \rangle$ that is minimal w.r.t. inclusion (so it contains no proper subset that is also a weak repair). Condition (i) states that weak repairs only include actions that change the database, and may be formally stated as $(\{+a \mid a \in \mathcal{I}\} \cup \{-a \mid a \in \mathcal{A}t \setminus \mathcal{I}\}) \cap \mathcal{U} = \emptyset$, or equivalently as $\mathcal{I} \circ \alpha \neq \mathcal{I}$ for every $\alpha \in \mathcal{U}$. Condition (ii) simply states that weak repairs make the database consistent w.r.t. η.

None of these conditions takes into account the operational nature of AICs, however, since they ignore the actions in the heads of the rules in η. For this purpose, one needs to consider the more sophisticated notion of founded (weak) repairs [5]. The intuition behind these is that they should contain only actions that are motivated (founded) by the application of some rule. An update action α is *founded*[1] w.r.t. $\langle \mathcal{I}, \eta \rangle$ and a set of update actions $\mathcal{U}$ if there is a rule $r \in \eta$ such that $\alpha \in \mathsf{head}(r)$ and $\mathcal{I} \circ \mathcal{U} \models L$ for every literal $L \in \mathsf{body}(r) \setminus \{\mathsf{lit}(\alpha)^D\}$. Quoting [5], "if $\mathcal{U}$ is to enforce r, then it must contain α" – if α is removed from $\mathcal{U}$, then all literals in the body of r are true and the rule is violated. A set of update actions $\mathcal{U}$ is *founded* w.r.t $\langle \mathcal{I}, \eta \rangle$ if every action in $\mathcal{U}$ is founded w.r.t. $\langle \mathcal{I}, \eta \rangle$ and $\mathcal{U}$. A set of update actions $\mathcal{U}$ is a founded (weak) repair for $\langle \mathcal{I}, \eta \rangle$ if (i) $\mathcal{U}$ is a weak repair for $\langle \mathcal{I}, \eta \rangle$ and (ii) $\mathcal{U}$ is founded w.r.t. $\langle \mathcal{I}, \eta \rangle$.

It is important to stress that founded repairs are minimal weak repairs that are founded. Indeed, there are founded weak repairs that do not contain any founded repair as subset (see [5] for an example). Also, being founded does not imply being a weak repair, so these two tests must be performed independently.

Founded repairs, however, sometimes exhibit unexpected properties, such as *circularity of support* [5] – e.g. they contain two actions α and β such that α is founded by means of a rule r whose body only holds because of β, and β is founded by means of a rule r' whose body only holds because of α –, and it is therefore interesting to consider a more complex type of repairs: justified repairs. In order to define these, we need some auxiliary notions. The set of *non-updatable literals* of a rule r is defined as $\mathsf{nup}(r) = \mathsf{body}(r) \setminus (\mathsf{lit}(\mathsf{head}(r)))^D$, were lit is extended to sets in the obvious way. A set of update actions $\mathcal{U}$ is closed for rule r if $\mathsf{nup}(r) \subseteq \mathsf{lit}(\mathcal{U})$ implies $\mathsf{head}(r) \cap \mathcal{U} \neq \emptyset$, and $\mathcal{U}$ is closed for η if $\mathcal{U}$ is closed for every rule in η.

An update action $+a$ (resp. $(-a)$) is a *no-effect action* w.r.t. $\mathcal{I}$ and $\mathcal{J}$ if $a \in \mathcal{I} \cap \mathcal{J}$ (resp. $a \notin (\mathcal{I} \cup \mathcal{J})$) – in other words, both $\mathcal{I}$ and $\mathcal{J}$ are unaffected by the action. The set of all no-effect actions w.r.t. $\mathcal{I}$ and $\mathcal{J}$ is denoted by $\mathsf{ne}(\mathcal{I}, \mathcal{J})$. Given a database $\mathcal{I}$ and a set of AICs η, a consistent set of update actions $\mathcal{U}$ is a *justified action set* for $\langle \mathcal{I}, \eta \rangle$ if $\mathcal{U}$ is a minimal set of update actions containing $\mathsf{ne}(\mathcal{I}, \mathcal{I} \circ \mathcal{U})$ and closed for η. In that case, the set $\mathcal{U} \setminus \mathsf{ne}(\mathcal{I}, \mathcal{I} \circ \mathcal{U})$ is a *justified weak repair* for $\langle \mathcal{I}, \eta \rangle$. Being closed for η implies being a weak repair for $\langle \mathcal{I}, \eta \rangle$, so this terminology is consistent with the previous usage of the latter term.

In spite of the minimality requirement in the definition of justified weak repair, there *are* justified weak repairs that contain a justified repair as a proper subset; this is because the minimality involved in this definition is within a different universe. All justified weak repairs are founded, but not conversely: indeed, these repairs successfully avoid circularity of support.

We will use the following alternative characterization of justified weak repair: a weak repair $\mathcal{U}$ for $\langle \mathcal{I}, \eta \rangle$ is justified if (i) $\mathcal{U} \cap \mathsf{ne}(\mathcal{I}, \mathcal{I} \circ \mathcal{U}) = \emptyset$ and (ii) $\mathcal{U} \cup \mathsf{ne}(\mathcal{I}, \mathcal{I} \circ \mathcal{U})$ is a justified action set. Indeed, taking $\mathcal{W} = \mathcal{U} \cup \mathsf{ne}(\mathcal{I}, \mathcal{I} \circ \mathcal{U})$, it can easily be checked that $\mathsf{ne}(\mathcal{I}, \mathcal{I} \circ \mathcal{W}) = \mathsf{ne}(\mathcal{I}, \mathcal{I} \circ \mathcal{U})$: the only differences between

[1] This equivalent characterization of founded action, which can be found in [7], is slightly different from that in [5], and simpler to use in practice.

$\mathcal{I}$ and $\mathcal{I} \circ \mathcal{W}$ must originate from $\mathcal{U}$ by definition of $\mathsf{ne}(\mathcal{I},\mathcal{I} \circ \mathcal{U})$. Therefore $\mathcal{W} \setminus \mathsf{ne}(\mathcal{I},\mathcal{I} \circ \mathcal{W}) = \mathcal{U}$.

The major problem with computing repairs for inconsistent databases lies in the complexity of deciding whether such repairs exist. Given $\mathcal{I}$ and η, the problem of deciding whether there exists a weak repair, a repair or a founded weak repair for $\langle \mathcal{I}, \eta \rangle$ is NP-complete (on the size of η), whereas deciding whether there is a founded repair, a justified weak repair or a justified repair for η is Σ_P^2-complete (again on the size of η).[2] In the special case where all AICs are *normalized* – they have only one action in their head – the last two problems also become NP-complete. Due to these ultimately bad complexity bounds, techniques to lower the size of the problem can be extremely useful in practice. The goal of this paper is to discuss how a set of AICs η can be divided into smaller sets such that the computation of (simple, founded, justified) repairs can be computed for each of those sets and the results combined in polynomial time.

3 Independent AICs

In this section, we introduce a notion of independence between active integrity constraints. The goal is the following: given a set of AICs η, to partition it in distinct independent sets $\eta_1, \ldots, \eta_n$ such that the search for repairs for a database $\mathcal{I}$ and η can be parallelized among the η_i. We define independent sets of AICs in such a way that (simple, founded, justified) repairs for the different sets can be combined into a (simple, founded, justified) repair for $\langle \mathcal{I}, \eta \rangle$.

The basic concept is that of independent AICs. Two AICs are independent if they do not share any atoms between their literals, so that applicability of one does not affect applicability of the other.

Definition 1.

1. *The* atom *underlying a literal L is $|L|$, defined as $|a| = |\mathsf{not}\ a| = a$.*
2. *Let r_1 and r_2 be two AICs, where r_1 is $L_1, \ldots, L_n \supset \alpha_1, \ldots, \alpha_p$ and r_2 is $M_1, \ldots, M_m \supset \beta_1, \ldots, \beta_q$. Then r_1 and r_2 are* independent, *$r_1 \perp\!\!\!\perp r_2$, if $\{|L_1|, \ldots, |L_n|\} \cap \{|M_1|, \ldots, |M_m|\} = \emptyset$.*
3. *Let η_1 and η_2 be sets of AICs. Then η_1 and η_2 are* independent, *$\eta_1 \perp\!\!\!\perp \eta_2$, if $r \perp\!\!\!\perp s$ whenever $r \in \eta_1$ and $s \in \eta_2$.*

Two comments are in place regarding this definition. First, the notion of independence does not take into account the actions in the rules (the "active" part of the AICs); this aspect will be dealt with in Section 5. Second, this concept only depends on the active integrity constraints themselves, and not on the underlying database. This issue has positive practical implications, as we will see later.

This notion of independence captures the spirit of parallelization, as the next lemmas state. Throughout the remainder of this section, let $\mathcal{I}$ be a database, η_1, η_2 be independent sets of AICs and $\eta = \eta_1 \cup \eta_2$.

[2] The size of $\mathcal{I}$ does not affect the complexity bounds for these problems [5].

Lemma 1. *Let $\mathcal{U}_1$ and $\mathcal{U}_2$ be sets of update actions such that every action in $\mathcal{U}_i$ corresponds to a literal (or its dual) in a rule in η_i,*[3] *and take $\mathcal{U} = \mathcal{U}_1 \cup \mathcal{U}_2$. For every literal L such that $L \in \mathsf{body}(r)$ with $r \in \eta_i$, $\mathcal{I} \circ \mathcal{U} \models L$ iff $\mathcal{I} \circ \mathcal{U}_i \models L$. In particular, for every $r \in \eta_i$, $I \circ \mathcal{U} \models r$ iff $\mathcal{I} \circ \mathcal{U}_i \models r$.*

Proof. Let $L \in \mathsf{body}(r)$ for some $r \in \eta_1$. If $\alpha \in \mathcal{U}_2$, then $|\mathsf{lit}(\alpha)| \neq |L|$ because $\eta_1 \perp\!\!\!\perp \eta_2$, whence $\mathcal{I} \circ \mathcal{U} \models L$ iff $\mathcal{I} \circ \mathcal{U}_1 \models L$ (note that $\mathcal{I} \circ \mathcal{U} = (\mathcal{I} \circ \mathcal{U}_1) \circ \mathcal{U}_2$). The result for rules is a straightforward consequence. The argument for $\mathcal{U}_2$ is similar. □

Lemma 2. *Let $\mathcal{U}_1$ and $\mathcal{U}_2$ be weak repairs for $\langle \mathcal{I}, \eta_1 \rangle$ and $\langle \mathcal{I}, \eta_2 \rangle$, respectively, such that the actions in $\mathcal{U}_i$ are all duals of literals in the body of some rule in η_i. Then $\mathcal{U} = \mathcal{U}_1 \cup \mathcal{U}_2$ is a weak repair for $\langle \mathcal{I}, \eta \rangle$.*

Proof. We first show that $\mathcal{U}$ is a consistent set of actions containing only essential actions. For consistency, note that the set of atoms underlying the actions in $\mathcal{U}_1$ is disjoint from that of the atoms underlying the atoms in $\mathcal{U}_2$, from the hypothesis and the fact that $\eta_1 \perp\!\!\!\perp \eta_2$; hence, if $+\alpha$ and $-\alpha$ were both in $\mathcal{U} = \mathcal{U}_1 \cup \mathcal{U}_2$ for some a, this would mean that $+\alpha, -\alpha \in \mathcal{U}_i$ for some i, whence $\mathcal{U}_i$ would be inconsistent. Furthermore, if $\alpha \in \mathcal{U}_i$ then α must change the state of $\mathcal{I}$ (since $\mathcal{U}_i$ is a weak repair for $\langle \mathcal{I}, \eta_i \rangle$), so $\mathcal{U}$ consists only of essential update actions.

Finally, we show that $\mathcal{U}$ is a weak repair. Without loss of generality, let $r \in \eta_1$. Then $\mathcal{I} \circ \mathcal{U}_1 \models r$, since $\mathcal{U}_1$ is a weak repair for $\langle \mathcal{I}, \eta_1 \rangle$, and by Lemma 1 $\mathcal{I} \circ \mathcal{U} \models r$. □

The hypothesis that the actions in each $\mathcal{U}_i$ are all duals of literals in the body of some rule in η_i is essential: if it were not required, then $\mathcal{U}_1$ could "break" satisfaction of some rule in η_2 or reciprocally, or there might be inconsistencies from joining $\mathcal{U}_1$ and $\mathcal{U}_2$. Although this hypothesis could be weakened, it is actually a (very) reasonable assumption: no reasonable algorithm for computing weak repairs should include actions that do not affect the semantics of the integrity constraints that should hold, since this verification can be done very efficiently.

If $\mathcal{U}_1$ and $\mathcal{U}_2$ are repairs, we get the following stronger result.

Lemma 3. *If $\mathcal{U}_1$ and $\mathcal{U}_2$ are repairs for $\langle \mathcal{I}, \eta_1 \rangle$ and $\langle \mathcal{I}, \eta_2 \rangle$, respectively, then $\mathcal{U} = \mathcal{U}_1 \cup \mathcal{U}_2$ is a repair for $\langle \mathcal{I}, \eta \rangle$.*

Proof. By Lemma 2, $\mathcal{U}$ is a weak repair for $\langle \mathcal{I}, \eta \rangle$. For any $\mathcal{U}' \subsetneq \mathcal{U}$, define $\mathcal{U}'_i = \mathcal{U}' \cap \mathcal{U}_i$ for $i = 1, 2$. Note that one of the inclusions $\mathcal{U}'_i \subseteq \mathcal{U}_i$ must be strict; without loss of generality, assume that $\mathcal{U}'_1 \subsetneq \mathcal{U}_1$. Since $\mathcal{U}_1$ is a repair, this means that $\mathcal{U}'_1$ cannot be a weak repair, hence there is a rule $r \in \eta_1$ such that $\mathcal{U}'_1 \not\models r$. By Lemma 1 $\mathcal{U}'_1 \cup \mathcal{U}'_2 \not\models r$, hence $\mathcal{U}' = \mathcal{U}'_1 \cup \mathcal{U}'_2$ cannot be a weak repair for $\langle \mathcal{I}, \eta \rangle$. Therefore $\mathcal{U}$ is a repair for $\langle \mathcal{I}, \eta \rangle$. □

The converse result also holds: if we split the actions in a weak repair $\mathcal{U}$ according to whether they affect rules in η_1 or η_2, we get weak repairs for those sets of AICs.

---

[3] Formally, $\{|\mathsf{lit}(\alpha)| \mid \alpha \in \mathcal{U}_i\} \subseteq \{|L| \mid \exists r \in \eta_i . L \in \mathsf{body}(r)\}$.

Lemma 4. *Let $\mathcal{U}$ be a weak repair for $\langle \mathcal{I}, \eta \rangle$. Then*

$$\mathcal{U}_i = \{\alpha \in \mathcal{U} \mid \exists r \in \eta_i.\mathsf{lit}(\alpha)^D \in \mathsf{body}(r)\}$$

are weak repairs for $\langle \mathcal{I}, \eta_i \rangle$. Furthermore, $\mathcal{U} = \mathcal{U}_1 \cup \mathcal{U}_2$ if the actions in $\mathcal{U}$ are all duals of literals in the body of some rule in η.

Proof. Assume that $\mathcal{U}$ is a weak repair for $\langle \mathcal{I}, \eta \rangle$ and let $\mathcal{U}_i$ be as stated. Since $\mathcal{U}$ is a weak repair for $\langle \mathcal{I}, \eta \rangle$, $\mathcal{I} \circ \mathcal{U} \models r$ for every rule $r \in \eta_i$. By Lemma 1, $\mathcal{I} \circ \mathcal{U}_i \models r$. Therefore $\mathcal{U}_i$ is a weak repair for $\langle \mathcal{I}, \eta_i \rangle$.

If the actions in $\mathcal{U}$ are all duals of literals in the body of some rule in η, then they occur in the body of a rule in η_1 or η_2, so they will all occur either in $\mathcal{U}_1$ or $\mathcal{U}_2$, whence $\mathcal{U} = \mathcal{U}_1 \cup \mathcal{U}_2$. □

The stated equality can be made to hold in the general case by adding the actions that do not affect any rule to either $\mathcal{U}_1$ or $\mathcal{U}_2$; however, this is not an interesting situation, and we will not consider it any further.

If $\mathcal{U}$ is minimal, then the same result can be made stronger.

Lemma 5. *If $\mathcal{U}$ is a repair for $\langle \mathcal{I}, \eta \rangle$, then $\mathcal{U}_1$ and $\mathcal{U}_2$ as defined above are repairs for $\langle \mathcal{I}, \eta_1 \rangle$ and $\langle \mathcal{I}, \eta_2 \rangle$, respectively, and furthermore $\mathcal{U} = \mathcal{U}_1 \cup \mathcal{U}_2$.*

Proof. By Lemma 4, each $\mathcal{U}_i$ is a weak repair for $\langle \mathcal{I}, \eta_i \rangle$. Suppose that $\mathcal{U}_1' \subsetneq \mathcal{U}_1$ is also a weak repair for $\langle \mathcal{I}, \eta_1 \rangle$. By Lemma 1, $\mathcal{U}' = \mathcal{U}_1' \cup \mathcal{U}_2$ is a weak repair for $\langle \mathcal{I}, \eta \rangle$ with $\mathcal{U}' \subsetneq \mathcal{U}$, which is absurd. Therefore $\mathcal{U}_1'$ is not a weak repair, hence $\mathcal{U}_1$ is a repair. The case for $\mathcal{U}_2$ is similar. Finally, $\mathcal{U}$ cannot contain actions that are not duals of literals in the body of rules in η, since these can always be removed without affecting the property of being a weak repair; therefore $\mathcal{U}_1 \cup \mathcal{U}_2 = \mathcal{U}$. □

These results also hold if we consider founded or justified (weak) repairs.

Lemma 6. *Let $\mathcal{U}_1$ and $\mathcal{U}_2$ be founded w.r.t. $\langle \mathcal{I}, \eta_1 \rangle$ and $\langle \mathcal{I}, \eta_2 \rangle$, respectively. Then $\mathcal{U} = \mathcal{U}_1 \cup \mathcal{U}_2$ is founded w.r.t. $\langle \mathcal{I}, \eta \rangle$.*

Proof. In order for $\mathcal{U}$ to be founded w.r.t. $\langle \mathcal{I}, \eta \rangle$, every action in $\mathcal{U}$ must be founded w.r.t. $\langle \mathcal{I}, \eta \rangle$ and $\mathcal{U}$. Let $\alpha \in \mathcal{U}$ and assume that $\alpha \in \mathcal{U}_1$ (the case when $\alpha \in \mathcal{U}_2$ is similar).

Since $\mathcal{U}_1$ is founded w.r.t. $\langle \mathcal{I}, \eta_1 \rangle$, there is a rule $r \in \eta_1$ such that $\alpha \in \mathsf{head}(r)$ and $\mathcal{I} \circ \mathcal{U}_1 \models L$ for every $L \in \mathsf{body}(r) \setminus \{\mathsf{lit}(\alpha)^D\}$. By Lemma 1, $I \circ \mathcal{U} \models L$ for every such L. Since $\eta_1 \subseteq \eta$, this means that α is founded w.r.t. $\langle \mathcal{I}, \eta \rangle$ and $\mathcal{U}$. □

Corollary 1. *If $\mathcal{U}_1$ and $\mathcal{U}_2$ are founded (weak) repairs, then $\mathcal{U}$ is also a founded (weak) repair.*

Proof. Consequence of Lemmas 2, 3 and 6. □

Lemma 7. *Let $\mathcal{U}$ be founded w.r.t. for $\langle \mathcal{I}, \eta \rangle$. Then $\mathcal{U}_1$ and $\mathcal{U}_2$ as defined in Lemma 4 are such that $\mathcal{U} = \mathcal{U}_1 \cup \mathcal{U}_2$ and each $\mathcal{U}_i$ is founded w.r.t. $\langle \mathcal{I}, \eta_i \rangle$.*

Proof. Let $\alpha \in \mathcal{U}_1$; since $\mathcal{U}$ is founded w.r.t. $\langle\mathcal{I},\eta\rangle$, there is a rule $r \in \eta$ such that $\alpha \in \mathsf{head}(r)$ and $\mathcal{I}\circ\mathcal{U} \models L$ for every $L \in \mathsf{body}(r) \setminus \{\mathsf{lit}(\alpha)^D\}$. But if $\alpha \in \mathsf{head}(r)$, then necessarily $r \in \eta_1$; and in that case $\mathcal{I}\circ\mathcal{U}_1 \models L$ for every $L \in \mathsf{body}(r) \setminus \{\mathsf{lit}(\alpha)^D\}$ by Lemma 1. Therefore α is founded w.r.t. $\langle\mathcal{I},\eta_1\rangle$ and $\mathcal{U}_1$, whence $\mathcal{U}_1$ is founded w.r.t. $\langle\mathcal{I},\eta_1\rangle$. The case when $\alpha \in \mathcal{U}_2$ is similar.

By definition of founded set, all actions in $\mathcal{U}$ must necessarily be in either $\mathcal{U}_1$ or $\mathcal{U}_2$, so $\mathcal{U} = \mathcal{U}_1 \cup \mathcal{U}_2$. □

Corollary 2. *If $\mathcal{U}$ is a (weak) founded repair, then $\mathcal{U}_1$ and $\mathcal{U}_2$ are also (weak) founded repairs.*

Proof. Consequence of Lemmas 4, 5 and 7. □

Lemma 8. *Let $\mathcal{U}_1$ and $\mathcal{U}_2$ be justified (weak) repairs for $\langle\mathcal{I},\eta_1\rangle$ and $\langle\mathcal{I},\eta_2\rangle$, respectively. Then $\mathcal{U} = \mathcal{U}_1 \cup \mathcal{U}_2$ is a justified (weak) repair for $\langle\mathcal{I},\eta\rangle$.*

Proof. We begin by making some observations that will be used recurrently throughout the proof.

(a) For $i = 1,2$, $\mathsf{ne}(\mathcal{I},\mathcal{I}\circ\mathcal{U}) \subseteq \mathsf{ne}(\mathcal{I},\mathcal{I}\circ\mathcal{U}_i)$, since $\mathcal{U}_i \subseteq \mathcal{U}$.
(b) For $i = 1,2$, $\mathsf{ne}(\mathcal{I},\mathcal{I}\circ\mathcal{U}_i) \subseteq (\mathsf{ne}(\mathcal{I},\mathcal{I}\circ\mathcal{U}) \cup \mathcal{U}_{3-i})$: $\mathcal{U}$ can only change literals that changed either by $\mathcal{U}_1$ or by $\mathcal{U}_2$. In particular, since $\eta_1 \perp\!\!\!\perp \eta_2$, if $\mathsf{nup}(r) \subseteq \mathsf{lit}(\mathsf{ne}(\mathcal{I},\mathcal{I}\circ\mathcal{U}_i))$ for some $r \in \eta_i$, then $L \in \mathsf{lit}(\mathsf{ne}(\mathcal{I},\mathcal{I}\circ\mathcal{U}))$; and if $\alpha \in \mathsf{head}(r)$ for some $r \in \eta_i$ and $\alpha \in \mathsf{ne}(\mathcal{I},\mathcal{I}\circ\mathcal{U}_i)$, then $\alpha \in \mathsf{ne}(\mathcal{I},\mathcal{I}\circ\mathcal{U})$.
(c) For $i = 1,2$, if $L \in \mathsf{body}(r)$ with $r \in \eta_i$ and $L \in \mathsf{lit}(\mathcal{U})$, then $L \in \mathsf{lit}(\mathcal{U}_i)$: since every justified weak repair is founded [5], $\mathcal{U}_i$ only contains actions in the heads of rules of η_i, and the thesis follows from $\eta_1 \perp\!\!\!\perp \eta_2$.

We first show that $\mathcal{U} \cup \mathsf{ne}(\mathcal{I},\mathcal{I}\circ\mathcal{U})$ is closed for η. Let $r \in \eta_1$; the case when $r \in \eta_2$ is similar. Suppose $\mathsf{nup}(r) \subseteq \mathsf{lit}(\mathcal{U} \cup \mathsf{ne}(\mathcal{I},\mathcal{I}\circ\mathcal{U}))$, and let $L \in \mathsf{nup}(r)$. If $L \in \mathsf{lit}(\mathcal{U})$, then $L \in \mathsf{lit}(\mathcal{U}_1)$ by (c), hence $\mathsf{nup}(r) \subseteq \mathsf{lit}(\mathcal{U}_1 \cup \mathsf{ne}(\mathcal{I},\mathcal{I}\circ\mathcal{U}))$, whence $\mathsf{nup}(r) \subseteq \mathsf{lit}(\mathcal{U}_1 \cup \mathsf{ne}(\mathcal{I},\mathcal{I}\circ\mathcal{U}_1))$ by (a). But $\mathcal{U}_1 \cup \mathsf{ne}(\mathcal{I},\mathcal{I}\circ\mathcal{U}_1)$ is closed for η_1, so $\mathsf{head}(r) \cap (\mathcal{U}_1 \cup \mathsf{ne}(\mathcal{I},\mathcal{I}\circ\mathcal{U}_1)) \neq \emptyset$. By $\mathcal{U}_1 \subseteq \mathcal{U}$ and (b), also $\mathsf{head}(r) \cap (\mathcal{U} \cup \mathsf{ne}(\mathcal{I},\mathcal{I}\circ\mathcal{U})) \neq \emptyset$.

To check minimality, suppose that $\mathcal{U}' \subsetneq \mathcal{U}$ is such that $\mathcal{U}' \cup \mathsf{ne}(\mathcal{I},\mathcal{I}\circ\mathcal{U})$ is closed for η and take $\mathcal{U}'_i = \mathcal{U}' \cap \mathcal{U}_i$ for $i = 1,2$. Note that one of the inclusions $\mathcal{U}'_i \subseteq \mathcal{U}_i$ must be strict. Without loss of generality, assume this is the case when $i = 1$, and take $r \in \eta_1$. If $\mathsf{nup}(r) \subseteq \mathsf{lit}(\mathcal{U}'_1 \cup \mathsf{ne}(\mathcal{I},\mathcal{I}\circ\mathcal{U}_1))$, then $\mathsf{nup}(r) \subseteq \mathsf{lit}(\mathcal{U}' \cup \mathsf{ne}(\mathcal{I},\mathcal{I}\circ\mathcal{U}))$, consequence of $\mathcal{U}'_1 \subseteq \mathcal{U}'$ and (b). Since $\mathcal{U}' \cup \mathsf{ne}(\mathcal{I},\mathcal{I}\circ\mathcal{U})$ is closed for η and $\eta_1 \subseteq \eta$, it follows that $\mathsf{head}(r) \cap (\mathcal{U}' \cup \mathsf{ne}(\mathcal{I},\mathcal{I}\circ\mathcal{U})) \neq \emptyset$. By definition of $\mathcal{U}_1$ and (a), it follows that $\mathsf{head}(r) \cap (\mathcal{U}'_1 \cup \mathsf{ne}(\mathcal{I},\mathcal{I}\circ\mathcal{U}_1)) \neq \emptyset$. Then $\mathcal{U}'_1 \cup \mathsf{ne}(\mathcal{I},\mathcal{I}\circ\mathcal{U}_1)$ is closed for η_1, contradicting minimality of $\mathcal{U}_1$.

Hence $\mathcal{U}$ is a justified weak repair for $\langle\mathcal{I},\eta\rangle$. By Lemma 3, if $\mathcal{U}_1$ and $\mathcal{U}_2$ are both justified repairs for $\langle\mathcal{I},\eta_1\rangle$ and $\langle\mathcal{I},\eta_2\rangle$, respectively, then $\mathcal{U}$ is also a justified repair for $\langle\mathcal{I},\eta\rangle$. □

Lemma 9. *Let $\mathcal{U}$ be a justified (weak) repair for $\langle\mathcal{I},\eta\rangle$. Then $\mathcal{U}_1$ and $\mathcal{U}_2$ as defined in Lemma 4 are such that $\mathcal{U} = \mathcal{U}_1 \cup \mathcal{U}_2$ and each $\mathcal{U}_i$ is a justified (weak) repair for $\langle\mathcal{I},\eta_i\rangle$.*

Proof. Again note that properties (a), (b) and (c) from the previous proof hold. We begin by showing that $\mathcal{U}_1 \cup \mathsf{ne}(\mathcal{I}, \mathcal{I} \circ \mathcal{U}_1)$ is closed under η_1. Take $r \in \eta_1$ and suppose that $\mathsf{nup}(r) \subseteq \mathsf{lit}(\mathcal{U}_1 \cup \mathsf{ne}(\mathcal{I}, \mathcal{I} \circ \mathcal{U}_1))$. Then $\mathsf{nup}(r) \subseteq \mathsf{lit}(\mathcal{U} \cup \mathsf{ne}(\mathcal{I}, \mathcal{I} \circ \mathcal{U}))$ by $\mathcal{U}_1 \subseteq \mathcal{U}$ and (b). Since $\mathcal{U} \cup \mathsf{ne}(\mathcal{I}, \mathcal{I} \circ \mathcal{U})$ is closed for η, it follows that $\mathsf{head}(r) \cap (\mathcal{U} \cup \mathsf{ne}(\mathcal{I}, \mathcal{I} \circ \mathcal{U})) \neq \emptyset$. By construction of $\mathcal{U}_1$ and (a), we conclude that $\mathsf{head}(r) \cap (\mathcal{U}_1 \cup \mathsf{ne}(\mathcal{I}, \mathcal{I} \circ \mathcal{U}_1)) \neq \emptyset$. The case for $\mathcal{U}_2$ is similar.

To check minimality, suppose that $\mathcal{U}_1' \subsetneq \mathcal{U}_1$ is such that $\mathcal{U}_1' \cup \mathsf{ne}(\mathcal{I}, \mathcal{I} \circ \mathcal{U}_1)$ is closed for η_1 and take $\mathcal{U}' = \mathcal{U}_1' \cup \mathcal{U}_2$. Let $r \in \eta$ and assume $\mathsf{nup}(r) \subseteq \mathsf{lit}(\mathcal{U}' \cup \mathsf{ne}(\mathcal{I}, \mathcal{I} \circ \mathcal{U}))$; there are two possible cases.

- Suppose $r \in \eta_1$ and let $L \in \mathsf{nup}(r)$. Note that $L \in \mathsf{lit}(\mathcal{U}_2)$ is impossible, since $\eta_1 \perp\!\!\!\perp \eta_2$. Therefore $\mathsf{nup}(r) \subseteq \mathsf{lit}(\mathcal{U}_1' \cup \mathsf{ne}(\mathcal{I}, \mathcal{I} \circ \mathcal{U}))$, whence by (a) $\mathsf{nup}(r) \subseteq \mathsf{lit}(\mathcal{U}_1' \cup \mathsf{ne}(\mathcal{I}, \mathcal{I} \circ \mathcal{U}_1))$, and therefore $\mathsf{head}(r) \cap (\mathcal{U}_1' \cup \mathsf{ne}(\mathcal{I}, \mathcal{I} \circ \mathcal{U}_1)) \neq \emptyset$. From $\mathcal{U}_1' \subseteq \mathcal{U}'$ and (b), also $\mathsf{head}(r) \cap (\mathcal{U}' \cup \mathsf{ne}(\mathcal{I}, \mathcal{I} \circ \mathcal{U})) \neq \emptyset$.
- Suppose $r \in \eta_2$ and let $L \in \mathsf{nup}(r)$. Since $L \in \mathsf{lit}(\mathcal{U}_1')$ is impossible, it follows that $L \in \mathcal{U}_2 \cup \mathsf{ne}(\mathcal{I}, \mathcal{I} \circ \mathcal{U})$, and since $\mathcal{U}_2 \subseteq \mathcal{U}$ we conclude that $\mathsf{nup}(r) \subseteq \mathsf{lit}(\mathcal{U} \cup \mathsf{ne}(\mathcal{I}, \mathcal{I} \circ \mathcal{U}))$. Since $\mathcal{U} \cup \mathsf{ne}(\mathcal{I}, \mathcal{I} \circ \mathcal{U})$ is closed for η (which contains η_2), it follows that $\mathsf{head}(r) \cap (\mathcal{U} \cup \mathsf{ne}(\mathcal{I}, \mathcal{I} \circ \mathcal{U})) \neq \emptyset$, and since $\mathsf{head}(r)$ does not contain actions in $\mathcal{U}_1$ necessarily $\mathsf{head}(r) \cap (\mathcal{U}_2 \cup \mathsf{ne}(\mathcal{I}, \mathcal{I} \circ \mathcal{U})) \neq \emptyset$, whence $\mathsf{head}(r) \cap (\mathcal{U}' \cup \mathsf{ne}(\mathcal{I}, \mathcal{I} \circ \mathcal{U})) \neq \emptyset$.

In either case, from $\mathsf{nup}(r) \subseteq (\mathcal{U}' \cup \mathsf{ne}(\mathcal{I}, \mathcal{I} \circ \mathcal{U}))$ one concludes that $\mathsf{head}(r) \cap (\mathcal{U}' \cup \mathsf{ne}(\mathcal{I}, \mathcal{I} \circ \mathcal{U})) \neq \emptyset$, whence $\mathcal{U}' \cup \mathsf{ne}(\mathcal{I}, \mathcal{I} \circ \mathcal{U})$ is closed for η, contradicting minimality of $\mathcal{U}$. This is absurd, so $\mathcal{U}_1$ is a justified weak repair. Again the case for $\mathcal{U}_2$ is similar.

Since justified weak repairs are founded, Lemma 7 guarantees that $\mathcal{U} = \mathcal{U}_1 \cup \mathcal{U}_2$. Furthermore, if $\mathcal{U}$ is a justified repair for $\langle \mathcal{I}, \eta \rangle$, then each $\mathcal{U}_i$ is a justified repair for $\langle \mathcal{I}, \eta_i \rangle$ by Lemma 5. □

The practical significance of the results in this section is a parallelization algorithm: if $\eta = \eta_1 \cup \eta_2$ with $\eta_1 \perp\!\!\!\perp \eta_2$, then all (simple, founded, justified) repairs for $\langle \mathcal{I}, \eta \rangle$ can be expressed as unions of (simple, founded, justified) repairs for $\langle \mathcal{I}, \eta_1 \rangle$ and $\langle \mathcal{I}, \eta_2 \rangle$ by Lemmas 5, 7 and 9, so one can search for these repairs instead and combine them in at the end; Lemmas 3, 6 and 8 guarantee that no spurious results are obtained. The next section expands on these ideas, and discusses how η can be adequately split.

4 Finding Independent Sets of AICs

The results in the previous section show that splitting a set of AICs η into two independent sets η_1 and η_2 allows one to parallelize the search for repairs of a database $\mathcal{I}$, by searching independently for repairs for $\langle \mathcal{I}, \eta_1 \rangle$ and $\langle \mathcal{I}, \eta_2 \rangle$. In this section we address a complementary issue: how can one find these sets? We begin by formulating the results in the previous section in a more general way.

Definition 2. *A* partition *of a set of AICs η is a set $\boldsymbol{\eta} = \{\eta_1, \ldots, \eta_n\}$ such that $\eta = \cup_{i=1}^n \eta_i$ and $\eta_i \perp\!\!\!\perp \eta_j$ for $i \neq j$.*

Theorem 1. *Let $\boldsymbol{\eta}$ be a partition of η.*

1. *If $\mathcal{U}$ is a simple/founded/justified (weak) repair for $\langle \mathcal{I}, \eta \rangle$, then there exist sets $\mathcal{U}_1, \ldots, \mathcal{U}_n$ with $\mathcal{U} = \cup_{i=1}^{n} \mathcal{U}_i$ such that $\mathcal{U}_i$ is a simple/founded/justified (weak) repair for $\langle \mathcal{I}, \eta_i \rangle$.*
2. *If $\mathcal{U}_i$ is a simple/founded/justified (weak) repair for $\langle \mathcal{I}, \eta_i \rangle$ for $i = 1, \ldots, n$ and $\mathcal{U} = \cup_{i=1}^{n} \mathcal{U}_i$, then $\mathcal{U}$ is a simple/founded/justified (weak) repair for $\langle \mathcal{I}, \eta \rangle$.*

Proof. By induction on n. For $n = 1$, the results are trivial. Assume that the result is true for n; applying the induction hypothesis to $\eta_1, \ldots, \eta_n$, on the one hand, and the adequate lemma from Section 3 to $\eta' = \bigcup_{i=1}^{n}$ and η_{n+1}, yields the result for $\eta_1, \ldots, \eta_{n+1}$, since $\eta' \perp\!\!\!\perp \eta_{n+1}$. □

To find a partition of η (actually, the best partition of η), we will define an auxiliary relation on AICs. Two AICs r_1 and r_2 are *dependent*, $r_1 \not\perp\!\!\!\perp r_2$, if there exist literals $L_1 \in \mathsf{body}\,(r_1)$ and $L_2 \in \mathsf{body}\,(r_2)$ such that $|L_1| = |L_2|$.

Lemma 10. *Let $\boldsymbol{\eta}$ be a partition of η. Then η_i is closed under $\not\perp\!\!\!\perp$ for every i, i.e. for every rule $r, r' \in \eta$, if $r \in \eta_i$ and $r \not\perp\!\!\!\perp r'$, then $r' \in \eta_i$.*

Proof. Let r be a rule in η_i and let $r' \in \eta$ be such that $r \not\perp\!\!\!\perp r'$. Since $\boldsymbol{\eta}$ is a partition of η, $r' \in \eta_k$ for some k. But $i \neq k$ would contradict $\eta_i \perp\!\!\!\perp \eta_k$ (since r and r' are not independent), hence $i = k$. Therefore η_i is closed under $\not\perp\!\!\!\perp$. □

This relation is reflexive and symmetric, so its transitive closure $\not\perp\!\!\!\perp^{+}$ is an equivalence relation. This equivalence relation defines the best partition of η.

Theorem 2. *The quotient set $^{\eta}/_{\not\perp\!\!\!\perp^{+}}$ is a partition of η. Furthermore, for any other partition $\boldsymbol{\eta}'$ of η, if $\eta'_i \in \boldsymbol{\eta}'$, there exists $\eta_j \in {}^{\eta}/_{\not\perp\!\!\!\perp^{+}}$ such that $\eta_j \subseteq \eta'_i$.*

Proof. Let $^{\eta}/_{\not\perp\!\!\!\perp^{+}} = \{\eta_1, \ldots, \eta_n\}$. By definition of quotient set, $\bigcup_{i=1}^{n} \eta_i = \eta$. By definition of $\not\perp\!\!\!\perp$, $\eta_i \perp\!\!\!\perp \eta_j$. Given η'_i as in the statement of the theorem and choosing $r \in \eta'_i$ and observing that η'_i is closed under $\not\perp\!\!\!\perp$ (Lemma 10) and $[r]$ is the minimal set containing r and closed under $\not\perp\!\!\!\perp$, it follows that $[r] \subseteq \eta'_i$. □

Furthermore, $^{\eta}/_{\not\perp\!\!\!\perp^{+}}$ can be computed efficiently and in an incremental way.

Theorem 3. *Let η be a set of AICs such that every rule in η contains at most k literals in its body. Then $^{\eta}/_{\not\perp\!\!\!\perp^{+}}$ can be computed in $\mathcal{O}\,(k \times |\eta|)$.*

Proof. Consider the undirected graph whose nodes are both the rules in η and the atoms occurring in those rules, and where there is an edge between an atom and a rule if that atom occurs in that rule. This graph has at most $k \times |\eta|$ nodes and can be constructed in $\mathcal{O}\,(k \times |\eta|)$ time; it is a well-known fact that its connected components can again be computed in $\mathcal{O}\,(k \times |\eta|)$ time, and the rules in each component coincide precisely with the equivalence classes in $^{\eta}/_{\not\perp\!\!\!\perp^{+}}$. □

Three important remarks are due. First, k typically does not grow with η and is usually small, so essentially this algorithm is linear in the number of AICs. Also, the algorithm is independent of the underlying database, which is useful since the database typically changes more often than η. Finally, if one wishes to add new rules to η one can reuse the existing partition for η as a starting point, which makes the algorithm incremental.

5 Stratified Active Integrity Constraints

In this section, we show how to define a finer relation among active integrity constraints that will allow an incremental construction of these repairs that can again substantially reduce the time required to find them.

Throughout this section we assume a fixed set of AICs η, so all definitions are within the universe of this set.

Definition 3. *Let r_1 and r_2 be active integrity constraints. Then $r_1 \prec r_2$ (r_1 precedes r_2) if $\{|\mathsf{lit}(\alpha)| \mid \alpha \in \mathsf{head}(r_1)\} \cap \{|L| \mid L \in \mathsf{body}(r_2)\} \neq \emptyset$.*

Intuitively, r_1 precedes r_2 if ensuring r_1 may affect applicability of r_2. In particular, $r_1 \prec r_2$ implies $r_1 \not\perp r_2$.

By definition of AIC, $\prec$ is a reflexive relation. Let $\preceq$ be its transitive closure (within η) and $\approx$ be the equivalence relation induced by $\preceq$, i.e. $r_1 \approx r_2$ iff $r_1 \preceq r_2$ and $r_2 \preceq r_1$. It is a well-known result that $\langle {}^{\eta}/_{\approx}, \preceq \rangle$ is a partial order, where $[r_1] \preceq [r_2]$ iff $r_1 \preceq r_2$.

Definition 4. *Let $\eta_1, \eta_2 \subseteq \eta$ be closed under $\approx$. Then $\eta_1 \prec \eta_2$ (η_1 precedes η_2) if (i) some rule in η_1 precedes some other rule in η_2, but (ii) no rule in η_2 precedes a rule in η_1.*[4]

In particular, if $\eta_1 \prec \eta_2$ then η_1 and η_2 must be disjoint. Note that, if η_1 and η_2 are distinct minimal sets closed under $\approx$ (i.e. elements of $\eta/\approx$), then $\eta_1 \preceq \eta_2$ iff $\eta_1 \prec \eta_2$.

This stratification allows us to search for weak repairs as follows: if $\eta_1 \prec \eta_2$, then we can look for weak repairs for $\eta_1 \cup \eta_2$ by first looking for weak repairs for η_1 and then extending these to $\eta_1 \cup \eta_2$.

Lemma 11. *Let $\eta_1, \eta_2 \subseteq \eta$ with $\eta_1 \prec \eta_2$, $\mathcal{I}$ be a database and $\mathcal{U}$ be a set of update actions such that all actions in $\mathcal{U}$ occur in the head of some rule in $\eta_1 \cup \eta_2$. Let $\mathcal{U}_i$ be the restriction of $\mathcal{U}$ to the actions in the heads of rules in η_i. If $\mathcal{U}$ is a weak repair for $\langle \mathcal{I}, \eta_1 \cup \eta_2 \rangle$, then $\mathcal{U}_1$ and $\mathcal{U}_2$ are weak repairs for $\langle \mathcal{I}, \eta_1 \rangle$ and $\langle \mathcal{I} \circ \mathcal{U}_1, \eta_2 \rangle$, respectively.*

Proof. Since $\eta_1 \prec \eta_2$, (a) actions in the head of a rule in η_2 cannot change literals in the body of rules in η_1 and in particular (b) $\mathcal{U}_1$ and $\mathcal{U}_2$ are disjoint.

By (a), $\mathcal{I} \circ \mathcal{U}_1 \models r$ iff $\mathcal{I} \circ \mathcal{U} \models r$ for every $r \in \eta_1$, so $\mathcal{U}_1$ is a weak repair for $\langle \mathcal{I}, \eta_1 \rangle$. By (b), $\mathcal{I} \circ \mathcal{U} = \mathcal{I} \circ (\mathcal{U}_1 \cup \mathcal{U}_2) = (\mathcal{I} \circ \mathcal{U}_1) \circ \mathcal{U}_2$, hence $\mathcal{U}_2$ is a weak repair for $\langle \mathcal{I} \circ \mathcal{U}_1, \eta_2 \rangle$. □

[4] Formally: $\eta_1 \prec \eta_2$ if (i) $r_1 \prec r_2$ for some $r_1 \in \eta_1$ and $r_2 \in \eta_2$, but (ii) $r_2 \not\prec r_1$ for every $r_1 \in \eta_1$ and $r_2 \in \eta_2$.

Lemma 12. *In the conditions of Lemma 11, if $\mathcal{U}$ is founded w.r.t. $\langle \mathcal{I}, \eta_1 \cup \eta_2 \rangle$, then $\mathcal{U}_1$ and $\mathcal{U}_2$ are founded w.r.t. $\langle \mathcal{I}, \eta_1 \rangle$ and $\langle \mathcal{I} \circ \mathcal{U}_1, \eta_2 \rangle$, respectively.*

Proof. (i) Let $\alpha \in \mathcal{U}_1$. Since $\mathcal{U}$ is founded w.r.t. $\langle \mathcal{I}, \eta_1 \cup \eta_2 \rangle$, there is a rule $r \in \eta_1 \cup \eta_2$ such that $\alpha \in \mathsf{head}(r)$ and $\mathcal{I} \circ \mathcal{U} \models L$ for every $L \in \mathsf{body}(r) \setminus \{\mathsf{lit}(\alpha)^D\}$. Since $\eta_1 \prec \eta_2$, necessarily $r \in \eta_1$. By (b) from the previous proof, $\mathcal{I} \circ \mathcal{U}_1 \models L$ for every $L \in \mathsf{body}(r) \setminus \{\mathsf{lit}(\alpha)^D\}$, whence α is founded w.r.t. $\langle \mathcal{I}, \eta_1 \rangle$ and $\mathcal{U}_1$. Thus $\mathcal{U}_1$ is founded w.r.t. $\langle \mathcal{I}, \eta_1 \rangle$.

(ii) Let $\alpha \in \mathcal{U}_2$. Again there must be a rule $r \in \eta$ such that $\alpha \in \mathsf{head}(r)$ and $\mathcal{I} \circ \mathcal{U} \models L$ for every $L \in \mathsf{body}(r) \setminus \{\mathsf{lit}(\alpha)^D\}$, and as before necessarily $r \in \eta_2$. Since $I \circ \mathcal{U} = (\mathcal{I} \circ \mathcal{U}_1) \circ \mathcal{U}_2$, it follows that α is founded w.r.t. $\langle \mathcal{I} \circ \mathcal{U}_1, \eta_2 \rangle$ and $\mathcal{U}_2$, hence $\mathcal{U}_2$ is founded w.r.t. $\langle \mathcal{I} \circ \mathcal{U}_1, \eta_2 \rangle$. □

Corollary 3. *If $\mathcal{U}$ is a founded weak repair for $\langle \mathcal{I}, \eta \rangle$, then $\mathcal{U}_1$ and $\mathcal{U}_2$ are founded weak repairs for $\langle \mathcal{I}, \eta_1 \rangle$ and $\langle \mathcal{I} \circ \mathcal{U}_1, \eta_2 \rangle$, respectively.*

Proof. Immediate consequence of Lemmas 11 and 12. □

Lemma 13. *In the conditions of Lemma 11, if $\mathcal{U}$ is a justified weak repair for $\langle \mathcal{I}, \eta_1 \cup \eta_2 \rangle$, then $\mathcal{U}_1$ and $\mathcal{U}_2$ are justified weak repairs for $\langle \mathcal{I}, \eta_1 \rangle$ and $\langle \mathcal{I} \circ \mathcal{U}_1, \eta_2 \rangle$, respectively.*

Proof. We first make some remarks that will be relevant throughout the proof.

(a) $\mathcal{I} \circ \mathcal{U}_1 \models L$ iff $\mathcal{I} \circ \mathcal{U} \models L$ for every literal $L \in \mathsf{body}(r)$ with $r \in \eta_1$, as argued in the proof of Lemma 11.
(b) $\mathsf{ne}(\mathcal{I}, \mathcal{I} \circ \mathcal{U}) \subseteq \mathsf{ne}(\mathcal{I}, \mathcal{I} \circ \mathcal{U}_1)$, as in the proof of Lemma 8.
(c) $\mathsf{ne}(\mathcal{I}, \mathcal{I} \circ \mathcal{U}_1) \subseteq \mathsf{ne}(\mathcal{I}, \mathcal{I} \circ \mathcal{U}) \cup \mathcal{U}_2$, since actions in $\mathcal{U}_2$ may not affect literals in the body of rules in η_1 (this would contradict $\eta_1 \prec \eta_2$). In particular, if $\mathsf{nup}(r) \subseteq \mathsf{lit}(\mathsf{ne}(\mathcal{I}, \mathcal{I} \circ \mathcal{U}_1))$ for some $r \in \eta_1$, then $L \in \mathsf{lit}(\mathsf{ne}(\mathcal{I}, \mathcal{I} \circ \mathcal{U}))$; and if $\alpha \in \mathsf{head}(r)$ for some $r \in \eta_1$ and $\alpha \in \mathsf{ne}(\mathcal{I}, \mathcal{I} \circ \mathcal{U}_1)$, then $\alpha \in \mathsf{ne}(\mathcal{I}, \mathcal{I} \circ \mathcal{U})$.

(i) Let $r \in \eta_1$ be such that $\mathsf{nup}(r) \subseteq \mathsf{lit}(\mathcal{U}_1 \cup \mathsf{ne}(\mathcal{I}, \mathcal{I} \circ \mathcal{U}_1))$. From $\mathcal{U}_1 \subseteq \mathcal{U}$ and (c), one gets $\mathsf{nup}(r) \subseteq \mathsf{lit}(\mathcal{U} \cup \mathsf{ne}(\mathcal{I}, \mathcal{I} \circ \mathcal{U}))$; since $\mathcal{U}$ is closed under η, $\mathsf{head}(r) \cap (\mathcal{U} \cup \mathsf{ne}(\mathcal{I}, \mathcal{I} \circ \mathcal{U})) \neq \emptyset$. By definition of $\mathcal{U}_1$ and (c), also $\mathsf{head}(r) \cap (\mathcal{U}_1 \cup \mathsf{ne}(\mathcal{I}, \mathcal{I} \circ \mathcal{U}_1)) \neq \emptyset$, whence $\mathcal{U}_1 \cup \mathsf{ne}(\mathcal{I}, \mathcal{I} \circ \mathcal{U}_1)$ is closed under η_1.

For minimality, suppose that $\mathcal{U}_1' \subsetneq \mathcal{U}_1$ is such that $\mathcal{U}_1' \cup \mathsf{ne}(\mathcal{I}, \mathcal{I} \circ \mathcal{U}_1)$ is closed under η_1 and take $\mathcal{U}' = \mathcal{U}_1' \cup \mathcal{U}_2$. We show that $\mathcal{U}' \cup \mathsf{ne}(\mathcal{I}, \mathcal{I} \circ \mathcal{U})$ is closed under η. Assume $\mathsf{nup}(r) \subseteq \mathsf{lit}(\mathcal{U}' \cup \mathsf{ne}(\mathcal{I}, \mathcal{I} \circ \mathcal{U}))$; there are two possible cases.

- $r \in \eta_1$: since $\eta_1 \prec \eta_2$, no literal in $\mathsf{nup}(r)$ can occur in $\mathsf{lit}(\mathcal{U}_2)$; therefore, from (b) it follows that $\mathsf{nup}(r) \subseteq \mathsf{lit}(\mathcal{U}_1' \cup \mathsf{ne}(\mathcal{I}, \mathcal{I} \circ \mathcal{U}_1))$, and thus $\mathsf{head}(r) \cap (\mathcal{U}_1' \cup \mathsf{ne}(\mathcal{I}, \mathcal{I} \circ \mathcal{U}_1)) \neq \emptyset$. From $\mathcal{U}_1' \subseteq \mathcal{U}'$ and (c), also $\mathsf{head}(r) \cap (\mathcal{U}' \cup \mathsf{ne}(\mathcal{I}, \mathcal{I} \circ \mathcal{U})) \neq \emptyset$.
- $r \in \eta_2$: from $(\mathcal{U}_1' \cup \mathcal{U}_2) \subseteq \mathcal{U}$, we conclude that $\mathsf{nup}(r) \subseteq \mathsf{lit}(\mathcal{U} \cup \mathsf{ne}(\mathcal{I}, \mathcal{I} \circ \mathcal{U}))$, hence $\mathsf{head}(r) \cap (\mathcal{U} \cup \mathsf{ne}(\mathcal{I}, \mathcal{I} \circ \mathcal{U})) \neq \emptyset$ because $\eta_2 \subseteq \eta$ and $\mathcal{U}$ is closed for η. But $\mathsf{head}(r)$ cannot contain actions in $\mathcal{U}_1$, so $\mathsf{head}(r) \cap (\mathcal{U}_2 \cup \mathsf{ne}(\mathcal{I}, \mathcal{I} \circ \mathcal{U})) \neq \emptyset$, whence $\mathsf{head}(r) \cap (\mathcal{U}' \cup \mathsf{ne}(\mathcal{I}, \mathcal{I} \circ \mathcal{U})) \neq \emptyset$.

In either case, from $\mathsf{nup}(r) \subseteq (\mathcal{U}' \cup \mathsf{ne}(\mathcal{I}, \mathcal{I} \circ \mathcal{U}))$ one concludes that $\mathsf{head}(r) \cap (\mathcal{U}' \cup \mathsf{ne}(\mathcal{I}, \mathcal{I} \circ \mathcal{U})) \neq \emptyset$, whence $\mathcal{U}' \cup \mathsf{ne}(\mathcal{I}, \mathcal{I} \circ \mathcal{U})$ is closed for η, which contradicts $\mathcal{U}$ being a justified weak repair for $\langle \mathcal{I}, \eta \rangle$. This is absurd, therefore $\mathcal{U}_1$ is a justified weak repair for $\langle \mathcal{I}, \eta_1 \rangle$.

(ii) Denote by $\mathcal{N}$ the set $\mathsf{ne}(\mathcal{I} \circ \mathcal{U}_1, \mathcal{I} \circ \mathcal{U}_1 \circ \mathcal{U}_2)$. To show that $\mathcal{U}_2$ is a justified weak repair for $\langle \mathcal{I} \circ \mathcal{U}_1, \eta_2 \rangle$, we need to show that $\mathcal{U}_2 \cup \mathcal{N}$ is closed for η_2 and that it is the minimal such set containing $\mathcal{N}$. Note that (d) $\mathcal{N} = \mathcal{U}_1 \cup \mathsf{ne}(\mathcal{I}, \mathcal{I} \circ \mathcal{U})$, since $I \circ \mathcal{U}_1$ is "between" $\mathcal{I}$ and $\mathcal{I} \circ \mathcal{U}$ (as $\mathcal{U}_1 \subseteq \mathcal{U}$).

First we show that $\mathcal{U}_2 \cup \mathcal{N}$ is closed for η_2. Let $r \in \eta_2$ and assume that $\mathsf{nup}(r) \subseteq \mathsf{lit}(\mathcal{U}_2 \cup \mathcal{N})$. By (d), $\mathsf{nup}(r) \subseteq \mathsf{lit}(\mathcal{U}_2 \cup \mathcal{U}_1 \cup \mathsf{ne}(\mathcal{I}, \mathcal{I} \circ \mathcal{U})) = (\mathcal{U} \cup \mathsf{ne}(\mathcal{I}, \mathcal{I} \circ \mathcal{U}))$, whence $\mathsf{head}(r) \cap (\mathcal{U} \cup \mathsf{ne}(\mathcal{I}, \mathcal{I} \circ \mathcal{U})) \neq \emptyset$ because $\mathcal{U} \cup \mathsf{ne}(\mathcal{I}, \mathcal{I} \circ \mathcal{U})$ is closed under η. By construction of $\mathcal{U}_2$ and (d), also $\mathsf{head}(r) \cap (\mathcal{U}_2 \cup \mathcal{N})$, whence $\mathcal{U}_2 \cup \mathcal{N}$ is closed under η_2.

Now let $\mathcal{U}_2' \subsetneq \mathcal{U}_2$ be such that $\mathcal{U}_2' \cup \mathcal{N}$ is closed for η_2 and take $\mathcal{U}' = \mathcal{U}_1 \cup \mathcal{U}_2'$. We show that $\mathcal{U}' \cup \mathsf{ne}(\mathcal{I}, \mathcal{I} \circ \mathcal{U})$ is closed under η. Let $r \in \eta$ be such that $\mathsf{nup}(r) \subseteq \mathsf{lit}(\mathcal{U}' \cup \mathsf{ne}(\mathcal{I}, \mathcal{I} \circ \mathcal{U}))$. Yet again, there are two cases to consider.

- $r \in \eta_1$: since $\mathcal{U}' \subseteq \mathcal{U}$, also $\mathsf{nup}(r) \subseteq \mathsf{lit}(\mathcal{U} \cup \mathsf{ne}(\mathcal{I}, \mathcal{I} \circ \mathcal{U}))$, whence $\mathsf{head}(r) \cap (\mathcal{U} \cup \mathsf{ne}(\mathcal{I}, \mathcal{I} \circ \mathcal{U})) \neq \emptyset$ because $\mathcal{U} \cup \mathsf{ne}(\mathcal{I}, \mathcal{I} \circ \mathcal{U})$ is closed for η. But actions in $\mathsf{head}(r)$ may not occur in $\mathcal{U}_2$, hence $\mathsf{head}(r) \cap (\mathcal{U}' \cup \mathsf{ne}(\mathcal{I}, \mathcal{I} \circ \mathcal{U})) \neq \emptyset$ since $(\mathcal{U} \setminus \mathcal{U}') \subseteq \mathcal{U}_2$.
- $r \in \eta_2$: by (d), $\mathcal{U}' \cup \mathsf{ne}(\mathcal{I}, \mathcal{I} \circ \mathcal{U}) = \mathcal{U}_1 \cup \mathcal{U}_2' \cup \mathsf{ne}(\mathcal{I}, \mathcal{I} \circ \mathcal{U}) = \mathcal{U}_2' \cup \mathcal{N}$, whence $\mathsf{head}(r) \cap (\mathcal{U}_2' \cup \mathcal{N}) \neq \emptyset$ because $\mathcal{U}_2' \cup \mathcal{N}$ is closed for η_2, which amounts to saying that that $\mathsf{head}(r) \cap (\mathcal{U}' \cup \mathsf{ne}(\mathcal{I}, \mathcal{I} \circ \mathcal{U})) \neq \emptyset$.

In either case, $\mathsf{head}(r) \cap (\mathcal{U}' \cup \mathsf{ne}(\mathcal{I}, \mathcal{I} \circ \mathcal{U})) \neq \emptyset$, so $\mathcal{U}' \cup \mathsf{ne}(\mathcal{I}, \mathcal{I} \circ \mathcal{U})$ is closed under η, contradicting the fact that $\mathcal{U}$ is a justified weak repair for $\langle \mathcal{I}, \eta \rangle$. Therefore $\mathcal{U}_2$ is a justified weak repair for $\langle \mathcal{I} \circ \mathcal{U}_1, \eta_2 \rangle$. □

Lemmas 11, 12 and 13 are analogue to Lemmas 4, 7 and 9, respectively. Interestingly, the analogue of Lemma 5 does not hold in this setting: it may happen that $\mathcal{U}$ is a repair, but $\mathcal{U}_1$ is a weak repair. The reason is that there may be a repair for $\langle \mathcal{I}, \eta_1 \rangle$ such that there is no (weak) repair for $\langle \mathcal{I} \circ \mathcal{U}_1, \eta_2 \rangle$.

Example 1. Let $\mathcal{I} = \emptyset$ and consider the following active integrity constraints.

$$r_1 : \mathsf{not}\ a \supset +a \qquad r_4 : a, \mathsf{not}\ b, \mathsf{not}\ c, d \supset -d$$

$$r_2 : \mathsf{not}\ b, c \supset +b \qquad r_5 : a, \mathsf{not}\ b, \mathsf{not}\ c, \mathsf{not}\ d \supset +d$$

$$r_3 : b, \mathsf{not}\ c \supset +c$$

Taking $\eta_1 = \{r_1, r_2, r_3\}$ and $\eta_2 = \{r_4, r_5\}$, one has $\eta_1 \prec \eta_2$. Furthermore, $\{+a\}$ and $\{+a, +b, +c\}$ are weak repairs for $\langle \mathcal{I}, \eta_1 \rangle$, the first of which is a repair. However, the only repair for $\langle \mathcal{I}, \eta_1 \cup \eta_2 \rangle$ is $\{+a, +b, +c\}$, which is not the union of $\{+a\}$ with a repair for $\langle \mathcal{I} \circ \{+a\}, \eta_2 \rangle$.

However, if both steps succeed then we can combine their results as before.

Lemma 14. *Let $\eta_1, \eta_2 \subseteq \eta$ with $\eta_1 \prec \eta_2$, $\mathcal{I}$ be a database, and $\mathcal{U}_1$ and $\mathcal{U}_2$ be sets of update actions such that all actions in $\mathcal{U}_i$ occur the head of some rule in η_i. If $\mathcal{U}_1$ is a weak repair for $\langle \mathcal{I}, \eta_1 \rangle$ and $\mathcal{U}_2$ is a weak repair for $\langle \mathcal{I} \circ \mathcal{U}_1, \eta_2 \rangle$, then $\mathcal{U} = \mathcal{U}_1 \cup \mathcal{U}_2$ is a weak repair for $\langle \mathcal{I}, \eta_1 \cup \eta_2 \rangle$.*

Proof. Since $\eta_1 \prec \eta_2$, the hypothesis over $\mathcal{U}_2$ imply that (a) actions in $\mathcal{U}_2$ cannot change literals in the body of rules in η_1 and in particular (b) $\mathcal{U}_1$ and $\mathcal{U}_2$ are disjoint.

Take $r \in \eta_1$. Then $\mathcal{I} \circ \mathcal{U}_1 \models r$, whence $\mathcal{I} \circ \mathcal{U} \models r$ by (a).

Take $r \in \eta_2$. Then $(\mathcal{I} \circ \mathcal{U}_1) \circ \mathcal{U}_2 \models r$, and by (b) $(\mathcal{I} \circ \mathcal{U}_1) \circ \mathcal{U}_2 = \mathcal{I} \circ \mathcal{U}$.

Therefore $\mathcal{U}_1 \circ \mathcal{U}_2$ is a weak repair for $\langle \mathcal{I}, \eta_1 \cup \eta_2 \rangle$. □

Lemma 15. *In the conditions of Lemma 14, if $\mathcal{U}_1$ is a repair for $\langle \mathcal{I}, \eta_1 \rangle$ and $\mathcal{U}_2$ is a repair for $\langle \mathcal{I} \circ \mathcal{U}_1, \eta_2 \rangle$, then $\mathcal{U}$ is a repair for $\langle \mathcal{I}, \eta_1 \cup \eta_2 \rangle$.*

Proof. By Lemma 14, $\mathcal{U}$ is a weak repair for $\langle \mathcal{I}, \eta_1 \cup \eta_2 \rangle$. Suppose $\mathcal{U}$ is not a repair; then there is $\mathcal{U}' \subsetneq \mathcal{U}$ such that $\mathcal{U}'$ is also a weak repair for $\langle \mathcal{I}, \eta_1 \cup \eta_2 \rangle$.

Take $\mathcal{U}'_1 = \mathcal{U}' \cap \mathcal{U}_1$ and $\mathcal{U}'_2 = \mathcal{U}' \cap \mathcal{U}_2$; by Lemma 11, $\mathcal{U}'_1$ is a weak repair for $\langle \mathcal{I}, \eta_1 \rangle$ and $\mathcal{U}'_2$ is a weak repair for $\langle \mathcal{I} \circ \mathcal{U}_1, \eta_2 \rangle$. But at least one of the inclusions $\mathcal{U}'_1 \subseteq \mathcal{U}_1$ and $\mathcal{U}'_2 \subseteq \mathcal{U}_2$ must be strict, contradicting the hypothesis that $\mathcal{U}_1$ and $\mathcal{U}_2$ are both repairs. Therefore $\mathcal{U}$ is a repair for $\langle \mathcal{I}, \eta_1 \cup \eta_2 \rangle$. □

In this setting, the condition that $\mathcal{U}_1$ and $\mathcal{U}_2$ be repairs is sufficient but not necessary, as illustrated by the example above – unlike in Lemma 3 earlier.

Lemma 16. *In the conditions of Lemma 14, if $\mathcal{U}_1$ is founded w.r.t. $\langle \mathcal{I}, \eta_1 \rangle$ and $\mathcal{U}_2$ is founded w.r.t. $\langle \mathcal{I} \circ \mathcal{U}_1, \eta_2 \rangle$, then $\mathcal{U}$ is founded w.r.t. $\langle \mathcal{I}, \eta_1 \cup \eta_2 \rangle$.*

Proof. Take $\alpha \in \mathcal{U}_1$. Since $\mathcal{U}_1$ is founded w.r.t. $\langle \mathcal{I}, \eta_1 \rangle$, there is a rule $r \in \eta_1$ such that $\alpha \in \mathsf{head}(r)$ and $\mathcal{I} \circ \mathcal{U}_1 \models L$ for every $L \in \mathsf{body}(r) \setminus \{\mathsf{lit}(\alpha)^D\}$. By (b) from the proof of Lemma 14, also $\mathcal{I} \circ \mathcal{U} \models L$ for every $L \in \mathsf{body}(r) \setminus \{\mathsf{lit}(\alpha)^D\}$, whence α is founded w.r.t. $\langle \mathcal{I}, \eta_1 \cup \eta_2 \rangle$ and $\mathcal{U}$.

Take $\alpha \in \mathcal{U}_2$. Since $\mathcal{U}_2$ is founded w.r.t. $\langle \mathcal{I} \circ \mathcal{U}_1, \eta_2 \rangle$, there is a rule $r \in \eta_2$ such that $(\mathcal{I} \circ \mathcal{U}_1) \circ \mathcal{U}_2 \models L$ for every $L \in \mathsf{body}(r) \setminus \{\mathsf{lit}(\alpha)^D\}$, and since $(\mathcal{I} \circ \mathcal{U}_1) \circ \mathcal{U}_2 = \mathcal{I} \circ \mathcal{U}$ this implies that α is founded w.r.t. $\langle \mathcal{I}, \eta_1 \cup \eta_2 \rangle$ and $\mathcal{U}$.

Therefore $\mathcal{U}$ is founded w.r.t. $\langle \mathcal{I}, \eta_1 \cup \eta_2 \rangle$. □

As before, Lemmas 14, 15 and 16 can be combined in the following corollary.

Corollary 4. *In the conditions of Lemma 14, if $\mathcal{U}_1$ is a founded (weak) repair for $\langle \mathcal{I}, \eta_1 \rangle$ and $\mathcal{U}_2$ is a founded (weak) repair for $\langle \mathcal{I} \circ \mathcal{U}_1, \eta_2 \rangle$, then $\mathcal{U}$ is a founded (weak) repair for $\langle \mathcal{I}, \eta_1 \cup \eta_2 \rangle$.*

Lemma 17. *In the conditions of Lemma 14, if $\mathcal{U}_1$ is a justified weak repair for $\langle \mathcal{I}, \eta_1 \rangle$ and $\mathcal{U}_2$ is a justified weak repair for $\langle \mathcal{I} \circ \mathcal{U}_1, \eta_2 \rangle$, then $\mathcal{U}$ is a justified weak repair for $\langle \mathcal{I}, \eta_1 \cup \eta_2 \rangle$.*

Proof. First observe that properties (a–d) of the proof of Lemma 13 all hold in this context. Define $\mathcal{N} = \mathsf{ne}(\mathcal{I} \circ \mathcal{U}_1, \mathcal{I} \circ \mathcal{U}_1 \circ \mathcal{U}_2)$ as in that proof.

To see that $\mathcal{U} \cup \mathsf{ne}(\mathcal{I}, \mathcal{I} \circ \mathcal{U})$ is closed for $\langle \mathcal{I}, \eta \rangle$, let $r \in \eta_1 \cup \eta_2$ be such that $\mathsf{nup}(r) \subseteq \mathsf{lit}(\mathcal{U} \cup \mathsf{ne}(\mathcal{I}, \mathcal{I} \circ \mathcal{U}))$. We need to consider two cases.

- If $r \in \eta_1$, then $\mathsf{nup}(r) \subseteq \mathsf{lit}(\mathcal{U}_1 \cup \mathsf{ne}(\mathcal{I}, \mathcal{I} \circ \mathcal{U}_1))$ by (a) and (b), and since $\mathcal{U}_1 \cup \mathsf{ne}(\mathcal{I}, \mathcal{I} \circ \mathcal{U}_1)$ closed for η_1 this implies that $\mathsf{head}(r) \cap (\mathcal{U}_1 \cup \mathsf{ne}(\mathcal{I}, \mathcal{I} \circ \mathcal{U}_1)) \neq \emptyset$, whence also $\mathsf{head}(r) \cap (\mathcal{U} \cup \mathsf{ne}(\mathcal{I}, \mathcal{I} \circ \mathcal{U})) \neq \emptyset$ by $\mathcal{U}_1 \subseteq \mathcal{U}$ and (c).
- If $r \in \eta_2$, then by equality (d) we have $\mathcal{U} \cup \mathsf{ne}(\mathcal{I}, \mathcal{I} \circ \mathcal{U}) = \mathcal{U}_2 \cup \mathcal{U}_1 \cup \mathsf{ne}(\mathcal{I}, \mathcal{I} \circ \mathcal{U}) = \mathcal{U}_2 \cup \mathcal{N}$; then $\mathsf{nup}(r) \subseteq \mathsf{lit}(\mathcal{U}_2 \cup \mathcal{N})$, whence $\mathsf{head}(r) \cap (\mathcal{U}_2 \cup \mathcal{N}) \neq \emptyset$ because $\mathcal{U}_2 \cup \mathcal{N}$ is closed for η_2, and the latter condition is precisely $\mathsf{head}(r) \cap (\mathcal{U} \cup \mathsf{ne}(\mathcal{I}, \mathcal{I} \circ \mathcal{U})) \neq \emptyset$.

In either case $\mathcal{U} \cup \mathsf{ne}(\mathcal{I}, \mathcal{I} \circ \mathcal{U})$ is closed for r, whence $\mathcal{U} \cup \mathsf{ne}(\mathcal{I}, \mathcal{I} \circ \mathcal{U})$ is closed for $\eta_1 \cup \eta_2$.

For minimality, let $\mathcal{U}' \subseteq \mathcal{U}$ be such that $\mathcal{U}' \cup \mathsf{ne}(\mathcal{I}, \mathcal{I} \circ \mathcal{U})$ is closed for $\eta_1 \cup \eta_2$ and take $\mathcal{U}'_i = \mathcal{U}' \cap \mathcal{U}_i$ for $i = 1, 2$. We show that $\mathcal{U}'_1 = \mathcal{U}_1$ and $\mathcal{U}'_2 = \mathcal{U}_2$.

- Let $r \in \eta_1$ be such that $\mathsf{nup}(r) \subseteq \mathsf{lit}(\mathcal{U}'_1 \cup \mathsf{ne}(\mathcal{I}, \mathcal{I} \circ \mathcal{U}_1))$. Since $\mathcal{U}'_1 \subseteq \mathcal{U}'$, from (c) and the fact that $\mathsf{nup}(r) \cap \mathsf{lit}(\mathcal{U}_2) = \emptyset$ (because $\eta_1 \prec \eta_2$) we conclude that $\mathsf{nup}(r) \subseteq \mathsf{lit}(\mathcal{U}' \cup \mathsf{ne}(\mathcal{I}, \mathcal{I} \circ \mathcal{U})$, whence $\mathsf{head}(r) \cap (\mathcal{U}' \cup \mathsf{ne}(\mathcal{I}, \mathcal{I} \circ \mathcal{U})) \neq \emptyset$. By (b) and the fact that $\mathsf{head}(r) \cap \mathcal{U}_2 = \emptyset$, also $\mathsf{head}(r) \cap (\mathcal{U}'_1 \cup \mathsf{ne}(\mathcal{I}, \mathcal{I} \circ \mathcal{U}_1)) \neq \emptyset$. Therefore $\mathcal{U}'_1 \cup \mathsf{ne}(\mathcal{I}, \mathcal{I} \circ \mathcal{U}_1)$ contains $\mathsf{ne}(\mathcal{I}, \mathcal{I} \circ \mathcal{U}_1)$ and is closed for η_1; since $\mathcal{U}_1 \cup \mathsf{ne}(\mathcal{I}, \mathcal{I} \circ \mathcal{U}_1)$ is the minimal set with this property and $\mathcal{U}_1 \cap \mathsf{ne}(\mathcal{I}, \mathcal{I} \circ \mathcal{U}_1) = \emptyset$, it follows that $\mathcal{U}'_1 = \mathcal{U}_1$.
- Let $r \in \eta_2$ be such that $\mathsf{nup}(r) \subseteq \mathsf{lit}(\mathcal{U}'_2 \cup \mathcal{N})$. From (d) and the equality $\mathcal{U}'_1 = \mathcal{U}_1$ established above, $\mathsf{nup}(r) \subseteq \mathsf{lit}(\mathcal{U}' \cup \mathsf{ne}(\mathcal{I}, \mathcal{I} \circ \mathcal{U}))$, whence $\mathsf{head}(r) \cap (\mathcal{U}' \cup \mathsf{ne}(\mathcal{I}, \mathcal{I} \circ \mathcal{U})) \neq \emptyset$. Again by (d) and $\mathcal{U}'_1 = \mathcal{U}_1$ this amounts to saying that $\mathsf{head}(r) \cap (\mathcal{U}'_2 \cup \mathcal{N}) \neq \emptyset$. Therefore $\mathcal{U}'_2 \cup \mathcal{N}$ contains $\mathcal{N}$ and is closed for η_2, whence as before necessarily $\mathcal{U}'_2 = \mathcal{U}_2$.

Therefore $\mathcal{U}' = \mathcal{U}$, hence the set $\mathcal{U} \cup \mathsf{ne}(\mathcal{I}, \mathcal{I} \circ \mathcal{U})$ is the minimal set containing $\mathsf{ne}(\mathcal{I}, \mathcal{I} \circ \mathcal{U})$ and closed for $\eta_1 \cup \eta_2$. Therefore $\mathcal{U}$ is a justified weak repair for $\langle \mathcal{I}, \eta_1 \cup \eta_2 \rangle$. □

Lemmas 11, 12 and 13 allow us to split the search for (weak) repairs into smaller steps, while Lemmas 14, 15, 16 and 17 allow us to combine the results. However, $\langle {}^{\eta}/_{\approx}, \preceq \rangle$ is in general not a total order. Therefore, to obtain (weak) repairs for η, we need to be able to combine weak repairs of sets η_1 and η_2 that are not related via $\prec$ (see example below).

Let η_1, η_2 be two such sets, and consider a weak repair $\mathcal{U}$ for $\langle \mathcal{I}, \eta_1 \cup \eta_2 \rangle$. By Lemma 11, restricting $\mathcal{U}$ to the actions in $\eta_1 \cap \eta_2$ yields a weak repair $\mathcal{U}'$ for $\langle \mathcal{I}, \eta_1 \cap \eta_2 \rangle$; furthermore, restricting $\mathcal{U}$ to the actions in $(\eta_1 \cup \eta_2) \setminus (\eta_1 \cap \eta_2)$ yields a weak repair for $\langle \mathcal{I} \circ (\eta_1 \cap \eta_2), (\eta_1 \cup \eta_2) \setminus (\eta_1 \cap \eta_2) \rangle$. This allows us to restrict ourselves, without loss of generality, to the analysis of the situation where $\eta_1 \cap \eta_2 = \emptyset$. Since in this case the application of rules in η_1 does not affect the semantics of rules in η_2 and vice-versa, the proofs of Lemmas 2, 3, 6 and 8 can be straightforwardly adapted[5] to prove the following result.

[5] Although these lemmas assume that $\eta_1 \perp\!\!\!\perp \eta_2$, the key argument is that applying rules in $\mathcal{U}_1$ does not affect the semantics of rules in η_2 and conversely, which still remains true if η_1 and η_2 are closed under $\approx$ and neither $\eta_1 \prec \eta_2$ nor $\eta_2 \prec \eta_1$.

Lemma 18. *Let $\eta_1, \eta_2 \subseteq \eta$ be closed under $\approx$ and such that $\eta_1 \not\prec \eta_2$ and $\eta_2 \not\prec \eta_1$. Let $\mathcal{U}$ be a weak repair for $\langle \mathcal{I}, \eta_1 \cup \eta_2 \rangle$ such that $\mathcal{U}$ only consists of actions in the heads of rules in $\eta_1 \cup \eta_2$. Define $\mathcal{U}_i$ to be the restriction of $\mathcal{U}$ to the actions in the heads of rules in η_i. Then:*

1. *each $\mathcal{U}_i$ is a weak repair for $\langle \mathcal{I}, \eta_i \rangle$;*
2. *if $\mathcal{U}$ is a repair for $\langle \mathcal{I}, \eta_1 \cup \eta_2 \rangle$, then $\mathcal{U}_i$ is a repair for $\langle \mathcal{I}, \eta_i \rangle$;*
3. *if $\mathcal{U}$ is founded w.r.t. $\langle \mathcal{I}, \eta_1 \cup \eta_2 \rangle$, then $\mathcal{U}_i$ is founded w.r.t. $\langle \mathcal{I}, \eta_i \rangle$;*
4. *if $\mathcal{U}$ is a justified (weak) repair for $\langle \mathcal{I}, \eta_1 \cup \eta_2 \rangle$, then $\mathcal{U}_i$ is a justified (weak) repair for $\langle \mathcal{I}, \eta_i \rangle$.*

Example 2. To understand how these results can be applied, consider the following set of AICs η.

$$r_1 : a, b \supset -a \mid -b \qquad r_4 : a, \mathsf{not}\ b, \mathsf{not}\ e \supset +e$$
$$r_2 : \mathsf{not}\ a, c \supset +a \qquad r_5 : d, e, \mathsf{not}\ f \supset +f$$
$$r_3 : b, c, d \supset -d$$

The precedence relation between these rules, omitting the reflexive edges, can be summarized in the following diagram.

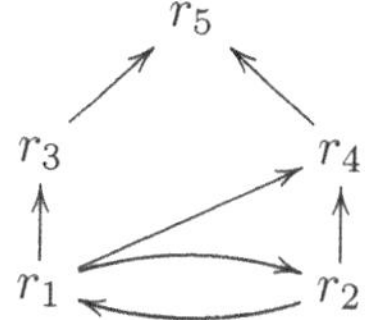

The equivalence classes are $\eta_1 = \{r_1, r_2\}$, $\eta_2 = \{r_3\}$, $\eta_3 = \{r_4\}$ and $\eta_4 = \{r_5\}$, with (direct) precedence relation $\eta_1 \preceq \eta_2 \preceq \eta_4$ and $\eta_1 \preceq \eta_3 \preceq \eta_4$. In order to find e.g. a founded weak repair for $\langle \mathcal{I}, \eta \rangle$, we would:

1. find all founded weak repairs for $\langle \mathcal{I}, \{r_1, r_2\} \rangle$;
2. extend each such $\mathcal{U}$ to founded weak repairs for $\langle \mathcal{I} \circ \mathcal{U}, \{r_3\} \rangle$ and $\langle \mathcal{I} \circ \mathcal{U}, \{r_4\} \rangle$, using Lemma 16;
3. for each pair of weak repairs $\mathcal{U}_2$ for $\langle \mathcal{I}, \{r_1, r_2, r_3\} \rangle$ and $\mathcal{U}_3$ for $\langle \mathcal{I}, \{r_1, r_2, r_4\} \rangle$ such that $\mathcal{U}_2$ and $\mathcal{U}_3$ coincide on the actions from heads of rules in $\{r_1, r_2\}$ (i.e. $-a$, $+a$ and $-b$), find weak repairs for $\langle \mathcal{I} \circ (\mathcal{U}_2 \cup \mathcal{U}_3), \{r_5\} \rangle$, using Lemma 18.

In the last step, we are using the fact that any weak repair $\mathcal{U}$ for $\langle \mathcal{I}, \eta \rangle$ must contain a weak repair $\mathcal{U}'$ for $\langle \mathcal{I}, \{r_1, r_2, r_3, r_4\} \rangle$; in turn, this can be split into a weak repair $\mathcal{U}_1$ for $\langle \mathcal{I}, \{r_1, r_2\} \rangle$ and weak repairs $\mathcal{U}_2'$ for $\langle \mathcal{I} \circ \mathcal{U}_1, \{r_3\} \rangle$ and $\mathcal{U}_3'$ for $\langle \mathcal{I} \circ \mathcal{U}_1, \{r_4\} \rangle$; defining $\mathcal{U}_2 = \mathcal{U}_2' \cup \mathcal{U}_1$ and $\mathcal{U}_3 = \mathcal{U}_3' \cup \mathcal{U}_1$, we must have $\mathcal{U}' = \mathcal{U}_1 \cup \mathcal{U}_2' \cup \mathcal{U}_3' = \mathcal{U}_2 \cup \mathcal{U}_3$. Lemma 12 guarantees that this algorithm finds all founded weak repairs for $\langle \mathcal{I}, \eta \rangle$.

6 Conclusions

We introduced independence and precedence relations among active integrity constraints that allow parallelization and sequentialization of the computation of repairs for inconsistent databases. These two processes allow us to speed up the process of finding these repairs: the advantages of parallelization are well-known, whereas the sequentialization herein presented allows a complex problem to be split in several small (and simpler) problems. Since size is a key issue in the search for repairs of a database – this being an NP- or Σ_P^2-comlete problem – it is in general much more efficient to solve several small problems than a single one as big as all of those taken together. Furthermore, the relations proposed are well-behaved w.r.t. the different kinds of repairs considered in the denotational semantics for AICs [5], so these results apply to all of them.

Using all the results presented in this paper, the strategy for computing repairs for a set η of AICs can be summarized as follows.

1. Compute ${}^{\eta}/_{\not\perp\!\!\!\perp^{+}}$
2. For each $\eta_i \in {}^{\eta}/_{\not\perp\!\!\!\perp^{+}}$
   (a) Compute $\eta_i / \approx$
   (b) Find (founded/justified) weak repairs for the minimal elements of $\eta_i / \approx$
   (c) For each non-minimal element η_j, find its (founded/justified) weak repairs by (i) combining the weak repairs for its predecessors, (ii) applying each result to $\mathcal{I}$, with result $\mathcal{I}'$, and (iii) computing (founded/justified) weak repairs for $\langle \mathcal{I}', \eta_j \rangle$ (as in the example at the end of the last section).

   This yields all (founded/justified) (weak) repairs for each element of ${}^{\eta}/_{\not\perp\!\!\!\perp^{+}}$.
3. Combine these (weak) repairs into a single (founded/justified) (weak) repair for η.

The only catch regards the situation depicted in Example 1: if one is interested in computing repairs, then one may restrict the search in the outer cycle to repairs. However, in step 2, whenever a repair cannot be extended when moving upwards in ${}^{\eta_i}/_{\approx}$, one must also consider weak repairs including that repair, since the end result may be a repair for the larger set. Also, if one does not want founded or justified repairs, the precedence relation cannot be used. The applicability of these techniques is summarized in Table 1.

In the worst case scenario, the set ${}^{\eta}/_{\not\perp\!\!\!\perp^{+}}$ will be a singleton (so there will be no parallelization) and likewise for ${}^{\eta}/_{\approx}$ (so there will be no sequentialization). However, in practical settings these are extremely unlikely situations: in typical databases concepts are built from more primitive ones, suggesting that the structure of these sets will be quite rich. Since finding repairs is an NP-complete or Σ_p^2-complete problem, this division can play a key role in making this search process much faster.

Work is in progress to implement these optimizations in order to obtain a more precise understanding of their benefits.

Table 1. Applicability of parallelization and stratification techniques to the different kinds of repairs

Type	Parallelization	Stratification
weak repairs	yes	no
repairs	yes	no
founded weak repairs	yes	yes
founded repairs	yes	yes†
justified weak repairs	yes	yes
justified repairs	yes	yes†

† may require computation of weak repairs

Acknowledgements. The author wishes to thank Patrícia Engrácia, Graça Gaspar and Isabel Nunes for their input and fruitful discussions on the topic of active integrity constraints. A special word of thanks goes to the anonymous referees, who provided me with very useful pointers that contributed to a much more comprehensive section on related work; and also for suggesting a simplification of the original proof of Theorem 3.

References

1. Abiteboul, S.: Updates, a new frontier. In: Gyssens, M., Paredaens, J., Van Gucht, D. (eds.) ICDT 1988. LNCS, vol. 326, pp. 1–18. Springer, Heidelberg (1988)
2. Abiteboul, S., Hull, R., Vianu, V.: Foundations of Databases. Addison-Wesley (1995)
3. Beeri, C., Vardi, M.Y.: The implication problem for data dependencies. In: Even, S., Kariv, O. (eds.) ICALP 1981. LNCS, vol. 115, pp. 73–85. Springer, Heidelberg (1981)
4. Caroprese, L., Greco, S., Sirangelo, C., Zumpano, E.: Declarative semantics of production rules for integrity maintenance. In: Etalle, S., Truszczyński, M. (eds.) ICLP 2006. LNCS, vol. 4079, pp. 26–40. Springer, Heidelberg (2006)
5. Caroprese, L., Truszczyński, M.: Active integrity constraints and revision programming. Theory Pract. Log. Program. 11(6), 905–952 (2011)
6. Chomicki, J.: Consistent query answering: Five easy pieces. In: Schwentick, T., Suciu, D. (eds.) ICDT 2007. LNCS, vol. 4353, pp. 1–17. Springer, Heidelberg (2006)
7. Cruz-Filipe, L., Engrácia, P., Gaspar, G., Nunes, I.: Computing repairs from active integrity constraints. In: Wang, H., Banach, R. (eds.) TASE 2013, pp. 183–190. IEEE (2013)
8. Eiter, T., Gottlob, G.: On the complexity of propositional knowledge base revision, updates, and counterfactuals. Artif. Intell. 57(2-3), 227–270 (1992)
9. Flesca, S., Greco, S., Zumpano, E.: Active integrity constraints. In: Moggi, E., Scott Warren, D. (eds.) PPDP, pp. 98–107. ACM (2004)
10. Kakas, A.C., Mancarella, P.: Database updates through abduction. In: McLeod, D., Sacks-Davis, R., Schek, H.-J. (eds.) VLDB 1990, pp. 650–661. Morgan Kaufmann (1990)
11. Katsuno, H., Mendelzon, A.O.: On the difference between updating a knowledge base and revising it. In: Allen, J.F., Fikes, R., Sandewall, E. (eds.) KR 1991, pp. 387–394. Morgan Kaufmann (1991)

12. Marek, V.W., Truszczynski, M.: Revision programming, database updates and integrity constraints. In: Gottlob, G., Vardi, M.Y. (eds.) ICDT 1995. LNCS, vol. 893, pp. 368–382. Springer, Heidelberg (1995)
13. Mayol, E., Teniente, E.: Addressing efficiency issues during the process of integrity maintenance. In: Bench-Capon, T.J.M., Soda, G., Tjoa, A.M. (eds.) DEXA 1999. LNCS, vol. 1677, pp. 270–281. Springer, Heidelberg (1999)
14. Naqvi, S.A., Krishnamurthy, R.: Database updates in logic programming. In: Edmondson-Yurkanan, C., Yannakakis, M. (eds.) PODS 1988, pp. 251–262. ACM (1988)
15. Przymusinski, T.C., Turner, H.: Update by means of inference rules. J. Log. Program. 30(2), 125–143 (1997)
16. Teniente, E., Olivé, A.: Updating knowledge bases while maintaining their consistency. VLDB J. 4(2), 193–241 (1995)
17. Widom, J., Ceri, S. (eds.): Active Database Systems: Triggers and Rules For Advanced Database Processing. Morgan Kaufmann (1996)
18. Winslett, M.: Updating Logical Databases. Cambridge Tracts in Theoretical Computer Science. Cambridge University Press (1990)

Belief Merging in Dynamic Logic of Propositional Assignments

Andreas Herzig[1], Pilar Pozos-Parra[2], and François Schwarzentruber[3]

[1] Université de Toulouse, CNRS, IRIT, France
[2] Universidad Juárez Autónoma de Tabasco, Mexico
[3] ENS Rennes, IRISA, France

Abstract. We study syntactical merging operations that are defined semantically by means of the Hamming distance between valuations; more precisely, we investigate the Σ-semantics, Gmax-semantics and max-semantics. We work with a logical language containing merging operators as connectives, as opposed to the metalanguage operations of the literature. We capture these merging operators as programs of Dynamic Logic of Propositional Assignments DL-PA. This provides a syntactical characterisation of the three semantically defined merging operators, and a proof system for DL-PA therefore also provides a proof system for these merging operators. We explain how PSPACE membership of the model checking and satisfiability problem of star-free DL-PA can be extended to the variant of DL-PA where symbolic disjunctions that are parametrised by sets (that are not defined as abbreviations, but are proper connectives) are built into the language. As our merging operators can be polynomially embedded into this variant of DL-PA, we obtain that both the model checking and the satisfiability problem of a formula containing possibly nested merging operators is in PSPACE.

Keywords: belief merging, belief change, dynamic logic.

1 Introduction

To merge a vector of belief bases $E = \langle B_1, \cdots, B_n \rangle$ means to build a new belief base $\Delta(E)$. In the literature, E is called a *profile*, and $\Delta(E)$ is sometimes called the fusion of E. Much efforts were spent on the characterisation of 'good' merging operations Δ by means of rationality postulates [14–16]. Beyond such families of abstract belief merging operations satisfying the postulates, several *concrete* operations were also introduced and studied in the literature. Some are syntax-based and others are semantic. The former are also called 'formula-based', and the latter are called 'model-based' or 'distance-based'. An example of the former is the MCS operation [2], where each element B_i of E is viewed as a set of formulas that is not closed under logical consequence and where the construction of $\Delta(E)$ is based on the extraction of maximal consistent subsets of each B_i of E. Such operations are syntax dependent: they do not guarantee that the merging of logically equivalent profiles leads to merged bases that are logically equivalent.[1]

[1] Two profiles E and E' are logically equivalent if for every B_i in E there is a logically equivalent B'_j in E' and the other way round, for every B'_i in E' there is a logically equivalent B_j in E.

C. Beierle and C. Meghini (Eds.): FoIKS 2014, LNCS 8367, pp. 381–398, 2014.

In contrast, syntax independence is guaranteed by the semantic merging operations, whose most prominent are Δ_Σ, $\Delta_{\max}$, and Δ_{Gmax} [19, 20]. These operations work on valuations of classical propositional logic. Indeed, even when the elements of the input profile are presented as formulas or sets thereof, the merging procedure starts by computing their models. The output set of valuations is sometimes transformed into a formula characterising the set, which can always be done because these operations are presented in terms of a finite set of propositional variables.

Contrasting with the existing literature, the present paper studies concrete semantic merging operations from a syntactic perspective: given a vector of formulas E, our aim is to obtain a syntactical representation of the merged belief base $\Delta(E)$, for Δ being Δ_Σ, $\Delta_{\max}$, or Δ_{Gmax}. As we have already said above, when the language is finite then it is easy to construct a formula representing $\Delta(E)$: it suffices to take the disjunction of the formulas describing the models of $\Delta(E)$, where each of these model descriptions is a conjunction of literals. Is there a better, more direct way of building a syntactic representation? In this paper we propose a powerful yet simple logical framework: Dynamic Logic of Propositional Assignments, abbreviated DL-PA [1]. DL-PA is a simple instantiation of Propositional Dynamic Logic PDL [7, 8]. Just as PDL, its language is built with two ingredients: atomic formulas and atomic programs. In both logics, atomic formulas are propositional variables. While PDL has abstract atomic programs, the atomic programs of DL-PA are assignments of propositional variables to either true or false, respectively noted $p{\leftarrow}\top$ and $p{\leftarrow}\bot$. The assignment $p{\leftarrow}\top$ corresponds to an update by p, while the assignment $p{\leftarrow}\bot$ corresponds to an update by $\neg p$. Complex programs π are built from atomic programs by the standard PDL program operators of sequential composition, nondeterministic composition, finite iteration (the so-called Kleene star), and test. Just as PDL, DL-PA has formulas of the form $\langle\pi\rangle\varphi$ and $[\pi]\varphi$, where π is a program and φ is a formula. The former expresses that φ is true after *some* possible execution of π, and the latter expresses that φ is true after *every* possible execution of π. For example, the DL-PA formula $\langle p{\leftarrow}\top \cup p{\leftarrow}\bot\rangle\varphi$ captures the propositional quantification $\exists p.\varphi$, illustrating that DL-PA naturally captures Quantified Boolean Formulas (QBF). It is shown in [1] that DL-PA formulas can be reduced to equivalent Boolean formulas. Just as for QBFs, the original formula is more compact than the equivalent Boolean formula. Star-free DL-PA has the same mathematical properties as the QBF reasoning problems; in particular, model checking, satisfiability and validity are all PSPACE complete. We believe DL-PA to be a more natural and flexible tool than QBF to reason about domains involving dynamics due to its more elaborate account in terms of programs.

Our main contributions are polynomial embeddings of semantic belief merging operators into DL-PA: to every profile E and merging operation Δ we associate a DL-PA formula $\varphi(\Delta, E)$, and we prove that the merged profile $\Delta(E)$ has the same models as $\varphi(\Delta, E)$. Then $\varphi(\Delta, E)$ may then be reduced to a Boolean formula, thus providing a syntactical representation of $\Delta(E)$ in propositional logic. A further contribution of our paper is a presentation of merging in terms of a recursive language with several merging operators Δ^σ in the object language, one operator per semantics σ. This contrasts with the usual presentations in terms of metalanguage operations (where we systematically use the term operat*or* for connectives in the object language, while we reserve the term operat*ion* for functions from the metalanguage).

The paper is organized as follows. In Section 2 we give the basic notation for propositional logic and recall the semantic definitions of the concrete merging operations Δ_Σ, Δ_{Gmax}, and $\Delta_{\max}$. In Section 3 we take a more syntactical stance: instead of viewing Δ as an operation in the metalanguage, we introduce a recursive language with families of n-ary merging operators in the object language and reformulate the above concrete merging operations in that language. In Section 4 we recall DL-PA. In Section 5 we embed the three merging operations into DL-PA. Section 6 concludes.

2 Background

We recall some standard notations and conventions for propositional logic, in particular distances between its valuations, as well as the definitions of the three concrete Boolean merging operators we are interested in.

2.1 Propositional Logic

Boolean formulas are built by means of the standard connectives $\neg$, $\vee$, etc. from a countable set of *propositional variables* $\mathbb{P} = \{p, q, \ldots\}$. We will in particular use the exclusive disjunction $\oplus$. We denote them by letters such as A, B, C; in particular, we use B, B_1, B_2, etc. for *Boolean belief bases*, which we identify with Boolean formulas.

Contrasting with that, *modal formulas*—to be defined in the next section—will be denoted by φ, ψ, etc. For a given Boolean formula A, the set of variables occurring in A is noted $\mathbb{P}_A$. For example, $\mathbb{P}_{p\vee\neg q} = \{p, q\}$.

A *valuation* associates a truth value to each propositional variable. We identify valuations with subsets of $\mathbb{P}$ and use v, v_1, v_2, etc. to denote them. The set of all valuations is $\mathbb{V} = 2^{\mathbb{P}}$. Sometimes it will be convenient to view v as a function from Boolean formulas into the set of truth values $\{0, 1\}$ and to write $v(p) = 1$ when $p \in v$ and $v(p) = 0$ when $p \notin v$.

Given a valuation v and a Boolean formula A, the truth value $v(A) \in \{0, 1\}$ is determined in the usual way. When $v(A) = 1$ then we say that v is an *A-valuation*. For example, $\{p, q\}$ is a $\neg p\vee\neg r$ valuation. The set of all A-valuations is denoted $||A||$. For example, $||p|| = \{v \in \mathbb{V} : p \in v\}$ and $||p \vee q|| = \{v \in \mathbb{V} : p \in v \text{ or } q \in v\} = ||p|| \cup ||q||$.

2.2 Distances

The *Hamming distance* between two valuations v_1 and v_2 is the cardinality of the symmetric difference between v_1 and v_2:

$$\begin{aligned} \mathsf{d_H}(v_1, v_2) &= \mathsf{card}((v_1 \setminus v_2) \cup (v_2 \setminus v_1)) \\ &= \mathsf{card}(\{p \in \mathbb{P} : v_1(p) \neq v_2(p)\}). \end{aligned}$$

So $\mathsf{d_H}(v_1, v_2)$ is the number of all those p such that $v_1(p) \neq v_2(p)$. For example, the Hamming distance between $\emptyset$ and $\{p, q\}$ is $\mathsf{card}(\emptyset \cup \{p, q\}) = 2$, and the Hamming distance between $\{p, q\}$ and $\{q, r, s\}$ is $\mathsf{card}(\{r, s\} \cup \{p\}) = \mathsf{card}(\{p, r, s\}) = 3$. Note that the Hamming distance might be infinite; for instance, $\mathsf{d_H}(\emptyset, \mathbb{P}) = \infty$.

The definition of Hamming distance can be extended to a distance between a valuation v and a set of valuations $V \subseteq \mathbb{V}$ as follows:

$$\mathsf{d_H}(v, V) = \begin{cases} 0 & \text{if } V = \emptyset \\ \min(\{\mathsf{d_H}(v, v') \ : \ v' \in V\}) & \text{otherwise} \end{cases}$$

This leads to the definition of the Hamming distance between a valuation and a Boolean formula as $\mathsf{d_H}(v, B) = \mathsf{d_H}(v, ||B||)$. For example:

$$\begin{aligned}
\mathsf{d_H}(\{p, q\}, p \wedge \neg p) &= 0 \\
\mathsf{d_H}(\{p, q\}, p \wedge q) &= 0 \\
\mathsf{d_H}(\{p, q\}, \neg p \vee q) &= 0 \\
\mathsf{d_H}(\{p\}, \neg p \vee q) &= 0 \\
\mathsf{d_H}(\{p, q\}, \neg p \vee \neg q) &= 1 \\
\mathsf{d_H}(\{p, q\}, \neg p \wedge \neg q) &= 2 \\
\mathsf{d_H}(\{p, q\}, \neg p \vee \neg r) &= 0 \\
\mathsf{d_H}(\{p, q\}, (\neg p \vee \neg r) \wedge \neg q) &= 1
\end{aligned}$$

Lemma 1. *For every valuation* v*,* $\mathsf{d_H}(v, B) \leq \mathsf{card}(\mathbb{P}_B)$*.*

Proof. Let v be a valuation. If $||B|| = \emptyset$ then $\mathsf{d_H}(v, B) = \mathsf{d_H}(v, ||B||) = \mathsf{d_H}(v, \emptyset) = 0$ and the lemma is correct. Otherwise, let $v' \in ||B||$. Without loss of generality, we can assume that for all $p \notin \mathbb{P}_B$, $v(p) = v(p')$. Thus, $\mathsf{d_H}(v, v') \leq \mathsf{card}(\mathbb{P}_B)$. By definition of $\mathsf{d_H}(v, B)$ we have $\mathsf{d_H}(v, B) = \mathsf{d_H}(v, ||B||) \leq \mathsf{d_H}(v, v') \leq \mathsf{card}(\mathbb{P}_B)$.

Finally, the Hamming distance between a valuation v and a vector of Boolean belief bases $\langle B_1, \ldots, B_n \rangle$ is defined to be the vector of the distances:

$$\mathsf{d_H}(v, \langle B_1, \ldots, B_n \rangle) = \langle \mathsf{d_H}(v, B_1), \cdots, \mathsf{d_H}(v, B_n) \rangle$$

For example:

$$\begin{aligned}
\mathsf{d_H}(\{p\}, \langle \neg p \vee \neg q \rangle) &= \langle 1 \rangle \\
\mathsf{d_H}(\{p, q\}, \langle \neg p \vee \neg r, (\neg p \vee \neg r) \wedge \neg q \rangle) &= \langle 0, 1 \rangle \\
\mathsf{d_H}(\{p, q\}, \langle \neg p \vee q, \neg p \vee \neg q, \neg p \wedge \neg q \rangle) &= \langle 0, 1, 2 \rangle
\end{aligned}$$

2.3 Various Merging Operations

A *profile*, typically noted E, is a vector of belief bases: $E = \langle B_1, \cdots, B_n \rangle$. The traditional definition of a belief merging operation is as a mapping Δ associating to every profile E a new belief base $\Delta(E)$. Such operations have been defined in several different ways and that is why we indicate a particular definition σ by a superscript and write $\Delta^\sigma(E)$. Throughout the present paper we suppose that there is no preference between the belief bases of a profile: we assume that $\Delta^\sigma(\varphi_1, \cdots, \varphi_n)$ is equivalent to $\Delta^\sigma(\varphi_{k_1}, \cdots, \varphi_{k_n})$, for every permutation $\langle \varphi_{k_1}, \cdots, \varphi_{k_n} \rangle$ of $\langle \varphi_1, \cdots, \varphi_n \rangle$. The reader may therefore view the vector as a set. We stick to the vector notation for two reasons: first, it is common in the merging literature, and second, it better fits the object language operators to be introduced in the next section.

Perhaps the best starting point is the merging operation that is based on minimisation of the sum of the Hamming distances to each belief base B_i of E, abbreviated Δ^{Σ}. It associates to every profile E the set of valuations such that the sum of the distances to the elements of E is minimal. Formally:

$$\Delta^{\Sigma}(E) = \Big\{v \in \mathbb{V} \ : \ \text{there is no } v' \in \mathbb{V} \text{ such that } \sum \mathsf{d_H}(v', E) < \sum \mathsf{d_H}(v, E)\Big\}.$$

For example:

$$\Delta^{\Sigma}(p, \neg p \vee q) = \{v \ : \ p, q \in v\} = ||p \wedge q||$$
$$\Delta^{\Sigma}(p \wedge q, \neg p \wedge \neg q) = 2^{\mathbb{P}} = ||\top||$$

Beyond Δ^{Σ} we consider other concrete merging operations: the Gmax merging operation Δ^{Gmax} and the max merging operator $\Delta^{\max}$. Their definitions are based on other minimisations. We do not give them here; instead, they will be presented in the next section in terms of object language operators.

Merging can also be done under integrity constraints. This leads to more general operations $\Delta^{\sigma}_{\psi}(E)$ where the formula ψ is an integrity constraint that the merged belief base should satisfy. The unconstrained $\Delta^{\sigma}(E)$ can then be identified with $\Delta^{\sigma}_{\top}(E)$. Then the Δ^{Σ} operation becomes:

$$\Delta^{\Sigma}_{\psi}(E) = \Big\{v \in ||\psi|| \ : \ \text{there is no } v' \in ||\psi|| \text{ such that } \sum \mathsf{d_H}(v', E) < \sum \mathsf{d_H}(v, E)\Big\}.$$

For example, $\Delta^{\Sigma}_{\neg r}(p, \neg p \vee q) = ||p \wedge q \wedge \neg r||$ and $\Delta^{\Sigma}_{p}(p \wedge q, \neg p \wedge \neg q) = ||p||$.

Observe that in the above definitions $\Delta_C(E)$ is a set of valuations. In contrast, the merging postulates to be given below are defined in terms of formulas: as already mentioned, papers on merging operations typically identify the set $\Delta_C(E)$ with the Boolean formula characterising it.

2.4 The Postulates for Merging with Integrity Constraints

We briefly recall the principles for merging operations that were introduced by Konieczny and Pino Pérez. We here present the version of [14], in a slightly adapted version because there, belief bases are considered to be finite sets of formulas (which are however often identified with their conjunction).

Let Δ be an mapping assigning to each belief profile E and integrity constraint C a belief base $\Delta_C(E)$. Δ is a merging operation if and only if it satisfies the following postulates.

(IC0) $\Delta_C(E) \to C$ is valid.
(IC1) If C is satisfiable then $\Delta_C(E)$ is satisfiable.
(IC2) If $C \wedge (\bigwedge E)$ is satisfiable then $\Delta_C(E) \leftrightarrow \bigwedge E$ is valid.
(IC3) For $E = \langle B_1, \cdots, B_n \rangle$ and $E' = \langle B'_1, \cdots, B'_n \rangle$, if $C \leftrightarrow C'$ and $B_i \leftrightarrow B'_i$ are valid for $1 \leq i \leq n$ then $\Delta_C(E) \leftrightarrow \Delta_{C'}(E')$ is valid.

(IC4) If $\Delta_C(\langle B, B'\rangle) \wedge B$ is satisfiable then $\Delta_C(\langle B, B'\rangle) \wedge B'$ is satisfiable.
(IC5) $\Delta_C(E) \wedge C' \to \Delta_{C \wedge C'}(E)$ is valid.
(IC6) If $\Delta_C(E) \wedge C'$ is satisfiable then $\Delta_{C \wedge C'}(E) \to \Delta_C(E)$ is valid.

In the above postulates, 'satisfiable' means 'propositionally satisfiable' and 'valid' means 'propositionally valid'.

The operations Δ^{Σ} and Δ^{Gmax} satisfy all the postulates, while the max merging operator $\Delta^{\max}$ does not. Nonetheless, many authors in the literature consider that the latter is an interesting merging operator.

3 A Modal Framework for Merging Operators

The Δ^{σ} are not logical connectives of the object language: they are part of the metalanguage. We highlight that by saying that they are opera*tions*. The merging opera*tors* to be introduced now are connectives of the object language, just as the Boolean operators $\neg$ and $\vee$ are.[2] For that reason we also write them differently as $\blacktriangle^{\sigma}$: for each semantics σ we have an object language operator $\blacktriangle^{\sigma}$.[3] It is an advantage of such a move that many things can then be proved in a formal, rigorous way inside a logical system, as opposed to lines of argument in natural language texts. Moreover, it also allows to take advantage of mathematical results such as complexity upper bounds and theorem proving methods for the logic.

If merging operators are in the object language, we have enough flexibility to nest merging operators and even talk about different semantics in the same formula, as illustrated by the well-formed formula $\blacktriangle^{\sigma_1}_{\blacktriangle^{\sigma_2}(p,q)}(p, p \vee q)$. To motivate this, consider a company whose productivity is declining and whose shareholders desire to implement a motivation policy in order to change the workers' conditions. They then have to merge the desires of every worker, while preserving several kinds of integrity constraints: job security, working environment, salary costs, job satisfaction. These different criteria have to be merged in their turn.

Formulas involving one or more kinds of merging operators may be given as an input to a reasoner. Observe that when we define the length of the input for the reasoner then one occurrence of a merging operator counts for 1 and certainly not for the length of the disjunction describing the corresponding set of valuations (as would be the case in the metalinguistic presentation).

3.1 Language

Our logical language $\mathcal{L}_{\blacktriangle}$ is defined by the following grammar:

$$\varphi ::= p \mid \neg\varphi \mid \varphi \vee \varphi \mid \blacktriangle^{\sigma}_{\varphi}(\varphi, \cdots, \varphi)$$

[2] While the term 'merging operator' is customary in the literature, our terminology is in line with that of abstract algebra.

[3] More precisely, we do not have a single operator but a family of operators $\blacktriangle^{\sigma,n}(.)$ that is parametrized by the length n of the profile vector. We abstract away from this here.

where p ranges over the set of propositional variables $\mathbb{P}$ and where σ ranges over the set of symbols $\{\Sigma, \text{Gmax}, \text{max}\}$. The informal reading of the formula $\blacktriangle^{\sigma}_{\psi}(E)$ is "the profile E has been merged (with merging semantics σ) under the constraint ψ".

Abusing language a bit, when the profile is $E = \langle \varphi_1, \ldots, \varphi_n \rangle$ then instead of $\blacktriangle^{\sigma}_{\psi}(E)$ we write $\blacktriangle^{\sigma}_{\psi}(\varphi_1, \cdots, \varphi_n)$.

The function $\mathbb{P}$ associating to a formula the set of its propositional variables naturally extends to our language; in particular we have $\mathbb{P}_{\blacktriangle^{\sigma}_{\psi}(\varphi_1,\cdots,\varphi_n)} = \mathbb{P}_{\psi} \cup \big(\bigcup_{1\leq i\leq n} \mathbb{P}_{\varphi_i}\big)$.

In the rest of the present section we introduce the truth conditions for the three merging operators $\blacktriangle^{\Sigma}$, $\blacktriangle^{\text{Gmax}}$, and $\blacktriangle^{\text{max}}$. Clearly, when the profile $E = \langle B_1, \ldots, B_n \rangle$ and the constraint C are Boolean then we expect the interpretation of the merging operator $\blacktriangle^{\sigma}$ under semantics σ to coincide with the merging operation Δ^{σ} defined in Section 2.3. In formulas, we expect the equality $\Delta^{\sigma}_{C}(E) = ||\blacktriangle^{\sigma}_{C}(E)||$ to hold for Boolean C and E.

3.2 The Σ-Semantics

The interpretation of $\blacktriangle^{\Sigma}$ is the set of valuations such that the sum of the distances to the elements of E is minimal. Formally:

$$||\blacktriangle^{\Sigma}_{\psi}(E)|| = \Big\{v \in ||\psi|| \ : \ \text{there is no } v' \in ||\psi|| \text{ such that } \sum \mathsf{d}_{\mathsf{H}}(v', E) < \sum \mathsf{d}_{\mathsf{H}}(v, E)\Big\}.$$

The definition of the Hamming distance d_{H} is as in Section 2.2. The function $||\cdot||$ is the interpretation we are currently defining by induction over the formulas of $\mathcal{L}_{\blacktriangle}$. The integer $\sum \mathsf{d}_{\mathsf{H}}(v, E)$ is the sum of the elements of the vector $\mathsf{d}_{\mathsf{H}}(v, E)$.

For example, $||\blacktriangle^{\Sigma}_{\top}(p\land q, \neg p\land\neg q)|| = ||\top|| = 2^{\mathbb{P}}$.

3.3 The Gmax-Semantics

The interpretation of $\blacktriangle^{\text{Gmax}}$ is as follows:

$$||\blacktriangle^{\text{Gmax}}_{\psi}(E)|| = \Big\{v \in ||\psi|| \ : \ \text{there is no } v' \in ||\psi|| \text{ such that } \mathsf{d}^{\text{sort}}_{\mathsf{H}}(v', E) <_{\text{lex}} \mathsf{d}^{\text{sort}}_{\mathsf{H}}(v, E)\Big\}$$

where $\mathsf{d}^{\text{sort}}_{\mathsf{H}}(v, E) = \text{sort}(d(v, \varphi_1), \ldots, d(v, \varphi_n))$ is the list that is obtained from the vector $\langle d(v, \varphi_1), \ldots, d(v, \varphi_n) \rangle$ by sorting it in descending order and where $<_{\text{lex}}$ is the lexicographical order between sequences of integers of the same length.

For example, $||\blacktriangle^{\text{Gmax}}_{\top}(p\land q, \neg p\land\neg q)|| = \{v \ : \ v(p) \neq v(q)\} = ||p\oplus q||$ because

$$\mathsf{d}^{\text{sort}}_{\mathsf{H}}(v, \langle p\land q, \neg p\land\neg q\rangle) = \begin{cases} \langle 2, 0 \rangle & \text{if } v(p) = v(q) \\ \langle 1, 1 \rangle & \text{otherwise.} \end{cases}$$

3.4 The max-Semantics

The interpretation of $\blacktriangle^{\text{max}}$ is as follows:

$$||\blacktriangle^{\text{max}}_{\psi}(E)||_{\text{max}} = \Big\{v \in ||\psi|| \ : \ \text{there is no } v' \in ||\psi|| \text{ such that } \max \mathsf{d}_{\mathsf{H}}(v', E) < \max \mathsf{d}_{\mathsf{H}}(v, E)\Big\}$$

where $\max \mathsf{d_H}(v, E)$ is the maximum of all the distances $\mathsf{d_H}(v, \varphi_i)$ between v and the elements φ_i of E.

For example, $||\blacktriangle^{\max}_{\top}(p{\wedge}q, \neg p{\wedge}\neg q)|| = \{v : v(p) \neq v(q)\} = ||p{\oplus}q||$ because for the valuations v such that $v(p) \neq v(q)$ we have that $\mathsf{d_H}(v, \langle p{\wedge}q, \neg p{\wedge}\neg q\rangle)$ equals $\langle 1, 1\rangle$ (and therefore the maximum of that vector is 1), while for the v such that $v(p) = v(q)$ the distance $\mathsf{d_H}(v, \langle p{\wedge}q, \neg p{\wedge}\neg q\rangle)$ is either $\langle 0, 2\rangle$ or $\langle 2, 0\rangle$ (and therefore the maximum is 2).

We recall that the max-semantics does not satisfy Konieczny and Pino Pérez's merging postulates. We also note that for the empty integrity constraint we have $||\blacktriangle^{\mathrm{Gmax}}_{\top}(E)|| \subseteq ||\blacktriangle^{\max}_{\top}(E)||$ for every profile E.

4 DL-PA: Dynamic Logic of Propositional Assignments

In this section we define syntax and semantics of dynamic logic of propositional assignments DL-PA and state complexity results. The star-free fragment of DL-PA was introduced in [9], where it was shown that it embeds Coalition Logic of Propositional Control [10–12]. The full logic with the Kleene star was further studied in [1].

4.1 Language

The language of DL-PA is defined by the following grammar:

$$\pi ::= p{\leftarrow}\top \mid p{\leftarrow}\bot \mid \pi;\pi \mid \pi{\cup}\pi \mid \varphi? \mid \pi^*$$
$$\varphi ::= p \mid \top \mid \bot \mid \neg\varphi \mid \varphi{\vee}\varphi \mid \langle\pi\rangle\varphi$$

where p ranges over the set of propositional $\mathbb{P}$. So the *atomic programs* of the language of DL-PA are of the form $p{\leftarrow}\top$ and $p{\leftarrow}\bot$. The operators of sequential composition (";"), nondeterministic composition ("$\cup$"), unbounded iteration ("$(.)^*$", the so-called Kleene star), and test ("(.)?") are familiar from Propositional Dynamic Logic PDL.

The *length* of a formula φ, denoted $|\varphi|$, is the number of symbols used to write down φ, without "$\langle$", "$\rangle$", parentheses and commas. For example, $|q{\wedge}r| = |\neg(\neg q{\vee}\neg r)| = 6$ and $|\langle q{\leftarrow}\top\rangle(q{\wedge}r)| = 2{+}6 = 8$. The length of a program π, denoted $|\pi|$, is defined in the same way. For example, $|p{\leftarrow}\bot; p?| = 5$.

We abbreviate the logical connectives $\wedge$, $\rightarrow$, $\leftrightarrow$, and $\oplus$ in the usual way. Moreover, $[\pi]\varphi$ abbreviates $\neg\langle\pi\rangle\neg\varphi$. Several program abbreviations are familiar from PDL. First, skip abbreviates $\top?$ ("nothing happens"). Second, the loop "while A do π" can be expressed as the DL-PA program $(A?;\pi)^*; \neg A?$. Third, for $n \geq 0$, the n-th iteration of π is defined inductively as:

$$\pi^0 = \mathsf{skip}$$
$$\pi^{n+1} = \pi^n; \pi$$

Let us now introduce the assignment of literals to variables by means of the following abbreviations that are proper to DL-PA:

$$p{\leftarrow}q = (q?; p{\leftarrow}\top) \cup (\neg q?; p{\leftarrow}\bot)$$
$$p{\leftarrow}\neg q = (q?; p{\leftarrow}\bot) \cup (\neg q?; p{\leftarrow}\top)$$

The former assigns to p the truth value of q, while the latter assigns to p the truth value of $\neg q$. The length of $p{\leftarrow}q$ is $(2+1+3)+1+(3+1+3) = 14$. That of $p{\leftarrow}\neg q$ is 14, too.

The *star-free fragment* of DL-PA is the subset of the language made up of formulas without the Kleene star "$(.)^*$".

4.2 Semantics of DL-PA

DL-PA programs are interpreted by means of a (unique) *relation between valuations*. The atomic programs $p{\leftarrow}\top$ and $p{\leftarrow}\bot$ update valuations in the obvious way, and complex programs are interpreted just as in PDL by mutual recursion. Table 1 gives the interpretation of the DL-PA connectives.

Table 1. Interpretation of the DL-PA connectives

$$\begin{aligned}
||p{\leftarrow}\top|| &= \{\langle v_1, v_2\rangle \ : \ v_2 = v_1 \cup \{p\}\} \\
||p{\leftarrow}\bot|| &= \{\langle v_1, v_2\rangle \ : \ v_2 = v_1 \setminus \{p\}\} \\
||\pi;\pi'|| &= ||\pi|| \circ ||\pi'|| \\
||\pi \cup \pi'|| &= ||\pi|| \cup ||\pi'|| \\
||\pi^*|| &= \bigcup_{k \in \mathbb{N}_0} (||\pi||)^k \\
||\varphi?|| &= \{\langle v, v\rangle \ : \ v \in ||\varphi||\} \\
||p|| &= \{v \ : \ p \in v\} \\
||\top|| &= \mathbb{V} \ = \ 2^{\mathbb{P}} \\
||\bot|| &= \emptyset \\
||\neg\varphi|| &= 2^{\mathbb{P}} \setminus ||\varphi|| \\
||\varphi \vee \psi|| &= ||\varphi|| \cup ||\psi|| \\
||\langle\pi\rangle\varphi|| &= \{v \ : \ \text{there is } v_1 \text{ s.t. } \langle v, v_1\rangle \in ||\pi|| \text{ and } v_1 \in ||\varphi||\}
\end{aligned}$$

Two formulas φ_1 and φ_2 are *formula equivalent* if $||\varphi_1|| = ||\varphi_2||$. Two programs π_1 and π_2 are *program equivalent* if $||\pi_1|| = ||\pi_2||$. In that case we write $\pi_1 \equiv \pi_2$. For example, the program equivalence $\pi; \mathsf{skip} \equiv \pi$ holds. A formula φ is DL-PA *valid* if it is formula equivalent to $\top$, i.e., if $||\varphi|| = 2^{\mathbb{P}}$. It is DL-PA *satisfiable* if it is not formula equivalent to $\bot$, i.e., if $||\varphi|| \neq \emptyset$. For example, the formulas $\langle p{\leftarrow}\top\rangle\top$ and $\langle p{\leftarrow}\top\rangle\varphi \leftrightarrow \neg\langle p{\leftarrow}\top\rangle\neg\varphi$ are DL-PA valid. Other examples of DL-PA validities are $\langle p{\leftarrow}\top\rangle p$ and $\langle p{\leftarrow}\bot\rangle\neg p$.

In DL-PA, all the program operators can be eliminated: for every formula φ there is a formula equivalent φ' such that no program operator occurs in φ' [1, Theorem 1]. For example, $\langle p{\leftarrow}\top^*\rangle r$ is equivalent to $p \vee \langle p{\leftarrow}\top\rangle r$ and $\langle p{\leftarrow}\top; q{\leftarrow}\top\rangle r$ is equivalent to $\langle p{\leftarrow}\top\rangle\langle q{\leftarrow}\top\rangle r$. This contrasts with PDL, where this is not the case. Once all the program operators have been eliminated, modal operators only contain atomic programs. The latter are both serial and deterministic modal operators and therefore distribute over negation and disjunction. They can finally be eliminated when they face a propositional variable, according to the following equivalences:

$$\langle p{\leftarrow}\top\rangle q \leftrightarrow \begin{cases} \top & \text{if } q = p \\ q & \text{otherwise} \end{cases}$$

$$\langle p{\leftarrow}\bot\rangle q \leftrightarrow \begin{cases} \bot & \text{if } q = p \\ q & \text{otherwise} \end{cases}$$

All together, we have a complete set of reduction axioms: every formula reduces to a Boolean formula [1, Theorem 2].

Theorem 1. *For every* DL-PA *formula* φ *there is a Boolean formula* φ' *such that* $\varphi \leftrightarrow \varphi'$ *is* DL-PA *valid.*

For example, for different propositional variables r and p, the formula $\langle p{\leftarrow}q\rangle(p \vee r)$ is successively equivalent to $\langle p{\leftarrow}q\rangle p \vee \langle p{\leftarrow}q\rangle r$ and to $q \vee r$.

It is proved in [9] that both model and satisfiability checking are PSPACE complete for the star-free fragment of DL-PA.

Observe that if p does not occur in φ then both $\varphi \to \langle p{\leftarrow}\top\rangle\varphi$ and $\varphi \to \langle p{\leftarrow}\bot\rangle\varphi$ are valid. This is due to the following semantical property that we will use later.

Proposition 1. *Suppose* $\mathbb{P}_\varphi \cap P = \emptyset$*, i.e., none of the variables in* P *occurs in* φ*. Then* $v \cup P \in ||\varphi||$ *iff* $v \setminus P \in ||\varphi||$.

In the rest of the paper we write $||\varphi||_{\mathsf{DL\text{-}PA}}$ in order to distinguish the interpretation of DL-PA formulas from the interpretation of the merging language.

4.3 Some Useful DL-PA Expressions

Table 2 collects some DL-PA expressions that are going to be convenient abbreviations.[4]

The program $\mathsf{vary}(P)$ nondeterministically changes the truth value of some of the variables in P. Its length is linear in the cardinality of P. So the program $\mathsf{vary}(\mathbb{P}_A); A?$ accesses all A-valuations that preserve the values of all those variables not occurring in A. Satisfiability of the Boolean formula A can be expressed in DL-PA by the formula $\langle\mathsf{vary}(\mathbb{P}_A); A?\rangle\top$ or the equivalent $\langle\mathsf{vary}(\mathbb{P}_A)\rangle A$. The program $\mathsf{flip}^1(P)$ changes the truth value of exactly one of the variables in P. The programs $\mathsf{flip}^{\leq m}(P)$ flip the truth value of at most m of the variables in P. The lengths of $\mathsf{flip}^m(P)$ and $\mathsf{flip}^{\leq m}(P)$ are quadratic in n. The formula $\mathsf{H}(\varphi, \geq d)$ is true in all those valuations whose Hamming distance to φ is d.

[4] An *expression* is a formula or a program. When we say that two expressions are equivalent we mean program equivalence if we are talking about programs, and formula equivalence otherwise.

Table 2. Some useful DL-PA expressions, for $P = \{p_1, \ldots, p_n\}$, where $m \le n$ in $\mathsf{flip}^m(P)$ and $\mathsf{flip}^{\le m}(P)$, and where $d \le \mathsf{card}(\mathbb{P}_\varphi)$ in $\mathsf{H}(\varphi, d)$

$$\mathsf{vary}(P) = (p_1{\leftarrow}\top \cup p_1{\leftarrow}\bot); \cdots; (p_n{\leftarrow}\top \cup p_n{\leftarrow}\bot)$$

$$\mathsf{flip}^m(P) = \begin{cases} \mathsf{skip} & \text{if } m = 0 \\ (p_1{\leftarrow}\neg p_1 \cup \cdots \cup p_n{\leftarrow}\neg p_n); \mathsf{flip}^{m-1}(P) & \text{if } m \ge 1 \end{cases}$$

$$\mathsf{flip}^{\le m}(P) = \begin{cases} \mathsf{skip} & \text{if } m = 0 \\ (\mathsf{skip} \cup \mathsf{flip}^1(P)); \mathsf{flip}^{\le m-1}(P) & \text{if } m \ge 1 \end{cases}$$

$$\mathsf{H}(\varphi, d) = \begin{cases} \varphi & \text{if } m = 0 \\ \neg\langle \mathsf{flip}^{\le d-1}(\mathbb{P}_\varphi)\rangle\varphi \wedge \langle \mathsf{flip}^d(\mathbb{P}_\varphi)\rangle\varphi & \text{if } m \ge 1 \end{cases}$$

For example:

$$\begin{aligned}
\mathsf{H}(p, 1) &= \neg\langle \mathsf{flip}^{\le 0}(\{p\})\rangle p \wedge \langle(\mathsf{flip}^1(\{p\}))\rangle p \\
&\leftrightarrow \neg p \wedge \langle p{\leftarrow}\neg p\rangle p \\
&\leftrightarrow \neg p \wedge \neg p \\
&\leftrightarrow \neg p \\
\mathsf{H}(\neg p \vee q, 0) &\leftrightarrow \neg p \vee q \\
\mathsf{H}(\neg p \vee q, 1) &\leftrightarrow \neg(\neg p \vee q) \wedge \langle p{\leftarrow}\neg p\rangle(\neg p \vee q) \\
&\leftrightarrow p \wedge \neg q \wedge (p \vee q) \\
&\leftrightarrow p \wedge \neg q \\
\mathsf{H}(\neg p \vee q, 2) &= \neg\langle(\mathsf{skip} \cup p{\leftarrow}\neg p); \mathsf{skip}\rangle(\neg p \vee q) \wedge \langle p{\leftarrow}\neg p; p{\leftarrow}\neg p\rangle(\neg p \vee q) \\
&\leftrightarrow \neg((\neg p \vee q) \vee (p \vee q)) \wedge (\neg p \vee q) \\
&\leftrightarrow \bot
\end{aligned}$$

Lemma 2. *The following hold:*

1. $\langle v_1, v_2\rangle \in ||\mathsf{vary}(P)||$ *iff* $(v_1 \setminus v_2) \cup (v_2 \setminus v_1) \subseteq P$.
2. $\langle v_1, v_2\rangle \in ||\mathsf{flip}^1(P)||$ *iff* $\langle v_1, v_2\rangle \in ||\mathsf{vary}(P)||$ *and* $\mathsf{card}(v_1 \dot{-} v_2) = 1$.
3. $\langle v_1, v_2\rangle \in ||\mathsf{flip}^{\le m}(P)||$ *iff* $\langle v_1, v_2\rangle \in ||\mathsf{vary}(P)||$ *and* $\mathsf{card}(v_1 \dot{-} v_2) \le m$.
4. $v \in ||\mathsf{H}(\varphi, d)||$ *iff* $\mathsf{d_H}(v, \varphi) = d$.

Note that $\mathsf{flip}^m(P)$ is nothing but the m-th iteration of $\mathsf{flip}^1(P)$, so one variable might be switched twice and therefore $\langle v_1, v_2\rangle \in ||\mathsf{flip}^m(P)||$ does not in general imply that the Hamming distance between v_1 and v_2 is m.

5 Embedding Merging Operators into DL-PA

In this section, we define a translation $tr(.)$ by induction over the formulas of our merging language $\mathcal{L}_\blacktriangle$. To every formula φ of our merging language $\mathcal{L}_\blacktriangle$ we associate a DL-PA formula $tr(\varphi)$. The Boolean part is translated as follows:

$$tr(p) = p$$
$$tr(\neg\varphi) = \neg tr(\varphi)$$
$$tr(\varphi \vee \psi) = tr(\varphi) \vee tr(\psi)$$

In the following three subsections we give the inductive cases of the definition of $tr(.)$ for $\blacktriangle^{\Sigma}$, $\blacktriangle^{\mathrm{Gmax}}$ and $\blacktriangle^{\max}$. We then prove that the translation is correct and that it gives us an algorithm to reason in $\mathcal{L}_{\blacktriangle}$ from an algorithm to reason in DL-PA. The reader may observe that our encodings are not particularly sophisticated and follow the semantic definitions in a fairly straightforward manner.

5.1 Embedding the Σ-Semantics

Let us define the translation for the $\blacktriangle^{\Sigma}$ as follows. Given a profile $E = \langle\varphi_1, \ldots, \varphi_n\rangle$, we define:

$$tr(\blacktriangle^{\Sigma}_{\psi}(E)) = tr(\psi) \wedge \bigvee_{\langle d_1,\ldots,d_n\rangle, d_k \leq \mathsf{card}(\mathbb{P}_{\varphi_k})} \Big(\big(\bigwedge_{i\leq n} \mathsf{H}(tr(\varphi_i), d_i)\big) \wedge \neg\langle \mathsf{vary}(\mathbb{P}_E)\rangle\big(tr(\psi) \wedge \bigvee_{\langle d'_1,\ldots,d'_n\rangle, \sum_{k\leq n}(d'_k) < \sum_{k\leq n}(d_k)} \bigwedge_{i\leq n} \mathsf{H}(tr(\varphi_i), d'_i)\big)\Big).$$

Intuitively, the translation does the following: first, the integrity constraint is required to be true (by $tr(\psi)$), second, it is checked that there is some vector $\langle d_1, \ldots, d_n\rangle$ of integers such that the Hamming distance from the present valuation to each $tr(\varphi_i)$-valuation is d_i (by $\mathsf{H}(tr(\varphi_i), d_i)$) and such that one cannot go to another valuation (by $\neg\langle\mathsf{vary}(\mathbb{P}_E)\rangle$) satisfying the constraint and whose sum of distances to the $tr(\varphi_i)$-valuations is smaller. As we are going to show, every model of the formula $tr(\blacktriangle^{\Sigma}_{\psi}(E))$ is indeed a model of the merged profile.

For example, $tr(\blacktriangle^{\Sigma}_{\top}(p, \neg p \vee q)))$ is

$$\begin{aligned}\top \wedge \Big(&(\mathsf{H}(p,0) \wedge \mathsf{H}(\neg p \vee q, 0) \wedge \neg\langle\mathsf{vary}(\{p,q\})\rangle(\top \wedge \bot) \vee \\ &(\mathsf{H}(p,0) \wedge \mathsf{H}(\neg p \vee q, 1) \wedge \neg\langle\mathsf{vary}(\{p,q\})\rangle(\top \wedge \mathsf{H}(p,0) \wedge \mathsf{H}(\neg p \vee q, 0)) \vee \\ &(\mathsf{H}(p,1) \wedge \mathsf{H}(\neg p \vee q, 0) \wedge \neg\langle\mathsf{vary}(\{p,q\})\rangle(\top \wedge \mathsf{H}(p,0) \wedge \mathsf{H}(\neg p \vee q, 0)) \vee \\ &(\mathsf{H}(p,1) \wedge \mathsf{H}(\neg p \vee q, 1) \wedge \neg\langle\mathsf{vary}(\{p,q\})\rangle(\top \wedge ((\mathsf{H}(p,0) \wedge \mathsf{H}(\neg p \vee q, 0)) \vee \cdots))\Big)\end{aligned}$$

which is equivalent to

$$\begin{aligned}&(p \wedge (\neg p \vee q) \wedge \top) \vee \\ &(p \wedge (p \wedge \neg q) \wedge \bot) \vee \\ &(\neg p \wedge (\neg p \vee q) \wedge \bot) \vee \\ &(\neg p \wedge (p \wedge \neg q) \wedge \cdots),\end{aligned}$$

i.e., to $p \wedge q$.

Here is another example:

$$
\begin{aligned}
tr(\blacktriangle^{\Sigma}_{\top}(p\wedge q, \neg p\wedge\neg q)) \leftrightarrow \top \wedge\ & (\mathsf{H}(p\wedge q, 0) \wedge \mathsf{H}(\neg p\wedge\neg q, 2)) \vee \\
& (\mathsf{H}(p\wedge q, 1) \wedge \mathsf{H}(\neg p\wedge\neg q, 1)) \vee \\
& (\mathsf{H}(p\wedge q, 2) \wedge \mathsf{H}(\neg p\wedge\neg q, 0)) \\
\leftrightarrow\ & (p\wedge q) \vee \\
& (\neg p\wedge q) \vee (p\wedge\neg q) \vee \\
& (\neg p\wedge\neg q) \\
\leftrightarrow\ & \top
\end{aligned}
$$

5.2 Embedding the Gmax-Semantics

The embedding of the Gmax-operator is in the same spirit as that of the previous operator. Given a profile $E = \langle\varphi_1, \ldots, \varphi_n\rangle$, we define:

$$
\begin{aligned}
tr(\blacktriangle^{\mathrm{Gmax}}_{\psi}(E)) = tr(\psi) \wedge & \bigvee_{\langle d_1,\ldots,d_n\rangle,\ d_k\leq \mathsf{card}(\mathbb{P}_{\varphi_k})} \Big(\big(\bigwedge_{i\leq n} \mathsf{H}(tr(\varphi_i), d_i)\big) \wedge \\
& \neg\langle \mathsf{vary}(\mathbb{P}_E)\rangle\big(tr(\psi) \wedge \bigvee_{\langle d'_1,\ldots,d'_n\rangle, \mathrm{sort}(d'_1,\ldots,d'_n)<_{\mathrm{lex}} \mathrm{sort}(d_1,\ldots,d_n)} \bigwedge_{i\leq n} \mathsf{H}(tr(\varphi_i), d'_i)\big)\Big).
\end{aligned}
$$

Intuitively, the translation checks the integrity constraints and checks for the vector characterising the Hamming distances to the φ_i-valuations that there exists no other valuation $tr(\psi)$ whose distance vector is smaller according to the sorted lexicographic ordering.

Table 3 contains an example. Another example is $tr(\blacktriangle^{\mathrm{Gmax}}_{\top}(p\wedge q, \neg p\wedge\neg q))$, which reduces to $p \leftrightarrow q$.

5.3 Embedding the max-Semantics

In a first try we have:

$$
\begin{aligned}
tr(\blacktriangle_{\psi}(E)) = tr(\psi) \wedge & \bigvee_{\langle d_1,\ldots,d_n\rangle,\ d_k\leq \mathsf{card}(\mathbb{P}_{\varphi_k})} \Big(\big(\bigwedge_{i\leq n} \mathsf{H}(tr(\varphi_i), d_i)\big) \wedge \\
& \neg\langle \mathsf{vary}(\mathbb{P}_E)\rangle(tr(\psi) \wedge \bigvee_{\langle d'_1,\ldots,d'_n\rangle, \max_{k\leq n}(d'_k)<\max_{k\leq n}(d_k)} \bigwedge_{i\leq n} \mathsf{H}(tr(\varphi_i), d'_i))\Big).
\end{aligned}
$$

This can actually be made more concise, and our official definition of the translation is as follows:

$$
\begin{aligned}
tr(\blacktriangle_{\psi}(E)) = tr(\psi) \wedge & \bigvee_{d,\ d\leq \max_{k\leq n}(\mathsf{card}(\mathbb{P}_{\varphi_k}))} \Big(\big(\bigwedge_{i\leq n} \langle \mathsf{flip}^{\leq d}(\mathbb{P}_{\varphi_i})\rangle tr(\varphi_i)\big) \wedge \\
& \neg\langle \mathsf{vary}(\mathbb{P}_E)\rangle\big(tr(\psi) \wedge \bigwedge_{i\leq n} \langle \mathsf{flip}^{\leq d-1}(\mathbb{P}_{\varphi_i})\rangle tr(\varphi_i)\big)\Big).
\end{aligned}
$$

Table 3. Example: translation of the Gmax merging of the profile $\langle p, p, \neg p \rangle$ under the empty integrity constraint $\top$

$$
\begin{aligned}
& tr(\blacktriangle^{\mathrm{Gmax}}_{\top}(p, p, \neg p)) \\
& \quad = \top \wedge (\mathsf{H}(p,0) \wedge \mathsf{H}(p,0) \wedge \mathsf{H}(\neg p,0) \wedge \neg\langle \mathsf{vary}(\{p\})\rangle(\top \wedge \bot)) \vee \\
& \qquad\quad (\mathsf{H}(p,0) \wedge \mathsf{H}(p,0) \wedge \mathsf{H}(\neg p,1) \wedge \neg\langle \mathsf{vary}(\{p\})\rangle(\top \wedge \mathsf{H}(p,0) \wedge \mathsf{H}(p,0) \wedge \mathsf{H}(\neg p,0))) \vee \\
& \qquad\quad (\mathsf{H}(p,0) \wedge \mathsf{H}(p,1) \wedge \mathsf{H}(\neg p,0) \wedge \neg\langle \mathsf{vary}(\{p\})\rangle(\cdots)) \vee \\
& \qquad\quad (\mathsf{H}(p,0) \wedge \mathsf{H}(p,1) \wedge \mathsf{H}(\neg p,1) \wedge \neg\langle \mathsf{vary}(\{p\})\rangle(\cdots)) \vee \\
& \qquad\quad (\mathsf{H}(p,1) \wedge \mathsf{H}(p,0) \wedge \mathsf{H}(\neg p,0) \wedge \neg\langle \mathsf{vary}(\{p\})\rangle(\cdots)) \vee \\
& \qquad\quad (\mathsf{H}(p,1) \wedge \mathsf{H}(p,0) \wedge \mathsf{H}(\neg p,1) \wedge \neg\langle \mathsf{vary}(\{p\})\rangle(\cdots)) \vee \\
& \qquad\quad (\mathsf{H}(p,1) \wedge \mathsf{H}(p,1) \wedge \mathsf{H}(\neg p,0) \wedge \neg\langle \mathsf{vary}(\{p\})\rangle(\cdots)) \vee \\
& \qquad\quad (\mathsf{H}(p,1) \wedge \mathsf{H}(p,1) \wedge \mathsf{H}(\neg p,1) \wedge \neg\langle \mathsf{vary}(\{p\})\rangle(\cdots)) \\
& \quad \leftrightarrow (p \wedge p \wedge \neg p \wedge \neg\bot) \vee \\
& \qquad (p \wedge p \wedge p \wedge \neg\bot) \vee \\
& \qquad (p \wedge \neg p \wedge \neg p \wedge \neg\bot) \vee \\
& \qquad (p \wedge \neg p \wedge p \wedge \neg\bot) \vee \\
& \qquad (\neg p \wedge p \wedge \neg p \wedge \neg\bot) \vee \\
& \qquad (\neg p \wedge p \wedge p \wedge \neg\bot) \vee \\
& \qquad (\neg p \wedge \neg p \wedge \neg p \wedge \neg\top) \vee \\
& \qquad (\neg p \wedge \neg p \wedge p \wedge \neg\bot) \\
& \quad \leftrightarrow p
\end{aligned}
$$

Intuitively, the integrity constrained is enforced and it is checked for some integer d that first, each φ_i in the profile has distance at most d and second, that there is no other valuation that both satisfies the integrity constraint and is strictly less than d away from each φ_i.

5.4 Correction of the Translations

Theorem 2. *Let φ be an $\mathcal{L}_{\blacktriangle}$ formula. Then* $||\varphi|| = ||tr(\varphi)||_{\text{DL-PA}}$.

Proof. The proof is by induction on the form of φ. The only interesting case is that of merging operators. Let us consider the case of $\blacktriangle^{\Sigma}$. We prove in detail that $||\blacktriangle^{\Sigma}_{\psi}(E)|| = ||tr(\blacktriangle^{\Sigma}_{\psi}(E))||_{\text{DL-PA}}$.

Let $v \in \mathbb{V}$ be a valuation. We have $v \in ||\blacktriangle^{\Sigma}_{\psi}(E)||$ iff $v \in ||\psi||$ and there is no other ψ-valuation v' such that $\sum \mathsf{d}_{\mathsf{H}}(v', E) < \sum \mathsf{d}_{\mathsf{H}}(v, E)$. The latter is the case iff $v \in ||\psi||$ and there are $\langle d_1, \ldots, d_n \rangle$ such that

1. $\mathsf{d}_{\mathsf{H}}(v, \varphi_i) = d_i$ for every i, and
2. there is no ψ-valuation v' and vector $\langle d'_1, \cdots, d'_n \rangle$ such that $\mathsf{d}_{\mathsf{H}}(v', \varphi_i) = d'_i$ for every i and $\sum d_i < \sum d'_i$.

By induction hypothesis, $v \in ||\psi||$ iff $v \in ||tr(\psi)||_{\mathsf{DL\text{-}PA}}$ and $v \in ||\varphi_i||$ iff $v \in ||tr(\varphi_i)||_{\mathsf{DL\text{-}PA}}$. Therefore $\mathsf{H}(\varphi_i, d_i)$ equals $\mathsf{H}(tr(\varphi_i), d_i)$.

We note that by Lemma 1 it is in order to only consider the d_i such that $d_i \leq \mathsf{card}(\mathbb{P}_{\varphi_i})$. By Lemma 2, Item 1 means that $v \in ||\mathsf{H}(tr(\varphi_i), d_i)||_{\mathsf{DL\text{-}PA}}$ for every i.

Item 2 means that the formula

$$\bigvee_{\langle d'_1,\cdots,d'_n\rangle, \sum_k(d'_k)<\sum_k(d_k)} \bigwedge_{i\leq n} \mathsf{H}(tr(\varphi_i), d'_i)$$

is unsatisfiable. According to Lemma 2 and Proposition 1, all the relevant valuations are accessed by the program $\mathsf{vary}(\mathbb{P}_E)$. Therefore Item 2 is equivalent to

$$v \in ||\neg\langle\mathsf{vary}(\mathbb{P}_E)\rangle(tr(\psi) \wedge \bigvee_{\langle d'_1,\cdots,d'_n\rangle, \sum_{k\leq n}(d'_k)<\sum_k(d_k)} \bigwedge_{i\leq n} \mathsf{H}(tr(\varphi_i), d'_i))||_{\mathsf{DL\text{-}PA}}.$$

Putting things together, items 1 and 2 are equivalent to $v \in ||tr(\blacktriangle^{\Sigma}_{\psi}(\varphi_1, \cdots, \varphi_n))||_{\mathsf{DL\text{-}PA}}$.

It follows from the above theorem that the merging of the Boolean profile $\langle B_1, \cdots, B_n\rangle$ under the Boolean constraint C equals $||tr(\blacktriangle^{\sigma}_{C}(B_1, \cdots, B_n))||_{\mathsf{DL\text{-}PA}}$.

The length of $tr(\varphi)$ is however exponential in the length of φ. Nevertheless, if we consider 'big disjunctions' such as $\bigvee_{\langle d_1,\ldots,d_n\rangle, d_k\leq \mathsf{card}(\mathbb{P}_{\varphi_k})}$, $\bigvee_{\langle d'_1,\ldots,d'_n\rangle, \sum_k(d'_k)<\sum_k(d_k)}$ etc. to be connectives of the object language—i.e., as symbolic disjunctions that are parametrised by sets and that are not defined as abbreviations, but are proper connectives—then the length of $tr(\varphi)$ is still polynomial in the length of φ. For instance, the length of

$$\bigvee_{\langle d'_1,\cdots,d'_n\rangle, \sum_k(d'_k)<\sum_k(d_k)} \bigwedge_{i\leq n} \mathsf{H}(\varphi^{\Sigma}_i, d'_i)$$

is $O(n)$ plus the length of $\mathsf{H}(\varphi^{\Sigma}_i, d'_i)$.

Corollary 1. *Both model checking and satisfiability checking of $\mathcal{L}_{\blacktriangle}$-formulas is in PSPACE.*

Proof. First we give the argument why both model and satisfiability checking are PSPACE-complete for the star-free fragment of DL-PA if we allow symbolic disjunctions in DL-PA formulas. We do so by adapting the proof of PSPACE membership of [9]: in order to check whether $\bigvee_{\langle d'_1,\ldots,d'_n\rangle, \sum_k(d'_k)<\sum_k(d_k)} \psi$ is true at a valuation v we backtrack and test all the choices $\langle d'_1, \ldots, d'_n\rangle$ such that $\sum_k(d'_k) < \sum_k(d_k)$. This backtrack process can be implemented as an algorithm that only uses a polynomial amount of memory. By Theorem 2 we then reduce polynomially model (satisfiability) checking of $\mathcal{L}_{\Delta}$ formulas to model (satisfiability) checking of a DL-PA-formulas, where 'big disjunctions' are viewed as being symbolic.

Note that the language of DL-PA is more succinct than that of Boolean formulas: although every formula of DL-PA is equivalent to a Boolean formula, equivalent Boolean formulas can be exponentially bigger. So SAT techniques for propositional logic do not provide interesting decision procedures for $\mathcal{L}_{\blacktriangle}$.

6 Conclusion

We have defined a single language $\mathcal{L}_{\blacktriangle}$ in which all merging operators are in the object language: they are considered to be modal operators and can be nested. This differs with other approaches such as [18] and [5]. As far as we know, the only similar approach is [17], where the merging operator (as well as the comma separating the elements of profiles) are considered to be in the object language.

We have then embedded this language into Dynamic Logic of Propositional Assignments, DL-PA. This has enabled us to give syntactic counterparts to the most popular semantically defined merging operations. Using the reduction principles of DL-PA we can therefore rewrite formulas to Boolean formulas. As our examples show, such formulas may be quite long; in particular, they typically contain a lot of disjunctions. They can however often be simplified by means of standard syntactical operations. This provides interesting syntactical representations of merged belief bases.

The logic DL-PA actually provides a sort of assembler language for merging operators. Its use avoids the design of specific tools implementing merging operators. Unfortunately, no efficient reasoning mechanisms for DL-PA exist up to now, and it would be interesting to have such tools. (It could also be based on Binary Decision Diagrams as in [5].) As we have seen, if we want the embeddings to be polynomial then such tools should be able to handle 'big disjunctions' and 'big conjunctions'.

The star-free fragment of DL-PA into which we have mapped various merging operators has PSPACE complexity (both model checking and satisfiability). This induces a result for our merging language $\mathcal{L}_{\blacktriangle}$, which is new because $\mathcal{L}_{\blacktriangle}$ authorizes arbitrary nesting of merging operators. It is possible that the translated formulas however have patterns that are less complex.

As to future work, a first perspective is to study the mathematical properties of merging operators in more detail. One example is the behaviour of iterated merging operators (which is a research project similar to that for iterated belief revision, see e.g. [3].) Reasoning should be considerably facilitated by the help of a DL-PA reasoner. For instance, suppose we want to know whether the operator $\blacktriangle_{\top}^{\max}$ is associative. We may run the following experimental protocol: first, choose some Boolean formulas A, B, C and write down the formula $\blacktriangle_{\top}^{\max}(A, \blacktriangle_{\top}^{\max}(B, C)) \leftrightarrow \blacktriangle_{\top}^{\max}(\blacktriangle_{\top}^{\max}(A, B), C)$; second, translate this formula into DL-PA; third, run a DL-PA reasoner. Note however that one cannot use the theorem proving procedure for DL-PA because it only works for formula instances and not for formula schemas. (This is related to the fact that the rule of uniform substitution does not preserve validity in DL-PA, which generally fails in dynamic logics).

Our embeddings are somewhat simpler than the embeddings of belief change operations into QBF as done in [4] since DL-PA is a logic of programs. The same argument applies to embeddings of merging problems into MSO. Our approach may also be useful to capture semantics of merging: one may think in particular of new semantics requiring loops, which can be directly captured in DL-PA by the Kleene-star operator, whereas the encoding as a QBF will most probably be trickier.

A second perspective is to focus on embeddings of other existing operations. We did not succeed yet in embedding other approaches to merging such as [13] and syntax-based operations such as MCS of [2]. Note that in principle this might however be feasible: while the Hamming distance is a semantical notion, the function $\mathbb{P}_\varphi$ is purely syntactic.

There exist also tentatives to define merging operations in first order logic [6]. In the long run, we may plan to extend DL-PA with first order constructions in order to capture those merging operations.

Acknowledgements. We wish to thank the three FOIKS reviewers for their critical comments and pointers to relevant work that we had not considered at the time of the submission.

References

1. Balbiani, P., Herzig, A., Troquard, N.: Dynamic logic of propositional assignments: A well-behaved variant of PDL. In: Kupferman, O. (ed.) Logic in Computer Science (LICS), New Orleans, June 25-28, IEEE (2013), `http://www.ieee.org/`
2. Baral, C., Kraus, S., Minker, J., Subrahmanian, V.S.: Combining knowledge bases consisting of first-order theories. Computational Intelligence 8, 45–71 (1992)
3. Darwiche, A., Pearl, J.: On the logic of iterated belief revision. Artificial Intelligence 89(1), 1–29 (1997)
4. Delgrande, J.P., Schaub, T., Tompits, H., Woltran, S.: On computing belief change operations using quantified boolean formulas. Journal of Logic and Computation 14(6), 801–826 (2004)
5. Gorogiannis, N., Hunter, A.: Implementing semantic merging operators using binary decision diagrams. International Journal of Approximate Reasoning 49(1), 234–251 (2008)
6. Gorogiannis, N., Hunter, A.: Merging first-order knowledge using dilation operators. In: Hartmann, S., Kern-Isberner, G. (eds.) FoIKS 2008. LNCS, vol. 4932, pp. 132–150. Springer, Heidelberg (2008)
7. Harel, D.: Dynamic logic. In: Gabbay, D.M., Günthner, F. (eds.) Handbook of Philosophical Logic, vol. II, pp. 497–604. D. Reidel, Dordrecht (1984)
8. Harel, D., Kozen, D., Tiuryn, J.: Dynamic Logic. MIT Press (2000)
9. Herzig, A., Lorini, E., Moisan, F., Troquard, N.: A dynamic logic of normative systems. In: Walsh, T. (ed.) International Joint Conference on Artificial Intelligence (IJCAI), IJCAI/AAAI, Barcelona, pp. 228–233 (2011), Erratum at `http://www.irit.fr/~Andreas.Herzig/P/Ijcai11.html`
10. van der Hoek, W., Walther, D., Wooldridge, M.: On the logic of cooperation and the transfer of control. J. of AI Research (JAIR) 37, 437–477 (2010)
11. van der Hoek, W., Wooldridge, M.: On the dynamics of delegation, cooperation and control: A logical account. In: Proc. AAMAS 2005 (2005)
12. van der Hoek, W., Wooldridge, M.: On the logic of cooperation and propositional control. Artif. Intell. 164(1-2), 81–119 (2005)
13. Konieczny, S.: On the difference between merging knowledge bases and combining them. In: KR, pp. 135–144 (2000)
14. Konieczny, S., Pérez, R.P.: Logic based merging. Journal of Philosophical Logic 40(2), 239–270 (2011)
15. Konieczny, S., Pérez, R.P.: On the logic of merging. In: Proc. 6th Int. Conf. on Principles of Knowledge Representation and Reasoning (KR 1998), pp. 488–498. Morgan Kaufmann (1998)
16. Konieczny, S., Pino Pérez, R.: Merging with integrity constraints. In: Hunter, A., Parsons, S. (eds.) ECSQARU 1999. LNCS (LNAI), vol. 1638, pp. 233–244. Springer, Heidelberg (1999)

17. Lang, J., Marquis, P.: Reasoning under inconsistency: A forgetting-based approach. Artificial Intelligence 174(12), 799–823 (2010)
18. Liberatore, P., Schaerf, M.: Brels: A system for the integration of knowledge bases. In: KR, pp. 145–152. Citeseer (2000)
19. Lin, J., Mendelzon, A.: Knowledge base merging by majority. In: Pareschi, R., Fronhoefer, B. (eds.) Dynamic Worlds: From the Frame Problem to Knowledge Management, Kluwer Academic (1999)
20. Revesz, P.Z.: On the semantics of theory change: arbitration between old and new information. In: Proceedings of the Twelfth ACM SIGACT-SIGMOD-SIGART Symposium on Principles of Database Systems, PODS 1993, pp. 71–82. ACM, New York (1993)

Incremental Maintenance of Aggregate Views

Abhijeet Mohapatra and Michael Genesereth

Stanford University
{abhijeet,genesereth}@stanford.edu

Abstract. We propose an algorithm called *CReaM* to incrementally maintain materialized aggregate views with user-defined aggregates in response to changes to the database tables from which the view is derived. CReaM is *optimal* and guarantees the self-maintainability of aggregate views that are defined over a single database table. For aggregate views that are defined over multiple database tables and do not contain all of the non-aggregated attributes in the database tables, CReaM speeds up the time taken to update a view as compared to prior view maintenance techniques. The speed up in the time taken to update a materialized view with n tuples is either $\frac{n}{\log n}$ or $\log n$ depending on whether the materialized view is indexed or not. For other types of aggregate views, CReaM updates the view in no more time than that is required by prior view maintenance techniques to update the view.

1 Introduction

In data management systems, views are derived relations that are computed over database tables (which are also known as extensional database relations or *edbs*). Views are materialized in a database to support efficient querying of the data. A materialized view becomes out-of-date when the underlying edb relations from which the view is derived are changed. In such cases, the materialized view is either recomputed from the edb relations or the changes in the edb relations are *incrementally* propagated to the view to ensure the correctness of the answers to queries against the view. Prior work presented in [5, 21, 27] shows that incrementally maintaining a materialized view can be significantly faster than recomputing the view from the edb relations especially if the size of the view is large compared to the size of the changes.

Several techniques [1–6, 10, 14–16, 18, 21–23, 26–28, 30–32, 34] have been proposed to incrementally maintain views in response to changes to the edb relations. However, only a small fraction of the prior work on incremental view maintenance [10, 14, 15, 21, 24, 27, 28] addresses the maintenance of views that contain aggregates such as sum and count. The techniques proposed in [10, 14, 21, 24] incrementally maintain views that have only one aggregation operator. Furthermore, the incremental maintenance algorithms presented in [10, 14, 15, 21, 24, 27, 28] support only a fixed set of built-in aggregate operators (min, max, sum, and count).

In contrast, we present an algorithm to incrementally maintain views with *multiple* aggregates each of which could be *user-defined*. As our underlying query language, we extend Datalog using tuples and sets, and express aggregates as predicates over sets. We note that, in Datalog, predicates are sets of tuples. Therefore, there are no *duplicates*.

C. Beierle and C. Meghini (Eds.): FoIKS 2014, LNCS 8367, pp. 399–414, 2014.

In Section 2, we discuss the specification of user-defined aggregates in our language. In Section 3, we present differential rules to correctly characterize the changes in edb relations to aggregate views. In Section 4, we present an algorithm *CReaM* to *optimize* the maintenance of a special class of materialized aggregate views that do not contain all of the non-aggregated attributes in the underlying edb relations. In Section 5, we establish the optimality of CReaM and the theoretical results on the performance of CReaM which are summarized in Table 1. In addition, we show that by materializing auxiliary views, CReaM guarantees the *self-maintainability* [17] of aggregate views that are defined over a single edb relation. This property is desirable when access to the edb relations is restricted or when the edb relations are hypothetical such as in a LAV integration scenario [33]. In Section 6, we compare our work to prior work on incremental maintenance of aggregate views.

Before we discuss our proposed solution, we illustrate the problem of incrementally maintaining aggregate views using a running example. We use our running example, which is based on the Star Wars universe, in examples throughout the paper.

Table 1. Performance summary of the CReaM algorithm

Speed up in the time taken to update a materialized aggregate view with n tuples as compared to prior techniques		
	Physical design of the database	
	Non-optimized	Optimized
Aggregate view over single edb relation	≥ 1	≥ 1
Aggregate view over multiple edb relations (single update)	$\log n$	$\frac{n}{\log n}$
Aggregate view over multiple edb relations ($k > 1$ updates)	$\log n$	$\frac{n}{k}$

Running Example: Suppose that there are tournaments in the Star Wars universe on different planetary systems. The tournament results are recorded in an edb relation, say tournament(V, D, L). A tuple $(V, D, L) \in$ tournament *iff* V has defeated D on the planet L. For instance, if Yoda has defeated Emperor Palpatine at Dagobah then the tuple (yoda, palpatine, dagobah) is in the extension of the edb relation tournament. We use the extension of tournament that is presented in Table 2 in examples throughout the paper.

Table 2. Extension of the edb relation *tournament* and the view *victories* in the Star Wars Universe

tournament		
Victor	**Defeated**	**Location**
yoda	vader	dagobah
yoda	palpatine	dagobah
vader	yoda	tatooine
yoda	palpatine	tatooine

victories	
Victor	**Wins**
yoda	2
vader	1
yoda	1

We define and materialize a view, say victories(V, W) to record the number of victories W achieved by a character V on a planet. For instance, Yoda has two victories in Dagobah and one victory in Tatooine. Therefore the tuples (yoda, 2) and (yoda, 1) $\in$ victories. The extension of victories that corresponds to the extension of the edb relation tournament is presented in Table 2.

Suppose that a new tournament match is played in Tatooine and that Darth Vader defeats Emperor Palpatine in this match. The new tournament match at Tatooine causes an *insert* to the tournament relation. In response to the insert to the tournament relation, the tuple (vader, 1) $\in$ victories must be updated to (vader, 2) to ensure the correctness of the answers to queries against the view. Now, suppose that the previous tournament match between Yoda and Palpatine at Tatooine is invalidated. In this case, the tuple (yoda, palpatine, tatooine) is *deleted* from the tournament relation. In response to this deletion, the tuple (yoda, 1) must be deleted from the materialized view victories.

2 Preliminaries

As our underlying language, we use the extension of Datalog that is proposed in [20]. We introduce tuples and sets as first-class citizens in our language. A tuple is an ordered sequence of Datalog constants or sets. A set is either empty or contains Datalog constants or tuples. For example, the tuple (yoda, vader, dagobah) and the sets $\{\}$, $\{$(yoda, vader, dagobah)$\}$ and $\{$(yoda, $\{1, 2\}$)$\}$ are legal in our language. We introduce the *setof* operator in our language to represent sets as follows.

Definition 1. *Suppose that $\phi(\bar{X}, \bar{Y})$ is a conjunction of subgoals. For every binding of values in $\bar{X}$, the subgoal setof($\bar{Y}$, $\phi(\bar{X}, \bar{Y})$, S) evaluates to true on a database D if $S = \{\bar{Y} \mid \phi(\bar{X}, \bar{Y})\}$ for D.*

We illustrate the construction of sets in our language using the following example.

Example 1. Consider the running example that we presented in Section 1. Suppose we would like to compute the set of characters who were defeated by Yoda at Dagobah. In our language, we compute the desired set using the following query.

$$q_1(S) \text{ :- setof}(D, \text{tournament(yoda}, D, \text{dagobah)}, S)$$

In query q_1, the set $S = \{D \mid \text{tournament(yoda}, D, \text{dagobah)}\}$. The evaluation of the query q_1 on the extension of tournament that is presented in Table 2 results in the answer tuple $q_1(\{\text{vader, palpatine}\})$.

Construction of Multisets: In Example 1, we computed the set of characters who were defeated by Yoda at Dagobah. In addition to generating sets, we could leverage the setof operator to effectively aggregate *multisets* as illustrated in the following example.

Example 2. Suppose we would like to count the total number of victories achieved by Yoda. Since Yoda defeats Emperor Palpatine at multiple locations, we would have to compute the cardinality of the multiset of people who were defeated by Yoda i.e. $\{$palpatine, vader, palpatine$\}$ to compute Yoda's total number of victories. Consider a query q_2 which is defined as follows.

$$q_2(C) \text{ :- setof}((D, L), \text{tournament(yoda}, D, L), S), \text{ count}(S, C)$$

We assume that the predicate count(X, Y) computes the cardinality Y of the set X. We discuss the representation of user-defined aggregates in our language shortly. In query

q_2, the set $S = \{(D, L) \mid \text{tournament}(\text{yoda}, D, L)\}$. The evaluation of q_2 on the extension of tournament relation (Table 2) computes the cardinality of the set {(palpatine, dagobah), (vader, dagobah), (palpatine, tatooine)} thus mimicking the computation of the cardinality of the multiset {palpatine, vader, palpatine}. As a result, the answer tuple $q_2(3)$ is generated.

We use the '|' operator in our language to represent the decompositions of a set. We represent the decomposition of a set S into an element $X \in S$ and the subset $S_1 = S \setminus \{X\}$ as $\{X \mid S_1\}$. For example, $\{3 \mid \{1, 2\}\}$ represents the decomposition of the set of numbers $\{1, 2, 3\}$ into 3 and the subset $\{1, 2\}$. We define the predicate *member* in our language to check the membership of an element in a set. The member predicate has the signature member(X, S), where X is a Datalog constant or a tuple and S is a set. If $X \in S$ then member(X, S) is true, otherwise it is false. The member predicate can be defined in our language using the decomposition operator '|' operator as follows.

$$\text{member}(X, \{X \mid Y\})$$
$$\text{member}(Z, \{X \mid Y\}) \text{ :- } \text{member}(Z, Y)$$

In addition, we use $\cup$ and $\setminus$ operators in our language to represent set-union and set-difference respectively. We note that we can define $\cup$ and $\setminus$ operators in our language using the member predicate and the decomposition operator '|' although we do not define them as such in this paper.

Aggregation over sets: In our language, user-defined aggregates are defined as predicates over sets. A user-defined aggregate could either be defined (a) in a stand-alone manner using the member predicate, the decomposition operator '|', the set-union $\cup$ and the set-difference $\setminus$ operators, and the arithmetic operators or (b) as a view over other aggregates. For instance, we can compute the cardinality of a set in our language by inductively defining an aggregate, say count(X, C), as follows.

$$\text{count}(\{\}, 0)$$
$$\text{count}(\{X \mid Y\}, C) \text{ :- } \text{count}(Y, C_1),\ C = C_1 + 1$$

The first rule specifies the base case of the induction i.e. the cardinality of an empty set is 0. The second rule decomposes a set S into an element X and the subset Y and computes the cardinality of S by leveraging the cardinality of Y. In addition, we can define an aggregate such as *average* modularly by leveraging the definitions of the aggregates count(X, C) and sum(X, S) as follows.

$$\text{average}(X, A) \text{ :- } \text{sum}(X, S),\ \text{count}(X, C),\ A = \frac{S}{C}$$

3 Maintenance of Aggregate Views

In the previous section, we discussed the specification of aggregates as predicates over sets in our language. Consider the running example that we presented in Section 1.

Suppose we would like to query the number of victories W achieved by a character V on a planet. We represent this query in our language as follows.

$$q(V, W) \text{ :- } \text{setof}(D, \text{tournament}(V, D, L), S),\ \text{count}(S, W)$$

For every binding of the variables V and L in the query q, W = cardinality of $\{D \mid \text{tournament}(V, D, L)\}$. In the Star Wars universe, this is equivalent to computing the number of victories W achieved by a character V on a planet. Since the answers to queries are computed under *set semantics*, the distinct numbers of victories are generated by the query q. To efficiently compute the answer to the query q, we can leverage the materialized view victories from our running example (in Section 1). The materialized view victories(V, W) is defined as follows.

$$\text{victories}(V, W) \text{ :- } \text{setof}(D, \text{tournament}(V, D, L), S),\ \text{count}(S, W)$$

When changes are made to the edb relation tournament, we must maintain the materialized view victories to ensure the correctness of answers to the query q.

Maintenance of Views that Contain Sets: Consider a view v in our language which is defined over the formula $\phi(\bar{X}, \bar{Y}, \bar{Z})$ using the aggregation predicate *agg* as follows.

$$v(\bar{X}, A) \text{ :- } \text{setof}(\bar{Y},\ \phi(\bar{X}, \bar{Y}, \bar{Z}),\ W),\ \text{agg}(W, A)$$

Since aggregates are defined as predicates over sets in our language, we can rewrite the definition of the view v using the following two rules, one of which contains a setof subgoal while the other does not.

$$v(\bar{X}, A) \text{ :- } u(\bar{X}, W),\ \text{agg}(W, A)$$
$$u(\bar{X}, W) \text{ :- } \text{setof}(\bar{Y},\ \phi(\bar{X}, \bar{Y}, \bar{Z}),\ W)$$

We note that if the definition of v contains k setof subgoals instead of one, we can rewrite the definition of v using $k + 1$ rules where only k of the rules contain setof subgoals. Since prior view maintenance techniques [13] already maintain views that do not contain sets, we focus *only* on the maintenance of views that contain setof subgoals. As a first step, we leverage differential relational calculus to incrementally propagate the changes in the edb relations to the views through *differential rules*. Then, in Section 4, we propose an algorithm called CReaM that applies these differential rules to optimally maintain materialized aggregate views.

Differential Rules: In differential relational calculus, a database is represented as a set of edb relations and views $r_1, r_2, \ldots, r_k$ with arities $d_1, d_2, \ldots, d_k$. Each relation r_i is a set of d_i-tuples [9]. The changes to a relation r_i in the database consist of insertions of new tuples and deletions of existing tuples. The new state of a relation r_i after applying a change is represented as r_i'. An update to an existing tuple can be modeled as a deletion followed by an insertion. The insertion of new tuples into a relation r_i and the deletion of existing tuples from r_i are represented as the differential relations r_i^+ and r_i^- respectively. Prior work in [23] presents a set of *differential rules* to compute the differentials (v^+ or v^-) of a non-aggregate view v. We extend the framework that is presented in [23] to compute the differentials of aggregate views.

There are two possible ways in which a view v can be defined in our language using the setof operator over the formula $\phi(\bar{X}, \bar{Y}, \bar{Z})$.

1. The view v is defined as $v(\bar{X}, \bar{Z}, W)$:- setof($\bar{Y}$, $\phi(\bar{X}, \bar{Y}, \bar{Z})$, W). In this case, all of the variables of ϕ that are bound outside the setof subgoal are passed to the view v.
2. The view v is defined as $v(\bar{X}, W)$:- setof($\bar{Y}$, $\phi(\bar{X}, \bar{Y}, \bar{Z})$, W). In this case, not all of the variables of ϕ that are bound outside the setof subgoal are passed to the view v.

We consider the above two cases separately and present differential rules to compute the differentials of the view v in each case.

Case 1: Suppose that a view v is defined over a conjunction of subgoals $\phi(\bar{X}, \bar{Y})$ as follows.

$$v(\bar{X}, W) \text{ :- setof}(\bar{Y},\ \phi(\bar{X}, \bar{Y}),\ W)$$

Suppose we define a view u over $v(\bar{X}, W)$ as $u(\bar{X})$:- $v(\bar{X}, W)$. The view u can be maintained using the differential rules that are proposed in [23]. In this case, the following differential rules correctly compute the differential relations $v^+(\bar{X}, W)$ and $v^-(\bar{X}, W)$ in response to the changes to $\phi(\bar{X}, \bar{Y})$.

$$v^+(\bar{X}, W) \text{ :- setof}(\bar{Y},\ \phi^+(\bar{X}, \bar{Y}),\ W),\ \neg u(\bar{X}) \qquad (\Delta_1)$$
$$v^+(\bar{X}, W \cup W') \text{ :- setof}(\bar{Y},\ \phi^+(\bar{X}, \bar{Y}),\ W),\ v(\bar{X}, W') \qquad (\Delta_2)$$
$$v^+(\bar{X}, W' \setminus W) \text{ :- setof}(\bar{Y},\ \phi^-(\bar{X}, \bar{Y}),\ W),\ v(\bar{X}, W') \qquad (\Delta_3)$$
$$v^-(\bar{X}, W) \text{ :- setof}(\bar{Y},\ \phi^-(\bar{X}, \bar{Y}),\ W),\ v(\bar{X}, W) \qquad (\Delta_4)$$
$$v^-(\bar{X}, W) \text{ :- setof}(\bar{Y},\ \phi^+(\bar{X}, \bar{Y}),\ \_),\ v(\bar{X}, W) \qquad (\Delta_5)$$
$$v^-(\bar{X}, W) \text{ :- setof}(\bar{Y},\ \phi^-(\bar{X}, \bar{Y}),\ \_),\ v(\bar{X}, W) \qquad (\Delta_6)$$

In the above differential rules, '_' represents *don't care* variables. We prove the correctness of the differential rules Δ_1– Δ_6 in the following theorem.

Theorem 1. *The differential rules Δ_1– Δ_6 correctly maintain a view containing a setof subgoal where all the variables that are bound outside the setof are passed to the head.*

Proof. Consider a view v that is defined over a conjunction of subgoals $\phi(\bar{X}, \bar{Y})$ as follows.

$$v(\bar{X}, W) \text{ :- setof}(\bar{Y},\ \phi(\bar{X}, \bar{Y}),\ W)$$

In the definition of v, the variables that are bound outside the setof subgoal i.e. $\bar{X}$ are passed to the head. Hence the view v contains exactly one tuple $(\bar{X}, W)$ for every distinct value of $\bar{X}$. Suppose a tuple, say $\phi(\bar{x}, \bar{y})$, is **inserted**. Either the view v does not contain any tuple $v(\bar{X}, W)$ where $\bar{X} = \bar{x}$ or v contains a tuple $v(\bar{x}, w)$. For correctly maintaining the view v, the tuple $(\bar{x}, \{\bar{y}\})$ is inserted into the view v in the former case. The differential rule Δ_1 handles this case. In the latter case, the tuple $v(\bar{x}, w)$

is updated to $v(\bar{x}, w \cup \{\bar{y}\})$ to correctly update the view v. The differential rules Δ_2 and Δ_5 capture this update.

Suppose a tuple, say $\phi(\bar{x}, \bar{y})$, is **deleted**. Either v contains the tuple $v(\bar{x}, \{\bar{y}\})$ or v contains the tuple $v(\bar{x}, W)$ where $\{\bar{y}\} \subset W$. For correctly maintaining the view v, the tuple $(\bar{x}, \{\bar{y}\})$ is deleted from the view v in the former case. The differential rule Δ_4 captures this deletion. In the latter case, the tuple $v(\bar{x}, w)$ is updated to $v(\bar{x}, w \setminus \{\bar{y}\})$ to correctly update the view v. The differential rules Δ_3 and Δ_6 capture this update. □

Case 2: Now, suppose that a view v is defined over a conjunction of subgoals $\phi(\bar{X}, \bar{Y}, \bar{Z})$ as follows.

$$v(\bar{X}, W) \text{ :- setof}(\bar{Y},\ \phi(\bar{X}, \bar{Y}, \bar{Z}),\ W)$$

In addition, suppose that we define a view u over $v(\bar{X}, W)$ as $u(\bar{X})$:- $v(\bar{X}, W)$. In the definition of v, the set W is computed as $\{\bar{Y} \mid \phi(\bar{X}, \bar{Y}, \bar{Z})\}$ for every binding of $\bar{X}$ and $\bar{Z}$. Since $\bar{Z}$ is not passed to the view v, a view tuple, say $v(x, w)$, potentially has multiple derivations. In this case, we compute the differentials $v^+(\bar{X}, W)$ and $v^-(\bar{X}, W)$ as follows.

$$v^+(\bar{X}, W) \text{ :- setof}(\bar{Y},\ \phi^+(\bar{X}, \bar{Y}, \bar{Z}),\ W),\ \neg u(\bar{X}) \qquad (\Gamma_1)$$

$$v^+(\bar{X}, W \cup W') \text{ :- setof}(\bar{Y},\ \phi^+(\bar{X}, \bar{Y}, \bar{Z}),\ W),\ \text{setof}(\bar{Y}',\ \phi(\bar{X}, \bar{Y}', \bar{Z}),\ W'),\ \neg v(\bar{X}, W \cup W') \qquad (\Gamma_2)$$

$$v^+(\bar{X}, W' \setminus W) \text{ :- setof}(\bar{Y},\ \phi^-(\bar{X}, \bar{Y}, \bar{Z}),\ W),\ \text{setof}(\bar{Y}',\ \phi(\bar{X}, \bar{Y}', \bar{Z}),\ W'),\ \neg v(X, W' \setminus W) \qquad (\Gamma_3)$$

$$v^-(\bar{X}, W) \text{ :- setof}(\bar{Y},\ \phi^-(\bar{X}, \bar{Y}, \bar{Z}),\ W),\ v(\bar{X}, W),\ \neg\text{setof}(\bar{Y}',\ \phi'(\bar{X}, \bar{Y}', \bar{Z}),\ W) \qquad (\Gamma_4)$$

$$v^-(\bar{X}, W) \text{ :- setof}(\bar{Y},\ \phi^+(\bar{X}, \bar{Y}, \bar{Z}),\ \_),\ \text{setof}(\bar{Y}',\ \phi(\bar{X}, \bar{Y}', \bar{Z}),\ W),\ \neg\text{setof}(\bar{Y}'',\ \phi'(\bar{X}, \bar{Y}'', \bar{Z}),\ W) \qquad (\Gamma_5)$$

$$v^-(\bar{X}, W) \text{ :- setof}(\bar{Y},\ \phi^-(\bar{X}, \bar{Y}, \bar{Z}),\ \_),\ \text{setof}(\bar{Y}',\ \phi(\bar{X}, \bar{Y}', \bar{Z}),\ W),\ \neg\text{setof}(\bar{Y}'',\ \phi'(\bar{X}, \bar{Y}'', \bar{Z}),\ W) \qquad (\Gamma_6)$$

We prove the correctness of the differential rules Γ_1– Γ_6 in the following theorem.

Theorem 2. *The differential rules Γ_1– Γ_6 correctly maintain a view containing a setof subgoal where all the variables that are bound outside the setof are not passed to the head.*

The proof of Theorem 2 is similar to the proof of Theorem 1, except that before deleting a tuple from the view we check for alternate derivations of the tuple in the updated subgoals. In addition, a tuple is inserted into the view only if the view does not contain an alternate derivation of the tuple to be inserted.

4 Efficient Incremental Maintenance

In the previous section, we extended the differential rules that are presented in [23] to incrementally compute the differentials of views containing setof subgoals. In this section, we leverage the differential rules (from Section 3) to *optimally* maintain views containing setof subgoals. As a first step, we present an example where the differential rules are leveraged to incrementally maintain views containing setof subgoals.

Example 3. Consider a materialized view *dominates* which is defined over the tournament relation from our running example (in Section 1) as follows.

$$\text{dominates}(V, W) \text{ :- } \text{setof}(D, \text{tournament}(V, D, L), W)$$

The extension of the view dominates that corresponds to the extension of tournament (in Table 2) is presented below.

dominates	
Victor	**Defeated**
yoda	{palpatine, vader}
vader	{yoda}
yoda	{palpatine}

Suppose that Yoda defeats Darth Vader at Tatooine in a new tournament match. This match results in the insertion of the tuple (yoda, vader, tatooine) into the tournament relation i.e. (yoda, vader, tatooine) $\in$ tournament$^+$. Since the non-aggregated variable L in tournament is not passed to the view dominates, we apply the differential rules Γ_1– Γ_6 to incrementally compute the differentials of the view dominates. By applying the differential rules Γ_2 and Γ_5 on the differential tournament$^+$ and the relations tournament and dominates, we derive the differentials dominates$^-$(yoda, {palpatine}) and dominates$^+$(yoda, {palpatine, vader}). The computed differentials correspond to updating the tuple (yoda, {palpatine}) $\in$ dominates to the tuple (yoda, {palpatine, vader}).

We note that in Example 3, the differential rules Γ_2 and Γ_5 access tournament's extension in addition to the differential tournament$^+$ to maintain the view dominates. A tuple $(V, W) \in$ dominates could potentially have multiple derivations in tournament because V could defeat the *same* set of characters W at multiple planetary systems. Hence, additional accesses to the extensions of edb relations are required to maintain materialized views using differential rules.

Alternatively, we could maintain the count of the different derivations of tuples to optimize the maintenance of aggregate views. The *counting algorithm*, which is presented in [14], leverages this idea to optimize the maintenance of views where the tuples in the view have multiple derivations in the edb relations. Suppose that in Example 3, we maintain the count of the different derivations of a tuple.

dominates		
Victor	**Defeated**	**Number of Derivations**
yoda	{palpatine, vader}	1
vader	{yoda}	1
yoda	{palpatine}	1

Now suppose that we delete a tuple, say (yoda, vader, dagobah) from tournament's extension. In this case, we decrease the count of the tuple (yoda, {palpatine, vader}) $\in$ dominates from 1 to 0 (thereby deleting it from the view) and increase the count of the

dominates		
Victor	**Defeated**	**Number of Derivations**
vader	{yoda}	1
yoda	{palpatine}	2

tuple (yoda, {palpatine}) from 1 to 2.

When the tuple (yoda, vader, dagobah) is deleted from tournament's extension, we do not have to access tournament's extension to incrementally maintain the materialized view dominates. However, consider a scenario where we delete the tuple (yoda, palpatine, dagobah) instead of the tuple (yoda, vader, palpatine) from tournament's extension. In this scenario, unless we access tournament's extension, we *cannot* correctly update the materialized view dominates because we do not have sufficient information to determine whether the existing tuple (yoda, {palpatine}) $\in$ dominates is to be deleted or the tuple (yoda, {palpatine, vader}) $\in$ dominates is to be updated.

Incremental Maintenance Using CReaM[1]: Consider the materialized view dominates that we presented in Example 3. Suppose we rewrite the definition of dominates using an auxiliary view v_a as follows.

$$\text{dominates}(V, W) \text{ :- } v_a(V, L, W)$$
$$v_a(V, L, W) \text{ :- setof}(D, \text{tournament}(V, D, L), W)$$

In addition, suppose that we materialize the auxiliary view v_a and maintain the counts of the derivations of a tuple in the view dominates. The extension of the auxiliary view v_a is presented below.

v_a		
Victor	**Location**	**Defeated**
yoda	dagobah	{palpatine, vader}
vader	tatooine	{yoda}
yoda	tatooine	{palpatine}

Now, suppose that we delete the tuple (yoda, palpatine, dagobah) from the extension of tournament. Since all of the non-aggregated variables of tournament are passed to the auxiliary view v_a, we can incrementally maintain v_a using the differential rules Δ_1– Δ_6 (from Section 3). We note that Δ_1– Δ_6 only access the extension of a view and the differentials of the edb relations over which the view is defined. Thus, we are able to compute the differentials v_a^-(yoda, dagobah, {palpatine, vader}) and v_a^+(yoda, dagobah, {vader}) without accessing the extension of tournament.

Since the modified definition of the view dominates *does not* contain setof subgoals, we use the counting algorithm [14] to incrementally maintain the count of the tuple derivations in the view dominates in a subsequent step. The updated extension of the view dominates is presented below.

[1] The algorithm has been named CReaM because it **C**ounts the tuple derivations in a view, **Re**writes the view using auxiliary views and **M**aintains the auxiliary views.

dominates		
Victor	**Defeated**	**Number of Derivations**
vader	{yoda}	1
yoda	{palpatine}	2

We now propose an algorithm called CReaM to incrementally maintain views containing setof subgoals. The CReaM algorithm is presented in Figure 1. In Step 1 of the algorithm, the supplied view is rewritten using an auxiliary view which is materialized in a subsequent step. In Step 3 of the algorithm, the number of derivations of the tuples in the supplied view is maintained. The *incremental maintenance* of the view v is carried out in Step 4 of the algorithm by computing the differentials of the auxiliary view which was created in Step 1 of the algorithm.

CReaM Algorithm

Input: 1. Materialized view $v(\bar{X}, W)$ defined as:

$$v(\bar{X}, W) \text{ :- setof}(\bar{Y}, \phi(\bar{X}, \bar{Y}, \bar{Z}), W),$$

2. Differentials $\phi^+(\bar{X}, \bar{Y}, \bar{Z})$ and $\phi^-(\bar{X}, \bar{Y}, \bar{Z})$

Step 1: **Rewrite** the view v using an auxiliary view v_a which contains all of the non-aggregated variables

$$v(\bar{X}, W) \text{ :- } v_a(\bar{X}, \bar{Z}, W)$$
$$v_a(\bar{X}, \bar{Z}, W) \text{ :- setof}(\bar{Y}, \phi(\bar{X}, \bar{Y}, \bar{Z}), W)$$

Step 2: **Materialize** the auxiliary view v_a

Step 3: Maintain the **count** of the tuple derivations in the view v

Step 4: Apply the differential rules Δ_1- Δ_6 over $\phi^+(\bar{X}, \bar{Y}, \bar{Z})$ and $\phi^-(\bar{X}, \bar{Y}, \bar{Z})$ to compute $v_a^+(\bar{X}, \bar{Z}, W)$ and $v_a^-(\bar{X}, \bar{Z}, W)$
Use $v_a^+(\bar{X}, \bar{Z}, W)$ and $v_a^-(\bar{X}, \bar{Z}, W)$ to incrementally update the counts of v's tuples using [9]

Fig. 1. Algorithm to *optimally* maintain views containing setof subgoals

We note that the CReaM algorithm incrementally maintains a view whose definition contains a single setof subgoal. However, when the supplied view definition contains multiple setof subgoals and aggregate predicates, we can incrementally maintain the view using CReaM as follows. Suppose a materialized view v contains k setof subgoals $\{s_i\}$ and m aggregate predicates $\{a_i\}$. First, we rewrite the definition of v using k auxiliary predicates, say $\{t_i\}$ where each t_i is defined as t_i :- s_i. Next, we maintain the counts of the tuple derivations in v and incrementally compute the differentials of t_i by applying CReaM to the extensions of the auxiliary predicates $\{t_i\}$ and the differentials of the edb relations. Since the modified definition of v does not contain setof subgoals, we use the counting algorithm that is presented in [14] to incrementally maintain the materialized view v.

In the following theorem, we establish that CReaM correctly maintains views that contain setof subgoals.

Theorem 3. *CReaM correctly maintains a materialized view containing setof subgoals.*

Proof. Consider a view v in our language which is defined using k setof subgoals $s_1, s_2, \ldots, s_k$ as v :- $s_1, s_2, \ldots, s_k$. Suppose that we introduce k auxiliary views v_{a_1},

$v_{a_2}, \ldots, v_{a_k}$ where each v_{a_i} is defined as v_{a_i} :- s_i. In the definition of the auxiliary view v_{a_i}, all of the variables that are bound outside the setof subgoal s_i are passed to the view. By replacing the setof subgoals using the auxiliary views, we can rewrite the definition of the view v as v :- $v_{a_1}, v_{a_2}, \ldots, v_{a_k}$. Since the modified definition of the view v does not contain setof subgoals, we can correctly maintain it by applying the counting algorithm [14]. In addition, we can leverage the rules Δ_1– Δ_6 to correctly compute the differentials of the auxiliary views $\{v_{a_i}\}$ by Theorem 1. □

In the next section, we discuss the performance of CReaM and prove that it optimally maintains materialized views containing setof subogals.

5 Performance of CReaM

Previous aggregate view maintenance algorithms [10, 14, 15, 21, 24, 27, 28] do not materialize and maintain additional views. Instead, the algorithms leverage differential relational algebra [10, 14, 15, 27, 28] or maintain the count of tuple derivations [10, 14, 15, 21, 24] to efficiently maintain aggregate views. In this section, we show that by rewriting, and maintaining additional auxiliary views, CReaM speeds up the time taken to incrementally maintain a view containing setof subgoals in comparision to previous view maintenance algorithms. As our underlying cost model, we assume that the time taken by an algorithm to incrementally maintain a view is proportional to the number of tuple accesses that are required by the algorithm to maintain the view. As a first step, we discuss the performance of CReaM when a single tuple is changed in the underlying edb relations. We then discuss the performance of CReaM with respect to multiple tuple updates.

Consider a view v which is defined as $v(\bar{X}, W)$:- setof$(\bar{Y}, \phi(\bar{X}, \bar{Y}, \bar{Z}), W)$. Suppose that the extensions of v and ϕ consist of n_v and n_ϕ tuples respectively. The extension of the view v and the differential ϕ^+ and ϕ^- are provided as inputs to the CReaM algorithm (see Figure 1). In Steps 1 and 2, CReaM rewrites the definition of v using an auxiliary view v_a and materializes v_a. Suppose that the number of tuples in v_a is n_{v_a}. Then, $n_v \leq n_{v_a} \leq n_\phi$. In Step 3, CReaM materializes the count of the tuple derivations in v. Steps 1, 2, and 3 of the CReaM algorithm are pre-processing steps that are executed before ϕ is updated. Therefore, in our analysis, we only consider the time taken to execute Step 4 of CReaM as the time that is required to incrementally maintain v.

In Step 4, CReaM maintains v_a using the differential rules Δ_1– Δ_6. The time required to update v_a using Δ_1– Δ_6 is equal to the time required to compute the join of the view v_a and the differentials ϕ^+ and ϕ^-. When the materialized view v_a is indexed on the attributes $\bar{X}$ and $\bar{Z}$, the time required to compute the join is O$(\log n_{v_a})$, otherwise it is O(n_{v_a}). In a subsequent step, CReaM leverages the differentials v_a^+ and v_a^- that were previously computed using Δ_1– Δ_6 to update the extension of the view v using the counting algorithm [14]. Since the view v is a projection of the view v_a, the time required to incrementally maintain v in response to the differentials v_a^+ and v_a^- is either O$(\log n_v)$ or O(n_v) depending on whether the attribute $\bar{X}$ in the view v is indexed or not. Therefore, the time taken by CReaM to update a view v with n_v tuples in response to a single tuple update in the underlying edb relations is either O$(\log n_{v_a})$ or O(n_{v_a}) depending on whether the physical design of the database is optimized or not.

Suppose that we did not materialize the auxiliary view v_a. In this case, we cannot update v without accessing the extension of ϕ. Suppose that the differential of ϕ consists of a single tuple $\phi(\bar{x}, \bar{y}, \bar{z})$. We need to recompute the set $S_{\bar{x},\bar{z}} = \{\bar{Y} \mid \phi(\bar{x}, \bar{Y}, \bar{z})\}$ to incrementally update v. We analyze the time required to update v under two possible scenarios depending on whether ϕ consists of a single edb relation or ϕ is a conjunction of edb relations. In the first case, the time required to incrementally compute $S_{\bar{x},\bar{z}}$ is either $O(\log n_\phi)$ or $O(n_\phi)$ depending on whether an index exists on the attributes $\bar{X}$ and $\bar{Z}$ in ϕ or not. However, in the second case, when ϕ is a conjunction of edb relations, we need to recompute ϕ and update it before recomputing the set $S_{\bar{x},\bar{z}}$. In this case, the cost of recomputing and updating ϕ is dominated by the cost of computing the join of the edb relations which is either $O(n_\phi)$ or $O(n \times \log n_\phi)$ depending on whether the physical design of the database is optimized or not.

When there are multiple (say k) changes to ϕ, the time required by CReaM to incrementally update the view $v(\bar{X}, W)$:- setof$(\bar{Y}, \phi(\bar{X}, \bar{Y}, \bar{Z}), W)$ with n_v tuples is either $O(k \times \log n_v)$ or $O(k \times n_v)$ depending on whether the views v_a and v are indexed or not. However, when v_a is not materialized and ϕ is a conjunction of multiple edb relations, we have to recompute the sets over ϕ. This requires $O(n \times \log n)$ time. Therefore, the speed up in the time to update v using CReaM in comparison to previous view maintenance algorithms is by a factor of $\frac{n}{k}$ when v_a and v are indexed.

We note that if the view v is defined as $v(\bar{X}, \bar{Z}, W)$:- setof$(\bar{Y}, \phi(\bar{X}, \bar{Y}, \bar{Z}), W)$, we can update v without accessing or computing the extension of ϕ. In this case, the time taken to update v is the same as is required by CReaM. The summary of CReaM's performance is presented in Table 1.

Next, we show that when the supplied materialized view and the auxiliary views that are materialized by CReaM are indexed, the time taken by CReaM to incrementally maintain a view is *optimal.*

Theorem 4. *The time taken by CReaM to maintain a materialized view containing a setof subgoal is optimal when the supplied materialized view and the auxiliary views are indexed.*

Proof. Suppose that a materialized view v containing a setof subgoal is supplied as an input to CReaM. In addition, suppose that v consists of n tuples. When v and the auxiliary view (that is materialized by CReaM) are indexed, CReaM maintains v in $O(\log n)$ time. If we prove that $\Omega(\log n)$ time is required to incrementally maintain an extension of a view with n tuples, then we would establish the optimality of CReaM.

To prove the lower bound, we reduce the problem of incrementally maintaining the *partial sums* of an array of n numbers to the problem of incrementally maintaining an extension of a view with n tuples. Prior work in [7, 8, 25] have independently proven that the maintenance of partial sums of an array of n numbers requires $\Omega(\log n)$ time. Consider an array of n numbers $\{a_i\}$. The partial sums problem maintains the sum $\sum_{i=0}^{k} a_i$ for every k $(1 \leq k \leq n)$ subject to updates of the form $a_i = a_i + x$, where x is a number. We reduce the instance of the partial sums problem over the array $\{a_i\}$ to an instance of the view maintenance problem in time that is polynomial in n as follows. Consider an instance of the view maintenance problem where we have two edb relations $r(A, B)$ and $s(B, C)$. The extension of $r(A, B)$ consists of the set of $n \times (n-1)$ tuples,

$\{(i,j) \mid 1 \leq j \leq i \leq n\}$. The extension of $s(A,B)$ consists of the set of n tuples, $\{(i,a_i) \mid 1 \leq i \leq n\}$. Suppose that we materialize n views $v_1, v_2, \ldots, v_n$ over $r(A,B)$ and $s(B,C)$ where each v_i is defined as $v_i(S)$:- setof$((B,C), r(i,B)\,\&\, s(B,C), W)$, sum$(W,S,2)$. In the definition of v_i, the aggregate sum$(W,S,2)$ computes the sum of the 2nd component of the tuples $\in W$.

When an array value a_i is updated to $a_i + x$, we update the tuple $(i,a_i) \in s(B,C)$ to the tuple$(i, a_i + x)$. Since we can compute the partial sum $\sum_{i=0}^{k} a_i$ by finding the value s which is in the extension of v_k, the problem of maintaining the partial sums of the array $\{a_i\}$ reduces to the problem of incrementally maintaining the views $v_1, v_2, \ldots, v_k$. Therefore, if the number of tuples in an extension of a view that contains a setof subgoal is $O(n)$, then $\Omega(\log n)$ time is required to incrementally maintain the view. □

Self-maintenance of Aggregate Views: In Theorem 4, we prove that CReaM optimally maintains views containing setof subgoals. Now, we show that by materializing auxiliary views, CReaM guarantees the self-maintainability [17] of aggregate views that are defined over single edb relations. In other words, the extension of an edb relation does not have to be accessed to incrementally maintain an aggregate view that is defined over the relation. The property of self-maintainability is desirable when access to the edb relations is restricted or when the edb relations themselves are hypothetical (such as in a LAV integration scenario [33]).

Consider a view v which is defined over an edb relation ϕ using the aggregation predicate *agg* as $v(\bar{X}, A)$:- setof$(\bar{Y}, \phi(\bar{X}, \bar{Y}, \bar{Z}), W)$, agg$(W, A)$. To incrementally maintain v, CReaM rewrites the definition of v using an auxiliary view v_a as follows.

$$v(\bar{X}, A) \text{ :- } v_a(\bar{X}, \bar{Z}, W),\ \text{agg}(W, A)$$
$$v_a(\bar{X}, \bar{Z}, W) \text{ :- } \text{setof}(\bar{Y}, \phi(\bar{X}, \bar{Y}, \bar{Z}), W)$$

CReaM materializes the view v_a and incrementally computes the diffentials v_a^+ and v_a^- by applying the differential rules Δ_1– Δ_6. The differential rules Δ_1– Δ_6 compute the join of the extension of the view v_a and the differentials of ϕ i.e., ϕ^+ and ϕ^-. In a subsequent step, CReaM leverages the differentials of v_a to incrementally maintain the view v using the algorithm presented in [14]. By materializing the auxiliary view v_a, CReaM is able to maintain v without accessing the extension of ϕ, thereby, making the view v self-maintainable.

6 Related Work

The problem of incrementally maintaining views has been extensively studied in the database community [1–6, 10, 14–16, 18, 21–23, 26–28, 30–32, 34]. A survey of the view maintenance techniques is presented in [13]. The view maintenance algorithms proposed in [2, 10, 14, 15, 18, 23, 26, 27] leverage differential relational algebra to incrementally maintain views in response to changes to the underlying edb relations. For instance, the prior work presented in [23] incrementally computes the differentials (or changes) of views by applying a set of differential rules over the extensions of edb relations and their differentials.

However, only a small fraction of the prior work on incremental view maintenance [10, 14, 15, 21, 24, 27, 28] addresses the maintenance of aggregate views. The techniques proposed in [10, 14, 21, 24] incrementally maintain views having only one aggregation operator. Furthermore, the incremental maintenance algorithms presented in [10, 14, 15, 21, 24, 27, 28] can support only a fixed set of built-in aggregate operators (such as min, max, sum, and count).

The problem of incremental view maintenance is closely related to the problem of self-maintainability [11, 12, 17, 19]. A view is self-maintainable if it can be incrementally maintained using the extension of the view and the changes to the edb relations. The view maintenance algorithms that are presented in [14, 15, 19] derive efficient self-maintenance expressions as well for certain types of updates to edb relations. Our view maintenance algorithm, CReaM, guarantees the self-maintenance of an aggregate view that is derived over a single edb relation by materializing auxiliary views.

Our work differs from prior work on incrementally maintaining aggregate views in two ways. First, we propose an algorithm called CReaM that *optimally* maintains aggregate views. For the special class of aggregate views where all of the non-aggregated attributes of the underlying edb relations are passed to the view, CReaM speeds up the time taken to incrementally update the view in comparison to previous view maintenance algorithms [10, 14, 15, 21, 24] by a factor that is at least logarithmic in the size of the extension of the view. Second, we can extend the CReaM algorithm to maintain views that contain *user-defined* aggregates. To maintain views with user-defined aggregates we rewrite the supplied view definitions using auxiliary views that contain setof subgoals and apply the CReaM algorithm to maintain the auxiliary views. Then, we apply prior maintenance algorithms [13] to maintain views whose definitions do not contain sets.

We note that even though CReaM optimally maintains aggregate views, we could further optimize the maintenance of views that contain *monotonic* aggregates [29] when new tuples are *inserted* to the edb relations. When new elements are inserted to a set that is aggregated by a monotonic aggregate, the aggregate value either always increases or decreases. For example, the aggregate sum is monotonic over the domain of postive numbers. Therefore, if we have a view v that is defined as $v(A)$:- setof$(B, r(A, B), W)$, sum(W, S), $S > 10$ and a tuple $t \in$ extension of v, the tuple t can never be changed by insertions into the relation $r(A, B)$.

7 Conclusion

We propose an algorithm called CReaM that incrementally maintains materialized aggregate views in response to changes to edb relations by materializing auxiliary views. By materializing auxiliary views, CReaM guarantees the self-maintainability of aggregate views that are defined over a single database table. CReaM optimally maintains views containing setof subgoals and speeds up the time taken to update materialized aggregate views with n tuples that are defined over multiple edb relations and do not contain all of the non-aggregated attributes in the edb relations either by a factor of $\frac{n}{\log n}$ or $\log n$ depending on whether the supplied materialized view is indexed or not. For other types of aggregate views, CReaM updates the view in no more time than that is required by prior view maintenance techniques to update the view.

References

1. Blakeley, J.A., Coburn, N., Larson, P.A.: Updating derived relations: Detecting irrelevant and autonomously computable updates. ACM TODS (1989)
2. Blakeley, J.A., Larson, P.A., Tompa, F.W.: Efficiently updating materialized views. SIGMOD (1986)
3. Buneman, O.P., Clemons, E.K.: Efficiently monitoring relational databases. ACM TODS (1979)
4. Ceri, S., Widom, J.: Deriving production rules for incremental view maintenance. VLDB (1991)
5. Colby, L.S., Kawaguchi, A., Lieuwen, D.F., Mumick, I.S., Ross, K.A.: Supporting multiple view maintenance policies. SIGMOD (1997)
6. Dong, G., Topor, R.W.: Incremental evaluation of datalog queries. ICDT (1992)
7. Fredman, M.L.: A lower bound on the complexity of orthogonal range queries. J. ACM (1981)
8. Fredman, M.L.: The complexity of maintaining an array and computing its partial sums. J. ACM (1982)
9. Gallaire, H., Minker, J., Nicolas, J.M.: Logic and databases: A deductive approach. ACM Computing Surveys (1984)
10. Griffin, T., Libkin, L.: Incremental maintenance of views with duplicates. SIGMOD (1995)
11. Gupta, A., Jagadish, H.V., Mumick, I.S.: Data integration using self-maintainable views. EDBT (1996)
12. Gupta, A., Jagadish, H.V., Mumick, I.S.: Maintenance and self-maintenance of outerjoin views. NGITS (1997)
13. Gupta, A., Mumick, I.S.: Maintenance of materialized views: Problems, techniques, and applications. In: Materialized Views (1999)
14. Gupta, A., Mumick, I.S., Subrahmanian, V.S.: Maintaining views incrementally. SIGMOD (1993)
15. Gupta, H., Mumick, I.S.: Incremental maintenance of aggregate and outerjoin expressions. Information Systems (2006)
16. Harrison, J.V., Dietrich, S.W.: Maintenance of materialized views in a deductive database: An update propagation approach. In: Workshop on Deductive Databases, JICSLP (1992)
17. Huyn, N.: Efficient view self-maintenance. Views (1996)
18. Kuchenhoff, V.: On the efficient computation of the difference between consecutive database states. DOOD (1991)
19. Mohania, M., Kambayashi, Y.: Making aggregate views self-maintainable. ACM TKDE 32 (1999)
20. Mohapatra, A., Genesereth, M.: Reformulating aggregate queries using views. SARA (2013)
21. Mumick, I.S., Quass, D., Mumick, B.S.: Maintenance of data cubes and summary tables in a warehouse. SIGMOD (1997)
22. Nicolas, J.M.: Yazdanian: An outline of bdgen: A deductive dbms. Information Processing (1983)
23. Orman, L.V.: Differential relational calculus for integrity maintenance. ACM TKDE (1998)
24. Palpanas, T., Sidle, R., Cochrane, R., Pirahesh, H.: Incremental maintenance for non-distributive aggregate functions. VLDB (2002)
25. Păatraşcu, M., Demaine, E.D.: Tight bounds for the partial-sums problem. SODA (2004)
26. Qian, X., Wiederhold, G.: Incremental recomputation of active relational expressions. ACM TKDE (1991)
27. Quass, D.: Maintenance expressions for views with aggregation. Views (1996)

28. Quass, D., Mumick, I.S.: Optimizing the refresh of materialized view. Technical Report (1997)
29. Ross, K.A., Sagiv, Y.: Monotonic aggregation in deductive databases. PODS (1992)
30. Shmueli, O., Itai, A.: Maintenance of views. SIGMOD (1984)
31. Stonebraker, M.: Implementation of integrity constraints and views by query modification. SIGMOD (1975)
32. Tompa, F.W., Blakeley, J.A.: Maintaining materialized views without accessing base data. Information Systems (1988)
33. Ullman, J.D.: Principles of Database and Knowledge-Base Systems: Volume II (1989)
34. Wolfson, O., Dewan, H.M., Stolfo, S.J., Yemini, Y.: Incremental evaluation of rules and its relationship to parallelism. SIGMOD (1991)

Towards an Approximative Ontology-Agnostic Approach for Logic Programs

João Carlos Pereira da Silva[1] and André Freitas[2]

[1] Departamento de Ciência da Computação - Instituto de Matemática - Universidade Federal do Rio de Janeiro - Brazil
[2] DERI National University of Ireland, Galway
jcps@dcc.ufrj.br,
andre.freitas@deri.org

Abstract. Distributional semantics focuses on the automatic construction of a semantic model based on the statistical distribution of co-located words in large-scale texts. Deductive reasoning is a fundamental component for semantic understanding. Despite the generality and expressivity of logical models, from an applied perspective, deductive reasoners are dependent on highly consistent conceptual models, which limits the application of reasoners to highly heterogeneous and open domain knowledge sources. Additionally, logical reasoners may present scalability issues. This work focuses on advancing the conceptual and formal work on the interaction between distributional semantics and logic, focusing on the introduction of a distributional deductive inference model for large-scale and heterogeneous knowledge bases. The proposed reasoning model targets the following features: (i) an approximative ontology-agnostic reasoning approach for logical knowledge bases, (ii) the inclusion of large volumes of distributional semantics commonsense knowledge into the inference process and (iii) the provision of a principled geometric representation of the inference process.

Keywords: distributional semantics, logic programming, knowledge bases, distributional vector space, approximate deductive reasoning.

1 Introduction

Logical models provide a comprehensive system for representing concepts, objects, their properties and associations. In addition to the representation of conceptual abstractions, logical models provide a precise definition of logical inference, allowing new knowledge to become explicit from existing facts and rules.

Despite the fundamental importance of inference to the development of intelligent systems, experimental research over large-scale and heterogeneous knowledge bases shows evidence that logical models have limitations in the provision of inference models which can cope with the level of contextual complexity, vagueness, ambiguity and scale present in open domain/commonsense knowledge bases. The lack of properties such as robustness to inconsistencies, a more principled mechanism of semantic approximation and the ability to scale to large

C. Beierle and C. Meghini (Eds.): FoIKS 2014, LNCS 8367, pp. 415–432, 2014.

volume knowledge bases represents a solid barrier to the applicability of existing inference models into this scenario.

More recently, *distributional semantic models* (DSMs) [17] have emerged from the empirically supported evidence that semantic models automatically derived from statistical co-occurrence patterns on large corpora provide simplified but comprehensive semantic models. With the availability of large volumes of text on the Web, DSMs have the potential to become a fundamental element in addressing existing challenges for enabling a robust semantic interpretation by computers.

This work investigates the complementary aspects between distributional semantics and logic programming models, focusing on the analysis of approximate inference and querying on a *distributional vector space.* While logical models provide an expressive conceptual representation structure with support for inferences and expressive query capabilities, distributional semantics provides a complementary layer where the semantic approximation supported by large-scale comprehensive semantic models and the scalability provided by the vector space model (VSM) can address the trade-off between expressivity and semantic/terminological flexibility.

The contributions of this work concentrate on advancing the conceptual and formal work on the interaction between distributional semantics and logic, focusing on the investigation of a distributional deductive inference model for large-scale and heterogeneous knowledge bases. The proposed inference model targets the following features: (i) an approximative ontology-agnostic reasoning approach for logical knowledge bases, (ii) the inclusion of large volumes of distributional semantics commonsense knowledge into the inference process and (iii) the provision of a principled geometric representation of the inference process.

This work is organized as follows: section 2 describes a motivational scenario; section 3 provides a brief introduction to distributional semantics ; section 4 describes the logic model; section 5 describes the geometric model; section 6 connects the logical and geometrical models ; section 7 shows the combined distributional-logic inference process; section 8 presents a prototype of the proposed approach; section 9 describes related work and section 10 presents the conclusions and future work.

2 Motivational Scenario

Every knowledge or information artifact (from unstructured text to structured knowledge bases) maps to an implicit or explicit set of user intents and semantic context patterns. The multiplicity of contexts where open domain and commonsense knowledge bases can be used, defines the intrinsic semantic heterogeneity for these scenarios. Different levels of conceptual abstraction or lexical expressions in the representation of predicates and constants are examples where a semantic/terminological gap can strongly impact the inference process.

In the scenario below an user executes a *vocabulary-independent (ontology-agnostic) query* over a logic program Π. A query is *vocabulary independent* if the user is not aware of the terms and concepts inside Π.

Consider the query *'Is the father in law of Bill Clinton's daughter a politician?'* that can be represented as the logical query:

$$? - daughter\_of(X, bill\_clinton), politician(Y), father\_in\_law(Y, X)$$

Let us assume that the logic program Π contains facts and rules such as:

$$child\_of(chelsea\_clinton, bill\_clinton).$$
$$child\_of(marc\_mezvinsky, edward\_mezvinsky).$$
$$spouse(chelsea\_clinton, marc\_mezvinsky).$$
$$is\_a\_congressman(edward\_\ mezvinsky).$$
$$father\_in\_law(A,B) \leftarrow spouse(B,C),\ child\_of(C,A).$$

meaning that Chelsea is the child of Bill Clinton, Marc Mezvinsky is the child of Edward Mezvinsky, Chelsea is the spouse of Marc, Edward Mezvinsky is a congressman and A is father in law of B when the spouse of B is a child of A.

The inference over Π will not materialize the answer $X = chelsea\_clinton$ and $Y = edward\_mezvinsky$, because despite the statement and the rule describing the same sub-domain, there is no precise vocabulary matching between the query and Π.

In order for the reasoning to work, the approximation of the following terms would need to be established: $daughter\_of \sim child\_of$, $is\_a\_congressman \sim politician$. The reasoner should be able to semantically approximate vocabulary terms such as *daughter_of* and *child_of*, addressing the terminological gap required by this inference.

To close the semantic/vocabulary gap in a traditional deductive logic knowledge base it would be necessary to increase the size of Π to such an extent that it would contain all the facts and rules necessary to cope with any potential vocabulary difference. Together with the aggravation of the scalability problem, it would be necessary to provide a principled mechanism to build such a large scale and consistent set of facts and rules.

3 Distributional Semantics and Semantic Approximative Inference

In this work *distributional semantics* supports the definition of an approximative semantic interpretation for facts and rules in a logic program Π where constants and predicates are mapped to vectors in a *distributional vector space.* This section provides a brief introduction to distributional semantics and outlines the core principles and the rationale of the proposed approximative inference model.

3.1 Distributional Semantics

Distributional semantics is defined upon the assumption that the context surrounding a given word in a text provides important information about its meaning [17]. It focuses on the construction of a semantic model for a word based on the statistical distribution of co-located words in texts. These semantic models are naturally represented by Vector Space Models (VSMs) [17], where the meaning of a word can be defined by a weighted vector over a term space, which represents the association patterns of co-occurring words in a corpus.

The existence of large amounts of unstructured text on the Web brings the potential to create comprehensive distributional semantic models (DSMs). DSMs can be automatically built from large corpora, not requiring manual intervention on the creation of the semantic model. Additionally, its natural association with VSMs, where dimensional reduction approaches or data structures such as inverted list indexes, can provide a scalability benefit for the instantiation of these models.

These models can provide a more scalable solution to the problem of capturing commonsense semantic information, complementing existing manually created knowledge bases such as Cyc[1]. The computation of semantic relatedness measures between pairs of words is one instance in which the strength of distributional models and methods is empirically supported ([10],[6]).

3.2 The Distributional Inference Vector Space

The commonsense semantic knowledge embedded in a distributional model is used to semantically complement a logic program Π. In a traditional deductive system the inference process is defined by a sequence of exact substitution operations, where the symbols representing constants and predicates are exactly matched under the syntax of the representation language. In the proposed inference model the symbols have an associated *concept vector* representation which encodes its relation to other symbols based on the symbols' co-occurrence statistics in a large unstructured reference corpus. The concept vectors define a distributional vector space which can be used to represent and embed the logic program symbols in the space.

The embedding of logic programs in the distributional vector space allows the definition of a geometric interpretation for the inference process. The geometry allows the definition of a semantic heuristics which defines a direction for the exploration of the solution space.

The proposed inference model uses the lexical-semantic information embedded in a distributional-relational vector space (named τ-Space [7]) to compute a measure of semantic relatedness between logic program symbols in the space. The distributional semantic relatedness measure can be used to establish an approximate semantic equivalence between two predicates at a given context. The intuition behind this approach is that *two terms which are highly semantically related in a distributional model are likely to have a close (implicit) relation*[2].

This work expands on the existing abstraction of the τ-Space, defined in [7], introducing the notion of inference process over a τ-Space, articulating the connections between logical inference and the geometry defined by the τ-Space.

---

[1] `http://www.cyc.com/platform/opencyc`

[2] Distributional semantic models can be specialized to exclude certain types of semantic relatedness (such as antonyms or relations in a negation context).

4 Logic Model

4.1 Syntax

An alphabet $\mathcal{A}$ is formed by the following disjoint set of symbols: (i) *Predicates* which are represented by $P = \{p_1, \cdots p_m\}$; (ii) *Constants* which are represented by $E = \{e_1, \cdots, e_n\}$; (iii) *Variables* which are represented by upper case letters $\{X, Y, Z, \cdots\}$. Also, we have a fixed set of *connectives* represented by $\{',', \leftarrow, \neg\}$. A *term* t is either a constant or a variable. An *atom* at is an expression of the form $p(t_1, \cdots, t_n)$ where p is a predicate and $t_1, \cdots, t_n$ are terms. An atom $p(t_1, \cdots, t_n)$ is *grounded* whenever $t_1, \cdots, t_n$ are all constants. A (grounded) literal is an (grounded) atom or a negated (grounded) atom.

A *clause* cl is an expression of the form: $head(cl) \leftarrow body(cl).$, where $head(cl)$ is an atom and $body(cl)$ is a conjunction of literals. A *grounded clause* is formed only by grounded literals.

A *logic program* Π is a set of clauses. We say that Π is a definite logic program when there is no negative atom in $body(cl)$. Otherwise, Π is a normal logic program. A *query* Q to Π is an expression of the form $?- q_1, \cdots, q_n.$ where $q_1, \cdots, q_n$ are literals. A *signature* Σ_x is a pair (P_x, E_x) where $P_x \subseteq P$ and $E_x \subseteq E$ are respectively the sets of all predicates and constants that appear in $x \in \{\Pi, at, Q\}$.

4.2 Semantics

Given the sets of constants E and predicates P, let $HU = E$ be the *Herbrand Universe* and HB be the *Herbrand Base* formed by all ground atoms that can be constructed using predicates and constants in P and E.

A Herbrand interpretation of a predicate p is any set $\Im(p) \subseteq HB$ such that all elements in $\Im(p)$ are of the form $p(e_1, \cdots, e_n)$, where for all $i \in [1, n], e_i \in HU$. A Herbrand interpretation $\Im$ *satisfies* a clause cl of the form $h \leftarrow b_1, \cdots, b_n, \neg b_{n+1}, \cdots, \neg b_m$ if $h \in \Im$ whenever each $b_1, \cdots, b_n \in \Im$ and each $b_{n+1}, \cdots, b_m \notin \Im$. A *Herbrand model* $\mathcal{M}(\Pi) = \bigcup_{p \in P_\Pi} \Im(p)$ of Π is a Herbrand interpretation that satisfies all clauses in Π. A Herbrand model $\mathcal{M}(\Pi)$ is *minimal* if no proper subset of $\mathcal{M}(\Pi)$ is also a model. A definite logic program Π has only one minimal Herbrand model, which we denote as $\mathcal{M}in(\Pi)$.

A set of atoms S is an answer set model of a normal logic program Π iff $S = \mathcal{M}in(\Pi^S)$ where Π^S is the definite logic program obtained from Π (the reduct of Π - [11]): (i) deleting all clauses that has $\neg at$ in its body such that $at \in S$ and (ii) deleting all negated atoms in the bodies of the remaining clauses. An answer set model S satisfies an atom at (resp., $\neg at$) when $at \in S$ (resp., $at \notin S$) which is denoted by $S \models at$ (resp., $S \models \neg at$).

5 Geometrical Model

5.1 τ-Space

The τ-*Space* [7] is a distributional structured vector space model that will be used to represent predicates and constants under a distributional semantic model. It is built from a *reference corpus* formed by a pair of sets (*Term*, *Context*) where $Term = \{k_1, \cdots, k_t\}$ is a set of terms and $Context = \{c_1, \cdots, c_t\}$ is a set of context windows in the corpus. For example, a given set of documents can be seen as a set of context windows and all terms that occur in those documents form the set of terms.

Term is used to define the basis $Term_{basis} = \{\overrightarrow{\mathbf{k}}_1, \cdots, \overrightarrow{\mathbf{k}}_t\}$ of unit vectors that spans the *term vector space* VS^{term}. In VS^{term}, a context window c_j is represented as:

$$\overrightarrow{\mathbf{c}}_j = \sum_{i=1}^{t} v_{i,j} \overrightarrow{\mathbf{k}}_i \tag{1}$$

where $v_{i,j}$ is 1 if term k_i appears in context window c_j and 0 otherwise.

The set *Context* is used to define the basis $Context_{basis} = \{\overrightarrow{\mathbf{c}}_1, \cdots, \overrightarrow{\mathbf{c}}_t\}$ of vectors that spans the *distributional vector space* VS^{dist}. A term x is represented in VS^{dist} as:

$$\overrightarrow{\mathbf{x}} = \sum_{j=1}^{t} w_j \overrightarrow{\mathbf{c}}_j \tag{2}$$

where

$$w_j = tf_j \times idf = \frac{freq_j}{count(c_j)} \times \log \frac{N}{n_{c_j}} \tag{3}$$

meaning that w_j is the product of the normalized term frequency tf_j (where $freq_j$ is the frequency of term x in the context window c_j and $count(c_j)$ is the number of terms inside c_j) and the inverse document frequency idf for the term x (where N is the total number of context windows in the reference corpus and n_{c_j} is the number of context containing the term x).

As consequence, a vector $\overrightarrow{\mathbf{x}} \in VS^{dist}$ can be mapped to VS^{term} by the following transformation:

$$\overrightarrow{\mathbf{x}} = \sum_{i=1}^{t} \sum_{j=1}^{t} w_j v_{i,j} \overrightarrow{\mathbf{k}}_i \tag{4}$$

We can see from the equations above that the set $C \subseteq Context$ where a term occurs defines the concept vectors associated with the term. This represents its meaning on the reference corpus. Since each concept vector is weighted according to the term distribution in the corpus, we can define the set $Context_{basis}$ in terms of $Term_{basis}$ where each dimension maps to a word in the corpus.

6 Linking the Logical and Geometrical Models

In this section, we will define the link between the geometrical (distributional) and logical models. The idea is that the former could provide a way to enrich the semantics and inference power of the latter, resulting in an approach that supports an *approximative semantic matching inference* process.

6.1 Mapping Predicates and Constants to Vectors

The signature of a given logic program Π can be translated into τ-Space vectors in the distributional vector space VS^{dist} as follows:

Definition 1. *Let $\{\vec{\mathbf{c}}_1, \cdots, \vec{\mathbf{c}}_t\}$ be the vectors basis that spans VS^{dist}. The vector representations of P and E in VS^{dist} are defined by:*

$$\vec{\mathbf{P}}_{VS^{dist}} = \{\vec{\mathbf{p}} : \vec{\mathbf{p}} = \sum_{i=1}^{t} v_i^p \vec{\mathbf{c}}_i, \text{ for each } p \in P\} \tag{5}$$

$$\vec{\mathbf{E}}_{VS^{dist}} = \{\vec{\mathbf{e}} : \vec{\mathbf{e}} = \sum_{i=1}^{t} v_i^e \vec{\mathbf{c}}_i, \text{ for each } e \in E\} \tag{6}$$

where v_i^e and v_i^p are defined by the weighting scheme over the distributional model. The weighting scheme will reflect the word co-occurrence pattern in the reference corpus.

Elements of a query Q with signature $\Sigma_Q = (P_Q, E_Q)$ are mapped to VS^{dist} in a similar way.

6.2 Semantic Relatedness of Predicates, Programs and Models

Consider, for example, two highly semantic related concepts represented by two syntactically different predicates, such as $daughter\_of$ and $child\_of$. In the unification process only syntactically identical predicates can be resolved. If we have stated facts/rules using the predicate $child\_of$, no query using predicate $daughter\_of$ would be answered.

In order to bring to logic programs the ability of semantically relate predicate symbols which use a meaningful natural language descriptor, we will define the notion of semantic relatedness between predicates as follows:

Definition 2. *Let p_1 and p_2 be predicate symbols with same arity and with normalized vector representations $\vec{\mathbf{p_1}}$ and $\vec{\mathbf{p_2}}$ in VS^{dist}. The semantic relatedness function $sr : P \times P \rightarrow [0, 1]$ is defined by the inner product between $\vec{\mathbf{p_1}}$ and $\vec{\mathbf{p_2}}$: $sr(p_1, p_2) = \vec{\mathbf{p_1}}.\vec{\mathbf{p_2}} = \cos(\theta)$ where θ is the angle between vectors $\vec{\mathbf{p_1}}$ and $\vec{\mathbf{p_2}}$.*

Definition 3. *Let p_1 and p_2 be predicates and $\eta \in [0, 1]$ be a threshold. We say that p_1 and p_2 are semantically related wrt η whenever $sr(p_1, p_2) > \eta$.*

The function sr allows us to extend the notion of semantic relatedness to logic programs, answer set models and unification procedure allowing the inference process to continue in cases where predicates are syntactically distinct. Initially, we use the semantic relatedness between predicate symbols to define the predicate substitution as follows:

Definition 4. *Let $P_1 = \{p_1, \cdots, p_n\}$ and $P_2 = \{p'_1, \cdots, p'_n\}$ be two sets of predicate symbols such that $\forall i \in [1, n], sr(p_i, p'_i) > \eta$. A predicate substitution of P_1 by P_2 wrt η is defined by $\lambda_\eta(P_1, P_2) = \{p_1/p'_1, \cdots, p_n/p'_n\}$. We denote $\lambda_\eta^{-1}(P_1, P_2) = \lambda_\eta(P_2, P_1) = \{p'_1/p_1, \cdots, p'_n/p_n\}$.*

Since the goal of this type of substitution is to allow that the inference process can continue despite the vocabulary differences, definition 4 does not allow the substitution of two different predicates p_i and p_j with a single predicate p'. This is done to preserve the logical semantics of the predicates, that is, both extensions of p_i and p_j. Otherwise, if p' could replace both p_i and p_j, we would have $\Im(p') = \{(c_1, \cdots, c_n)$ such that $(c_1, \cdots, c_n) \in (\Im(p_i) \cup \Im(p_j))\}$.

We associate to predicate substitutions a semantic relatedness measure:

$$sr_{subst}(\lambda_\eta(P_1, P_2)) = \frac{1}{n} * \sum_{i \in [1,n]} sr(p_i, p'_i) \tag{7}$$

which will be also used to define semantic relatedness between logic programs and Herbrand models.

Definition 5. *Let Π_1 and Π_2 be logic programs with signatures, respc., $\Sigma_{\Pi_1} = (P_{\Pi_1}, E_{\Pi_1})$ and $\Sigma_{\Pi_2} = (P_{\Pi_2}, E_{\Pi_2})$. We say that Π_1 and Π_2 are semantically related wrt a threshold η (or sr-logic programs wrt η) when there is some predicate substitution $\lambda_\eta(P_1, P_2)$ such that $\Pi_2 = \Pi_1 \cdot \lambda_\eta(P_1, P_2)$ where $P_1 = (P_{\Pi_1} \setminus P_{\Pi_2})$ and $P_2 = (P_{\Pi_2} \setminus P_{\Pi_1})$.*

Note that when $\Pi_2 = \Pi_1 \cdot \lambda_\eta(P_1, P_2)$, $\Pi_1 = \Pi_2 \cdot \lambda_\eta^{-1}(P_1, P_2) = \Pi_2 \cdot \lambda_\eta(P_2, P_1)$.

Definition 5 states that two sr-logic programs are different versions of the same program that use a set of different predicate symbols, which are semantically related from a natural language perspective. From the logical point of view, the answer set models of Π_1 are preserved in Π_2 (and vice-versa) in the sense that the extensions of all predicates in both programs are the same: different predicate symbols that are semantically related have the same extension. This can be shown as follows:

Proposition 1. *Let Π be a normal logic program, $S \subseteq HB_\Pi$ be a set of atoms. For any predicate substitution λ_η, $(\Pi^S \cdot \lambda_\eta) = (\Pi \cdot \lambda_\eta)^{S \cdot \lambda_\eta}$.*

Proof. ($\subseteq$): Suppose that $cl \in (\Pi^S \cdot \lambda_\eta)$ and $cl \notin (\Pi \cdot \lambda_\eta)^{S \cdot \lambda_\eta}$. Since $cl \in (\Pi^S \cdot \lambda_\eta)$, we have that $cl \cdot \lambda_\eta^{-1} \in \Pi^S$. One of the following cases can occur:

- $cl \cdot \lambda_\eta^{-1} \in \Pi$ when there is no occurrence of negative atoms in the body of cl. Then $cl \in \Pi \cdot \lambda_\eta$ and $cl \in (\Pi \cdot \lambda_\eta)^{S \cdot \lambda_\eta}$, a contradiction; or
- there is a clause $cl' \cdot \lambda_\eta^{-1} \in \Pi$ with negative atoms $\neg at_1 \cdot \lambda_\eta^{-1}, \cdots, \neg at_n \cdot \lambda_\eta^{-1}$ in the body that are all eliminated by S generating $cl \cdot \lambda_\eta^{-1}$. Since $\{at_1 \cdot \lambda_\eta^{-1}, \cdots, at_n \cdot \lambda_\eta^{-1}\} \subseteq S$, we have $\{at_1, \cdots, at_n\} \subseteq (S \cdot \lambda_\eta)$. So $cl \in (\Pi \cdot \lambda_\eta)^{S \cdot \lambda_\eta}$, which contradicts our hypothesis.

($\supseteq$): Suppose that $cl \in (\Pi \cdot \lambda_\eta)^{S \cdot \lambda_\eta}$ and $cl \notin (\Pi^S \cdot \lambda_\eta)$. Since $cl \in (\Pi \cdot \lambda_\eta)^{S \cdot \lambda_\eta}$, we can have one of the following cases:

- $cl \in (\Pi \cdot \lambda_\eta)$. Then $(cl \cdot \lambda_\eta^{-1}) \in \Pi$, and consequently $(cl \cdot \lambda_\eta^{-1}) \in \Pi^S$ or $cl \in (\Pi^S \cdot \lambda_\eta)$, contradicting our hypothesis; or

- there is a clause $cl' \in (\Pi \cdot \lambda_\eta)$ with negative atoms $\neg at_1 \cdot \lambda_\eta, \cdots, \neg at_n \cdot \lambda_\eta$ in the body that are all eliminated by $(S \cdot \lambda_\eta)$ generating cl. Hence $\{at_1, \cdots, at_n\} \subseteq S$ and since $(cl' \cdot \lambda_\eta^{-1}) \in \Pi$, we have that $(cl \cdot \lambda_\eta^{-1}) \in \Pi^S$, or, $cl \in (\Pi^S \cdot \lambda_\eta)$, contradicting our hypothesis.

Corollary 1. *Let Π_1 and Π_2 be sr-logic programs wrt η and S a set of atoms such that $P_S \subseteq P_{\Pi_1}$. Then $\Pi_1^S = (\Pi_2^{S \cdot \lambda_\eta(P_1,P_2)}) \cdot \lambda_\eta(P_2, P_1)$.*

Proof. Since Π_1 and Π_2 be sr-logic programs wrt η, by definition 5, we have $\Pi_1 = \Pi_2 \cdot \lambda_\eta(P_2, P_1)$. Given a set of atoms S, let $S' = S \cdot \lambda_\eta(P_1, P_2)$ and consequently $S = S' \cdot \lambda_\eta(P_2, P_1)$.

Then, we have: $\Pi_1^S = (\Pi_2 \cdot \lambda_\eta(P_2, P_1))^S = (\Pi_2 \cdot \lambda_\eta(P_2, P_1))^{S' \cdot \lambda_\eta(P_2,P_1)} = (\Pi_2^{S'}) \cdot \lambda_\eta(P_2, P_1) = (\Pi_2^{S \cdot \lambda_\eta(P_1,P_2)}) \cdot \lambda_\eta(P_2, P_1)$

Proposition 2. *Let Π_1 and Π_2 be sr-logic programs wrt η.*

$\mathcal{M}(\Pi_1)$ is an answer set model of Π_1 iff $\mathcal{M}(\Pi_2) = \mathcal{M}(\Pi_1) \cdot \lambda_\eta(P_1, P_2)$ is an answer set model of Π_2.

Proof. We have $\Pi_2 = \Pi_1 \cdot \lambda_\eta(P_1, P_2)$ and $\mathcal{M}(\Pi_1) = \mathcal{M}in(\Pi_1^{\mathcal{M}(\Pi_1)})$. So,

$$\mathcal{M}in(\Pi_2^{\mathcal{M}(\Pi_1)\cdot\lambda_\eta(P_1,P_2)}) = \mathcal{M}in((\Pi_1 \cdot \lambda_\eta(P_1, P_2))^{\mathcal{M}(\Pi_1)\cdot\lambda_\eta(P_1,P_2)}) =$$

$$= \mathcal{M}in((\Pi_1^{\mathcal{M}(\Pi_1)}) \cdot \lambda_\eta(P_1, P_2)) = \mathcal{M}in(\Pi_1^{\mathcal{M}(\Pi_1)}) \cdot \lambda_\eta(P_1, P_2) = \mathcal{M}(\Pi_1) \cdot \lambda_\eta(P_1, P_2)$$

The semantic relatedness sr_{prog} between logic programs Π_1 and Π_2 and the semantic relatedness sr_{models} between (answer set) models $\mathcal{M}(\Pi_1)$ and $\mathcal{M}(\Pi_2) = \mathcal{M}(\Pi_1) \cdot \lambda_\eta(P_1, P_2)$ are defined using the predicate substitution $\lambda_\eta(P_1, P_2)$ used to transform Π_1 in Π_2: $sr_{prog}(\Pi_1, \Pi_2) = sr_{models}(\mathcal{M}(\Pi_1), \mathcal{M}(\Pi_2)) = sr_{subst}(\lambda_\eta(P_1, P_2))$

The satisfiability of atoms expressed using a predicate symbol that does not belong to the signature of an answer set model is defined by:

Definition 6. *Let S be an answer set model of a logic program Π. Given a grounded atom $p(t_1, \cdots, t_n)$ such that $p \notin P_\Pi$, we say that:*

- *S sr-satisfies $(p(t_1, \cdots, t_n), \zeta)$ wrt η, denoted by $S \models_\eta (p(t_1, \cdots, t_n), \zeta)$ when there is a substitution $\lambda_\eta(\{p\}, \{p'\})$ for some $p' \in P_\Pi$ such that $S \models (p(t_1, \cdots, t_n) \cdot \lambda_\eta(\{p\}, \{p'\}))$ and ζ is the semantic relatedness measure associated to the predicate substitution $\lambda_\eta(\{p\}, \{p'\})$ as defined in equation (7) (i.e., $\zeta = sr_{subst}(\lambda_\eta(\{p\}, \{p'\}))$).*
- *S sr-satisfies $(\neg p(t_1, \cdots, t_n), \zeta)$ wrt η, denoted by $S \models_\eta (\neg p(t_1, \cdots, t_n), \zeta)$ when there is a substitution $\lambda_\eta(\{p\}, \{p'\})$ for some $p' \in P_\Pi$ such that $S \models (\neg p(t_1, \cdots, t_n) \cdot \lambda_\eta(\{p\}, \{p'\}))$ and ζ is the semantic relatedness measure associated to the predicate substitution $\lambda_\eta(\{p\}, \{p'\})$ as defined in equation (7) (i.e., $\zeta = sr_{subst}(\lambda_\eta(\{p\}, \{p'\}))$).*

Given a set of grounded literals Q such that $Q = Q_1 \cup Q_2$, $P_{Q_1} \cap P_{Q_2} = \emptyset$, $P_{Q_1} \subseteq P_\Pi$ and $P_{Q_2} \not\subseteq P_\Pi$, we say that

– *S sr-satisfies* (Q, ζ) *wrt* η*, denoted by* $S \models_\eta (Q, \zeta)$ *iff there is a substitution* $\lambda_\eta(P_{Q_2}, P')$ *for some* $P' \subseteq P_\Pi$ *such that* $S \models (Q \cdot \lambda_\eta(P_{Q_2}, P'))$ *and* ζ *is the semantic relatedness measure associated to the predicate substitution* $\lambda_\eta(P_{Q_2}, P')$ *as defined in equation (7) (i.e.,* $\zeta = sr_{subst}(\lambda_\eta(P_{Q_2}, P'))$*).*

7 Distributional-Logic Inference

In this section, we will present the combined distributional-logical inference process. The first step to answer Q is to order the literals in it according to a relevance order of elements in $P_Q \cup E_Q$.

Definition 7. *Let* Q *be a query. The relevance order of the literals in* Q *is the sequence of literals* $< l_1, l_2, \cdots, l_m >$ *such that: (i)* Q *is equivalent to* $\bigwedge_{i=1}^{m} l_i$*; (ii)* $\forall i \in [1, m-1], f_{relevance}(l_i) \geq f_{relevance}(l_{i+1})$

The function $f_{relevance}$ is a heuristic measure of specificity over the query symbols which gets the most specific constant or predicate (which we call *semantic pivot symbol*). The specificity can be defined as the IDF (Inverse Document Frequency) of a term over a reference corpus, as a function of the *lexical categories* associated with the term, or as a combination of the number of elements associated with x (where x is a predicate or constant). The rationale behind prioritizing the selection of a symbol with high specificity is that the algorithm prioritizes the hardest constraint in the query and selects the query element less prone to semantic ambiguity, vagueness and polysemy. Normally the first semantic pivot symbol selected in a query Q is a constant, if any exists.

The selection of a semantic pivot allows a reduction in the search space where just the elements of Π associated with the pivot at a given iteration are candidates for the semantic matching. In each iteration, a set of semantic pivots is defined, which propagates to other points in the τ-Space, following the topological relations defined by the syntactic structure of the atoms. The order of the sequence is unique, with regard to a $f_{relevance}$ function and the syntactic constrains of the query elements.

Once the order of literals in a query Q is fixed, to answer the ordered query Q over Π, first we use algorithm 1 to find all predicate substitutions wrt a given threshold η between the predicate symbols that appear in Π (P_Π) and all the predicate symbols q in Q such that $q \notin P_\Pi$ (P_{query}).

Each predicate substitution $\lambda_\eta(P_{query}, P'_\Pi)$ generated by algorithm 1 can be applied to Q resulting in a query $(Q \cdot \lambda_\eta(P_{query}, P'_\Pi))$ where all predicates belong to P_Π. So, we can answer this transformed query using any answer set solver.

Note that for each $\lambda_\eta(P_{query}, P'_\Pi) \in$ *Substitutions* we can calculate the score associated with that substitution $(sr_{subst}(\lambda_\eta(P_{query}, P'_\Pi)))$ since the semantic relatedness measure sr is stored whenever a substitution is found (line 13 in algorithm 1).

Algorithm 1. Distributional Predicate Substitution Algorithm - DPS

INPUT

- P_{Π}: The list of all predicate symbols that appear in a program Π
- P_{query}: The list of all predicate symbols q that appear in a query Q such that $q \notin P_{\Pi}$
- η: Threshold

OUTPUT

- *Substitutions*: A set with all predicate substitutions $\lambda_{\eta}(P_{query}, P'_{\Pi})$ where $P'_{\Pi} \subseteq P_{\Pi}$ and $|P'_{\Pi}| = |P_{query}|$

PROCEDURE $DPS(P_{\Pi}, P_{query}, \eta)$:

```
if P_query == [ ] then
    return ([ [ ] ])
else
    for all i ∈ [1, |P_query|] do
        X ← P_query(i)
        P'_query ← remove(X, P_query)
        Substitutions ← [ ]
        for all Y ∈ P_Π do
            if sr(X, Y) > η then
                P'_Π ← remove(Y, P_Π)
                Subst ← [ ]
                for all Z ∈ DPS(P'_Π, P'_query, η) do
                    Subst ← append(Z, [(X, Y, sr(X, Y))])
                    Substitutions ← append(Substitution, [Subst])
                end for
            end if
        end for
    end for
end if
return Substitutions
```

8 Prototype and Evaluation

A prototype of the proposed approach was built and it contains two modules: (i) the prolog module implemented using SWI-Prolog[3] which identifies if a predicate in a query belongs or not to the signature of a given normal logic program and does all predicate substitutions with the respective semantic relatedness measure; (ii) the τ-Space module, which was constructed using Explicit Semantic Analysis (ESA) as the distributional model built over Wikipedia 2006, where the Wikipedia articles were the context windows and TF/IDF was the weighting scheme.

The query is of the form (Q, η), where Q is a query and η is the desired threshold which has its value determined experimentally accordingly to the cor-

[3] www.swi-prolog.org/

pus that is used. The experimental threshold η was based on the semantic differential approach for ESA proposed in [6]. The approach was simplified to a ground threshold of 0.05.

The answers to (Q, η) are usual logic program answers with the scores corresponding to sr_{subst}. When the predicate q in the selected literal l of Q is identified as not belonging to P_Π, the τ-Space module is called and returns all predicate names of Π semantically related to q wrt η. Each one of these predicates (if any exists) replaces q in the query, which proceeds in the inference process as usual.

Example 1. Let Π be formed by:

child_of(chelsea_clinton, bill_clinton).
child_of(marc_mezvinsky,edward_mezvinsky).
spouse(chelsea_clinton, marc_mezvinsky).
is_a_congressman(edward_mezvinsky).
father_in_law(A,B) $\leftarrow$ *spouse(B,C),child_of(C,A).*

Suppose that we want to answer the query *"Is the father in law of Bill Clinton's daughter a politician?"* wrt a threshold $\eta = 0.05$:

?-*((daughter_of(X,bill_clinton),father_in_law(Y,X),politician(Y)),0.05).*

Since the predicate *daughter_of* does not appear in Σ_Π, we need to verify if there is a semantically related binary predicate to *daughter_of* wrt $\eta = 0.05$. As can be seen in table 1, only *child_of* is semantically related to *daughter_of* wrt η $(sr(child\_of, daughter\_of) = 0.054 > 0.05)$. Thus, we allow that these predicates unify and they have a *mgu* $(\{X/chelsea\_clinton\}, 0.054)$. The complete inference is shown in figure 1 and the score of the answer is $(0.054 + 0.06)/2 = 0.057$.

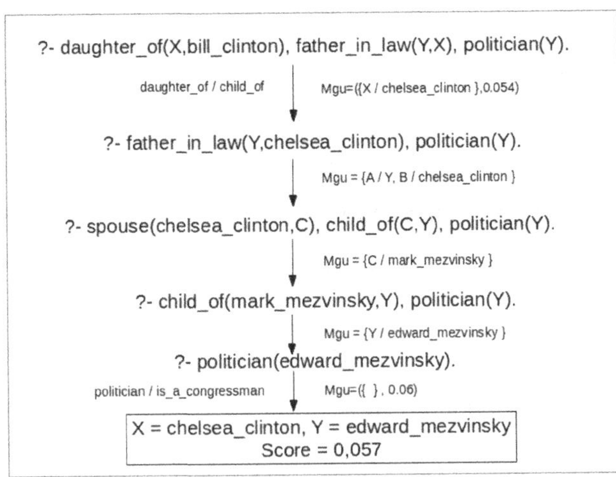

Fig. 1. Derivation for the question *"Is the father in law of Bill Clinton's daughter a politician?"*

Table 1. Semantic relatedness determined by the τ-Space module between the predicates in Q and Π, according to arity

sr	*child_of/2*	*spouse/2*	*father_in_law/2*	*is_a_congressman/1*
daughter_of/2	0.054	0.012	0.048	-
politician/1	-	-	-	0.06

As an approximative approach, we can have some undesirable answers as shows the following example:

Example 2. Suppose that we have Π' defined as:

$$\Pi \cup \{spouse(bill\_clinton, hilary\_clinton), child\_of(hilary\_clinton, hugh\_rodman)\}$$

where Π is the program defined in example 1.

Consider that we query Π' with *"Who is Bill Clinton's daughter ?"* using a threshold $\eta = 0.04$. In this case, we have two predicates semantically related to *daughter_of* obtaining the answers:

- $X = chelsea\_clinton$ with score 0.054, replacing *daughter_of* by *child_of*,
- $X = hugh\_rodman$ with score 0.048, replacing *daughter_of* by *father_in_law*.

These answers could be filtered using a higher threshold as input (in the example, 0.05), using a threshold over the final score or through a principled interaction mechanism (dialog/disambiguation system).

To illustrate the use of negation, we present the following example:

Example 3. Suppose we query Π' using $\eta = 0.04$ with:

?- ($\neg$ *daughter_of(chelsea_clinton,bill_clinton),0.04)*.

We obtain the following answers:

- **no** with score 0.054, replacing *daughter_of* by *child_of* (since $child\_of(chelsea\_clinton, bill\_clinton) \in \Pi'$),
- **yes** with score 0.048, replacing *daughter_of* by *father_in_law* (since $child\_of(hilary\_clinton, chelsea\_clinton) \notin \Pi'$).

As before, the answers could be filtered either manually or by adjusting the threshold.

To evaluate the semantic matching (τ-Space module) we used DBpedia[4], a heterogeneous and large-scale data set which consists of 45,767 predicates, 5,556,492 classes, 9,434,677 instances, as knowledge base. The relevance function used a combination of IDF over predicates, cardinality (number of associated constants to another constant) and a dice string similarity coefficient.

[4] http://dbpedia.org

To query this knowledge base we selected 18 queries (Table 2) extracted from the Question Answering over Linked Data (QALD)[5] 2011 test collection. The selected subset concentrates on queries with a vocabulary gap between query and knowledge base terms. Queries in the original test collection with a perfect vocabulary match and with functional operators were removed.

The approach achieved **avg. recall=0.935**, corroborating the hypothesis that distributional semantics provides a comprehensive semantic matching solution. **Average mean reciprocal rank (mrr) = 0.632** shows that most of the results are in the first two positions of the ranked list. Different from traditional approaches where the matching is done at a syntactical level, the semantic approximation implies that absolute precision will unlikely to be achieved in all cases since some level of ambiguity and vagueness is intrinsic to the vocabulary gap problem.

The **avg. precision=0.561** confirms that the distributional semantic relatedness measure is able to provide a selective semantic filtering mechanism. However, in the context of logic programs higher precision should be targeted by the use of more selective distributional models and by the introduction of disambiguation/dialog user feedback mechanisms. ESA is a distributional model which, by its construction, favours the broader class of semantic relatedness instead of the more constrained class of semantic similarity (such as taxonomic relations). The use of ESA favours recall and broader vocabulary independency over precision and assume that noisy inferences can be filtered out by a disambiguation mechanism. A relevant research direction is to improve precision by using distributional models with narrower context windows [3]. Enlarging the set of inferences can be problematic in large-scale knowledge bases, as for example, in the context of the Semantic Web. The composition with scalable and selective reasoning models (e.g. in Bonatti et al. [4]) should be investigated in order to minimize the impact of the additional inference process.

The average predicate distributional **matching time is 1,523 ms** in a core i5 8GB RAM machine. The τ-Space works as a *semantic best-effort* approximation [7] mechanism where there are no warranties of absolute precision but recall is close to 1. The distributional semantic relatedness measure provides a high selectivity rate over unrelevant results (shown by the precision value). These assumptions mean that in most cases the final result is found, but spurious inferences are present in the current distributional models. These spurious inferences can be eliminated by the provision of dialog mechanisms, where users can provide additional information in order to disambiguate the query.

The computational cost of the distributional semantic approximation concentrates on the cosine similarity operation for the semantic relatedness computation which can be performed at O(nlogn) time complexity using Locality Sensitive Hashing (LSH) techniques.

[5] www.sc.cit-ec.uni-bielefeld.de/qald-1

Table 2. Examples of prolog queries in the test collection. In the third column, (p,r,fm) represents (precision, recall, F-measure)

NL-query	Prolog-query	(p, r, fm)	vocabulary gap (query=dataset)
Who was the wife of Abraham Lincoln?	wife(X,abraham_lincoln)	(0.0305, 1,0.0592)	wife = spouse, President Lincoln = Abraham Lincoln
Who created English Wikipedia ?	created(X,english_wikipedia)	(1,1,1)	created = author, English Wikipedia = English Wikipedia
Who is the owner of Aldi?	owner(X,aldi)	(0.3333, 1, 0.5)	owns = key Person, Aldi = Aldi
How tall is Claudia Schiffer?	tall(claudia_schiffer,X)	(0,09090,1, 0.1667)	Claudia Schiffer = Claudia Schiffer, tall = height
Is Natalie Portman an actress?	actress(natalie_portman)	(1,1,1)	Natalie Portman = Natalie Portman, actress = Actor
Who wrote the book The Pillars of the Earth?	wrote(X,the_pillars_of_earth)	(0.5, 0.5, 0.5)	wrote = author, The Pillars of the Earth = The Pillars of the Earth
Who was Tom Hanks married to?	married_to(X,tom_hanks)	(0.75, 1, 0.8571)	Tom Hanks = Tom Hanks, married to = spouse
When was Lucas Arts founded?	founded(lucas_arts,X)	(1,1,1)	Lucas Arts = Lucas Arts, founded = foundation
Who is the daughter of Bill Clinton married to?	daugther_of(X,bill_clinton), married_to(X,Y)	(0.5, 0.5, 0.5)	Bill Clinton = Bill_Clinton, daughter = child, married to = spouse
Where did Abraham Lincoln die?	die(abraham_lincoln,X)	(0.0162, 1, 0.032)	Abraham Lincoln = Abraham Lincoln, die = death Place
Who is the mayor of New York City ?	mayor(X,new_york_city)	(0.2, 1, 0.3333)	New York City = New York City, mayor = leader Name
What is the profession of Frank Herbert ?	profession(frank_herbert,X)	(0.01428, 1, 0.0281)	Frank Herbert = Frank Herbert, profession = occupation
What did Bruce Carver die from ?	die(bruce_carver,X)	(0.1818, 1, 0.3077)	Bruce Carver = Bruce Carver, die = death Cause
Who designed the Brooklyn Bridge ?	designed(X,brooklyn_ridge)	(0.5, 1, 0.6667)	Brooklyn Bridge = Brooklyn Bridge, designed = designer
Give me all films produced by Hal Roach?	produced(hal_roach,X)	(0.98, 0.9722, 0.9761)	films = Film, produced = producer, Hal Roach = Hal Roach
When was Capcom founded ?	founded(capcom,X)	(1, 1, 1)	Capcom = Capcom, founded = foundation
Which albums contain the song Last Christmas?	contains(X,last_christmas)	(1, 0.8571, 0.9231)	music albums = album, contain = , song = single, Last Christmas = Last Christmas
Was U.S. president Jackson involved in a war ?	u_s_president(jackson), involked(jackson,war)	(1, 1, 1)	U.S. president = Presidents Of The United States, Jackson = Jackson, war = battle

9 Related Work

In [13], Lukasiewicz & Straccia presented probabilistic fuzzy dl-programs, which is a uniform framework that deals with uncertainty and fuzzy vagueness. Our work focus on the ontology mapping aspect (uncertainty) and in the use of a distributional semantic approach to align semantically equivalent terms. The common goal of both fuzzy/probabilistic and distributional approaches is the introduction of flexibility into the reasoning process. The main benefit of using distributional semantics is the use of large-scale unstructured or semi-structured information sources to complement the semantics of logic programs. One of the streghts of distributional semantic models is from the acquisitional perspective, where comprehensive semantic models can be automatically built from large-scale corpora.

Distributional semantic models are evolving in the direction of coping with better compositional principles, supporting the semantic interpretation of complex sentences/statements. Baroni et al. [3] provide an extensive discussion of state of the art approaches for compositional-distributional models. In this work the compositional model is given by the structure of the logical atoms in a logic program Π, which defines a set of vectors in the distributional vector space.

In [12], Grefenstette presented how elements of a quantifier-free predicate calculus can be modelled using tensors and tensor contraction. The basic elements, truth values and domains objects, are modelled as vectors and predicates and relations are modelled through high order tensors. Also, Boolean connectives are modelled using tensors and with the basic elements used to build a quantifier-free predicate calculus.

Research on schema matching/alignment [5] have extensively investigated semantic matching approaches for entities on different schemas. Different matching strategies are employed ranging from structural approaches to strategies based on linguistic resources [5]. Most of the approaches focusing on linguistic resources concentrate on the use of manually created resources such as WordNet. Distributional semantic models are still not extensively used in this context. Another difference between this work and schema alignment approaches is the context in which the semantic matching takes place, which here focuses on the query - knowledge base semantic matching.

Freitas et al. [8] and Novacek et al. [9] describe distributional approaches applied to Semantic Web Data. While Freitas et al. [8] focuses on a natural language query scenario, [9] Novacek et al. targets the description of a tensor-based model for RDF data and its evaluation on entity consolidation.

10 Conclusion and Future Work

This work presented a principled approximative inference model for large-scale and heterogeneous knowledge bases which adds the flexibility and the scale of commonsense-based semantic approximation of distributional semantics to logic programming models. The approach was formalized, a prototype was implemented and evaluated over a large knowledge base, achieving avg. recall=0.935, avg. mean reciprocal rank=0.632 and avg. precision=0.561. The proposed approach provides a provides a high recall and mean reciprocal rank semantic matching mechanism, under a semantic best-effort scenario (accurate approximation, but which demands a user interaction or post processing step).

Future work will concentrate on the implementation of a pre-processing strategy for natural language queries, the investigation of more constrained distributional models focussing on the improvement of precision and the study of the connection of our approach to synonymous theories in answer set programming proposed by Pearce and Valverde [16].

Acknowledgments. The work presented in this paper has been funded by Science Foundation Ireland under Grant No. SFI/08/CE/I1380 (Lion-2). João C. P. da Silva is a CNPq Fellow - Science without Borders (Brazil).

References

1. Abelson, H., Sussman, G.-J., Sussman, J.: Structure and Interpretation of Computer Programs. MIT Press, Cambridge (1985)
2. Baral, C.: Knowledge Representation, Reasoning, and Declarative Problem Solving. Cambridge University Press, Cambridge (2003)
3. Baroni, M., Bernardi, R., Zamparelli, R.: Frege in Space: A Program for Compositional Distributional Semantics. Linguistic Issues in Language Technologies (to appear)
4. Bonatti, P.A., Hogan, A., Polleres, A., Sauro, L.: Robust and Scalable Linked Data Reasoning Incorporating Provenance and Trust Annotations. Journal of Web Semantics 9(2), 165–201 (2011)
5. Shvaiko, P., Euzenat, J.: A survey of schema-based matching approaches. Journal on Data Semantics 4, 146–171 (2005)
6. Freitas, A., Curry, E., O'Riain, S.: A Distributional Approach for Terminology-Level Semantic Search on the Linked Data Web. In: 27th ACM Symposium On Applied Computing (SAC 2012). ACM Press (2012)
7. Freitas, A., Curry, E., Oliveira, J.G., O'Riain, S.: A Distributional Structured Semantic Space for Querying RDF Graph Data. International Journal of Semantic Computing 5(4), 433–462 (2012)
8. Freitas, A., Oliveira, J.G., Curry, E., O'Riain, S.: A Multidimensional Semantic Space for Data Model Independent Queries over RDF Data. In: Proceedings of the 5th International Conference on Semantic Computing, ICSC (2011)
9. Nováček, V., Handschuh, S., Decker, S.: Getting the Meaning Right: A Complementary Distributional Layer for the Web Semantics. In: Aroyo, L., Welty, C., Alani, H., Taylor, J., Bernstein, A., Kagal, L., Noy, N., Blomqvist, E. (eds.) ISWC 2011, Part I. LNCS, vol. 7031, pp. 504–519. Springer, Heidelberg (2011)
10. Gabrilovich, E., Markovitch, S.: Computing semantic relatedness using Wikipedia-based explicit semantic analysis. In: Proceedings of the 20th International Joint Conference on Artificial Intelligence, pp. 1606–1611 (2007)
11. Gelfond, M., Lifschitz, V.: Logic programs with classical negation. In: Logic Programming: Proc. of the Seventh International Conf., pp. 579–597 (1990)
12. Grefenstette, E.: Towards a Formal Distributional Semantics: Simulating Logical Calculi with Tensors. CoRR (2013)

13. Lukasiewicz, T., Straccia, U.: Description logic programs under probabilistic uncertainty and fuzzy vagueness. International Journal of Approximate Reasoning 50, 837–853 (2009)
14. Lukasiewicz, T., Straccia, U.: Managing uncertainty and vagueness in description logics for the Semantic Web 6, 291-308 (2008)
15. Lloyd, J.W.: Foundations of Logic Programming. Springer-Verlag New York, Inc., USA (1993)
16. Pearce, D., Valverde, A.: Synonymus Theories in Answer Set Programming and Equilibrium Logic. In: Proc. of 16th Eureopean Conference on Artificial Intelligence (ECAI 2004), pp. 388–392 (2004)
17. Turney, P.D., Pantel, P.: From frequency to meaning: vector space models of semantics. J. Artif. Int. Res. 37(1), 141–188 (2010)

Author Index